THE PAPER MUSEUM OF CASSIANO DAL POZZO

SERIES B ~ PART THREE

FOSSIL WOODS AND OTHER GEOLOGICAL SPECIMENS

THE PAPER MUSEUM OF CASSIANO DAL POZZO

A CATALOGUE RAISONNÉ

Drawings and Prints
in the Royal Library at Windsor Castle,
the British Museum, the Institut de France
and other Collections

General Editors
†Francis Haskell and Jennifer Montagu

Assistant General Editor
Henrietta McBurney

Series Editors
Series A ~ Antiquities and Architecture
Amanda Claridge
Series B ~ Natural History
David Freedberg

Patrons and Sponsors
Accademia dei Lincei
Académie des Inscriptions et Belles-Lettres
British Academy
Getty Grant Program
Ing. C. Olivetti & C
Royal Collection Trust

SERIES B ~ NATURAL HISTORY

PART THREE

FOSSIL WOODS
AND OTHER GEOLOGICAL SPECIMENS

by Andrew C. Scott and David Freedberg
with documentation provided by
Francesco Solinas

with contributions by
Jo Taylor

HARVEY MILLER PUBLISHERS

HARVEY MILLER PUBLISHERS
An Imprint of Brepols Publishers
Begijnhof 67, B-2300 Turnhout, Belgium

British Library Cataloguing in Publication Data
A catalogue record for this book is available
from the British Library

ISBN 1872501915

This work is published with the assistance of
The Getty Grant Program

The production of this volume
has been supported by The BG Group

BG

Monochrome and colour origination by Neue Schwitter AG, Basel, Switzerland
Printed by Grafiche Milani S.p.a. Milan, Italy

CONTENTS

THE CATALOGUE RAISONNÉ

SERIES A
ANTIQUITIES & ARCHITECTURE
Series Editor: Amanda Claridge

I
Ancient Mosaics and Wall Paintings

II
Early Christian and Medieval Antiquities
Two Volumes

III
Sarcophagi and other Reliefs
Three Volumes

IV
Statues and Busts

V
Classical and Later Metalwork, Glass and Terracotta
Three Volumes

VI
Classical Manuscript Illustrations

VII
Inscriptions

VIII
Tombs of Illustrious Italians at Rome

IX
Roman Topography and Architecture
Three Volumes

X
Renaissance and Later Architecture

SERIES B
NATURAL HISTORY
Series Editor: David Freedberg

I
Citrus Fruit

II
Fungi: 'Icones Fungorum'
Five Volumes

III
Fossil Woods and other Geological Specimens

IV
Birds
Two Volumes

V
Animals, Fish and other Fauna

VI
The 'Erbario Miniato' and other Flora
Two Volumes

VII
Flora: 'Syntaxis Plantaria'
Five Volumes

VIII
The 'Erbario Messicano'

IX
Minerals and Curiosities

ACKNOWLEDGEMENTS

Andrew Scott thanks Henrietta McBurney for her constant encouragement and unstinting commitment to the project. He would also like to thank Dr Margaret Collinson, Dr W. J. Van der Burgh and Dr Alan Hemsley for useful discussion concerning Tertiary plants; Dr E. Martinetto, Professor P. Ambrosetti, Dr O. Girotti and Dr E. Biondi for help concerning the Italian fossil sites, and Professor P. Ambrosetti, Dr Flavius Ikome, Dr Federico Venturi and Dr Giorgio Basilici for generous assistance in locating some of the field sites. Lynne Blything executed the diagrams (Figs 12–14, 29–31; Comp. fig. 51a) and Kevin de Souza assisted with photography. Fieldwork in 1998 was supported by a grant to Professor P. Ambrosetti, and in 1999 by Proctor and Gamble (Health and Beautycare) Ltd. and the Geology Department, Royal Holloway, University of London. Special thanks are due to Toshiba Information Systems (UK) Ltd who have contributed so generously towards this publication.

John Thackray and Ian Rolfe kindly read and commented on early drafts of this volume.

David Freedberg wishes to thank Andrew Scott for providing the scientific context for his introductory essay, Henrietta McBurney for her constant support and editorial guidance, Jo Taylor for her unflaggingly cheerful assistance on many topics, and Irina Oryshkevich for her extraordinary research assistance over many years; but gratitude is due above all to Francesco Solinas, whose initial work on the subject of this essay provided the stimulus and inspiration for everything written therein.

It should be noted here, both as a matter of form and as a tribute, that the bibliographic citation 'Gabrieli 1989' refers to the collection of Gabrieli's fundamental essays, originally written in the amazingly productive years between 1927 and 1941, on every aspect of Cesi and his fellow Linceans' lives. Gabrieli was the real pioneer in the study of every one of the fields touched upon here.

Andrew C. Scott, London
David Freedberg, New York
August 2000

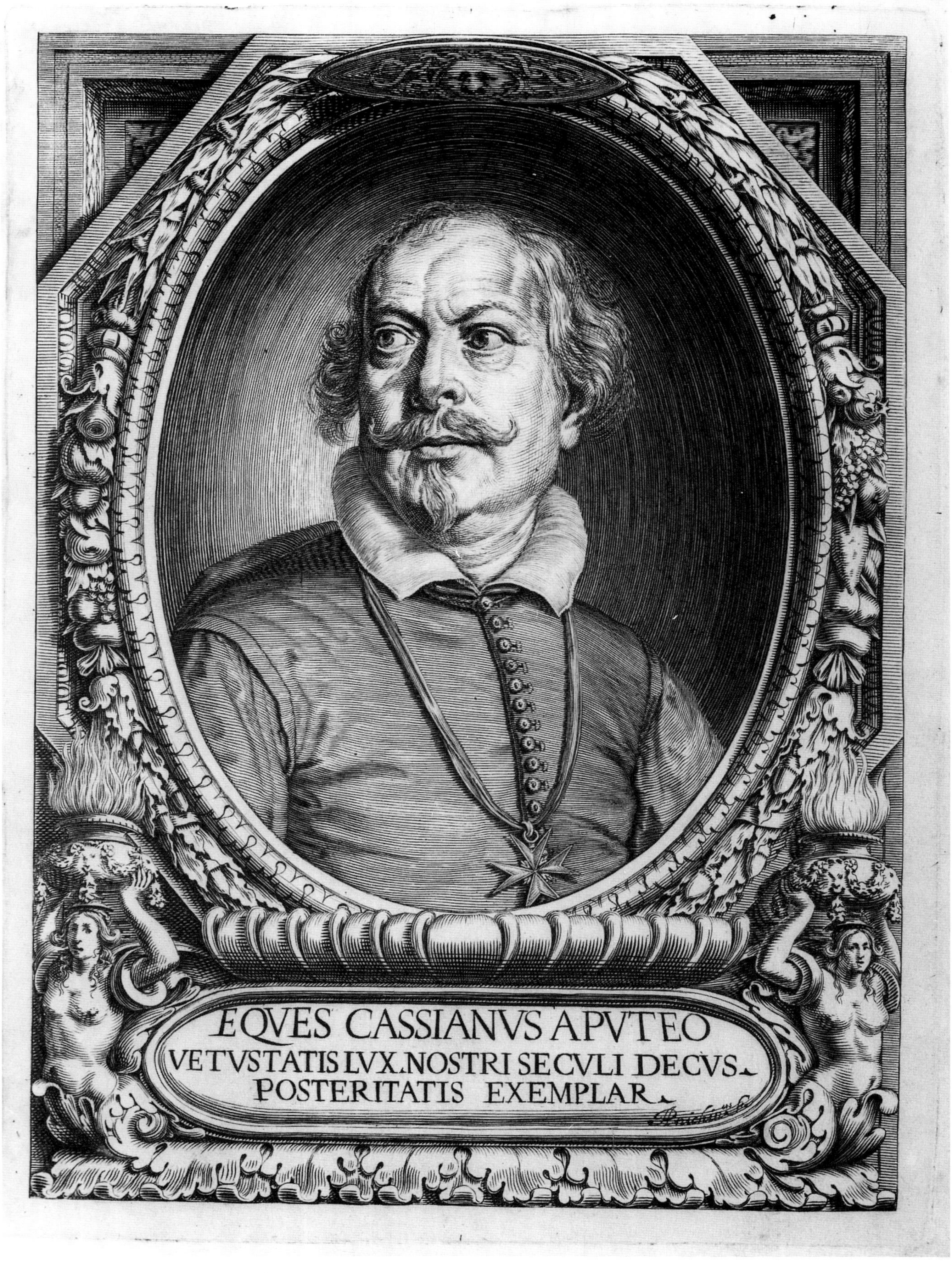

Fig. 1. Pietro Anichini, *Portrait of Cassiano dal Pozzo,* engraving, frontispiece to Dati's funerary oration, 1664

THE PAPER MUSEUM OF CASSIANO DAL POZZO

Francis Haskell and Henrietta McBurney[*]

CASSIANO DAL POZZO is celebrated today as one of the most important art patrons in seventeenth-century Italy [Fig. 1]. He commissioned more than forty paintings by Nicolas Poussin and owned many others by some of the finest artists of the day. A friend of Galileo and a correspondent of Peiresc, Cassiano was closely concerned with developments in the political and intellectual life of his time. Yet contemporary scholars admired him above all for his learning in the fields of antiquities and natural history and for the extraordinarily ambitious project which grew out of these interests, a project which he described as his *Museo Cartaceo* or 'Paper Museum'.

Born in Turin in 1588, Cassiano went as a young boy to Pisa, where he was placed in the care of his father's cousin, the Archbishop of Pisa, and received his university education in law (see page 13). It was here in Tuscany that his interests in the worlds of art and science were nurtured, partly through his cousin's connections at the Medici court, and partly in the intellectual circles centred around the University of Pisa and its Botanic Gardens. In 1612, after briefly pursuing a legal career in Siena, he moved to Rome, where he was to remain for the rest of his life.

In 1622 Cassiano was elected to the first scientific society in Europe, the *Accademia dei Lincei*, in which he was to play a very active role. Meanwhile he had entered the service of Francesco Barberini (1597–1679), who became a cardinal in 1623 after the election in that year of his uncle Maffeo as Pope Urban VIII. In 1625–6, Cassiano accompanied the cardinal on diplomatic missions to Paris and Madrid, in which cities he enriched his understanding of painting and developed a particular interest in the works of Leonardo da Vinci and the young Velázquez. On his return to Rome, Cassiano, together with his younger brother Carlo Antonio (1606–89), began to build up a collection of paintings, books, medals and drawings. After the death of Prince Federico Cesi (1585–1630), founder of the *Accademia dei Lincei*, he acquired a large number of Cesi's books and scientific instruments.

Cassiano also conceived the ambitious project of employing artists—many of them young and unknown, but some of great distinction—to make copies for him of all surviving traces of Roman civilization from household items to manuscripts, sculpture and mosaics. This 'Paper Museum' was then classified thematically in order to build up as complete a record as possible of ancient religion, customs, dress, architecture, theatrical spectacles and so on. Equally interested in the natural world

[*] This contribution has been abridged from the full version published in volume B.I of this series.

as in antiquities, Cassiano commissioned a similarly large number of drawings of geological specimens, citrus fruit, fungi, plants, fish and birds. In collaboration with scholars such as Giovanni Battista Ferrari and Federico Cesi, these natural history drawings were also classified, and in such a way that they anticipated modern systems of taxonomy.

Although unpublished, the Paper Museum was widely used in the seventeenth century as an instrument of study and research: scholars and dilettanti from all over Europe visited Cassiano in Rome or wrote to him asking for information, and in return sent him reports of strange natural occurrences and local antiquarian investigations.

When Cassiano died in 1657, the Paper Museum passed to his younger brother Carlo Antonio, who not only added to the collection, but rearranged parts of it. On the latter's death, the collection passed to his son Gabriele (1642–95), and on Gabriele's death to his son Cosimo Antonio (1684–1740). It was during Cosimo's custodianship that the family collections first began to be dispersed, mainly to pay off the large debts that he had incurred. The library, which included the Paper Museum, was the first to go: in 1703 it was sold for the sum of 4,000 *scudi* to Giovanni Francesco Albani, by then Pope Clement XI (reg. 1700–21), from whom, in 1714, it passed to his nephew, Cardinal Alessandro Albani (1692–1779), for the same sum. The Paper Museum thus joined the important collection of drawings and prints already belonging to the Albani family. Under Albani ownership, large parts of the dal Pozzo collection were reorganized and some were amalgamated into other sequences of drawings in their possession. In 1762 the greater part of Cardinal Albani's collection (about 200 folio volumes) was purchased for 14,000 *scudi* by King George III, through the agency of the architect James Adam. It was brought back to London in 1763 and incorporated into George III's extensive library in Buckingham House. By the time an inventory of the collection was drawn up at the end of the eighteenth century (Inventory A: see page 17) many of the dal Pozzo volumes had been reordered and rebound, probably under George III's librarian, Richard Dalton, in the 1780s. In some cases the drawings were lifted from their original mounting sheets, remounted on eighteenth-century paper and given decorative wash borders.

Towards the end of the eighteenth century, two volumes containing drawings of classical antiquities were separated from the main body of the material by Richard Dalton; these, known as the Franks volumes—after Augustus Wollaston Franks, the name of the last collector through whose hands they passed—are now in the British Museum. Another offshoot from the main collection is the group of four volumes of architectural drawings now in Sir John Soane's Museum. These were acquired by the Adam brothers at some stage between their leaving the library of the Albani and reaching that of George III, and were bought by Soane at the Adams' sale in 1818. The volumes of drawings in the Royal Botanic Library in Kew, and in the Institut de France, Paris, appear to have remained in the Albani Library until the nineteenth century.

In 1823, when most of George III's library was given to the nation by his son, George IV, and transferred to the British Museum, a number of dal Pozzo volumes of prints (and at least one volume of the drawings) were further separated from the main corpus; these were re-discovered in the British Museum in recent years. However, the main body of the Paper Museum was retained in the Royal Collection, and since 1834 has been housed (with the rest of the Royal Library) in Windsor Castle. Just after the First World War a large part of the natural history material (probably fifteen volumes) was dispersed from the Royal Library at the behest of the then Royal Librarian, Sir John Fortescue. The contents of these volumes, many of which were in the possession of a London art dealer called Jacob Mendelson just before the Second World War, are now scattered among private and public collections in England, Italy, America and Australia. A number of these drawings have in recent times been returned to the Royal Collection, either through gift or purchase. Other drawings from the Paper Museum have appeared in salerooms: in September 1987 a group of 107 natural history drawings was sold at Sotheby's, New York, and in 1990 another collection of 500 drawings after the antique was sold at Phillips, London. Since then other drawings have gradually been coming to light.

The decision to produce the present catalogue has its own history. The Paper Museum remained in relative obscurity after it left Italy in the mid eighteenth century until the late nineteenth century, when scholars began once more to consult Cassiano's drawings. It is to two German archaeologists—Friedrich Matz (1843–74), one of the founders of the great published corpus of Roman sarcophagus reliefs, and Adolph Michaelis (1835–1910)—that credit is due for the 'modern' revival of interest in the Paper Museum. Their work, and that of the British scholar, Thomas Ashby (1874–1931), provided the inspiration for the study by Cornelius Vermeule who, in two major articles in the 1960s, catalogued the drawings of sarcophagi and reliefs in the Royal Library and the British Museum. In the same period Sheila Somers Rinehart embarked on a comprehensive study of the Paper Museum; and in the 1970s Jean Goldman attempted a general artistic assessment of the drawings. The decision finally to produce a fully illustrated catalogue raisonné of the entire Paper Museum was made in the mid 1980s, when research brought international scholars to Windsor to inspect the visual counterparts of the documentary evidence for Cassiano's *Museo Cartaceo*.

However, it soon became clear that it would be impossible to adopt for this body of drawings the same principles as those used in the volumes devoted to the drawings of Holbein, Guercino, Canaletto and other great masters belonging to Her Majesty The Queen. Despite Cassiano's well-informed love of art and the fact that he had access to distinguished artists, including Pietro Testa, Pietro da Cortona and others of real stature, it is clear beyond any doubt that it was the documentary and not the artistic aspect of the enterprise that was important to him. The drawings in the present publication, therefore, are catalogued not by artist, but by subject matter. Viewed from this perspective the drawings readily divide into two categories: those concerned with Antiquities and Architecture (which we have called Series A) and those dealing with Natural History (Series B). The former group encompasses about 4,200 drawings and the latter some 2,700. The aim of this catalogue raisonné is to reconstruct in some measure the intellectual and artistic achievements once embodied in Cassiano's Paper Museum.

Francis Haskell, joint General Editor of the Catalogue Raisonné since its initiation, died in January 2000.

CHRONOLOGY

Events in the dal Pozzo Household	Year	Contemporary events	Publications
Cassiano dal Pozzo born in Turin	1588		
Moves to Pisa	1598		
Made Knight of the Order of S. Stefano	1599		
	1603	Cesi founds *Accademia dei Lincei*	
Birth of Carl Antonio dal Pozzo	1606		
Cassiano graduates from Pisa University; Carlo Antonio dal Pozzo, Archbishop of Pisa, dies	1607		
Cassiano appointed judge in Siena	1608		
	1610	Galileo moves to Pisa	
Leaves Tuscany	1611	Galileo becomes member of *Accademia dei Lincei*	
Arrives in Rome	1612		
	1614	Cesi records Palestrina mosaic	
	1616	Galileo censured	
Cassiano's father (Antonio dal Pozzo) dies	1619		
Carlo Antonio arrives in Rome	1620		
Cassiano elected to *Accademia dei Lincei*	1622	Francesco Barberini elected to *Accademia dei Lincei*	Olina, *Ucelliera*
Appointed to the Household of Francesco Barberini	1623	Maffeo Barberini elected Pope Urban VIII	Galileo, *Il Saggiatore*
	1624	Poussin arrives in Rome	
Travels to France: meets Peiresc	1625		
Travels to Spain; elected to *Accademia della Crusca*	1626		
Carlo Antonio marries Teodora Costa; brothers move to Via dei Chiavari	1626/7		
	1628		Faber, *Animalia Messicana*
	1629	Faber dies	
	1630	Cesi dies	Stelluti, *Persio Tradotto*
	1631		Campanella, *Atheismus*
	1632		Galileo, *Dialogo*
Cassiano buys Cesi's collection	1633	Galileo is condemned	Ferrari, *De Florum Cultura*
	1634-5		Bosio, *Roma Sotterranea*
Poussin begins planning *Seven Sacraments* for Cassiano	1635-6		
	1637	Peiresc dies	Stelluti, *Trattato del Legno Fossile Minerale*
Carlo Antonio nominated *Capitano di Fanteria*	1642		

CHRONOLOGY *(Continued)*

Events in the dal Pozzo Household	Year	Contemporary events	Publications
	1644	Urban VIII dies; Giovanni Battista Pamphilj elected Pope Innocent X	
	1645	Francesco Barberini in exile in Paris	
	1646		Ferrari, *Hesperides*
	1648	Francesco Barberini's return to Rome	
Cassiano's last documented commission from Poussin: *Pyramus and Thisbe*	1651/2		Hernandez, *Rerum Medicarum Novae Hispaniae Thesaurus*
	1655	Fabio Chigi elected Pope Alexander VII	
Carlo Antonio serves as *Conservatore* of Rome	1656-78		
Cassiano dies; Poussin's *Annunciation* executed, probably for his memorial. Carlo Antonio takes over the knight-hood in the Order of S. Stefano	1657		
	1664		Bellori, *Nota delli Musei* Dati, Delle lodi Commendatore...
	1670-94	Bellori holds post of 'papal antiquary'	
Carlo Antonio serves as *Console Capitolino*	1684		
	1687		Mabillon, *Iter Italicum*
Carlo Antonio dies	1689		
	1690-99		Ciampini, *Vetera Monumenta*
Gabriele dal Pozzo dies	1695		
	1700	Giovanni Francesco Albani elected Pope Clement XI	
Paper Museum sold to Pope Clement XI [Albani] by Cosimo Antonio dal Pozzo	1703		
Paper Museum passes to Cardinal Alessandro Albani	1714		
	1721	Michelangelo dei Conti elected Pope Innocent XIII	
Cosimo Antonio dal Pozzo dies	1740		
	1743	Winckelmann becomes librarian to Cardinal Albani	
Paper Museum bought by King George III	1762		

LOCATIONS OF DAL POZZO MATERIAL

Series A: Antiquities and Architecture

I. WINDSOR CASTLE: Royal Library	Folios	RL inventory	Ashpitel	Heaton-Smith	Michaelis	Robert
Bassi Relievi Antichi 1	95	8145-8255	A40	155	-	I
Bassi Relievi Antichi 2	66	8256-8322	A41	156	II	II
Bassi Relievi Antichi 3	70	8323-8400	A42	157	III	III
Bassi Relievi Antichi 4	79	8401-8483	A43	158	IV	IV
Bassi Relievi Antichi 5	69	8484-8551	A44	159	V	V
Bassi Relievi Antichi 6	69	8552-8628	A45	160	VI	VI
Bassi Relievi Antichi 7	69	8784-8860	A48	163	IX	-
Bassi Relievi Antichi 8	79	8702-8783	A47	162	VIII	VIII
Buste e Statue Antichi 9	77	8629-8701	A46	161	VII	VII
Bassi Relievi o Antichi (10)	120	7994-8113	A52	153	X	-
Bassi Relievi Antichi (pf. IXa)	30	8114-8144*	A53	154	IXa	IXa
Mosaici Antichi I	96	8928-9035	A50	165	XI	-
Mosaici Antichi II	103	9036-9221	A51	166	XII	-
Antichità Diverse	108	10189-10297	A9	184	XVIII	XVIII
Architec[tura] Civil[e]	150	10354-10504	A12	186	XV	-
Ancient Roman Architecture	50	10780-10846	A17	190	-	-
Buildings and Architectural Ornaments	12	10741-45, 10750-56	A16	189	-	-
Disegni di varie Antichità, Nettuno	220	11120-11521	A31	196	XIII	-
Tombs of Illustrious Italians at Rome	243	11715-11957	-	201	-	-
Terence	23	26645-26667	-	-	XXI	-
Il Vergilio nel Vaticano I	52	26668-26719	-	-	XXII	-
Il Vergilio nel Vaticano II	53	26720-26773	-	-	XXIII	-
Ancient Mosaics & Paintings (loose folios)	34	19201-19234	pf. 5	-	-	-
Architectural drawings (loose folios)	32	19243-19287	-	-	pf. 5	-
Bas-reliefs (loose folios)	2	11880-11881	-	-	-	-

* dispersed,4 in 'Franks' I (see British Museum), drawing 8144 rediscovered April 1999 (see A.I.132).

II. BRITISH MUSEUM	Folios	Drawing no.	Michaelis	Robert
'Franks' I	172	1-202	I	-
'Franks' II	134	203-523	II	-
'New' drawings	87	-	-	-

III. BRITISH LIBRARY	Folios	Shelf mark
Plans of Fortified Places ...	28	118.e.18

IV. SIR JOHN SOANE'S MUSEUM	Folios	Vol. no.	Ashby 1904 nos	Fairbairn 1998 nos
Montano I: *Architectu del Montani*	138	125	-	840–1027
Montano II: *Montan 1*	97	124	-	1028–1191
Montano III:*Tempj del Montani*	98	123	-	1192–1342
'Codex Coner': *Architect[ura] Civilis Andrea Coneri Antiquita Monume[nta] Rome*	165	115	1-165	–
Architect[ura] Civile Frammenti	88	119		488–533

V. DISPERSED	Folios	Phillips 1990 lot nos
ex-Stirling-Maxwell I - *Sculpture*	124	219-298
ex-Stirling-Maxwell II - *Architecture*	101	299-374

RL inventory = Royal Library inventory number
Michaelis = number given in Michaelis 1882
Robert = number given in Robert *ASR*
Ashpitel = number given in Ashpitel 1862
Heaton-Smith = number given by Evelyn Heaton-Smith
(worked in Royal Library between 1905 and 1916, under Sir John Fortescue)

LOCATIONS OF DAL POZZO MATERIAL

Series B: Natural History

I. WINDSOR CASTLE: Royal Library	**Folios**	**RL inventory**
Natural History: Birds I	44	dispersed
Natural History: Birds II	49	dispersed
Natural History: Birds III	54	dispersed
Natural History: Birds IV	44	dispersed
*Natural History: Fossils*V	45	25480-25535
Natural History: Fishes VI	61	dispersed
Natural History: Fruits VII	53	dispersed
Natural History: Plants VIII	74	dispersed
Natural History: Plants IX	67	dispersed
Natural History: Plants X	56	dispersed
Natural History: Plants XI	57	dispersed
Natural History: Plants I	53	dispersed
Natural History: Plants XIII	58	dispersed
Natural History: Fossils XIV	62	25536-25591
Natural History: Fossils XV	48	25592-25639
Natural History: Fossils XVI	57	25640-25696
Natural History: Fossils XVII	41	25697-25734
Natural History: Quadrupeds	25	dispersed
Natural History: Corals	16	dispersed
Con[cae] & Plant[ae]	29	dispersed
Natural History of Birds	92	27599-27690
Erbario Miniato	214	27691-27901
Erbar[io] Minia[to]/ 'Libellus dei Medicinalibus Indorum herbis...'	41	27902-27938

II. KEW: Library, Royal Botanic Gdns		**Folios**	**File no.**
Icones Fungorum Ineditorum	I	122	QG 3884
Icones Fungorum Ineditorum	II	123	QG 3884

III. PARIS: Institut de France		**Folios**	**MS nos**
Fungorum genera et species	I	216	968
Fungorum genera et species	II	206	969
Fungorum genera et species	III	177	970
Plantae et flores	I	129	974
Plantae et flores	II	108	975
Plantae et flores	III	193	976
Plantae et flores	IV	193	977
Plantae et flores	V	178	978

IV. DISPERSED, previously at Windsor (see I. above)

Located, as of 2000:

Courtauld Institute Galleries	Witt colln cat. nos: 3218, 3745, 3865
Fitzwilliam Museum	Accession nos: PD.42-1956, PD.43-1956
Victoria & Albert Museum	Accession nos: Circ.13-1925, E.731-735-1949, E.2776-2777-1962
Rijksmuseum, Amsterdam	Inv. no. 1966: 87
Private collections	
Ex-Herbert Boone	Sotheby's NY 1988, lot nos: 125-182. 99 drawings
Other	98 drawings

Unlocated:

Witt photos	6 drawings
Courtauld Inst. Microfilm	16 drawings
Ex-sale room	4 drawings

INVENTORY A

References to volumes containing dal Pozzo material

Antiquities and Architecture

[p. 3] *Terence*, illustrated with Figures. 23 D°

[p. 3] *Il Vergilio nel Vaticano* 2 Vols. V[ol] 1, 52 pages. V2, 54 D°

[p. 4] *Ancient Roman Architecture* 50 pages. Fol.

[p. 4] *Buildings & Architectural Ornaments.* 43 [*in pencil* 42] D° Fol.

[p. 4] *Antichità Diverse.* 173 Drawings. Fol.

[p. 4] *Mosaici Antichi.* 2V. Fol. V1–96 pages. V2–102 D°

[p. 4] *Bassi Relievi Antichi.* 9 Vols Fol. V1–95 pages V2–67 D° V3–70 D° V4–79 D° V5–67. V6–70. V7–69. V8–78. V9–120 D°

[p. 4] *Busti & Statue Antichi.* 76 pages

[p. 4] *Disegni di varie Antichità.* 220 pages. Fol.

[p. 4] *Architectu: Civile & Milit:* (containing a medley of Plans in different parts of Italy & elsewhere & entire Buildings) 147 pages

[p. 4] *Tombs of illustrious Italians at Rome etc.* MS: Fol. 243 D°

[pp. 166–7] Six Portfolios of Architecture
No. 5
1. 18 large coloured Drawings of the Mosaic Pavement at Palestrina, with 2 Copies of the same engraved in 1721. The Abbé Barthelmy says that the *Commandeur Pozzo fit copier en grand la Mosaique, mais par parties detachées, conservant dans les desseins les couleurs des differentes objets qu'elle representent, les Dessins etoient retracées en dix-huit morceaux, dont Suares, Evêque de Vaisons donna une description succincte dans son histoire imprimée a Rome en 1655.*
2. 2 Prints of separate parts of the same Pavements
3. 7 coloured Drawings of Mosaic
4. 43 Drawings of ancient Buildings in Rome [*with a later annotation in pencil* These are partly prepared for binding]

Natural History

[p. 3] Drawings of *Birds*. 4 Vols Fol. V1, 44 pages. V2, 49 D° V3, 54 D° V4, 44 D°

[p. 3] D°... of *Fruits*. 1 Vol Fol. 53 pages

[p. 3] D°... of *Fishes*. 1 Vol Fol. 61 DE

[p. 3] D°... of *Plants*. 6 Vols Fol. V1–74. V2–67. V3–56. V4–57 [*in pencil* 40 only]. V5–53. V6–58.

[p. 3] D°... of *Fossils*. 5 Vols Fol. V1–45. V2–62 [*in pencil* including] 12 of Index. V3–48. V4–57. V5–38 & 3 of Index.

[p. 3] Sixteen Drawings of *Plants*. Fol.

[p. 3] Twenty five Plates of Drawings in Water colours of *Quadrupeds*. Fol.

[p. 4] *Conc: & Plant*: Fol. 16 Shells & 13 Plants

[p. 4] Drawings of *Birds*. Fol. 92. Drawings

[p. 4] *Erbario Miniato*. 209 Drawings, with an Index

[p. 4] *Libellus de Medicinalibus Indorum herbis a Martino, Collegii Sanctae Crucis. Compostus: cum 37 figuris coloratis*. MS.

[p. 143] A Volume containing twenty-five pages of Drawings of *Animals*

[p. 143] A Volume (sixteen pages) containing drawings of *Corals*?

[p. 143] A Volume (92 pages) cont: Drawings of *Birds*. Small Fol. let.d *'Natural History of Birds'*

[N.B: on page 141 of Inventory A is a list of 17 of the above volumes labelled 'Natural History' with a running sequence of numbers assigned to them (for this sequence see 'Locations of dal Pozzo material Series B', I. Windsor Castle: Royal Library, p. 16)]

Inventory A is a manuscript inventory preserved in the Print Room, Royal Library, Windsor, and dating from the end of the eighteenth century: drawn up by, or under the direction of, George III's librarian, Richard Dalton.

SERIES B: NATURAL HISTORY
General Introduction to the Drawings

David Freedberg

THE IMPORTANCE of Cassiano's drawings of natural history cannot be overestimated. Largely commissioned by a group of Galileo's closest friends and supporters, they provide the visual evidence for crucial developments in the modern study of palaeontology, mycology, botany and zoology. Among them may be found some of the earliest illustrations ever produced with the aid of a microscope. Together they constitute one of the most ambitious attempts ever to classify the natural world through the use of visual description. Ultimately that attempt would fail, as the very men who commissioned these drawings also came to realize. But before they arrived at the conclusion that visual description in itself was not entirely adequate to revealing the mysteries of nature—and was even of limited use in analysing them—they provided the world with one of the most beautiful and comprehensive records of how the natural world looked at any period in history.

Made at the same time as Galileo's pioneering investigations in astronomy, mathematics and physics, these drawings and the documents associated with them have remained unknown to the vast majority of historians of science, and have been omitted from almost every treatment of the scientific revolution of the seventeenth century; and while there were certainly precedents for encyclopaedic collections of drawings of nature—commissioned by men like Conrad Gesner and Ulisse Aldrovandi—none could match this one in terms of size, comprehensiveness or accuracy. No one ever before had paid quite such meticulous attention to local flora, fauna or geology, or to the ever-increasing variety of species that were daily being reported and imported from the Americas. And none had been even remotely as rigorous in excluding the fabulous and the fantastic specimens that until then had so bedevilled the study of nature. For while there is no shortage of anomalous specimens in Cassiano's collection of drawings, each one of them reproduced something that had actually been seen and studied by one or other of his friends. Every specimen, in other words, actually occurred in nature, unlike the thousands of fantastic monsters, prodigies and wonders that populated the pages and sheets of Gesner and Aldrovandi, and so many other sixteenth-century natural historians. This was no longer a world in which the chief motor for explanation was either historical or the force of similitude (generally anthropomorphic): it was a world in which nothing could be taken as given unless it had been recorded and analysed by direct observation. It is not surprising that the first systematic use of both the telescope and the microscope should have emerged from the very same group of people who were responsible for the drawings commissioned in the present series, and who themselves went into the field to make detailed notes on everything they saw.

But who were these people? At first, they were the young men who had gathered round Federico Cesi, Prince of Acquasparta (1585–1630), when he founded the *Accademia dei Lincei* (Academy of Linceans) in 1603, at the age of eighteen [Fig. 2]. Its first members were Cesi's lifelong friend and follower, Francesco Stelluti (1577–1653), then the obsessive and ever-travelling Dutchman, Johannes Heck (1576–1618?), and the more obscure Anastasio de Filiis (1577–1608). Seven years later, in 1610, they invited the aged Giovanni Battista della Porta (1535?–1615) to join them: he was perhaps the most famous Italian student of nature, science and magic in the age before Galileo. In the following year the *Lincei* invited none other than Galileo Galilei (1564–1642) himself to join them, and he willingly accepted. This was the year in which Galileo made his famous demonstration of the capabilities of his telescope on the Janiculum in Rome, and when he was received in triumph and celebration by the Jesuits there. Later the same year, three talented Germans, Johannes Schreck (1576–1630), Johannes Faber (1574–1629) and Theophilus Müller (1576–1618) were invited to join the group.

Fabio Colonna (1566–1640), the distinguished botanist from Naples, became a member in 1612, but it would be six more years before two key figures in the active core of the group, Virginio Cesarini (1595–1624) and Giovanni Ciampoli (1590–1643), were added to their number. By then Heckius had gone mad and died, and Galileo had encountered the first serious opposition to his Copernican ideas. In 1622 Francesco Barberini, whose uncle would soon be elected Pope, joined them, and in the same year Cassiano dal Pozzo himself was elected a Lyncean. These were the chief players in the story, and they will feature time and again in the volumes in this series. They were the ones who went into the field, incessantly collecting botanical, zoological and geological specimens, who supported Galileo and edited the *Saggiatore* in the years in which he was discouraged by the threat of excommunication, and who put together the bulk of the drawings in the present corpus. And it was they who realized the prime importance of making visual records (or commissioning such records) of everything they saw and encountered.

Cassiano, of course, became a prime mover in all this. It was he who commissioned the beautiful drawings of birds and citrus fruit in the corpus and he who had the facsimile made of an old Aztec herbal (the so-called *Codex Badianus*) when he visited the Escorial in 1626. But—most importantly of all in the present context—it was he who in 1633 bought the contents of the great library formed by Cesi, thus preserving for posterity the visual heritage of the Lincean researches. The library contained not only books and instruments, but also the drawings of natural specimens which Cesi had commissioned from the earliest days of the academy onwards. But since we know that Cassiano continued to add to this inheritance, and that he himself commissioned natural history drawings in order to aid his own researches and to make them available to others, the following questions arise: which of the surviving

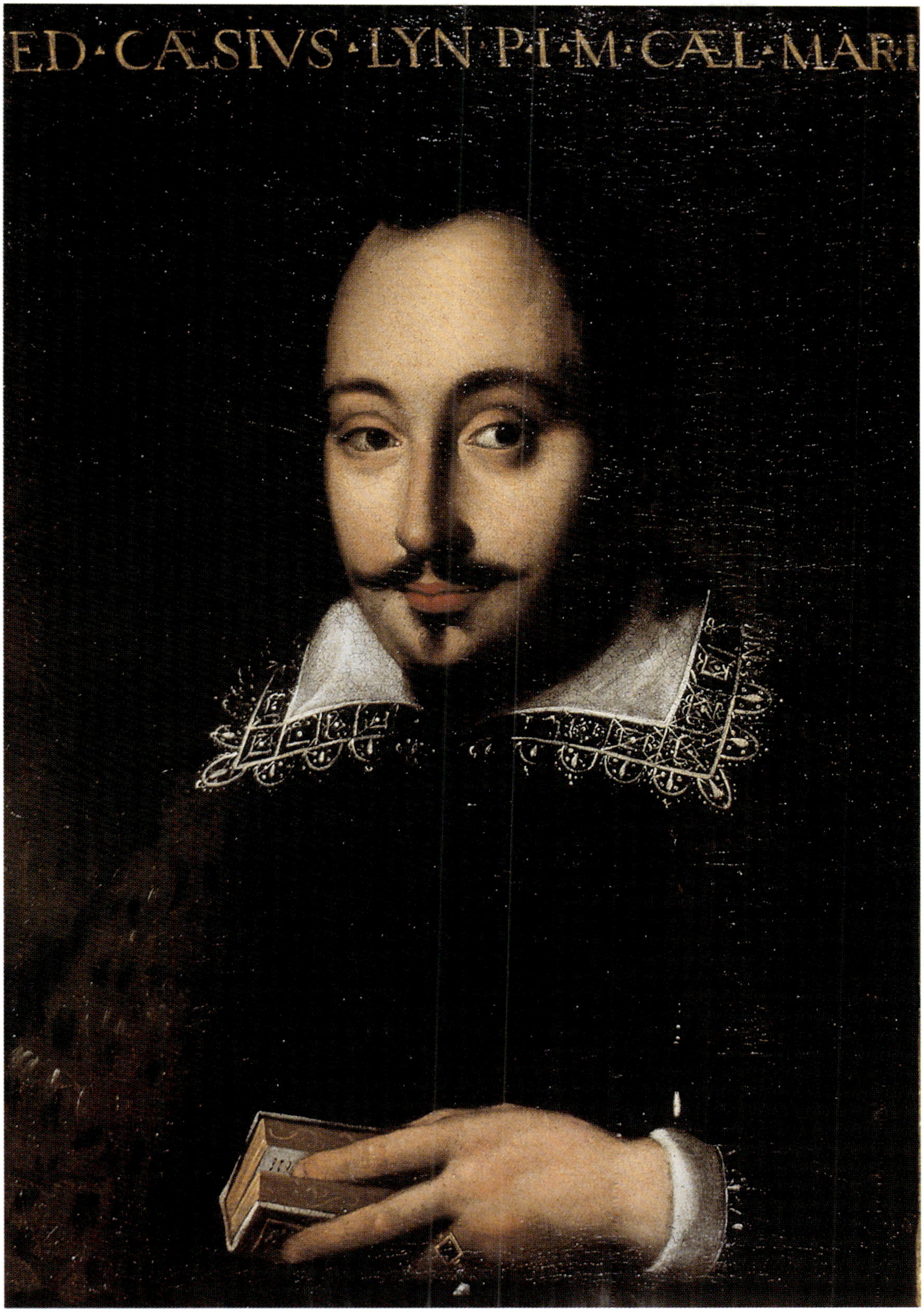

Fig. 2. *Federico Cesi*, attr. to Simon Vouet (?). Rome, Accademia dei Lincei

drawings come from Cesi's collection, and which were commissioned and collected by Cassiano himself? And who made them? These are difficult questions to answer, although there is a considerable amount of internal and external evidence to help us along the way.

At the time of writing, the natural history drawings that survive from Cassiano's Paper Museum number almost 3,000. They are chiefly to be found in the Royal Library, Windsor Castle, and in the Institut de France, Paris, but also in a number of other public and private collections. Since this is material that has not generally been known, and since it represents only a portion of what we know that Cesi and Cassiano commissioned and collected, it is very likely that more will come to light, even as the volumes in this Catalogue Raisonné appear.

The surviving drawings are preserved in a variety of bindings and as loose sheets. The numbers may be broken down as follows: in Windsor there are a total of 253 folios of drawings of fossil and mineral specimens bound into the five volumes entitled *Natural History: Fossils* [Fig. 3]. The so-called *Erbario Miniato* has 211 folios of drawings of plants and mushrooms, while that entitled *Libellus de medicinalibus Indorum herbis* (also known as the Mexican Herbal) has 41 folios showing Mexican plants with their Aztec names. The album entitled *Natural History of Birds* has 92 folios of ornithological drawings. The volumes described as *Plantae et flores,* numbered I–V (MSS 974–978 in the Institut de France), contain over 800 numbered folios of botanical specimens and the occasional drawing of insects and other animals (in MS 978). Although entitled *Plantae et flores* on their spines, these are presumably the volumes of the so-called *Syntaxis Plantaria* often referred to in the Lincean documents. The volumes *Fungorum genera et species,* numbered I–III (MSS 968–970 in the Institut de France) have 599 folios of fungi, while those entitled *Icones fungorum* in the Library of the Royal Botanic Gardens, Kew, contain a further 245 folios of the same subject, mostly copies of the drawings in Paris.

Some 480 loose sheets—showing animals, fish, crustacea, corals and sponges, citrus fruit, plants, mushrooms and birds—have been located so far. These were once bound into now dismembered volumes. For example, the original numbering of the drawings in the *Erbario Miniato* runs from 499 to 705, and the numbers of all of the loose drawings of plants so far located fall between 1 and 498. The spine of the *Erbario* carries the number VI, thus suggesting that it was part of a series of earlier volumes, at least one of which was presumably of plants, and the other of related natural history subjects. Even from these figures the enormous scale and ambition of the corpus is immediately apparent.

The Catalogue Raisonné will follow broadly the chief divisions already present in this material, along these lines: (i) citrus fruit; (ii) the five volumes of the *Icones Fungorum*, three in Paris and two in Kew; (iii) fossils and other geological specimens; (iv) birds; (v) animals, fish, and other fauna; (vi) the *Erbario Miniato* and other flora; (vii) the five volumes of the *Syntaxis Plantaria* in Paris; (viii) the *Erbario Messicano* and (ix) minerals and curiosities (see p. 6). Some subdivisions will be inevitable,

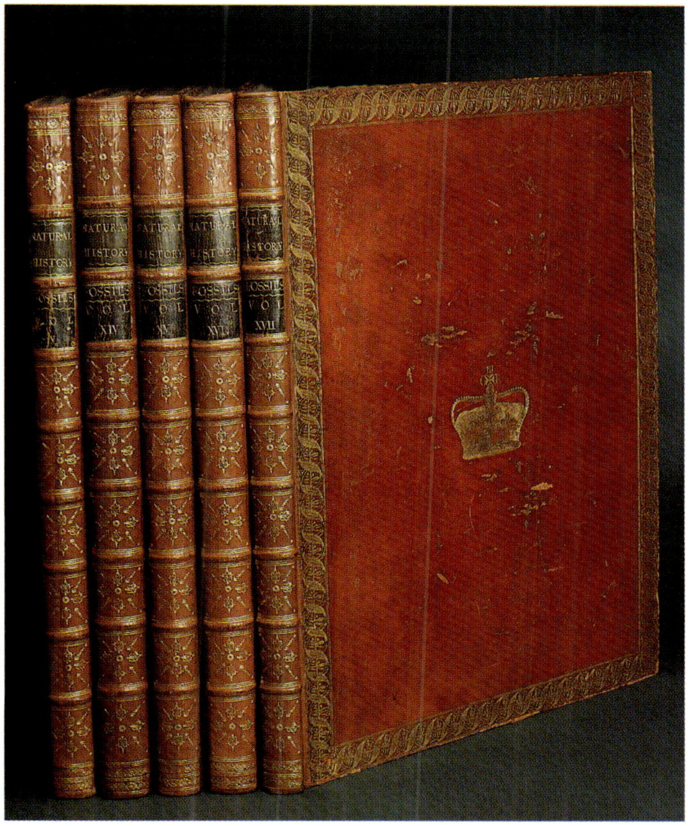

Fig. 3. The five volumes of *Natural History: Fossils* in George III bindings. Windsor, Royal Library

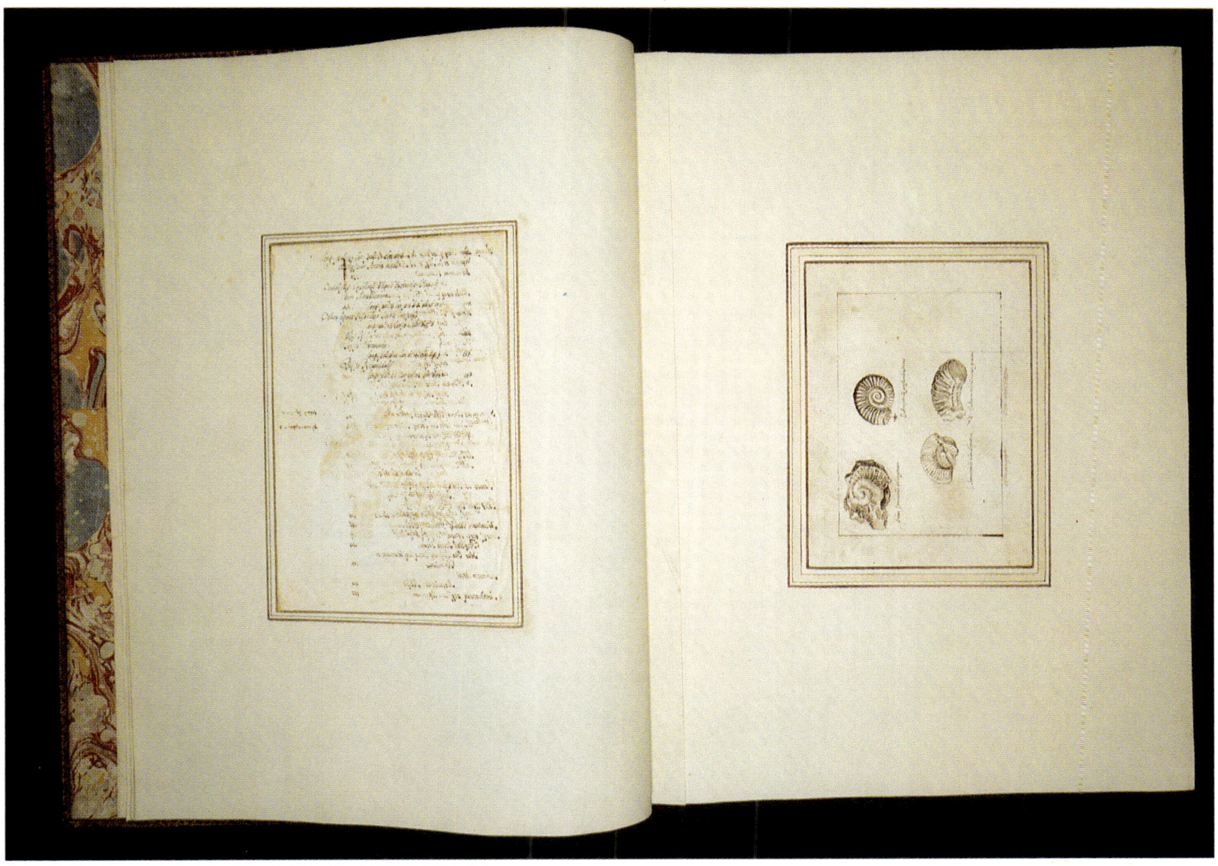

Fig. 4. Volume XIV of *Natural History: Fossils*, open to show 17th-century Index and page of ammonite drawings **(198)**

according to the quantity of drawings concerned and the internal logic of whatever may be discovered about their original or early organization.

The *Erbario Miniato* and Mexican Herbal in Windsor are in vellum bindings which date at least from the time they were in the Albani collection and which very probably are the bindings made for Cassiano; both have green wash fore-edges. The drawings in the *Erbario Miniato* are still on their original folios; those in the Aztec herbal are inlaid. The volumes entitled *Natural History: Fossils* and *Natural History of Birds* are in calf bindings specially made some time after the drawings they contain entered the Royal Collection in 1762; here the drawings are laid down. The volumes in the Institut de France are in eighteenth-century boards with leather spines, perhaps also made at the time they belonged to the Albani family. On the basis of internal evidence it would seem that the drawings in the Albani volumes reflect their earlier seventeenth-century ordering much more closely than the sometimes haphazard ordering in the volumes re-bound in the Royal Library in the late eighteenth century.

Many of the drawings carry inscriptions and numbers. These are in a variety of hands, and refer to many different things—the name of a specimen, the date of its discovery or its reproduction, and its possible occurrence in other sources, and so on. A large number of the fossil and fungus drawings and several of the botanical drawings contain inscriptions in Cesi's own hand. Especially in the case of the fossil drawings, these are often erased or cut off; fortunately Cassiano had had them copied by a scribe who reproduced these now truncated inscriptions elsewhere on the same sheet. This same scribal hand appears in the extensive indices now bound into the plant and fungi manuscripts in the Institut de France and in the plant and fossil volumes in Windsor [Fig. 4]. To what extent these indices reflect the original ordering of the folios they refer to (whether in Cassiano's or Cesi's collection) has yet to be clarified.

A similar uncertainty applies to the surviving indications of numbering. These are mainly in lower-case arabic numerals, but some are in roman numerals, and others take the form of lettering, occasionally capitals. Almost all are in brown ink in a small, often identical seventeenth-century hand. If there is any area that needs further research it is this; but, given the lacunose state of our knowledge about the original ordering and binding of the majority of these drawings, no conclusions may yet be reached. The numbers do not correlate in any helpful way with the order of reproductions in the printed books associated with them. The extent to which the numbers reflect the organization (or proposed organization) of the images in existing treatises, such as those on birds and citrus fruit (and perhaps others yet to be discovered), remains uncertain. In the case of the Antiquities drawings in Series A of the Catalogue Raisonné, one set of these numbers refers to an inventory of a group of drawings kept and acquired by Cassiano's brother, Carlo Antonio dal Pozzo; further research has to be done on the extent to which Carlo Antonio continued to expand the collection of natural history drawings.

The method of mounting drawings most commonly used by Cassiano was by inlay. A window a few millimetres smaller than the drawing sheet was cut out of the mount sheet, and the drawing pasted on it, with an overlap of about 5 mm. The overlap was burnished at the back so that the sheet could lie completely flat, and the drawing was often framed within double lines drawn in ink on the mount sheet. Because most of the drawings in the natural history series were at the time of their entry into the Royal Collection lifted from their seventeenth-century mount, trimmed and re-laid, with wash borders added, evidence of the original mounting method has disappeared. It is likewise possible that drawings from Cesi's collection were cut down when they were in the collection of Cassiano, his brother, or his heirs. Thus, with the passage of time, much valuable evidence has been cut off or erased, and we must look for other evidence for the original ordering and provenance of the drawings.

Sometimes the situation is very clear. For example, on the basis of their inscriptions, the occasional date, and a wide variety of external evidence (all of which will be discussed in the appropriate volumes in this series), it can be established that the drawings in the Institut de France were commissioned by Cesi himself. The drawings of fungi are the *Icones Fungorum* which he and his friends repeatedly discussed from the very beginning of their association onwards. The botanical drawings both in the so-called *Syntaxis Plantaria* in Paris and in the Windsor *Erbario Miniato* also carry Cesian inscriptions and can be related in many ways to the Lincean researches in the field. In the *Erbario Miniato* there are constant cross-references to discussions and to illustrations of the corresponding specimen in 'my Matthioli', presumably Cesi's own copy of an edition of the work of the great sixteenth-century botanist. The simple but proud inscriptions referring to the use of the microscope, chiefly in the Paris volumes, may be paralleled in Cesi's letters to his Lincean friends and in the occasional passage in their great compendium of South American flora and fauna, the *Rerum Medicarum Thesaurus Novae Hispaniae, sive Plantarum Animalium Mineralium Mexicanorum Historia* which was published in preliminary form in 1628 but in its complete form only in 1649/51. We shall return to this work, the so-called *Tesoro Messicano*, later.

Then there are the drawings which relate to other, chiefly Lincean publications. Those drawings now bound into Volumes XIV–XVII of the *Natural History: Fossils*, for example, were also commissioned by Cesi himself: a few were reproduced as etchings in the small *Trattato del Legno Fossile Minerale* published by Stelluti in 1637, and the group as a whole is repeatedly referred to there, as well as in Cesi's and Stelluti's correspondence and other papers still preserved in the Vatican Library, in the library of the Accademia dei Lincei in Rome, and in that of the École de Médecine in Montpellier.

If the fossil and fungus drawings, and most of the flower drawings, have a clearly Cesian provenance, many others were verifiably commissioned by Cassiano himself. As already noted, the volume in Windsor known as the *Libellus de Medicinalibus Indorum Herbis* is the facsimile which Cassiano had made of the Aztec herbal, the

Codex Badianus. This had been made over seventy years earlier (in 1552) by a converted Nahuatl Indian for the son of Philip II's viceroy, Antonio de Mendoza. It probably inspired Philip to send his personal physician, Francisco Hernández, to Central America to explore its botany and zoology. Hernández, subsequently known as the *Protomedicus* of the Indies, spent over six years there, from 1571 to 1577, gathering all kinds of information and many drawings about these matters. He brought this material back with him and deposited it in the Escorial, where it was seen by Cassiano in 1626.

Several further groups of drawings, now largely disbound from the volumes which once contained them, actually exceed most of the Cesian ones in terms of finish and refinement, and can only have been commissioned by Cassiano. Firstly, there are the drawings that he had commissioned in preparation for his own entry presentation to the Academy, the *Uccelliera* of 1622. These are by the same meticulous, skilful and attractive hand as the other bird drawings ordered by Cassiano in connection with the small unpublished ornithological treatises he wrote himself and which may still be found among his papers. Indeed ornithology seems to have been his particular area of interest and competence. Then come the fine drawings of citrus fruit, documented as commissioned and collected by Cassiano, and made in preparation for the illustrations of a book that eventually appeared in 1646, Giovanni Battista Ferrari's *Hesperides sive de malorum aureorum cultura*. Of all the drawings that came from Cassiano's Museum, this is by far the largest group ever to have been published.

In their almost complete suppression of any acknowledgement of his role in their production, both the *Uccelliera* and the *Hesperides* are typical of the natural history works published with the help of Cassiano. For although we know from his papers that he played a key role in their preparation and in the preliminary research for them, they were actually written by friends of his, in these instances the lawyer Giovanni Pietro Olina and the Jesuit professor of Hebrew and Rhetoric, Giovanni Battista Ferrari respectively; and Cassiano's name is scarcely mentioned in their pages. Both books name Vincenzo Leonardi as the author of the drawings on which their illustrations were based. We thus have the name of at least one artist in the corpus. Furthermore, one of the bird drawings has his signature on its verso, and it is reasonably clear that it is by the same hand as the best of the citrus drawings, documented as being by him. On stylistic grounds alone, there is much to be said for the view that many other of the finest animal drawings in the corpus which are not thus documented are also by Leonardi. Unfortunately his is the only name we have discovered so far which can definitely be associated with any drawing or group of drawings in the Natural History series.

Certainly a number of other possible candidates may be excluded. While we know that Heckius, Stelluti and Colonna all made drawings of natural history and other specimens, their hands appear not to be found among the drawings from Cassiano's Paper Museum. In the course of his travels, Heckius made a large number of rather crude drawings of what he saw, now preserved in his manuscripts in Rome,

Florence, Montpellier and elsewhere, but these are very different from anything in this series. So too is the occasional surviving drawing attributable to Stelluti, while for Colonna we have only the occasional reference to drawings which he made. It is certain that the other *Lincei*, especially Faber, Schreck and Cesarini, made use of drawings in their researches, but we have almost no evidence (except rarely in the *Tesoro Messicano*) of what they may have commissioned or even drawn themselves. Whatever the case, however, there can be absolutely no doubt of the importance that all these men attached to the graphic recording of everything they saw.

But why, specifically, were these drawings made? It was not just to have a visual record and proof of their manifold researches, or to make clear the errors and shortcomings of the ancient authorities, from Aristotle to Theophrastus, Pliny to Dioscorides. It was to help them in the complex taxonomic tasks in which they were constantly engaged. How much better to have a clear drawing than to describe something in words (always potentially ambiguous), or to have a multiplicity of different names for what might, after all, be the same thing. With the aid of drawings, people might at least attempt to ensure that one name was given to one thing, not one name to many things or many names to one. If they could not send a specimen around to their correspondents, or ask for one to be sent, they could at least circulate a drawing with a name attached to it.

The other reason for the production of these drawings was that they should serve as a means of assisting in the distinctions between species, of ensuring that a particular specimen was not assigned to the wrong species, or that it did not partake of more than one species. Nothing obsessed Cesi and the early Lynceans more than those phenomena which seemed to straddle the borderlines between classes, such as the fossil woods which are the subject of the present volume, or what they called species of ambiguous or doubtful nature. Indeed, if we turn to Cesi's vast attempt to classify all of the plant world, the *Tabulae Phytosophicae* appended to the *Tesoro Messicano* from 1628 on, or to his even more ambitious attempt to put all of nature in order, the outline plan still preserved in Naples (Biblioteca Nazionale) and entitled the *Speculum Rationis*, we find one overriding concern. This was the problem of how to deal with species that were of a middle nature, as Cesi put it, intermediate between one species and another, or apparently on the borderline between two species, or partaking of the character of two or more species (for further discussion of this problem see especially pp. 37–9, 48).

The problem of anomaly was evidently closely allied to this; and so in the *Tabulae Phytosophicae* we find a whole section on plants that were of a *mezza* or *mezzana natura* (the terms often used by Cesi and his correspondents), and in the *Speculum Rationis*, a division headed *De mediis naturis in universo*. Here would be included species of a doubtful or ambiguous nature, *de ancipiti natura, seu de ancipiti specie, seu ambiguis*. And the next division was that entitled *De Plantis imperfectis*, in which Cesi planned to discuss those plants with a defect or distinctive characteristic—*defectu vel nota peculiari*—or which were otherwise anomalous. One of Cesi's great breakthroughs was his clear-headed realization that the problem of the ambiguous or borderline or intermediate case would only be resolved by close attention to the reproductive systems of plants. In this respect he was one of Linnaeus's most important predecessors.

Very swiftly then, Cesi and his friends realized that in order to discover the truth in such matters, they would have to penetrate the mysteries of reproduction, and to this end they studied ferns and fungi of every description, anatomized the marine turtle and the bee, and began to make use of the microscope. Indeed, there are several puzzling drawings of eggs and foetuses, and the earliest illustrations ever made of botanical specimens with the aid of a microscope. These last appear in the volumes of plant drawings in the Institut de France where the newly-perfected instrument (a rudimentary example had been used previously by Giovanni Battista della Porta) was used to scrutinize the sori and sporophytes on the underside of ferns. It would be several years before the first *printed* illustrations made with the aid of the microscope appeared. These were the anatomy of the bee prepared by Cesi, Stelluti and Colonna in honour of the papal jubilee in 1625, and the enlargements of the seed-pod and seeds of the 'Chinese Rose' (*Hibiscus mutabilis* L.) in Giovanni Battista Ferrari's *De Florum Cultura* of 1633.

While it is clear that most of the drawings in this corpus were made for the purposes of research and taxonomy, it is just as clear that many were intended for eventual publication. Above all, this emerges in the case of the citrus drawings. Very many of these were published in Ferrari's *Hesperides*; and for this, as we have already noted, Cassiano himself did a considerable portion of the preliminary research, as well as organizing much of the necessary funding. In the case of the fossil drawings, only a few were eventually selected for reproduction in Stelluti's strangely minimal and evasive work on fossil woods of 1637, the *Trattato del Legno Fossile* (see especially pp. 56–63 below). One of the beautiful series of ornithological drawings was used in the *Uccelliera* of 1622, and others (as we may judge from the excellent treatises written by Cassiano himself and preserved in the *carteggio* now in the Library of the Accademia Nazionale dei Lincei in Rome) may well have been intended for further, never-realized publications in this domain. The same probably applies to the animal and flower drawings too. Certainly Cesi himself must have intended to publish at least part of his great series of fungus drawings, his *Icones Fungorum*, but this project, like so many others of his, never saw the light of day.

The one project to which this does not apply was the *Rerum Medicarum Thesaurus Novae Hispaniae, sive Plantarum Animalium Mineralium Mexicanorum Historia*. This was the great compendium of material on American plants and animals that first appeared in incomplete form in 1628 but was eventually published in full as late as 1649–51, long after Cesi had died and the Academy was effectively defunct. Moreover, its many illustrations bear no direct relation to the drawings in the present corpus. But its story is an instructive and telling one. Cesi had been preparing this work ever since he had heard of the Neapolitan doctor Nardo Antonio Recchi's summary of the massive collection of notes and illustrations gathered in Mexico by

Hernández between 1571 and 1577, and neglected ever since. Already in 1610 Cesi had proposed to publish Recchi's manuscript (which by then he had bought from Recchi's heirs), accompanied by a commentary by his fellow Lynxes. When Galileo made his epochal visit to Rome in 1611 he paid a special visit to Cesi's palace in the Via Maschera d'Oro to look at illustrations of over 500 Mexican plants (presumably those in Recchi's manuscripts), and when asked his opinion confessed that neither he nor anyone else could identify them correctly.

By the end of the decade Cesi had already prepared the set of over 800 woodcuts for the proposed work, several of which he then annotated himself; and finally, by 1628, a first edition, funded by Cardinal Franceso Barberini, appeared. This contained Schreck's edition of the botanical material, Faber's great and expansive commentary on Mexican animals, a set of important additions and annotations by Fabio Colonna, and the first thirteen tables of Cesi's own vast attempt to classify all of the botanical world, the *Tabulae Phytosophicae.* Then Cesi died, and had it not been for the devoted and patient efforts of Cassiano and Stelluti, none of this material would ever have appeared. But for years they kept the project alive, bringing it up to date, writing hither and thither for more information, and always seeking new sponsors for a revised and expanded version. Thus it was that the definitive edition of the *Tesoro Messicano* only appeared in 1651 (though several copies carry the dates of 1648 and 1649 too), long after the death of Cesi and of almost all his original *confrères*. This time the book was not subsidized by the now disgraced Barberini, but rather by the Spanish ambassador Alfonso de las Torres, who had been involved in the project by Cassiano. It contained the parts already printed in 1628, but with new front- and end-matter, as well as seven of Cesi's *Tabulae*, which the latter had apparently left incomplete but which were now edited and completed by Stelluti. It also included Hernández's *Liber Unicus* on Mexican animals (including birds, fishes and insects) and minerals. This Cassiano also had copied at the time of his visit to Madrid in 1626.

But the woodcuts in the *Tesoro Messicano* are positively crude and rudimentary and lacking in detail compared to the bulk of the surviving drawings in the Paper Museum. Presumably they were based on the now-lost illustrations that Hernández himself had made in South America, or on Recchi's copies of them. But at the same time as Cesi realized the importance of publishing Recchi's material as evidence of a world unknown to the ancient authorities, he also knew that it was possible to have much better illustrations made, with closer indications of colour, texture, and detail, both inner and outer, than had ever come from the Americas. He and his fellow Lynceans made it their business to find draughtsmen—and then engravers—who could convey with as great as possible precision and detail both the things they had examined with their own eyes and the remarkable specimens seen by reliable eyes abroad.

This is why these drawings—preserved in the Royal Library, the Institut de France and in many private and public collections—are so important. They, more than anything else, provide the visual evidence for the Lincean researches into natural history. Aside from Ferrari's *Hesperides,* also published long after Cesi's death, and written by someone who was never a *Linceo* himself, the published material offers only the palest shadow of the intensity and quality of their researches in this domain. The illustrations in works such as the *Trattato del Legno Fossile* and the *Tesoro Messicano* have nothing like the precision, refinement, and analytic intensity of the drawings. Over and over again they testify to the fact that they were based on direct observation, and on a kind of observational persistence that the printed illustrations hardly ever suggest. The drawings show the same specimen from one side and from another, in sections and in segments; they focus on specific parts of plants or animals or minerals, from the snout of a porcupine to the seed pods of plants, from the sulphuric florets and pyritic growths on fossil woods to the interiors of shells and the details of leaves and grasses, from the legs and beaks of flamingoes to their wings and feathers. They carry inscriptions about colours or actually show them; sometimes they record the date they were made, or the location. Repeatedly they bear witness to the Lincean desire to look ever more closely into the interior of things that until then had been described only from the outside. This was done either with the aid of the microscope or simply by picking apart and delicately dismantling the specimens which they had found or which had been sent to them by friends and colleagues.

Together these drawings represent the high-water mark of the belief in visual reproduction as a means of analysing and classifying the creations of nature, indeed of everything in the visible world. Other ambitious attempts had been made before to reproduce much of nature, and there would, of course, be more; but it is hard to think of any other collection to match the drawings from Cassiano's Paper Museum in terms of quantity, range, scope and, in so many cases, descriptive and artistic refinement. At the very height of the Linceans' engagement with drawing as a tool of natural history research, however, their own fellow and friend Galileo laid to rest the idea that the subjective qualities of things were capable of being recorded with the same objectivity as number, mass, volume, weight and size. And so the *Lincei* themselves began to lose faith in the possibilities of pictorial reproduction as an instrument of what we would now simply call scientific research. Cesi would bury himself in his efforts to classify the natural world not by means of drawing but in tabular form; and the basis of this effort, he realized, had to be geometrical and schematic, not densely descriptive and pictorial. But neither he nor any of his fellow Linceans ever felt the same hostility to picturemaking that Linnaeus did; and whatever scepticism they may have had about the possibilities of reproduction in visual form lies silent beside the observational intensity, descriptive beauty and analytic closeness of the drawings reproduced in the volumes of this corpus.

THE PALAEONTOLOGICAL RESEARCHES OF FEDERICO CESI AND THE FIRST ACADEMY OF LINCEANS[1]

David Freedberg

NOTHING in the early history of palaeontology prepares one for the intensity with which Federico Cesi set out to record and study the fossil remains in the Umbrian hills and villages around his native town of Acquasparta. Until now, the little we know about his work in this domain has had to be inferred from a few fragmentary indications in his correspondence and from the strangely inconclusive allusions to his research in the *Treatise on Fossil Wood* (*Trattato del Legno Fossile Minerale*) written by Francesco Stelluti and published in Rome in 1637.[2] Almost no historian of palaeontology has known of the extraordinary corpus of 199 drawings preserved in the Royal Library at Windsor and here published *in toto* for the first time.[3] Together they provide eloquent testimony to the intensity and texture of Cesi's palaeontological researches. They also flesh out the little information to be gathered from other sources about the work he planned to write on the subject, but which was cut short by his premature death in 1630.

It is true that there had been some illustrations of fossils in the pioneering works of earlier natural historians such as Encelius, Gesner and Aldrovandi,[4] and it is still possible to admire the sharpness of observation and insight shown in the fossil researches of writers such as Agricola, Cardano, Palissy and Cesalpino.[5] Mysterious and puzzling fossil phenomena such as aetites, *glossopetrae*[6] and the fabled bezoar had long occupied the attention of natural historians, but these were more the object of fantasies based on myth and fable than any kind of careful and realistic scrutiny.[7] There had never been anything to match what Cesi produced

Aristotle, Theophrastus, Celsus and Pliny had all mentioned a variety of fossil forms in the course of their discussions of stones, metals and minerals; and Cesi, well-read in the ancient natural historians as he was, knew these references well. He was also familiar with the work of his immediate predecessors, such as Gesner and Aldrovandi. But Gesner never produced the larger work he planned, based on the collections and findings of his friend Johannes Kentmann, though his small book of 1565 did contain illustrations of a few fossil forms. Like most of the other works which go under his name, Aldrovandi's 1648 *Musaeum Metallicum* was a posthumous production, based on the material he had collected in the second half of the sixteenth century, just before Cesi began his own investigations. Put together by Bartolomeo Ambrosino, the *Musaeum Metallicum* gives the reader a vivid, almost dizzying sense of the old encyclopedism—which Cesi would significantly renounce. As in the other works which go under the name of Aldrovandi, everything conceivably related to the ostensible subject of the book seems to be here. A vast and promiscuous quantity of information is brought together under a series of headings in which history, fable and what we would now call science were all, as Michel Foucault so splendidly described, inextricably mixed together.[8]

For each stone, fossil or metal in the *Musaeum Metallicum*, there were headings for its synonyms, etymology, nature and effect, differences (the technical Aristotelian term for the principles on which classification and therefore taxonomy had traditionally been based), locus, medical use, other uses, myth, fable, emblem, and so on. Fantastic fable, pure gossip, folk remedies, sheer superstition, tourist's report—all this was mixed in with sometimes useful and accurate information and observation. It was almost impossible, given such a farrago, to sort out what was of genuine natural historical value, and what was simply the stuff of myth and fable. Aldrovandi and his editors were random in the information they put down, and all too credulous.

There is almost none of this in the corpus illustrated in the present volume. Eschewing all superfluous information, Cesi simply annotated, with spare notes on colour, the meticulous and detailed drawings he had commissioned from his draughtsman. Nothing was shown that he had not seen and observed himself. Not for Cesi the tales and travellers' reports—only the colours of his specimens, their measurements, the locations in which they were found, and occasionally a date. Unlike several other parts of the Cesian corpus of drawings, here there was no drive to anthropomorphism and similitude as a basis for explanation. He simply made his draughtsman draw what he saw, as objectively as he could, and often from several angles. The descriptive density and sheer exactitude of these drawings is without precedent—as is the repetitive concentration on single specimens. It is as if Cesi set out to catalogue every single one of the fossils of the region of Umbria around Acquasparta, coldly and without superfluity. Moreover, he was determined to devise a whole new taxonomy for the particular fossil woods and concretions he found there, and to deal with the critical problem, of their apparently combinatory forms and double or even triple natures. The struggle to find an adequate terminology was also without precedent, resulting in a flood of new terms and variations of old ones that still defy satisfactory translation.

As Martin Rudwick has shown, the earlier writers lacked consistency in their views of fossil origins (if such writers knew they were dealing with fossils in the modern sense at all); but one precursor must be singled out above all others. He, of course, is Leonardo da Vinci. Although it is unlikely that Cesi knew of Leonardo's work in this area—long before the Linceans and their sixteenth-century predecessors—Leonardo made many perceptive observations on the subject of fossils, and marine fossils in particular. He set aside the old idea that Noah's flood (or some other such deluge) was responsible for the presence of fossils on mountain tops and concluded that the cause was to be found in the pushing-up of the earth, from beneath the waters that once covered it, rather than their being carried along by the water itself. He too studied fossil woods—one of the very few to have done so—and he took issue with the Aristotelian view of the generation of fossils, according to which they arose from some kind of inorganic 'seeds' within the rock. Such views would have been of much assistance to the Linceans in their research, had they but known of them. But they did not. Leonardo's researches were to remain buried in his manuscripts for some time yet.[9]

For many in the sixteenth century (and later), fossils were categorized under the rubric of 'things dug up' (the literal meaning of the word *fossiles*). Little if any distinction was made between the organic and the inorganic, and the term 'fossils' covered the whole gamut, from ancient statues recovered from the earth to mineralized plant and animal forms. Sometimes it may seem that in the work of Cesi and Cassiano dal Pozzo the two forms of *fossiles* came together; but the fact remains that no one before Cesi had ever set out to make a comprehensive visual record of the organic fossil remains of a particular region. Indeed there is nothing prior to the corpus of drawings reproduced in this volume that even remotely approaches the sustained quality and quantity of the drawings which Cesi commissioned of the fossils he found around the place of his birth.

Although we do not know the name of the artist, it is clear that these drawings were commissioned by Cesi himself. Many carry annotations in his own hand, giving the name he devised for each specimen and detailed information about the location in which it was found, its exact size, its appearance, and above all its colour.[10] Along with the contents of his library and the other drawings commissioned by Cesi, these drawings were bought from his widow by Cassiano in 1633. Even if the drawings were not bound together before—which most probably was the case—they were presumably bound into volumes at this stage, of the kind known from elsewhere in this corpus.[11] It was probably while they were in Cassiano's Paper Museum that a scribe prepared the surviving indices to these volumes[12] and made generally (but by no means always) accurate transcriptions, on the sheets themselves, of what Cesi had written there.[13]

The drawings passed from the dal Pozzo family collection to that of the Albani, and were bought for the collection of King George III in 1762.[14] The drawings catalogued here, having first been removed from their original mount sheets and laid down onto new ones,[15] were then bound into four large folio volumes. The spines of these volumes carry the label *Natural History: Fossils* and are numbered XIV, XV, XVI and XVII [Fig. 3][16] in the series of (probably twenty) volumes created by George III's librarian from some of the original volumes of natural history drawings in the Paper Museum.[17] The drawings in the first three of these volumes appear to have been retained in their original order. They are numbered 1–157 in pen and ink in a seventeenth-century hand, with four unnumbered sheets interspersed in the otherwise continuous series.[18] The coloured drawings of concretions in Volume XVII are separately numbered 1–38. The indices referred to above were mounted into the beginning of volumes XIV and XVII [see Fig. 4].[19]

Twelve of the drawings in the corpus represent the sites—often quite dramatic—of the fossilized woods and everlasting fires near the Umbrian villages of Dunarobba, Sismano and Rosaro (**1–6** and **12–16**), and Perrosciolo (**17**). Five show pieces of fossil wood actually embedded in the earth in these locations (**7–11**). The largest group is formed by the extraordinarily meticulous and closely observed drawings of fragments of fossilized wood (**18–133**). Equally compelling are the smaller groups showing aetites or 'eagle-stones'[20] (the nodular concretions which so fascinated early students of fossils because they were apparently 'pregnant', **134–145**);[21] the eight sheets with fossil animals embedded in clay or in one kind of concretion or another (**146–154**); the group of baked or burnt clays shown on **155–196**; and, last, the three splendid sheets of drawings of ammonites (**197–199**).

With the exception of the beautiful, subtle and closely observed colours of the 38 drawings of concretions in Volume XVII, the drawings are all in pen and ink. They are all of remarkably high quality and of exceptional technical accomplishment. It is possible that the bulk of them were made around the time of the only date on any of the drawings, 1611, when Cesi brought that piece of pyritized wood (**30**) to Rome.[22] Together they constitute the first extensive set of field drawings of a fossil site or set of geological features yet made. That there was always an intention to publish some of them is clear, both from the correspondence between Cesi and his friends, and from the frame-lines that surround 43 of them.[23] Similar frame-lines appear in all but two of the thirteen plates eventually derived from them in Stelluti's *Treatise*, that ultimately pale reflection of what Cesi intended.

New Light on the Question of Heat

In the hills opposite the small Cesi family castle in Acquasparta, and in the neighbouring region of Dunarobba, Rosaro and Sismano, one may still see the extraordinary remains of fossil woods that first gripped the attention of Cesi and his three young friends, Johannes Heckius (otherwise van Heeck or Heck), Francesco Stelluti and Anastasio de Filiis. In the catalogue below, many examples are reproduced of what they actually saw, and then had drawn. Embedded with other fossils in the ground, or still protruding as strange and forlorn trunks, these fossil remains seemed to Cesi and Stelluti to offer the key to two of the central problems of their scientific lives: the classification of the natural world, and the nature of light and heat. What exactly were these fossils: animal, vegetable or mineral? Why did they glow in the dark, or heat up when dipped in cold water, and what was the nature of the subterranean fires that burned beneath them? Most of the Linceans, but Cesi and Stelluti in particular, were absorbed by the problem of petrified woods, of underground fires, of ammonites, of concretions, of the marine fossils embedded in concretions, of combustible lignites, of pyrites of various kinds. Among the very first letters the Linceans wrote (as a group) to botanists and other natural historians is a revealing pair of letters to Adone Campello, the famous alchemist of Spoleto, in which they posed the problems of petrifaction and of the vitrification of metal.[24]

Even before Cesi went to Naples in 1604, to escape harassment by his father (who hated his scientific proclivities) and to explore the possibility of setting up a branch of his Lyceum or 'Lynceum' there, he was well acquainted with the work one of the most brilliant scientists of that city, Fabio Colonna. All his life Colonna was interested in fossils and in phenomena on the edges of plant, animal and mineral life. To his 1616 treatise on the *Murex* or sea-snail, entitled *Purpura*, Colonna appended a work that was not just pretty but a landmark in the history of palaeontology, his *De Glossopetris Dissertatio*. As with his *Phytobasanos* of 1592, he himself etched its plates, and in it he proved, long before Nicholas Steno, that the so-called *glossopetrae*, or 'tongue-stones', were fossilized sharks' teeth, often occurring within the shells of molluscs. In the same year Colonna also republished his important treatise of 1606 on marine and terrestrial animals, the *De Aquatilibus Alijsque Animalibus*, itself illustrated with more of his etchings of fossil remains.[25]

Although Colonna was away from Naples at the time of Cesi's visit in 1604, the nineteen year-old Prince could at least visit the aged and much-admired Giovanni Battista della Porta, one of the greatest authorities in all of Europe on natural magic, and a leader in many fields of natural history, including botany and mineralogy. Cesi also went to see other distinguished natural historians of Naples, notably Ferrante Imperato. Imperato's museum was known all over Europe, and among its many treasures were the fossils, on which both he and his son Francesco wrote their own much-respected works: Ferrante's own *Dell'Historia Naturale ... libri XXVIII. Nella quale ordinatamente si tratta della diversa condition' di miniere e pietre*, published in Naples in 1599; and Francesco's *De Fossilibus Opusculum* of 1610 (also published there).[26]

But what was it about the fossil woods, the concretions, the baked clays, the pyrites, the underground fires, that was so important to the Linceans? There can be no doubt that the issues of spontaneous combustion and the spontaneous generation of light—and the extent to which the generation of light had to be accompanied by heat or not—remained central to them. These issues were closely related to Galileo's concerns about the diffusion of light from the sun, the phases of Venus, and the whole question of the secondary light of the moon.[27] For Galileo, the question of whether light was a quantity or quality (as the Aristotelians claimed) was a critical one, because the quantitative view lay at the basis of his commitment to an atomistic view of the universe.[28] No wonder that the Linceans' correspondence is full of references to underground fires, to the generation of light without heat, and to the phenomenon of pieces of fossil wood that could catch fire even when damp and cold. These issues appear in Antonio Persio's hardly-known treatise, the *De Natura Ignis et Caloris* (of which a tiny summary was published in 1613),[29] in an important lecture delivered by Cesi's friend and fellow *Linceo*, Johannes Faber in 1622,[30] and in the works of the now almost forgotten friend of the Linceans, Fortunio Liceti (in whose writings the study of antiquity is so closely combined with that of fossil remains).[31] Right until his last days Galileo engaged with Liceti over issues of heat and light.[32]

The story of how Galileo demonstrated his telescope to a group of eight friends gathered together by Cesi on the Janiculum on 14 April 1611, is well-known;[33] but what happened the following morning has been told less often—and is of even greater relevance to the present story.

Just as dawn was breaking, with the telescope put away for the time being, Galileo invited Cesi and his friends to leave the terrace and come inside Monsignor Malvasia's villa. His old interlocutor and opponent Magini had sent him something from Bologna he wanted to show them. Even though they must by then have been exhausted and excited by their night-long study of the heavens, Galileo had one more experiment to share with them. He took them into a small room, and there, in that crepuscular light, opened a little box. To the amazement of those among them who had not yet heard about it, the stone inside (probably a calcined form of barium sulphate) glowed briefly with a cold light, and then began to fade. They called it the Bologna stone, or the *spongia solis*.[34] A few moments later, Galileo went out onto the terrace again, exposed the stone to the weak light of the dawn, closed the box, brought it inside, and opened it again. Once more it glowed for a while, and then petered out.[35]

The full implications of this experiment would not by any means have been clear to those present. But everyone there must have realized at least some of the implications for the new cosmology. After all, if the attenuated light of the sun at dawn could be weakly re-emitted by the stone in the box, why could not the same principle apply to the planets? And if this light was quantitative (and not the qualitative phenomenon the Aristotelians thought it to be) could it not be reflected differentially across space, thus accounting for the diversity in planetary brilliance? Furthermore, it was now much easier to grasp Galileo's proposal that the moon was not naturally luminescent, and that the light of its phases was not its own, but solar light reflected off the earth. Just like the Bologna stone, it could receive its light from the sun, absorb it, as it were, and transmit it coldly across the darkness of the heavens; it did not have to burn with its own fiery light in order to do so. And what *this* entailed was something much more substantial and momentous: namely that the earth was not the centre of the universe. One begins, dimly, to sense the controversial nature of the investigations in which the Linceans were involved.

This was one set of crucial issues in Cesi's mind when he first began to contemplate making an accurate record of the fossil woods around Acquasparta. As the drawings show, these remains were full of pyritic and sulphurous substances, and their formation was clearly related to the underground fires which burned in that region. Can it be a coincidence that 1611 is the only date to appear on any of the fossil drawings in Windsor, recording that a piece of contorted pyritized wood—or 'metallophyte', as Cesi called it—had been found by him and taken to Rome in that year? *Anno 1611 a nobis invent[a] et Romam allat[a]*, he proudly wrote on it.[36]

The term 'metallophyte' was a critical one for Cesi.[37] He defined it as a natural form that was 'doubtful whether plant or metal'. The term was 'invented by us, and which we have explained in several books'.[38] These books, unfortunately, were never written; later we shall see how much or how little of them may be reconstituted now. In any case, the metallophytes were only one of several kinds of fossils in which Cesi was interested, for a variety of reasons. The drawings in this volume are evidence of the Linceans' commitment to observation and experiment, as is their correspondence. But there was another and more crucial motive in all this. It first emerges in a remarkable draft manuscript which Cesi began to work on about the same time as Galileo did his seminal experiments on the Janiculum.

Classification and the 'mezzana natura'

In this manuscript[39] Cesi attempted to set out a general scheme for the ordering of all human knowledge—natural historical, physical, mathematical, and cultural. Entitled the *Speculum Rationis*, it was intended to give a foretaste of what Cesi called the frontispieces—or summaries—of his *Theatrum Naturae*. All too aware of the likelihood of being overwhelmed by the abundance of information which the Linceans were gathering, Cesi realized they needed some system for ordering that information (which, they believed, would in turn provide clues to the order of nature).[40] Cesi thus intended his frontispieces to serve as a kind of synoptic overview of the things both of nature and of culture. Not surprisingly, given the extraordinary ambition of such a scheme, only one set of these frontispieces was ever realized.[41]

The importance of the Naples manuscript lies not just in its outline of Cesi's scheme to discover the order of nature, but also in the evidence it offers for his belief that no system of classification, no provision of a taxonomy or an ordering grid, could ever be adequate unless it first confronted those things that threatened to confound such a system. The problem lay in the borderline cases, that is, those things which participated in more than one class, and could therefore not easily or satisfactorily be classified within the boxes provided for them. No test case (other than those of fungi and ferns, to which other volumes in this series will be devoted) occupied Cesi's attention more closely than the metallophytes from the hills and villages round Acquasparta. Not only did they raise questions about the nature of heat and light, they also dramatically posed the problem of how to establish criteria for classification that would not be subverted by apparently overlapping or intermediate classes—those phenomena that Cesi, repeatedly in this manuscript, described as being of a middle nature.

It is precisely this problem, that of the *mezzana natura*, that emerges so clearly from the drawings in Windsor. No one could fail to be struck by Cesi's preoccupation with it. Especially in the drawings of fossil woods, Cesi and his draughtsman went to extraordinary lengths to emphasize, both verbally and visually, where specimens showed an overlapping—and therefore ambiguous—nature. Even a superficial survey prompts the question of whether one is dealing with wood or with stone. And then, one asks oneself (just as Cesi and Stelluti did), what is one to make of the group of baked clays that are so frequent in these sheets? These baked clays contain a whole variety not just of metalliferous ingredients such as pyrites, but also of fossil animals and plants. Pyritic elements were crucial to the issues about spontaneous light and heat that Galileo was raising (one has only to read the relevant pages in Pliny to be aware of that);[42] while the combination with fossil animals and plants self-evidently raised the issue of the *mezzana natura*.

Then there were the aetites. These were stony forms that often contained molluscs, gastropods or bivalves. Even if one did not stumble over the definitional problem of animal *vs*. mineral, one might still want to know how such combinations arose in the first place. The case of the ammonites, much more frequently studied in the past than fossil woods, was by no means resolved either. A multiplicity of views about their origins was still current. No one even began to suspect that they were actually the shells of extinct cephalopods.

The inscriptions on the drawings catalogued here also declare the double or triple nature of these finds. They offer seemingly endless variations on the theme of 'stony on the inside, woody on the outside', and a plethora of terms such as *petrilignum*, *lapifactum*, *metallophyta*, *metalligo*, *lignimetallum*, *ligni metallificati*, *lapilignea metallificata*, *lapilignum* (*aliquantulum metallico*), *lapiferreum* and *cretilignum*. With varying degrees of precision, such terms convey something of the endless doubts about whether the fossilized woods were actually wood or clay, or how *much* of each. In both inscriptions and drawings, the degree of participation in one class or another was almost pedantically calibrated, as Cesi struggled to find appropriate terms for the phenomena he was seeking to describe.

Most of the drawings bear inscriptions in two hands. The first of these is the distinctive hand of Cesi himself. The second is the equally characteristic hand of the scribe whom Cassiano employed to transcribe Cesi's inscriptions. This seems to have been the same scribe who made the indices to the volumes and whose hand often appears in the surviving copies of Cassiano's own letters. The idea of the transcriptions was presumably to transform the working notes of Cesi into clearer and more elegant handwriting. These scribal transcriptions have a twofold merit. First, they are indeed often clearer than Cesi's own handwriting; and second, they often valuably restore to us words which were cut off when the drawings were prepared for mounting.

Unfortunately for us, the scribe's Latin was not so good. Gender endings often fail to agree, complicating the task of translating Cesi's already terse descriptions. Sometimes cases are patently incorrect. Prepositions are misused. Occasionally the transcription of what Cesi wrote is incorrect. The separation of parts of words—often Cesi's complex nouns for his fossil finds—does not help either (an egregious case is the *lapila teri lignum* in **38**; a misleading one is *An Fractuosum* in **39**). But the chief difficulty with the inscriptions has actually to do with Cesi himself—or rather with his strenuous, even heroic, efforts to provide adequate terms for different samples.

Often he fell back on old terms, themselves sometimes very rare. But more frequently he devised (or 'invented' as he was proudly accustomed to put it) new terms for the apparent combinations of animal, vegetable and mineral which so fascinated him about the fossil remains of Umbria.

Some of these new terms present little difficulty in understanding—at least in a broad sense—what Cesi intended by them. We grasp immediately what he was getting at with his favourite term, which he claimed to have invented, of 'metallophytes'. Many of his other terms convey the hybrid nature of his finds, such as *lapilignum*, *petrilignum*, *cretilignum*, and even the strangely hybrid (but rather touchingly descriptive) *litoxilum* (as in **20**, **24** and **43**). No one could fail to be moved by the classificatory poignancy of such terms, intended to separate what was probably inseparable. Cesi tried to introduce even further refinements to what we might now call his middle terms (terms to indicate combinations from two separate kingdoms, such as those just mentioned). A typical example is the odd case of the *lapisurgens* (also used in his *Tabulae Phytosophicae*), as well, perhaps, as that of the *lapi Resina* (**141**). And even though we can see what he was driving at with phrases such as *lapilignea metallificata*, *lapilignum aliquantulum metallico* or *ligni lapiferreum*, we still wonder what kind of precision was envisaged by terms such as *metalligo* (**111**), a term which will elude most modern interpreters, as it has us. Other refinements such as *lapiferreum* and *lignimetallum* are rather clearer.

Cesi seems to have been especially fond of a number of descriptive terms which he apparently thought of as reasonably precise. We may now be pained by their seeming lack of precision. For this reason they do not always receive consistent translation in this catalogue, but are adapted to their context. They include words such as *corticosus* (for 'bark-like'), *suberaceus* ('corky'), *coriaceus* ('leathery'), and a variety of terms for knotty excrescences, such as *[con]tuberatio*, and *tuberculosus* or *tuber[cul]ose*. Of course, there are all-purpose terms that recur with justified frequency, such as *ligneus* ('woody') and *lapideus* ('stone-like'). The term *nervosus* is a little—but not much—more problematic. Specimens are compacted (*compactus* or *pressus*), anfractuous (*anfractuosus*) or simply knotty (*nodosus*). In fact, there is no shortage of terms to describe qualities of compaction. But when we encounter terms such as *musculus*, for the 'sinews' or 'muscles' of a piece of wood (*costa*, a rib, is sometimes used too), we realize how limited was the technical vocabulary available to Cesi, with terms based on similitude rather than analysis. Lacking any better terminology, he could only explain these strange pieces of wood in terms of what they resembled. We should perhaps be grateful for the latitude such seemingly picturesque description allows us.

Cesi seems to have been especially fond of a few old terms as well. These are to be found on the drawings and in the correspondence (especially in the letter of 1 December 1624 discussed on pp. 44–5 below and translated in Appendix III a). Many such terms are found in the ancient writers, especially Theophrastus and Pliny: *gagate* (presumably for what we now call jet), *ebenus*, and of course the ubiquitous aetites. But the Italian words *litantrace* (a form of anthracite) and the rare word *acciavaccio* (also, presumably, some kind of anthracite) are more mysterious. Unusual words such as *pompholyx* (or *pompholigum*, in **180**) are discussed in the relevant entries.

The most puzzling recurrent term, however, is that of *lateritium*. It is puzzling not because we have no idea what it means, but because Cesi's exact meaning is not clear. Indeed, this may be the one term he used which he did not envisage as having any exact meaning: it may simply have had the broad sense of 'brick-like', its standard Latin use. One thing we *do* know is that Cesi did not use *lateritium* either to indicate the modern laterite, or to suggest that *lateritia* were just man-made bricks—though in some respects these came close to many of the examples actually shown. The term *lateritium* almost always appears on these drawings in the context of some kind of baked clay or another, and so the epithet 'brick-like'is not entirely off the mark. Cesi clearly used the term to indicate the brick-like texture or colour of the specimen concerned (baked, as he supposed, by the underground fires of the region). The trouble, though, is that Cesi used the word as a substantive, not as an adjective; and so we are left wondering whether he really thought the class of *lateritia* was indeed just that, a class. Possibly he did not; but to use a specific term such as this with such lax classificatory intent would have been very uncharacteristic of him. We are thus left with the problem, not entirely resolved in the catalogue entries, of how to translate this critical term. For the rest the inscriptions are straightforward enough, with the exception of the occasional mystifying place name (such as *Perrosciolo* in **17** and *Picchia* in **48**) or epithet (such as *Firmanus* in **62** and **153**—unless it is a proper name, which is quite possible).

All this is reflected in the drawings, which seem to document every subtle difference between the specimens of fossil woods. And the draughtsman attempted to match Cesi's verbal descriptions in pictorial terms. Measurements are nearly always given, and sometimes the same specimen is shown in different views (for example, **36, 45–6, 77**). Thanks to Cesi's faith in representation, we seem to see with our own eyes how wood is pyritized, baked, turned into stone, composed half of clay and half of wood, or how specimens that look just like a branch or knotty twig resemble nothing so much as pure stone. One can almost imagine the earth itself turning into these woody forms (as Stelluti later said that Cesi claimed). The drawings show not only the inexhaustible varieties of woodiness; they also show veins, undulating growth lines, grains, chips, metallic fibres, sulphurous speckles and twinkling pyritic elements.

In these drawings, even more than in their inscriptions, Cesi still clings to the old beliefs in surface description and subjectively perceived qualities. He was doing so in the years just preceding the *Saggiatore* of 1623 in which Galileo made it clear that such qualities could not tell one about the essential nature of things, or be of much use in the matter of classification.[43] Cesi realized that in order to classify things, to avoid the appearance of overlap, one had to be able to discover what was essential about them. And yet in these drawings it is impossible to escape the problem of overlapping and borderline forms. There was no more persistent an attempt at recovering the minutiae of 'difference', as the Aristotelians would then have put it, until the remarkable drawings of citrus fruit commissioned by Cassiano a few years later for his friend Giovanni Battista Ferrari.[44]

The breakdown of drawings into the groups indicated in this catalogue does not correspond with the way they were actually bound in Cesi's and later Cassiano's collections. How then did Cesi order these drawings? Although the names inscribed on them and the alphabetical indices at the beginning of each volume give us some idea of the principal divisions into which Cesi placed (or intended to place) his specimens, the differences between them were often so small, or dependent on secondary qualities (as they later came to be called),[45] that it is still difficult to know exactly how he planned to classify the abundance of the palaeontological material before him.[46]

Fortunately, the surviving documentary evidence, whether in the form of the little-known manuscripts of Rome and Naples,[47] or in the published letters of the *Lincei*, casts much supplementary light on Cesi and his palaeontological interests, particularly when read in the context of these drawings. In first place there is his manuscript now in Naples. Not everything in it can be directly related to the drawings, but many elements can be. It also provides a background to the many references to fossil finds in the Linceans' correspondence, as well as to the connection with Galileo.

The early sections of the Naples manuscript contain much that relates not only to Galilean astronomy, but also to his optics and atomism. Thus, in a draft there for his treatise entitled the *Coelispicium*, Cesi provides headings for comets, novae and sunspots, as well as for various problems relating to the transmission of light in the heavens. Then comes a section headed *Taumatombria*, 'on marvellous [celestial] rains', which discusses another general topic of central interest to Cesi—that of extraordinary liquefactions, saps, juices and resins. These were all to be related to the old theories about the petrifying juices instrumental in the formation of fossils, as well as to the even older issue of meteorites (which showed, as it was then thought, that things could be petrified even in the upper atmosphere).[48] Above all, this section pointed to the problem 'of the marvels of stolen light'—*de luce furtiva admiranda*—and 'nocturnal light and the art of the light-bearing stone',[49] in other words the *litheosphorus* (or *pietra lucifera*) from Bologna, which Galileo had demonstrated on 15 April 1611.[50]

Then follows the section that has perhaps the greatest bearing on the fossil treatise that Cesi never completed. Under the heading 'On middle natures in general'[51] came

> things of doubtful nature, or doubtful species, or ambiguous things. On [the nature of] ambiguous Nature. Two different natures joined in a single species ... species participating in two natures. A very remarkable corner of the Theatre of Nature, all the more hidden, concealed and obscure because passed over.[52]

For Cesi, such intermediate classes included not only amphibians and the 'whole class of oviparous quadrupeds', bats, giraffes, divers and so on, but also 'zoophytes', 'lithophytes' or 'metallophytes', and 'zoolithophytes', the terms we encounter time and again in his work.

In this same section, Cesi swiftly moved on to refer to 'imperfect plants', those somehow 'mutilated, deficient, inordinate'. Fungi were the most important in this group, and for the rest of his life Cesi devoted attention to fungi and ferns—both instances of the problems of crossover between classes and unclear principles of reproduction. These crossover forms also included the *lapides plantales figurandi*, 'stones figuring plant forms', another fossilized form that consistently interested Cesi and his friends.

The next section of the manuscript began with a revealing note: 'In this whole family of plants, monstrosities and marvels occur as a result of some particular defect or peculiar characteristic, arising rather from the rudiments of plants than from plants themselves'.[53] The problem was that often these imperfect plants showed no visible signs of seeds or reproductive organs. No wonder that here were further references to lithophytes, and zoolithophytes, as well as to the omnipresent metallophytes.

Always, even when it came to the abstract processes of classification, first-hand observation remained paramount. One almost senses the pride with which Cesi noted down the various 'workshops of nature' which he had visited. These *officinae naturae* (perhaps 'laboratories of nature' would be a more relevant translation) included the *lapificium Tiburtinum* (the travertine quarries of Tivoli), the *ignificium Puteolanum* (clearly the *Solfatara* at Pozzuoli, seen by Cesi when he went to Naples in 1604), and, of course, the *lignificium Acquaspartanum*, the site of the metallophytes and other ligneous fossils of Acquasparta illustrated in these volumes.[54]

The Naples manuscript thus introduces us to a number of terms that were to become the leitmotifs of Cesi's preoccupation with fossil wood. Let us see how these different themes played themselves out in Cesi's lifetime—and afterwards, for it was only in 1637, seven years after Cesi's death, that Stelluti eventually published a pale reflection of the great treatise on fossils which the prince had intended to produce.

The great excitement of 1611–12 arose from Galileo's discoveries with the telescope, which he had first demonstrated to the Linceans in April 1611; the other interest of that year lay in Galileo's second demonstration on the Janiculum on that famous morning.[55] On 10 June, Ferrante Imperato wrote from Naples to Faber in Rome enquiring about the 'little stones shown by Signor Galileo, with the quality of being able to catch and retain the light, certainly unknown to me', to which he added sceptically, 'I don't believe this to be a natural effect, but rather an artificial one'.[56] Cesi himself admitted to Galileo a few months later that the matter was a complicated and difficult one. On 7 April 1612, Della Porta expressed *his* frustration to Cesi about 'those small pieces of stone that retain light in the dark'.[57]

The little stones that caught the light and then mysteriously retained it in the dark, without giving off any heat, were examples of the *litheosphorus* referred to in Cesi's Naples manuscript and several of his letters to Galileo between 1611 and 1614.[58] Faber did some experiments with the 'Bologna stone' in December 1612,[59] while in the spring of 1613 Galileo sent Cesi a box of luciferous stones to work with and, as Cesi himself put it, to enjoy.[60] In 1640, long after Cesi had died, and he himself had been exiled, Galileo wrote a long letter to Prince Leopoldo de' Medici about 'chapter 4 of Fortunio Liceti's *Litheosphorus*'.[61] The debate about the spontaneous emission of light remained a central problem in Galileo's cosmology. In this context one can understand the attention the Linceans devoted to the pyritized fossils they found around Acquasparta, and the even more extraordinary phenomena of underground fires and wood that caught fire when dipped in water.

But the more general implications of such phenomena were best left to Galileo himself, for they touched on the crucial issue of Aristotelian atomism, which only he had the physical and mathematical knowledge to deal with; Cesi, for his part, turned to the problem of classification, and the 'intermediate' status of the metallophytes. It was on Cesi's mind at the same time that he and his closest friends, Stelluti, Faber and Cassiano, were working on the manuscripts of Galileo's *Saggiatore*. In a letter of January 1621, discussing the great botanist Bauhin's classification of plants in the 1618 *Phitopinax*, Cesi wrote to Faber that 'I would also like to know if there is anyone who has distinguished in orderly fashion and enumerated into their classes the fossils, and especially the metallized ones, and the half-metals; and similarly if any has synoptically reduced the sciences into trees and tables'.[62]

Of course Gesner, as well as the two members of the family Imperato, had attempted a classification of fossils; and in Gesner's book of 1565, there is even an illustration of Johann Kentmann's little cabinet, with the compartments in which he had neatly placed (and thereby classified) his fossil and gemmological specimens; but none of these men had investigated and worried about the issues as intensely as Cesi.[63]

Further Lincean researches

It was only from 1624 on that Cesi and his friends could devote more of their time to the issue of classification, and of fossils more generally, because their work on Galileo's *Saggiatore* had been completed. It contained Galileo's considered reference to the great controversy over the comets of 1618 and was the fullest statement of his Copernicanism before the publication of the *Dialogue on the Two Great World Systems* and the trial of 1633. At last they could return to their most cherished projects.

On 18 May 1624 Giovanni Battista Winther wrote to Faber from Acquasparta, where he had been staying at the Cesi palace along with Stelluti and Galileo, now resting from his labours on the *Saggiatore* and the ever-increasing intensity of his controversy with the Jesuits. Those were pleasant and reflective days for all of them, and it is not surprising that Winther's letter should have pondered, perhaps under Galileo's influence, the nature of petrifaction. This time too, the conversation took a rather more aesthetic turn than usual:

> Twice I have been in a very mountainous and clayey place, where a large quantity of very hard woods are to be found beneath the earth, black and perfumed, with very beautiful veins, regarded by our Prince Cesi as minerals; we've had many discussions about the site and the petrified substance of some of them.

Winther had read Cesalpino's chapter on ebony and discussed it with Cesi.[64] In the Windsor volumes, the interest in veining and ebony-like fossils is intense. How did such phenomena, apparently the result of underground fires and the burying of the wood, arise?

On 18 August 1624 Cesi himself wrote enthusiastically to Cassiano dal Pozzo that 'daily I discover wonderful things in the natural laboratories of these parts, about which I shall write to you in due course'.[65] He was referring, of course, to the same *naturae officinae* mentioned in the Naples manuscript, and in particular to the *lignificium Aquaspartanum*. Five days later Stelluti, still in Acquasparta, wrote to Galileo about more finds of extraordinary and unexpected fossil fuels. It was the height of summer, he said, but even in those very hot days, Cesi did not cease his studies and his 'very beautiful observations' of that puzzling intermediate class, the *legno-minerale*. Cesi had found:

> very large pieces up to eleven palms in diameter, and others with iron fibres in them, or similar metallic pieces within such woods, and others which secrete a kind of resin, like incense, even having a similar odour; and then too he has found a very large quantity of woods which have been turned into stone or pyritized,[66] of most extravagant forms; and if your Excellency passes by on his return to Florence, he will see, both with amazement and pleasure, all these woods, and where they originated, as well as some of the underground fires that are here.[67]

In the meantime Cesi had begun to turn his sights on the man who became one of their most important sponsors—and had already encouraged them considerably—Francesco Barberini, nephew of the new pope, Urban VIII. Recently appointed cardinal and elected a *Linceo*, he had not yet turned against Galileo and his friends. It was natural enough that Cesi should have thought of getting him interested in their fossil discoveries, and just as natural that he should have done so by sending him an artistic product made from a large piece of the *legno-minerale*. On 29 September, Cesi wrote to Barberini about a small table which he had succeeded in having made from 'a solid piece of the mineralized wood which I have found in these parts'. He continued: 'and the learned and virtuoso delight which you take in natural observations has encouraged me to send it to you, as something which can best show you the nature of these things, especially the variety of its veins and concretions'.[68]

By the beginning of December, Cardinal Francesco Barberini had become completely involved, and Cesi wrote to him at some length about the more complex phenomena he had found. Clearly he felt that he could take the cardinal into his confidence. The letter is a remarkable—indeed critical—one:

> In the little free time my troublesome affairs have left me, I have pursued my physical observations, particularly of the new type of 'middle nature' between plants and metals, the veined and varied mineralized wood discovered by me some years ago in this region, and which I have already brought to you. And after having studied and seen many differences in these things, which are so full of curiosities and such immeasurable excesses, that within the species of similar composite things they even exceed the proportion the cetaceans have among sensate creatures, to whom it seems that even the sacred texts attribute the notion of largeness.
>
> I have found a great variety of petrified objects, that is, some completely petrified, of the consistency of stone or brick, or mixed—part wood, part stone—or a similar condition, with internal and external combinations of many forms.
>
> That which gave me the greatest cause first for wonderment, then for reflection, was the discovery of metallized woods, and those which were similarly either completely or partially transformed into metal, and strangely altered. I believe this phenomenon to be something completely unexpected by natural historians, as the mixtures described above are most pleasing and at various stages of petrifaction. But for many the matter has remained very doubtful and difficult, being more of a natural problem than a scholastic one.
>
> Moreover I believe that the relationship between these terrestrial natures will open up to me the road to a full understanding of them, which until now has been poor and confused. Because apart from the aforesaid middle nature state, I have found many ambiguities among them, which display mixtures in various ways: of wood and earth, of wood and stone, of wood, metal and stone together, of wood and mineral juices, drawn together and participating equally in each substance, in a body that is almost neutral, or rather combined and reduced to quite apparent unity, whence the very obscure nature of those terrestrial liquids which we call bitumens may receive no little clarification.[69]

Cesi then went on to discuss a whole series of 'mixed' phenomena, many of which we recognize from the volumes of fossil drawings and from his correspondence: gagate, aetites,[70] anthracite, fossilized ebony, *acciavaccio* (probably a form of anthracite), *litantrace* (or lithanthrax: a low-grade anthracite), and finally the fragrant agalloch. Having already commented that he had found the gagate far from the Ganges, where it was supposed to arise, he concluded of agalloch that its origin was also unknown but 'its transport by floods is related' and that it was not likely to be 'so different from this wonderful new species of metallophyte'.[71]

How proud Cesi was of this discovery, about which he was continuing to learn so much! He noted that he was 'also attempting to arrive at, not only the full historical point of view, but also every natural consideration and discussion, in searching for causes and examining how much can be found about these subjects or concluded from them'.

But why this reticence? Barely perceptible it may be—but still unmistakable. What constraints underlie Cesi's words? The question will soon become more urgent (though never clamorous). It may seem that Cesi gives us no help with it here, but is there an allusion, in the passage that follows, to the controversy about substances pregnant with light, and to the whole issue, raised by Galileo himself, of the dissemination not of earthly but of celestial light?

> On such occasions, I have also observed furnaces kindled by nature, their manner and their fumes. I also saw, in this new class of things, the sulphurous efflorescence, as well as the saline, vitreous and crystalline ones; and I confirmed with particular satisfaction many of my theories, made over many years, about the essence and causes of these fires, which although very little known and hidden to human intellects, always seemed to me for the most part to be full of light.

Cesi promised to come in person and give the cardinal a summary of his views the following week. In the meantime he was sending a few small gifts, illustrations of what he was describing: 'a petrified trunk in which both substances are very clearly evident, a piece of simple fossilized ebony, and a more solid and variegated fragment of the type of the same substance in which are interposed metal veins, similar to those which appear in some additional pieces of a much more fragile and dissolved nature similar to lithanthrax'.[72]

This letter was immediately recognized as central to Cesi's work. In addition to several drafts for it,[73] no fewer than six manuscript copies survive, generally in the company of manuscript versions of Stelluti's *Trattato*.[74] The day after he wrote his letter, in a brief note to Cassiano, Cesi commented that 'I am a sending to Signor Cardinale a little indication of the significance of the things discovered by me *nella inferior Natura de' terranei*, along with some pieces of such objects'.[75]

It is hard, in rendering Cesi's letter to Francesco Barberini into English, to avoid the word 'fossilized'. One wants it, because it would so easily account for the phenomena described by Cesi. But although the term 'fossil' was certainly current, it was generally used in the sense of anything dug up from the earth, and without the evolutionary and palaeontological connotations it now has. Indeed, although there was nothing in the existing literature on fossils to offer an adequate explanation for the fossil woods of Acquasparta, Cesi himself never offered anything really convincing either—at least not in the few surviving indications of what he once intended to write.

Cesi's 'Tabulae Phytosophicae'

Following the dispatch of his letter to Cardinal Barberini, Cesi worked ever more strenuously on his great classificatory tasks, and in particular on the 'frontispieces' to his *Theatrum Naturae*. But his health was failing and family responsibilities pressed ever more heavily. Several years passed without any mention in his correspondence of the metallophytes, or of any related topic. Finally, in a letter to Stelluti of 10 November 1628, Fabio Colonna announced his receipt of the printed version of the very first of the frontispieces:

> The first page, a table of the whole of the *Theatre* and the basis of division—or better, distinction—has pleased me very much. The intelligence of the Prince is such that I hope it will be admired by all when they see the parts of these things being printed. To tell the truth I do not know what the stone animal–plant with the name of *Zoolithophytum* is. I would like to know about it, since it is something I have not yet seen—probably because it hasn't yet been sent to me. The same thing with the *Pianta-Metallo*—unless it is the branchy silver described by Imperato.[76] And certainly this is a very beautiful and new distinction.[77]

Five weeks later Colonna expressed his pleasure to Cesi himself:

> I have taken immense delight in the tables of Your Excellency, and learnt many things. First of all I have been overwhelmed by the composition of the whole, such as the observations on the peculiarity of the *Lithozoophyton*. For a while, I thought about whether I have anything that merits such a name, but nothing I have gives me complete satisfaction. One might, if one takes as a stone the same bone of the testacean, thus call that Zoophyte which sticks to rocks.[78]

Here Colonna went on to discuss various other species that could be called Zoophytes, species that seemed to be both animal and vegetable, or that bordered on mineral.[79] There was not much to satisfy him in the work of Aldrovandi (and here, significantly, he criticized the Aldrovandian illustrations of fossils while referring to his own more successful illustrations).[80] He had much to learn from Cesi, he acknowledged.

The 'table' or 'tables' to which Colonna refers in these and other letters of this period are the famous *Tabulae Phytosophicae*.[81] The first twelve and a half were printed in 1628 as a kind of grand appendix to the greatest Lincean project of all, their edition of Nardo Antonio Recchi's epitome of Francisco Hernández's massive account of the *materia medica* and the animals, plants and minerals of New Spain. The remaining seven and a half were added by Stelluti on the basis of Cesi's unpublished manuscripts, to the full edition of the *Rerum Medicarum Novae Hispaniae Thesaurus, seu Plantarum Animalium Mineralium Mexicanorum Historia*, or *Tesoro Messicano*, as the Lincean work was popularly called, which only appeared betweeen 1649 and 1651. In these frontispieces, devoted almost entirely to botanical matters, but with very long digressions, Cesi gave a specimen of his attempt to classify all of nature.

In the very first of these tables, Cesi placed his fossil finds within the larger scheme of natural things. In struggling to do so, he confronted the problem of one ambiguous or borderline species after another. To see these tables, however, simply as an attempt to find places for intermediate species in some form of the 'Great Chain of Being', or even as an effort to link the three great kingdoms of nature, animal, plant and mineral,[82] would be to sell it short. Cesi's was a comprehensive attempt to *understand* the order of nature, as well as to illustrate some of the fundamental forms *and* processes of life. One of the reasons Cesi worked so hard on the Acquaspartan finds in the last years of his life was his realization that the key to any system—as well as the critical problem with it—was the borderline case, that which partook of more than one nature and therefore was 'middle' and in some sense 'less perfect'.

At the head of everything came the perfect, infinite creator—God, the Cause of All Causes. Immediately below was the subdivision into the angels and separate souls. Beneath this came Man ('Homo'), endowed with reason and intellect. Then followed the heading *Brutum*, 'lacking in reason', divided into the more perfect and the less perfect. Here we begin to encounter forms directly relevant to the present discussion. At the top of the subdivisions of the less perfect was the *testafixum*, the 'fixed head', lacking all locomotive ability. Then came the *phytozoum*, an animal retaining something of a plant (a sharp and perceptive division, still present in the modern sense of the *phytozoa*), the *zoophytum*, partaking both of the nature of the plant and of the animal, and finally the *zoolithophytum*, a category which Cesi amplified by defining as a *'lapiplantanimal, [or 'mineral-plant-animal', a category] invented by us'.*

Beneath the 'brutes' on this same, extraordinarily dense page, lay the plants, Cesi's main concern in the tables that followed; and he also claimed to have devised (or 'invented', as he puts it) most of his categories for these. The first subdivision was the transitional one of *sensiplantae*, plants which 'in some way' approached the status of animal. After this came the 'more perfect' and the 'less perfect' plants. Cesi said he had distributed the whole group of perfect plants into classes, and asserted that he was the first to survey them from a physical and mathematical point of view. He claimed he was the first to divide the less perfect plants into classes; and that he was now offering a synopsis of both categories (or, as he tellingly says, a syntactic form of both).

Directly relevant, however, to the corpus of Cesi's drawings were the categories that came next, namely those of the metallophytes ('doubtful whether plants or metals, invented by us, and which we have explained in several books'); the lithophytes ('plants approaching the status of stone; and which we have been the first to discuss from the perspective of their physical history') and finally the *lapisurgentes*, stones on the verge of planthood. A separate and rather difficult subheading now followed, that of the *figulapis*, under which Cesi seems to have classified veined substances from the earth, concretions of various kinds and other phenomena which come from the earth, or are basically earthy. Here, in tortuous language, he referred to the 'excretions' and 'juices' from these concretions (presumably the resinous substances that he observed coming from some specimens from Acquasparta; cf. **7**). After these came the baked concretions resulting from the underground fires so often mentioned in the Lincean documents and letters.[83]

On this first page of the table, Cesi also dealt cursorily with the problem of petrifaction. In noting that forms changed gradually from one condition to another, and that the changes were a matter of process, he introduced a further complicating factor: that of change over time. This, of course, was a factor which threatened the whole notion of fixity upon which all systems of classification then depended. Cesi had already subverted these with his intermediate and transitional phenomena, clearly incapable of logical division and separation. No wonder that the problem of change over time was almost wholly absent from the remainder of the tables; no wonder, too, that the whole project on fossils stalled. But for this, as we shall see, there were other reasons as well.

The after-life of Cesi's researches

Cesi's friends knew about his fossil researches, even when they were not directly involved in them. In his expansive commentary on Mexican animals[84] in the *Tesoro Messicano*, Johannes Faber referred to Cesi's fossil work on several occasions. In a passage on crossover species, which he concluded with a discussion of the double nature of amber (it is both plant and mineral), Faber referred explicitly to the significance of Cesi's finds: 'Our most illustrious and Excellent Prince Cesi was the first to discover and reveal this middle nature between plants and minerals which he treats specifically in his books on metallophytes'. 'Last year', Faber continued, 'he sent different specimens of these from Acquasparta to Rome to the most illustrious Prince Francesco Cardinal Barberini'.[85]

In an earlier passage Faber had commented on the beauty and elegance of the furniture which Cesi had had made from fossilized wood from Acquasparta, as a gift to Cardinal Barberini. Everyone had been stupefied by what could be made from fossil woods. Cesi himself had been:

> the first to uncover and to see to the digging up of this substance. Not only was this sprinkled with the most beautiful flecks and waves, it was also adorned, as it were, with all the other qualities of such concretions. These were so attractive that they seemed to compete with—and even to outdo—the elegance of the veins of the *Arundo* or Indian shot. You could even say that these were tiger-like tables, since they glistened with such a dark yellow colour and were marked with wavy lines ... . But the interested reader will learn all these things much more clearly from Cesi's writings themselves, which will very soon be published, and in which many new things discovered and observed by him will be presented. This same Prince recalled to us the ancient use of working terebinthine wood into tables, caskets, small chests, and various other pieces of furniture.[86]

Cassiano mentioned just such pieces of furniture in an undated letter to Lorenzo Pignoria in Padua,[87] while among the curiosities listed in the post-mortem inventory of Cesi's museum are two inkwells and a rather crude and broken table made from fossil wood.[88] Time and again we are reminded how aesthetic and scientific approaches were linked then—and not just as a way of ensuring patronage, as must have been the motive with the gift to Cardinal Francesco Barberini.

In 1630 Francesco Stelluti published one of the most remarkable of the works to emerge from the Academy of the Linceans, his translation into Tuscan blank verse of the difficult Latin poet Persius.[89] In the notes, Stelluti used the slightest pretext in the poetry of Persius to expound on whatever scientific and natural historical issue was of importance to him and his colleagues.[90] The notes contained passages of great importance on subjects such as the history of the telescope and the microscope.[91] In them too, Stelluti referred to the aesthetic interest of the Acquaspartan fossil finds and of the furniture that could be made from them. As with the cedar used in some of the ancient beds mentioned by Persius, the wood found by Cesi ('a most learned writer and most diligent observer') 'has a hundred of the most beautiful and diverse forms of undulating waves, that are rather like camlet and other forms of watered and shot silk of the kinds that one uses today. When well seasoned it can be beautifully worked and acquires a lustre like ebony'.[92]

Stelluti recorded that for Cesi, the Acquaspartan wood was nothing but 'the trunks of trees, born beneath the earth where there are other minerals', wood which began as earth and gradually acquired:

> the form and nature of wood; and for this reason he wished to call it by the name of metallophyte, finding it of a middle nature between plants and minerals. This was first discovered and invented by Lord Cesi.

Here the term 'invented' is to be taken as meaning 'identified and thus named'. Stelluti recalls how:

> On many occasions I accompanied him when he went to observe the place where this wood originates, and not without amazement I saw that earth transformed into wood, and there I found large pieces of trunks up to thirty and more *palmi* in circumference.

The size of these pieces remained an important factor in the two friends' speculation about the origins of the fossil woods: in their view, as reported by Stelluti, they were so large that they could not possibly be naturally occurring life-forms. They must have been formed from the earth itself.

At this point Stelluti mentioned that Cesi was fully engaged in writing about the said wood,

> both from a historical and a physical point of view, with his observations about it, and a variety of figures and causes of these matters, such as one will be able to see in his books *De Metallophytis*, which will shortly be published, and in which he also writes about other relevant objects, and about the said [middle] Nature discovered by him, and in particular of stones not observed or

> described by anyone else, and other things no less marvellous than new and curious, never ceasing his consistent labour in contemplating things either unseen or neglected by others.[93]

Stelluti wrote these lines in the last days of Cesi's life; how poignant then his comment later on—referring to Cesi's discussions with Ferrante Imperato about a kind of *Cytisus* very similar to the ebony-like minerals he had discovered—where he announced that Cesi's planned books on the metallophytes 'ought soon to be printed'.[94] A vain hope! Nothing at all appeared for seven years, and even then what did appear, under Stelluti's—not Cesi's—name was but the palest semblance of what the prince had intended.

Peiresc and La Ferrière

Every project—not only the book on fossils—was thrown into disarray by Cesi's death. Stelluti later wrote that in taking him away before he could complete his books, it seemed that Nature herself envied his great knowledge.[95] The work on fungi never saw the light of day; the publication of the *Tesoro Messicano* was held up for almost twenty years; practically none of the Linceans' plant investigations ever appeared. The fact is that the events that led up to Galileo's trial in 1633 followed hard upon Cesi's death, which in turn had followed Faber's death in February 1629.

The fate of the Linceans' investigations into fossils and the problem of middle natures now lay in the hands of two men: Cassiano dal Pozzo and Francesco Stelluti. The latter in particular was unwilling to let the memory of Cesi's researches in this domain fade. Indeed he felt a responsibility not to allow this to happen, and he knew he could count on Cassiano, now the owner of Cesi's material, and the closest link with Cardinal Francesco Barberini. Despite the latter's abandonment of Galileo, Stelluti felt that they might still be able to benefit from his patronage, whether financial or moral. They must also have thought of invoking other support, for example from the Grand Dukes of Tuscany, who had remained faithful to Galileo. We know that in the course of 1635 they received as presents from Cesi's widow some rather fragile *tavoloni* of fossil wood. These tables had to be polished by an ebony-worker and then strengthened by the addition of ordinary wood.[96]

But at this point, a few years after Cesi's material had entered Cassiano's Paper Museum, another player came on the scene—that stimulating polymath, the great French antiquarian, Nicolas Claude Fabri de Peiresc. For several years he had been preparing a work of his own on shells and fossils. Naturally enough, as a friend of Cassiano, he had been kept abreast of the Linceans' work, but it was a few years after Cesi died that he decided to take a personal interest in the fossils of Acquasparta. On 28 February 1635 Peiresc wrote from Aix-en-Provence, to Cassiano in Rome, thanking him for :

> the letter of Prince Cesi and the little discourse [*discorzetto*] and the illustrations added to it of that fossil wood, the description of which pleased me a great deal. One day I will tell you my opinion, as soon as I have the opportunity to send you a small account of certain other petrified woods which are formed, and are found formed in various places in this country.

The letter from Cesi which Cassiano had sent to Peiresc may well be one of the two copies, now preserved in the library of the Bibliothèque de l'École de Médecine in Montpellier, of the letter to Francesco Barberini of 1 December 1624;[97] the illustrated *discorzetto*, as he called it, was almost certainly one of the several manuscript versions of Stelluti's *Trattato del Legno Fossile* which were already circulating at this time, two years before its actual publication.

Peiresc went on to ask Cassiano to send him:

> some fragments of those [pieces of fossilized wood] which have the crust on them; and those which have a mixture of metal; and those which have a mixture of resins and gums; and those which retain some of the nature of wood and of stone together; and above all those which retain the nature of clay, susceptible to fire and brick-like by nature. These seem to me easier to understand than the others, unless they contain mixtures of different natures together, and are as it were incorporated and entangled one within the other.[98]

Peiresc was describing in verbal form some of the very types of fossil woods, concretions, baked clays and aetites depicted in the Windsor drawings. So what did he know of Cesi's researches? Was it just what might be gleaned from Stelluti's *Trattato* and its illustrations, or was he indeed sent some sample, now lost, of Cesi's work? We do not know. In any case, Peiresc typically wasted no time in sending out one of his men to conduct an investigation *in situ*.

On 20 March 1635, Peiresc wrote to Cassiano with a letter of introduction for the aged Jacques de la Ferrière, the personal physician of the Cardinal-Archbishop of Lyon (brother of the much more famous Cardinal Richelieu). Peiresc knew him well and trusted him with the necessary investigations in Italy.[99] Peiresc intended to 'persuade him to make the journey to Acquasparta, so that he can see with his own eyes those fossil woods, or metallophytes; I am sure that he will find something more there'.[100]

At this point in the letter Peiresc wrote that he wanted to send La Ferrière to Sicily as well, in order to examine the different fossil remains there. He had been told about these a few years earlier by their mutual friend, Claude Menestrier.[101] If La Ferrière could not go himself, then Peiresc would send Menestrier; but it would be best of all if they went together. He knew that Cassiano would be interested in any attempt to discover the context in which the giant fossil bones and teeth of Sicily were found—in the earth, in the sand, accompanied by stones and rocks, included in the rocks excavated at the place, or mixed up with the matter of the rocks themselves, like other petrifications.[102] Cesi had already been sent some of these giant fossil remains by

Menestrier, while Peiresc's interest in the exact location of such remains (like Cesi's) is testimony to his scepticism about all explanations in terms of fable—or, indeed, any other non-physical account.[103]

On 29 March 1635 Peiresc wrote again to Cassiano about La Ferrière: 'I showed him the little book (*libretto*) by Your Lordship about the metallophytes of Todi and Acquasparta, and I made him promise to go there and see them personally'.[104] The 'little book' was presumably the *discorzetto* which Cassiano had sent to Peiresc a month earlier. That it was a manuscript version of Stelluti's *Trattato* emerges from Stelluti's letter to Cassiano of 5 May 1635, in which Stelluti described some fossil woods: 'one has more stone than wood, which you can tell from its great weight; the other has those cuts and cracks which I refer to in my discourse, and where you can see that white resin';[105] in the course of the previous year he had written *un trattatello a penna* on a subject 'completely new and ambiguous'—the very words which appear in the dedication to Francesco Barberini in the printed *Trattato del Legno Fossile Minerale* a couple of years later.[106]

By the beginning of May, La Ferrière was in Italy. He was going to see 'the metallophytes or fossil woods of Todi and Acquasparta' and to prepare his report on them.[107] But it all took much longer than Peiresc might have expected. At the end of May, Peiresc wrote to his friend Pierre Gassendi telling him that La Ferrière was undertaking this project; in August he wrote to Menestrier to the same effect.[108] The delay seems to have been caused by other geological and palaeontological matters that occupied La Ferrière's attention. First he investigated a number of other fossil remains—including the problematic giant's bones—around Rome, and then began preparing to go to Sicily with Menestrier himself.[109] In the meantime, Peiresc—anxious to confirm the correctness of his views about fossils over those of the 'two gentlemen' Cesi and Stelluti—was writing again to Cassiano about his interest in 'mixed things', and in the fossil woods of Acquasparta.[110]

By October La Ferrière, having been to Civitavecchia to look at some caves, finally reached Acquasparta.[111] We know this from an unexpected source, the heterodox and libertine French doctor, philosopher, politician and bibliophile, Gabriel Naudé, who along with Peiresc was preparing counter-arguments to Cesi's alleged views on the 'generation' of fossils. On 30 November 1635, Naudé wrote to Peiresc that La Ferrière had been in Acquasparta 'a little before me' (*peu auparavant moy*), and that Peiresc could expect a good account from him of the fossils there.[112]

Indeed he did, in the second of the two most important letters in this whole story, the first being Cesi's letter of 1 December 1624 to Francesco Barberini (see Appendix III a; for a full translation of La Ferrière's letter see Appendix III b). 'Monsieur', wrote La Ferrière in November 1635,

> Having promised you in all my previous letters to go to Acquasparta so as to acquaint you with the peculiarities of the fossil woods which are to be found near there, I was obliged to allow the great heat of summer and of the early autumn to pass, so that I left Caprarola only on 22 October, having first asked the Cavaliere dal Pozzo to write on my behalf to someone there of his

> acquaintance, which he did. I was taken to a place three miles from the aforementioned Acquasparta by someone who lives below the castle of Rosaro, close to which these woods are found.

Thus did La Ferrière begin his investigation on behalf of Peiresc. In his letter he described the exact location of the fossilized woods and tree-trunks, revealed by the washing away by strong rains of the earth in a ravine (e.g. Figs 1–17, 26). These trunks, he continued, seemed once to have been cut into large logs of eight to ten feet (2.5–3.0 m) in length. The wood was so strong that it could not be broken across the grain, 'though it splits easily lengthwise'.

At this point Cesi's large table made from fossil wood made another appearance. Having asked the *châtelain* of Acquasparta what was the largest piece ever found, he was told that Cesi had had a table made of the wood, which measured nine spans wide and ten long. The same *châtelain*, however, had never heard of the vein of rock forty or fifty feet (12–15 m) long referred to by both Cassiano and Peiresc. On the other hand, La Ferrière's guide did show him the spot where an underground fire had burned for seven years and where the shepherds, in warming themselves, had actually set fire to the soil. In this connection, La Ferrière specifically mentioned the treatise sent to Peiresc by Cassiano, and which he himself had seen. He 'suddenly remembered the treatise which the late Prince Cesi had written about this fossil wood, of which you have a copy, which states that there was a place where, the earth being naturally sulphurous, it had burnt for ten years and without doubt this was the place my guide had shown me'.

La Ferrière now embarked upon a description of the fossil woods and baked clays of Acquasparta and Rosaro. It could almost be an account of the relevant drawings in the Windsor volumes:

> ... I found that the soil broke off in what looked like terracotta scales resembling earthen pots or broken bricks, and there were many of them even at the bottom of the escarpment that had fallen and which cracked and splintered under my horse's hooves like broken pots. I also saw large lumps of this baked brick-coloured clay, but of rounded shape of about the size of an ox head or larger, which the water had caused to roll down quite a distance from their starting place, as well as pieces of the same size and shape as slag [*scoria ferri*], which is the waste of unusable iron that is thrown into forges in large quantities and which, when broken up with hammers, reveals the iron inside. I must confess to you that having wished to look for some piece in which one could see the chalk or clay turning into a something akin to wood, and then into wood itself, I could find none. But I did unearth more than a dozen well-formed pieces of wood, separated as if they had been cut, of which only the ends were visible, the remainder being buried in the earth. I was told that the late Prince Cesi had once harnessed eight pairs of oxen to a trunk to uproot or exhume it.

La Ferrière then suggested that the earth in which he found this wood must be sulphurous, and that underground fires caused the 'inner aridity of the soil' and made it crack, 'though I saw no smoke passing through, and I believe that the great drought of the season caused it and that it must rain for the fires to be rekindled, which even the locals rarely see'. This kindling of fire by moisture was one of the phenomena that most interested Cesi.

Here La Ferrière set out some of his own disagreements with Cesi about the origins and formation of fossil wood. Firstly he referred to the traditional views held by educated people of the region. They 'think that it was wood which had been buried since the Flood, or by an earthquake and it looks as if it had been cut and sawn in order to be fashioned into barrels or other wine containers'. These possibilities, he wrote, must be rejected, since the trunks had always been found lying on the ground, and were all rotten on the outside; much of the timber was therefore unusable. Yet Cesi's *tavolino*, made from the heart of the wood and not across the grain, was nine spans wide—so the tree from which the table was made was no common one, 'for in this part of the world we don't have trees of this size and diameter' (or weight, for that matter). How then had it arisen?

While La Ferrière only hinted at the antediluvian origin of such a tree, he considered Cesi's views on the subject to be mistaken. Peiresc and La Ferrière had read a copy of Cesi's 1624 letter to Francesco Barberini, in which Cesi had described these giant tree-trunks as petrified; and in his letter La Ferrière admitted 'it is easier for nature to turn wood into terracotta or stone' than vice versa. Yet he went on to state his belief that here stone or clay was being turned into wood, before ending lamely, 'I leave you to resolve this difficulty'.

> And as for what Prince Cesi says to prove the generation, that he had seen pieces which at one end were like hard chalk and at the other end like wood, I think he is wrong and that he has mistaken the beginning of petrifaction for the beginning of the generation of wood. Indeed, I have a piece given to me by the Cavaliere dal Pozzo which is like wood at one end and like a half-baked brick at the other. But it is easier for nature to turn wood into terracotta or stone than to turn earth into wood. We have even seen flesh which has been similarly petrified . . .. Therefore I don't think one can rely on Cesi's report as to how this came to be formed, having found two points which cast doubt upon it, namely the recumbent positions of the trunks and their extraordinary size. I leave the resolution of this difficulty to your judgement—to which I will always yield.[113]

Here indeed was a major difference of opinion about the generation of fossils and, in particular, fossil woods.

In his letters to Cassiano during the month of August, Peiresc had been expressing his own views about the organic origins of fossils.[114] Such views were clearly at odds with those of Cesi and Stelluti.[115] By the time La Ferrière made his journey to Acquasparta, there were already quite a number of people interested in, and being kept abreast of, the fossil finds around Acquasparta. Copies of Stelluti's discourse or *trattatello* were circulating in both Rome and Tuscany (as well as France) and disseminating what purported to be Cesi's views on the origins of these woods.[116] The Grand Dukes of Tuscany not only had furniture made of fossilized wood sent to them by Cesi's widow; they also owned a copy of the little treatise.

Even Galileo, now more a Medici protégé than ever, was involved. At just about the time La Ferrière was visiting the site in Rosaro, Stelluti wrote to Galileo expressing sympathy for his troubles, and tried to cheer him up with some news about the progress of an old Lincean project:

> Your Lordship ought to know that when Signor Balì Cioli [Secretary to Grand Duke Ferdinand II of Tuscany] was here recently [in May 1635][117] he went to visit our Lady the Duchess [Isabella Salviati, Cesi's widow] on several occasions, and when he left she gave him several pieces of the fossil wood that is born near Acquasparta ... and at the same time he wanted to know where this fossil wood was found, and how it was generated—since he had seen in the commentary to my *Persius* that Prince Cesi was writing about it. Her Ladyship the Duchess ordered me to write something about this, which I did, and which was sent to their Serene Highnesses along with a box of different pieces of the said wood.[118]

Everything thus points to the fact that Stelluti was writing his little treatise on the fossil woods by May 1635, probably at the instance of Cesi's widow.

In 1636 interest in the finds grew even further. In April Peiresc wrote to thank Cassiano for having sent him some pieces of petrified wood, at the suggestion of La Ferrière.[119] All through this year, and at the beginning of the next, Peiresc and Naudé were preparing their counterblast to Cesi's views (or at least those reported by Stelluti). It is not surprising to find Stelluti writing on 23 May 1637 to an unidentified correspondent in Rome that he had put down everything he knew

> about the said wood. In fact I had already done this in the shape of a small treatise to give to his Eminence Signor Cardinal Barberini, who had already had a small summary from Prince Cesi [referring again to the letter of 1 December 1624], and I gave it to Signor Balì Cioli ... But now I'm being asked by many, especially by the French and by other *curiosi oltramontani*, for a copy of the said treatise, and since I really can't make so many copies, and since I didn't want to destroy myself because of the many figures that I had to have engraved, I have decided to publish it, and in fact I finished it just this week.[120]

The 'Trattato del Legno Fossile Minerale'

The languorous full title, in a way, says everything—*Treatise on the newly discovered fossilized mineralized wood; in which is briefly summarized the diverse and mutable nature of the said wood, represented here with some illustrations which show its place of origin, the diversity of the undulations that are seen within it, and its such varied and wondrous forms; by Francesco Stelluti, member of the Academy of Linceans and native of Fabriano.* Yet there is not a word of the classificatory and taxonomic problems that so preoccupied Cesi. Instead there was a strong emphasis on the aesthetic interest of the fossils, and (as we also see clearly in La Ferrière's letters) a preoccupation with the geographical location of the fossils. The treatise was certainly brief: only eight pages of text and thirteen etched plates.

The dedication to Francesco Barberini confirms what Stelluti had already written to the unknown correspondent of 1635:

> Last year I gave your Eminence my hand-written treatise [*trattatello*] on the fossil wood which is indigenous to our region. I wrote about it in this way to satisfy the curiosity of many who wished to have a true and faithful account of such strange wood. And because afterwards many copies were requested of me, being unable to satisfy everyone, I decided to put the work into print, with the addition of other, rather curious figures, for the greater satisfaction of my readers: so that through visual means they might be more satisfied and convinced of all they will learn in the treatise. Obliged to publish it with engravings, I have wished to dedicate it to Your Eminence, both for the pleasure which you are wont to take from the contemplation of the secret parts of Nature;[121] and also that it may bring to the fore a testimony so worthy and so eminent as to assure others that I present nothing that is false or fantastic.
>
> I bring before your Eminence a matter which is in truth wholly new and ambiguous.[122]

But what did Stelluti have to say about this wholly new and ambiguous subject? Less, perhaps than we might expect; and rather different, perhaps, from what Cesi might have said. As so often with the Linceans, however, things were seldom what they seemed. Francesco Barberini, the man who had so coldly abandoned Galileo at the time of his trial four years earlier, would surely have recognized the design of this title-page, which had the same layout and motif of a crowned lynx as that of Galileo's *Letters on the Sunspots* of 1613, the real beginning of his rift with orthodoxy.[123] Let us examine the contents of the treatise on fossil wood a little more closely.

It begins with Stelluti's declaration that he intends to discuss the fossil wood discovered by Cesi, 'by nature a metallophyte, being part plant, part metal; if well considered in its entirety, it partakes much more of the plant than of metal'. He emphasizes its localized occurrence and its remarkable aesthetic qualities: the variety of its beautiful growth lines, and its many diverse forms. Its closest antique parallel, he writes—for one always had to begin with the classical references—was probably the cedar wood of Mauretania, from which the ancients also had made furniture. He reminds his readers that he had already mentioned these things in his translation from *Persius*.[124]

Stelluti then identifies the general location of the finds, in the area once occupied by the western arm of the ancient Tiberine Lake. It is all nicely illustrated by the map that forms the first plate of his book, showing the district extending from Todi to Amelia, and up past Acquasparta into the mountains to the east.[125] Suddenly he moved to the heart—and the most improbable aspect—of his discussion. 'From what I have been able to see', he asserts, 'the wood is not generated from the seed or root of any plant whatsoever, but only from a type of earth, containing much clay, which is slowly transformed into wood, nature thus operating until all the earth is converted into the said wood'.[126]

This odd and unlikely proposal is the central element of the Stellutian theory of the generation of fossil wood; and it is possible to see how Stelluti (and perhaps Cesi too) might have arrived at such a view. If we look at the Windsor drawings, the fossil wood often seems to emerge from the clay—but can Stelluti really have been so naïve? Was it perhaps a case of not being critical enough about what he thought he had seen and observed at first hand?

Stelluti elaborates his position on the basis of the subterranean fires or 'fumaroles' of the region:

> This [conversion into wood] I believe to take place with the assistance of the heat of subterranean fires which exist in these areas, and which, snaking underground, continuously emit very thick smoke, and sometimes flames, and especially when the weather is wet, with the further assistance of sulphurous and mineral waters. And when the heat is sufficient, the wood is browned, or lightly scorched, and remains similar to coal. If, then, the earthen material is not yet converted into wood, this fire bakes it, and it remains similar to the vases of terracotta baked in a kiln, or to bricks.

The actual observations are unassailably correct.[127] But Stelluti seems to be saying that the underground fires help transmute the clay into fossil wood, and that when the heat was more intense, any unconverted clay would be baked like terracotta. From observation he inferred that: 'it is clear that the essence of this wood is nothing other than earth, because I myself have seen pieces one part of which was hard clay, the other wood, and the rest similar to coal.'[128] So much for close observation! The Linceans were better at it than almost anyone else; but still they could draw a wrong conclusion, as Stelluti appeared to do here. There is indeed a lesson here about how empirical and observational evidence may be used to bolster a wholly incorrect hypothesis.

In support of the view that the wood could not have had plant origins, Stelluti emphasized that the fossilized trunks always lay prostrate and were of such a size and weight that they could never have stood up, as trees must. Moreover, these specimens never actually reproduced any part of a plant besides the trunk, and so (he reasoned) they could not originally have been in the plant, or more specifically, the arboreal, kingdom. 'The raw material', he believed he could demonstrate, 'is nothing other than clay-rich earth, in confirmation of which one may adduce its great weight and the fact that even a very small flake sinks to the bottom of the water...'

At the same time, however, Stelluti seemed all too aware of the opposite (and certainly more correct) theory: 'Neither can these pieces of wood be considered trunks nor [are they] fragments of other trees buried in this area, or which have fallen and been covered by earth and then given wave-like form by the mineral waters which have their springs there. Nor have they been created by subterranean fires' he said, 'no trees ever having been found formed by Nature like those in the following illustrations'.[129]

If he came so close, why then did Stelluti attempt to suppress, or sideline this much more plausible account of the origins of fossil wood (for all the continuing ignorance with regard to the growth lines of trees)? Instead, he continued to digress—about the peculiarities of the grain, the veining and the 'waves' (that is, the growth–lines) of the fossil woods; he described in some detail the different ways in which the fossil woods burnt; and then he discussed how some samples might seem to have wood inside and earth outside, and some the other way round. In this and many other ways, he commented, Nature makes jokes (this is a commonplace of natural history in those days) and makes us marvel at the wondrousness of its works: *scherzando così la natura per farci forse maravigliare dell'opere sue così rare*.[130]

Finally, before moving on to discuss the illustrations of his treatise, Stelluti produces what he seems to regard as his trump card: 'When there is no other evidence or proof of the origins of this wood, let this suffice: that, a quantity of damp earth, which was around a piece of this wood, having been removed and placed in a room in the Palace of Acquasparta ... it was found after some months to be wholly converted into wood.' Even a schoolchild could put forward the alternative explanation that upon drying the clay cracked and separated from the wood inside, and fell off. But according to Stelluti, Cesi was so amazed by this that he had 'no doubt whatsoever that the earth is the seed and mother of this wood, the earth of these parts being most suitable for its generation'.[131]

It is hard not to wonder why Stelluti protested so much in favour of this explanation of the generation of fossil woods; or why he continued to insist that Cesi's finds demonstrated that the specimens passed through various stages from shapeless pieces of earth through to those which were wholly wooden. On the face of it, he does not seem to have come very close to believing that the specimens Cesi found were in various stages of permineralization. True, many of the drawings in the present catalogue do indeed show pieces of wood with more earth, or less, actually clinging to them, some seeming to be pure earth, others mixed, and others pure wood, just as he described them in his book; but can he really have been satisfied with a conclusion like this? Can he and his colleagues really have trusted so naïvely to a *prima facie* explanation of what they were observing? Perhaps—but not certainly.

The illustrations in Stelluti's book are a very meagre selection from the massive visual record put together by Cesi. There are thirteen numbered plates (reproduced here as Figs 54–66): a map of the region, a composite site drawing, ten plates of fossil woods, and one of ammonites. Seven of the plates are based on fifteen of the drawings in Windsor.[132] For no immediately apparent reason, all the examples seem to be based on specimens from Sismano.[133] Their descriptions by Stelluti are disappointingly thin, but the drawings give us some sense of the chief preoccupations of this curiously enervated treatise.[134]

The first illustration is the map of the territory of Todi and Acquasparta, indicating the four main sites where metallophytes were discovered. The second, also a double-page spread, shows the site close to the Castle of Rosaro—the site described so closely by La Ferrière—where a large quantity of the fossil woods had been found, and where the fumaroles (nicely illustrated on the plate) had been burning for a period of ten years.

Then come four pages showing the meandering growth lines (the 'waves') of the fossil wood.[135] No drawings for these plates survive. They clearly demonstrate the old topoi of the artfulness and artifice of nature. But more significantly, Stelluti took these wavy lines to be evidence that the specimens could never have been real trees, because in that case they would have been regular concentric circles, rather than wandering waves or other patterns. In either event, he did not recognize that they were growth lines.

Then comes an illustration of a large specimen of fossil wood embedded in the earth [Fig. 60], and, on the next plate [Fig. 61], two more pieces of fossil wood, these being of the white, resinous kind already mentioned by Stelluti in his letter to Cassiano of May 1635—and also referred to by Cesi on several occasions as well. Of these specimens, he had earlier said that he believed the reason they were oval in section was because they were unable to grow upwards owing to the great weight of the earth in which they were embedded, and could thus not achieve the cylindrical form of ordinary tree-trunks.[136] This he took to be further proof of his view that they were actually generated from the earth itself (and were not trees which had been transported by water and deposited away from the site where they had grown).

Three pages of interesting specimens—of fossilized wood, and of the pieces of baked earth called laterite—follow, both here and in the corpus (**7** and **11** and **129, 169–170**). Stelluti describes the small piece of fossilized wood illustrated as *D* on Plate 10 (see **19**) as follows:

> Trunk *D* ... also excited great wonder: because its inner part was already totally petrified, and reddish in colour, and the outside covering was of a bark-like wood, and knotty, exactly as it has been drawn. The length of the entire trunk was six palms, and its width approximately one palm in diameter.

This kind of straightforward description of colour, texture, size, and degree of fossilization occurs throughout Stelluti's sparing comments on the plates: aside from an occasional reinforcement of the view that fossil woods are generated out of earth, there is hardly a genuine theoretical moment. One might, of course, say the same of the drawings in the corpus. But with the latter, one knew that Cesi was preparing a larger dissertation; here it is hard not to be disappointed by the scarcity of reflective comment.

Finally, on Plate 13, comes a page of ammonites (see **197–199**). Stelluti's text about them is significant, not because it adds anything to the discussion so far (indeed, it comes as a kind of distraction from the relatively concentrated discussion of fossil woods, concretions, and baked clays), or because it is in any way an accurate account of the origins of these fossils (he held that all fossils belong to the mineral kingdom and grow within rocks),[137] but because it reminds us of how much more Cesi himself planned to discuss, and what his own treatise, had it ever appeared, might have illuminated. The plate, Stelluti says, shows several stones—looking like snails or snakes turned in upon themselves—of the kind called Ammon's horn:

> I would always have believed them to be artificial, or else not natural, had I not discovered them like this, not only in the territory of Acquasparta, but also in my native territory of Fabriano ... And of their origins and nature His Highness Prince Cesi, mentioned above, had also in part written and was writing, when death prevented the finishing of the treatise which dealt with this material, the novelty of which would have made it most delightful and curious to read: because not only was he was writing of the generation of these stones, and wood, and eagle-stones, which were generated in such quantities in these places, but of all other stones known until now, and even of others not yet observed nor described by other authors; he having through long and diligent observation discovered how the pregnant eagle-stone is generated;[138] how in arboreal stones, which Imperato calls 'imbosked', are formed those small trees with branches and fronds so resembling their natural counterparts as to appear artificially painted.

Stelluti then goes on to mention a whole host of stones (by no means all fossils) which Cesi had been discussing, including the astroites and stellates, the petrified fungi, the 'mixed' (veined) marbles, the bucardites (the heart shell), lyncurium (or *lapis lyncis*), ostracites, fossilized sea-urchins, vermiculite, milleporite, and reteporic and frondeporic stones.[139] Many of these are to be counted among the *figulapides* referred to in the first of the *Tabulae Phytosophicae*. A few are illustrated in the Cesian manuscripts in the Institut de France,[140] while others appear in Volume V of the *Natural History: Fossils* in Windsor.[141]

Passages such as these reveal much more than the aesthetic dimensions of what passed for scientific description in those days. Even in their relatively simple way they go beyond the connections between art and nature, between *naturalia et artificialia* so favoured by *virtuosi* like Cassiano. (Cesi himself, like many of his friends loved the kinds of 'mixed marbles' illustrated in Volume V of the *Natural History: Fossils* and paralleled by Plates 6–9 in Stelluti's treatise.) In short, they remind us, of the fundamental Lincean emphases on observation, discovery, taxonomy and, above all, their interest in problems of generation and reproduction.

But the book ends in a lower key. After the tantalizing glimpse into the further areas of Cesi's palaeontological researches, Stelluti laments how Nature herself had shortened the days of his life 'for no other reason than that he should not come to discover the manner of her secret operations'. But how much more would Cesi have discovered had he lived? The indications, at least in the field of palaeontology, are not especially promising. Was he obliged to hold back in his researches on fossils because of his relationship with Galileo? It is hard to tell. Beyond doubt, however, is the energy with which Cesi brought to the surface a range of the hidden things of nature and for the first time made them the subject of scientific attention. He may not have found an adequate explanation of their origins, but the long-forgotten drawings—forgotten almost from the day they were made—offer the finest testimony to his achievement. And, seven years after his death, there seem to have been quite a number of people who were eager to have a copy of Stelluti's summary of Cesi's findings.

Manuscript versions of the 'Trattato'

Five manuscript versions of the *Trattato* survive, two in Montpellier (Bibliothèque de l'École de Médecine, MSS H.170 and H.173); two in the Vatican Library (MSS Barb. Lat. 4355 and Vat. Lat. 8258); and one in the Bibliothèque Nationale in Paris (MS It. 1684: see Concordance D, pp. 400–3 below). All but Barb. Lat. 4355 are bound together with a copy of Cesi's letter of 1 December 1624 to Francesco Barberini on the subject of his fossil discoveries. MS It. 1684 contains the original of La Ferrière's letter to Peiresc, and H.170 in Montpellier contains the only surviving copy of that letter.[142]

Since the documents discussed in this essay refer so frequently to manuscript versions of the *Trattato*, it is tempting to speculate who the respective owners of these particular manuscripts might have been. Peiresc's version may well have been MS It. 1684 in the Bibliothèque Nationale, where so many of his other papers (and his important copy of the printed version of the *Trattato*)[143] are also kept; the one La Ferrière took with him or consulted was perhaps Montpellier H.170, with its copy of his letter to Peiresc; Cassiano's could easily be Montpellier H.173 (ex-Albani 861); and Francesco Barberini's is certainly Barb. Lat. 4355.[144] Whether the Medicean copy of the *discorzetto* is the other surviving manuscript (Vat. Lat. 8258) is difficult to establish with any certainty. But in the absence of further corroborative or technical evidence, these associations—apart from Barberini's—must remain speculative. They are not critically significant here, however, except perhaps in two respects which happen also to support strongly the associations with Peiresc and Cassiano.

Both from the evidence presented earlier and from a comparison of the texts it is clear that all these manuscripts preceded the printed version of the *Trattato* (rather than being variant copies of it, as has sometimes been thought). Moreover, the earliest drafts among these manuscripts (all fairly close to the published version, apart from variations mentioned below) are almost certainly MS H.173 in

Montpellier and MS It. 1684 in the Bibliothèque Nationale (the copies here associated with Cassiano and Peiresc respectively).[145] How do we know that this was their sequence? These two manuscripts contain revealing differences from the *Trattato* itself, and from the other manuscripts.

Firstly, in describing the discovery of the important *pezzo ovato* found embedded in the earth (and illustrated on Plate 7 of the *Trattato*), both H.173 and It. 1684 refer (as does the published version) to the fact that the hole dug to reveal it had caved in, and that the owner refused to allow it to be opened up again. But while H.173 and It. 1684 add the clause 'even though the owner was offered payment to open up another hole',[146] the printed version (like the other manuscripts) adds: 'though I have been informed, while writing this account, that the waters have once more uncovered it'.[147] Thus the later manuscripts and the printed version were brought right up to date.

The other divergences between these two prime manuscripts and the printed *Trattato* are more important. In H.173 and It. 1684, the last three paragraphs printed in the *Trattato* appear in much shorter form and as just one paragraph; the digression on ammonites and figured stones is absent; and the ruminations on Cesi's premature death are reduced to these revealing lines about his project and its progress at the time of his death:

> In the final figure are engraved various kinds of petrified stones found where this wood occurs, among the petrified trunks mentioned above. And about their generation and nature—as well as on the wood and on the aetites of which there is so much in these places—our Prince Cesi of blessed memory was writing; and overcome by death was unable to complete his treatise on this subject; because of the three books he planned about it, I believe he had barely finished one.[148]

But perhaps most significant is the absence from the manuscript versions of a passage that appears only in the printed version. On the third folio of H.173 in Montpellier and the second folio of MS It. 1684 in Paris, Stelluti says, as he does on page 8 of his book, that 'it remains to discuss the illustrations that follow'. In the manuscripts he moves straight on to refer to the map of the territory of Acquasparta and Todi, and to the beds of fossil woods not far from the Castle of Rosaro. But in the book he inserts first his account of Cesi's observation in the Palace of Acquasparta of the 'conversion' of a quantity of damp earth to wood, whence Cesi is said to have derived the conclusion that 'the earth itself is seed and mother of this wood, the earth of these parts being most suitable for its generation'.[149.]

In other words, at some time between the early drafts of the *discorzetto* (written, at the latest, in February 1635 but perhaps in 1634 or 1633) and the published *Trattato* of 1637, Stelluti decided to insert his bizarre insistence on the generation of fossil woods from earth. What moved him, at this particular juncture, to emphasize the view that fossils did not have organic origins? And was that view genuinely Cesi's? We cannot yet answer these questions with certainty.

Cesi, Galileo, Naudé

We know from Stelluti—and to some extent from Faber too—that Cesi intended to produce three books on metallophytes and that he barely finished writing the first one.[150] These books would also have contained his views on other fossil forms, such as those mentioned in the letter to Francesco Barberini and in the paragraph at the end of the *Trattato* about ammonites and other figured stones. Beyond that we are in the dark. Stelluti's book gives almost no hint of Cesi's work on the problems of classification and their implications for palaeontology. The *Tabulae Phytosophicae* provide some indication of Cesi's sense that good classification is thwarted by the factor of time, and that this factor presents a major obstacle to traditional views of fossils. In all his work, Cesi realized—like Linnaeus long after him—that any attempt at classification has to be based upon the study of generative and reproductive principles.

But was Stelluti justified in insisting that Cesi believed that wood was generated from earth, that the seeds of the wood were actually earth? There is little in Cesi's own work to suggest this. Could it all have been a front, to divert attention away from the ever more obvious conclusion that over immensely long periods parts of the earth once covered by water had been pushed up and then revealed fossil remains of the vegetable and animal life of long past ages? Such views of fossils, of course, stood at odds with the traditional biblical accounts of both the Creation and the Deluge.

The events leading up to Galileo's condemnation in 1633 had made the whole matter of biblical interpretation an extremely tense one. The theologians alleged that Galileo had attempted to reconcile nature and the Bible in the most unacceptable ways, and that he twisted scripture to fit his new interpretations of nature and the cosmos. The problems began with the sunspot controversy of 1612–15, and continued through the debate over the 1618 comet and the publication of the *Saggiatore*, Galileo's wonderfully clear and pungent response to Orazio Grassi's views about it.

Grassi and his fellow Jesuits believed that Galileo was not sticking to his last[151] when he tried to interpret scripture to suit his own views. And this central aspect of the controversy around Galileo found an echo in the ominous, tense and bitter comments in Colonna's letter of 10 November 1628, when he first wrote to Stelluti about the opening table of the *Tabulae Phytosophicae* and praised Cesi's 'very beautiful and new division' of the metallophytes:[152]

> How it will be when it comes to the subdivisions of meteors and heaven and earth I really cannot imagine. I am not speaking of corporeal things, about which you need to curtail your intelligence—because if the scholastic theologians are of one opinion today, they will have another much harsher one tomorrow. *Et sapienti basta il motivo—and to the wise man the motive is enough.*[153]

At this time, Galileo had for the moment been forced into silence, defeated by the theologians and by the lack of support from his erstwhile friend, Maffeo Barberini, now Pope Urban VIII. Colonna's letter continued, even more ominously and not a little rancorously:

> As I wrote on another occasion to you, I think it would be a good idea to warn Galileo not to expatiate too much on scriptural matters, particularly with regard to the miracle of the furnace, in which the three [holy] children were seen to walk and to praise God; in case these men find the occasion to prohibit his work, in order to make themselves the prior inventors of all your inventions; as they were unable to do with regard to the movement of the earth, and the moon not shining in and of itself.[154]

Little upset Galileo more than the idea that others might usurp his work, or claim his discoveries as theirs; nor did he like being told what he could or could not say. But what is relevant here is the way Colonna moves directly from the Cesian researches to the vexed situation around Galileo. Indeed, as if aware of the intensely risky context in which they were all moving, Colonna makes a passing reference to the great Dominican heretic Tommaso Campanella; and then pointedly expresses his hope that God grant Galileo a long and healthy life, *per beneficio de' studiosi*.[155]

In this letter, Colonna was warning two fellow-Linceans of the risk that they might be silenced. Could this have been what really stood in the way of the full publication of Cesi's views on fossils? Cesi's findings threatened to stand at odds with the way the Bible was interpreted by those who claimed the sole right to do so: the theologians, and the other clerical opponents of Galileo, particularly his arch-enemy Grassi. The story of the Three Holy Children in the third chapter of the Book of Daniel, for example, was repeatedly adduced by Grassi at the time of the controversy over the comets as one of the proofs of the transparency of flames, against Galileo's own repeated insistence that the comet could not be a flaming substance (it was light reflected from gases passing through the atmosphere) precisely because it *was* transparent. And he went to elaborate lengths in the *Saggiatore* to demonstrate not just that fire was not transparent,[156] but that the story of the Three Holy Children in the Bible did not imply that either.[157] In the years immediately preceding the publication of his *Dialogue on the Two Great World Systems*, Galileo found that almost every aspect of natural history, including the problem of fossils, related to one or another aspect of biblical interpretation, which lay at the heart of the controversies that swirled about him. If his views could not be impugned from a mathematical—and what we would now call a scientific—point of view, then they were always vulnerable from the purely hermeneutic one.

The frustrating aspect of all this is that we do not know what progress Cesi actually made towards the understanding of fossils. Certainly oblivion came early to Cesi's work, and this might explain the extraordinary attacks on his purported views by Peiresc and Naudé. I have already commented on the scepticism Peiresc expressed about Cesi's alleged views on the generation of fossil woods in his letters to Cassiano; and on La Ferrière's agreement with Peiresc; but there is one further letter that is directly relevant to the present context.

On 2 August 1635, at a time when La Ferrière was in Italy but had yet to report from Acquasparta, Peiresc wrote to Cassiano and set out his views at length. It will be remembered, as one reads this, that for Stelluti one of the constant objections to the claim that the fossilized woods were indeed once living trees was the improbably large scale and weight of these woods, as well as the fact that most of the trunks were found lying horizontally. In this extraordinary letter Peiresc states that it is perfectly possible that there should once have existed giant forms in nature which have since become petrified (though he insisted that many fossils were actually giants' bones or parts of them). After all, he notes, one finds a variety of shells and other marine creatures, even the bodies of very large fish and whales, high up on mountains:

> The whole of antiquity, as well as the Holy Fathers, are in agreement that in earlier times the waters inundated the face of the earth inhabited by us, and left behind the remains of many shells, crabs, and other fishes, and marine plants and animals, which are found there and in the whole of Italy, France and Germany ... And these revolutions could easily have brought together very large fishes as well as small ones. [158]

Peiresc was going to send La Ferrière to Sicily, since

> in these times when one is so busy researching the true anatomies of things, the problem deserves to be clarified. And on this occasion one could also investigate the forms and qualities of fossil ebony and other figured things that might be found buried together with them, and conjecture more clearly whether they are the teeth of an elephant, or an hippopotamus, or another marine monster ... it would not be surprising if larger monsters [than crustaceans] frequented the place where fossil ebony is found, and more readily left their remains there than elsewhere.

In accounting for the presence of marine fossils on high mountains the story of the Deluge was made to fit together here with Aristotelian views of the role of cataclysms.

Peiresc continued by saying that since 'these things are not easy to write about to others in letters (given the near-impossibility of explaining them in words), in order to make plain to others one's concepts about these things ... if one cannot go to the sites oneself, it might be a good idea to send practical people who also have insight into the most recondite mysteries of nature'.[159] He thought Menestrier or someone like him would be good. We know the rest of the story: when La Ferrière went to Acquasparta, he saw the difficulties with Cesi's alleged views about the generation of fossil wood. And this, of course, was the sceptical line adopted in much more vigorous form by Naudé.

In his letter of 30 November to Peiresc (in which he mentioned that La Ferrière had been there a little while before him), Naudé discussed 'my little trip made especially' to see the fossil woods of Acquasparta:

> I am more uncertain than ever before about what one ought to believe, all the more since on the one hand the authority of Duke Cesi, who was well versed in these matters, and who had worked so hard to inform himself about them, together with the nature of the earth, which seems in certain parts to change itself into wood, causes one to believe that this wood is really fossilized and natural in this place. But on the other hand, when one considers the form of this wood which resembles trees, trunks, logs, and other ordinary pieces of wood such as we ourselves find, and that one finds on it the knots, branches, roots, eyes, and other ordinary parts of wood, and that one piece is buried here and another there all confusedly, and at the worst that they are even sawn and cut in different places so that most have rotted with the bark, and having all the other accidents of our ordinary wood, one can hardly conclude anything else except that all this is rather the effect of those changes which happen in nature according to the opinion of Aristotle, and which I believe to be true, since one can see many other examples, such as where once there was a bridge; once there was land where now there is water, and once an anchor was found on the top of a mountain. At least one cannot deny that even if the nature of this place had the property of producing this fossil wood, it did not have some upturned forest, and perhaps one would also say that it is this wood which changes into earth, and not the earth into wood.[160]

'In order to clarify these matters', Naudé continued, 'one will really have to offer a lengthier substantial explanation, which I cannot do at the moment because of the haste in which I'm writing to you'.[161] Hastily written, perhaps—but it was a long enough letter.

The metallophytes of Acquasparta had clearly become a problem of great interest to those in the know. In addition to his extraordinary connections within the *République des Lettres*, Naudé was not only the librarian of Cardinal Guidi di Bagno (who had also encouraged La Ferrière in his fossil researches),[162] but had occupied the same position in the service of Cardinals Antonio Barberini and Richelieu in France, and would go on to do so for Cardinal Mazarin and Queen Christina.[163] At first expressing himself with comparative moderation, Naudé adopted a truly sceptical, cynical, and derisive tone a few months later:

> One of the main reasons I had until now for believing that these were fossils was the authority of Prince Cesi, who had done so many experiments with them. But proposing this difficulty to M. de Bonnaire the other day, I was astonished when he made a joke of it, saying that he thought nothing at all of the judgement of this Prince, whom he knew well, and that he was the poorest man in the world, and that easily and without the slightest discretion he believed everything which one told him.[164]

Thus wrote Naudé to Peiresc, on 29 March 1636. He was still in Rieti, and planning a second trip to Acquasparta to see the fossils first-hand.

Many years later, in his *Mascurat* of 1649–50, Naudé did not hesitate to lump together every one of the actors in this saga of the fossils as instances of *hommes de bien et curieux* who suffered from being much too credulous:

> This great man, the late Monsieur Peiresc, had the same problem, while the Duke Federico Cesi, who was the head of the Academy of Linceans in Rome, in deferring too much to the judgment of others, trusted almost nothing to his own ... Just as there are some spirits who have the marvellous disposition to believe nothing, so too there are those who even more vigorously doubt nothing at all ... In this regard, I remember that when I was travelling in Italy, I had the curiosity to go to see a mine or quarry of fossil wood, or thought to be such, near the town of Acquasparta, and about which one of the Linceans named Stelluti, put together a book at the instruction of the said Duke Federico; but although the whole of this Academy and a certain Claude Menestrier of Besançon, who was said to be a great Naturalist, or rather a great Fabulist (because everything was fine with him), as well as the Doctor of the Cardinal of Lyon [La Ferrière], had judged that this was fossil wood, I found after observing it well, and having exhumed the batons, planks, staves, beams, trunks and various kinds of knotty, straight, curved, and twisted wood, with bark and without, horizontal in one spot, standing up in another, I recognized, as I said, that all this wood came from some destroyed and crushed forest with all its timber yards and warehouses that had been in it, as a result of the overturning of the higher and neighbouring terrain, in the furious tremors and agitations of the earthquake of 1618 ... and that after the lapse of time, they acquired a hardness, accompanied by a blackness, and certain veins, that can also be seen in oak wood once it has been buried for hundreds of years beneath the water, or in the earth, as one sees in the stakes of Venice and Amsterdam.[165]

His careful excavation and observation of the fossil woods of Acquasparta had put Naudé on the right track, but his huge jumps in logic made his correction of Cesi unpersuasive as well as arrogant. He recognized the origin of the fossil wood where another researcher might have surmised it; and, while the instance of oak piles was certainly relevant, he failed to show how eighteen years in the ground could produce the same effect as 'hundreds of years beneath the water, or in the earth' nor how it was that in 1618 there were trees there growing of a size the aged La Ferrière had never in his life seen.

Certainly Naudé's cynicism was breathtaking, and rather cavalier to boot. After all, Cesi's critical abilities were also spurred on by the very Aristotelianism which Naudé supported in his letter of 1635. Naudé himself was known as a libertine, a heterodox thinker, a supporter of Richelieu and Mazarin, one moment acerbic on behalf of the radical sceptics, another time keenly conservative and supportive of his established employers. Of course there was also some truth in what he said about the fossils (if not about Cesi). 'It is true that the views of Gabriel Naudé about the phenomenon of fossil wood are in many respects more exact and pertinent scientifically than the confused and often ingenuous observations, the fabulous hypotheses, the puerile enthusiasms expressed by the Linceans and by Stelluti in particular', a modern commentator has written.[166] But everything we know about Cesi, if not necessarily about his colleagues, suggests that he was not credulous. On the contrary.

One of the most moving aspects of Cesi's work was his scepticism about the traditional Aristotelian views of nature and its objects. However much he may have been steeped in those ideas, however much he may have been inclined to accept the charms of their appeal to the purely empirical, he resisted them, more and more. His researches on fossil woods led him to researches on other fossils. What his views were he may have had to suppress, since to suggest that the world was not made in a week, that there was no fundamental unity of species across time, at the same time as Galileo was in the deepest trouble for proving—over and over again—that the world was not the centre of the universe, would indeed have been dangerous.

We will never know exactly what Cesi felt about the origins of the fossils of Acquasparta, or whether, like Leonardo, he formulated some rather more general explanation about the presence of marine fossils on mountain tops and far from the oceans; but what we are left with in the case of the fossil drawings in Windsor is yet one more instance of the magnificent visual testimony to his extraordinary curiosity. Nothing, it seems, escaped the eyes of these Linceans, no wrinkle or muddy patch of nature or any of its specimens. In their commitment to the close visual recording of nature, they may have tried to record too much that was not strictly relevant to the project of classification, and they may indeed not have devoted enough energy to what in modern terms might be called the principles of scientific hypothesis in its relation to experiment (principles articulated so clearly by Galileo), but they were never uncritical. It may have been that all their energies were absorbed by the amassing of visual and verbal records of so much that had never been recorded before. But their attentiveness to the multifold abundance of nature, as well as their persistence in trying to reach what lay on the surface of things was unparalleled.

However frustrating it may be not to know what conclusions Cesi reached, we may console ourselves with the magnificent legacy of the drawings he commissioned. History may have failed to save his ideas from oblivion, but the extraordinarily detailed evidence on which he planned to base them has now belatedly begun to emerge from long obscurity. What new insights he may have gained, we do not know, but this part of his achievement is not in doubt: the first comprehensive study in visual form of the fossil remains of a particular region. There was nothing like this before, and nothing for a long time to come. Cesi's own contribution—massive, committed, full of remarkable natural historical intuition, and fierce in its observational intensity—prepared the way for a whole new approach to the use of illustration in the study of what we now call science.

NOTES

1. The author has expanded many of his observations in this essay, and set them in their broader Galilean context, in his forthcoming book on the natural historical researches of the Linceans, Freedberg forthcoming.

2. Stelluti 1637. A more literal translation of the title *Trattato del Legno Fossile Minerale* would be *A Treatise of Fossil Mineralized Wood*. For the full title see the bibliography and the translation, pp. 383–8 below. For what has been deduced so far on Cesi's work on fossils, see Bromehead 1947a, the many scattered references to it in Gabrieli 1989, Nicolò 1986 and the articles by Freedberg and Solinas listed in n. 3. An excellent summary of the Linceans' palaeontological researches (on the basis of the indications in the correspondence and Stelluti's little book) appeared in Gould 1998b.

3. The drawings were briefly discussed by Bromehead 1947a, and then by Solinas 1989a, Solinas 1989b, Freedberg 1989b and Freedberg 1993. No one, as far as the present author is aware, has ever mentioned the remarkable fossil drawings in Cesi's manuscripts in the Institut de France referred to in nn. 11 and 46, below.

4. See Encelius 1557, Gesner 1558, Gesner 1565, Aldrovandi 1648. The last work, of course, was a posthumous publication, based, as in the case of most of the natural historical publications that bear his name, on Aldrovandi's papers. From his will we learn that Aldrovandi was preparing a *Geologia ovvero De Fossilibus*, and much of his work on this subject—as on many others—was well known to scholars. Cesi's closest early friend, Johannes Heckius, was writing to Aldrovandi for natural historical information as early as 1603 (cf. Gabrieli 1989, pp. 1084–7). Then too, Aldrovandi's 1606 work on molluscs, crustaceans and so forth, was available—the *De Reliquis Animalibus Exanguibus, nempe de Mollibus Crustaceis, Testaceis et Zoophytis* (the last term is particularly significant for the story to be set out here). Of course, Mercati's *Metallotheca*, though only published in 1709, was long known in manuscript as well. For a comprehensive survey of the background, see Rudwick 1976, pp. 1–48.

5. See Agricola 1546, Agricola 1556, Cardanus 1550, Palissy 1580, Cesalpino 1596. For a good summary, see Rudwick 1976.

6. *Glossopetrae* were reputed to be the tongues of serpents but are now known to be fossilized sharks' teeth. Cf. also p. 34 above on Colonna's and then Steno's important works on this subject; for titles of works by Colonna, see the Bibliography; for Steno's work see n. 25, below.

7. In addition to the discussion in, for example, Rudwick 1985, Accordi 1981a and Accordi 1981b, see also the important contextualization of these fossil curiosities in terms of 17th-century collecting culture in Schnapper 1988.

8. On this aspect of the Aldrovandian approach, see especially Foucault 1973, Part I, 2.

9. On Leonardo's important contributions to the study and understanding of fossils, see the useful discussion in Rudwick 1976, pp. 39–41, as well as the excellent article, Gould 1997.

10. Over forty of the drawings of fossils still show evidence of these annotations. They are not always legible around the edges of the sheets where they were erased after being copied by a scribe when they entered Cassiano's collection.

11. It is not only the fossils drawings, of course, that are thought to have been bound under Cassiano's ownership, but also the *Erbario Miniato* and the *Erbario Messicano* in the Royal Library (to be catalogued in Parts VI and VIII of the present series), as well as the volumes on fungi and plants in the Bibliothèque de l'Institut de France in Paris, MSS 968–970 and MSS 974–978 (to be catalogued as Parts II and VII).

12. On these indices, see Appendix I and n. 15, below.

13. Since many of Cesi's own inscriptions were cut off or erased when the drawings were mounted in albums, either when they entered Cassiano's Paper Museum or at the time of George III, these scribal transcriptions, for all their spelling errors, are often quite useful. See also n. 10, above, and n. 15, below.

14. George III's acquisition of the drawings is described on p. 10 above.

15. A substantial number of drawings in which the iron-gall ink had 'burned' through the paper were restored after they reached Windsor in 1763: the holes in the paper were touched up with brown watercolour where the backing paper showed through. Each drawing was then framed by a decorative wash border, as in the case of many of the other drawings from the dal Pozzo–Albani collections at this time (see p. 10 above). The only sheets not treated in this way were those bearing the two indices to the corpus, compiled by Cassiano, which were mounted into the beginning of Volumes XIV and XVII. As these index folios (each measuring about 302 × 210 mm) had writing on both recto and verso, apart from the second folio of the index in Volume XVII, they were inlaid (as in Fig. 4).

16. Volumes XIV to XVII (each measuring 608 x 460 mm) were bound uniformly in calf, with gold tooling around the edges and George III's crest blocked in gold on the front and back boards [Fig. 3].

17. See Locations of dal Pozzo Material (p. 16). A fifth volume with the title *Natural History: Fossils* is numbered V, and contains very beautiful coloured drawings of rocks, gems, minerals, coloured marbles and other curiosities (such as bottle-imps). It will be catalogued in Series B, Part IX of this series. For a brief summary, see Napoleone 1989.

18. See Concordance C, p. 398, below.

19. For these 17th-century indices, see Appendix I, below.

20. For aetites, see n. 70, below.

21. Besides Bromehead 1947b, for the Linceans' interest in so-called 'pregnant' species, cf. also Freedberg 1992b, Freedberg 1996b and Freedberg/Baldini 1997.

22. See pp. 36 and 148; and **30** (Windsor, RL 25656).

23. Ruled around the images either in pen and ink or in black chalk, these frame-lines were evidently added at an early stage, before the numbers were written on the sheets.

Another mark that appears on some of the sheets is a small cross in pen and ink. Twelve sheets are marked in this way, one with two crosses by two separate specimens. Although it has been suggested that these crosses indicate images selected for engraving, the drawings thus marked do not relate to any of the drawings with frame-lines, nor, indeed, to any of the engravings in Stelluti's *Trattato*. Other possible explanations are that the crosses denoted specimens from a particular part of Cesi's collection, from another collection, or from a particular locality. The fact that these crosses do not appear in Cassiano's indices of the drawings (see n. 15, above, and Appendix I), reinforces the possibility that they may have been added by Cesi himself.

24. See the Linceans' letters of March 1604, and Campello's discursive reply from Spoleto of 10 April 1604 (Gabrieli 1959, nos. 10 and 12, *passim*; and 13, pp. 31–4).

25. For these works by Colonna, see the Bibliography. Steno published his findings in the *Canis Carchariae Caput Dissectum*, Copenhagen, 1667; available in Nicholas Steno, *The Earliest Geological Treatise*, trans. Arvel Garboe, London, 1958.

26. In the father's *ordinatamente* is foreshadowed the fundamentally classificatory purpose of the son's treatise. Little could they have known, in those days, how difficult it was to provide a fixed classification for species that so plainly revealed the problem of mutability over time. The literature on the Imperato family is now growing rapidly: a summary is provided by Accordi 1981a, while more up-to-date material is available in Findlen 1994.

27. On the generation of heat and light, cf., in addition to the material in Shea 1977 and Redondi 1987, the references to Psalm 18: 5–6 (Psalm 19 in the King James version) in the famous letter to Pietro Dini of 23 March 1615, in Galileo *Opere*, V, pp. 297–305, as well as the discussion in Blackwell 1991, p. 74 and Rossi 1978, pp. 45–6.

28. On the nature of light and the atomistic approach, see Redondi 1982, Redondi 1984 and Redondi 1987, as well as the important material (particularly as it relates to Faber) in De Renzi 1992–3, especially pp. 148–50.

29. See Persio 1613. On Antonio Persio (1542–1611), whose manuscripts became one of the major unfinished publication projects of the Linceans, see Gabrieli 1989, pp. 442–3 and 865–88. The manuscript of the *De Natura Ignis et Caloris* was ready for publication but blocked by the hostility of the censors (Rome, BANL, Archivio Linceo VI, fols. 13–347 and VII, fols. 4–370); on this see De Renzi 1992–93, p. 137, with earlier literature.

30. Johannes Faber, *Oratio qua Ignis & Metallorum Exemplo, quam parum sciamus demonstratur* (written 1622), now in Naples, BN, MS VIII.D.13; described in Gabrieli 1989, pp. 1183–4; published in full and well discussed in De Renzi 1992–3.

31. Liceti brings together the topics of fossil remains and the ancient world most strikingly in the oft-reprinted *De Lucernis Antiquorum Reconditis*, first published in Venice in 1621 (see Bibliography for this and other relevant books by Liceti combining investigations into light, heat and antiquity).

32. Cf. Redondi 1984 and Redondi 1987, pp. 20–27, as well as Liceti 1640a and 1640b.

33. On Galileo's demonstration of his telescope, cf., for example, the references supplied in Gabrieli 1959, pp. 161–2, as well as the longer account set out in Freedberg forthcoming.

34. The 'Bologna stone' had been found by a Bolognese alchemist on Monte Paderno near Bologna *c.* 1604, just about the time Cesi was beginning his own explorations of the fossil woods. See Gomez 1991 and the works cited in n. 35.

35. On this experiment with the luminous stone, and its significance, see especially Redondi 1984, and Redondi 1987, pp. 5–27; see also Galileo *Opere*, XX, p. 290; Gabrieli 1959, p. 215, n. 8; De Renzi 1992–3, pp. 64–5 and 136-7, and Findlen 1994, pp. 230–32.

36. The date of 1611 appears on **30**, Windsor RL 25656.

37. See above, pp. 38–40, as well as Stelluti 1637, *Thesaurus*, p. 905 and Stelluti 1630, p. 170.

38. Cf. pp. 38, 48 and nn. 56–61 below.

39. Naples, BN, MS XII.E.4, published in Gabrieli 1989, pp. 38–60.

40. For more on Cesi's 'frontispieces', see also Freedberg forthcoming.

41. These published frontispieces were the *Tabulae Phytosophicae* appended to the one project of the Linceans that was indeed published in something approaching its intended entirety, the *Tesoro Messicano* (the *Thesaurus* of 1649–51); discussed above on pp. 46–9.

42. Pliny, *Natural History*, Book 37, Caps. 25–8 and 83.

43. Cf. Freedberg 1994 on these matters.

44. See Freedberg/Baldini 1997.

45. See Freedberg 1994.

46. There are several other drawings in Volume V of the *Natural History: Fossils* in the corpus that came from Cassiano's Paper Museum (including ones originally belonging to Cesi) that testify to the Linceans' concern with borderline species. Then too the many folios in MSS 977–8, in the Institut de France in Paris, give some indication of the Linceans' earliest researches into fossils, including several in MS 978 (some engraved and some drawn) where the overlap between what we would call fossil remains and *artificialia* such as statues could not be clearer. These are discussed at greater length in Freedberg forthcoming.

47. For example, Rome, BANL, Archivio Linceo 2, and Naples, BN, MS XII.E.4, as well as the manuscripts in the Vatican Library cited in n. 74, below.

48. Naples, BN, MS XII.E.4, fols. 13r–15v.

49. Ibid., fol. 15r.

50. See n. 27 above for references; cf. also Galileo *Opere*, XX, p. 290.

51. Things of a 'middle nature' were distinguished from prodigies and monsters, which Cesi also acknowledged—clearly for once in this period—as presenting special problems for any clear system of classification.

52. Naples, BN, MS XII.E.4, fol. 15r.

53. Ibid., fol. 15v.

54. Ibid., fol. 16v.; cf. the useful discussion in Gabrieli 1989, p. 45.

55. Cf. n. 32 above, especially Gabrieli 1959, p. 215, n. 8.

56. Imperato in Naples to Faber in Rome, 10 June 1611, in: Gabrieli 1959, no. 65, p. 163.

57. Cesi to Galileo in Florence, 21 October 1611, in: Gabrieli 1959, no. 78, p. 176; G. B. Della Porta to Cesi, 7 April 1612, in: Gabrieli 1959, no. 113, p. 215. For Peiresc's views on the imperfect correlation of heat and light in phenomena that shine without warmth, see Reeves 1997, p. 87.

58. Cesi in Tivoli to Galileo in Florence, 21 October 1611, in: Gabrieli 1959, no. 78, pp. 175–6; Cesi in Rome to Galileo in Florence, 15 February 1613, in: Gabrieli 1959, no. 219, pp. 326–8; Cesi in Monticelli to Galileo in Florence, 30 May 1613, in: Gabrieli 1959, no. 242, pp. 357–8; Cesi in Rome to Galileo in Florence, 3 January 1614, in: Gabrieli 1959, no. 296, pp. 405–6. For an important discussion of the issues of luminescence and secondary light, with particular reference to its Neo-Stoic context, see Reeves 1997, pp. 84–90.

59. See *Des Bamberger Fürstbischofs Johann Gottfried von Aschhausen Gesandtschaftsreise nach Italien und Rom, 1612 und 1613*, ed. Ch. Haütle, Tübingen 1851, p. 92.

60. Cesi in Monticelli to Galileo in Florence, 30 May 1613, in: Gabrieli 1959, no. 242, p. 357.

61. Galileo *Opere*, VIII, pp. 471–555; cf. also Redondi 1987, pp. 20–25.

62. Cesi in Acquasparta to Faber in Rome, 20 January 1621, in: Gabrieli 1959, no. 588, pp. 732–3. Cf. Freedberg forthcoming on classification and fossils.

63. For a brief overview of 16th- and 17th-century classificatory attempts in the field of geology (though, typically, with no mention of Cesi), see Accordi 1981b.

64. Winther in Acquasparta to Faber in Rome, 20 January 1621, in: Gabrieli 1959, no. 588, pp. 732–3.

65. Cesi in Acquasparta to Cassiano in Rome, 18 August 1624, in: Gabrieli 1959, no. 772, pp. 933–4 (at p. 934).

66. *Inferriti* is the term Stelluti uses and which is translated here as 'pyritized'. For the secretion of resin, see **7**.

67. Stelluti in Acquasparta to Galileo in Florence, 23 August 1624, in: Gabrieli 1959, no. 773, pp. 934–5.

68. Cesi in Acquasparta to Francesco Barberini in Rome, 29 September 1624, in: Gabrieli 1959, no. 785, p. 948. Not surprisingly, it was in precisely this context—furniture made from fossil woods—that Cassiano referred to the Acquaspartan metallophytes in a letter to Lorenzo Pignoria in Padua (Montpellier, BEM, MS H.267, fol. 34).

69. Cesi in Acquasparta to Francesco Barberini in Rome, 1 December 1624, in Gabrieli 1959, no. 798, pp. 965–7 (at pp. 965–6). For the full text, see Appendix III.

70. On the aetites or eagle-stone, which can be defined as 'any hollow stone containing loose matter, a smaller stone or sand, which rattles when shaken', see Bromehead 1947b, p. 22. The eagle-stone was thus called because it was supposed to be found in eagles' nests and to be used by them as an aid to breeding. It was also believed to prevent miscarriage in women and help in an easy delivery, this talismanic power no doubt deriving from the view that the stones themselves got pregnant and gave birth. For a sceptical account see also Sir Thomas Browne's *Pseudodoxica Epidemica* (1646), Book II, Chapter V.

71. Cesi in Acquasparta to Francesco Barberini in Rome, 1 December 1624, in: Gabrieli 1959, no. 798, pp. 965–7. See Appendix IIIa for the text of this important letter.

72. Ibid., pp. 966–7. Cf. Figs 32–7, Comp. figs 21–4 for the fossil remains brought back from Umbria by Andrew Scott.

73. Including Rome, BANL, Archivio Linceo 2, fols. 57–60, 62.

74. Copies of Cesi's letter of 1 December 1624 to Francesco Barberini are found in Rome, BAV, MS Vat. Lat. 8258; Montpellier, BEM, MSS. H.170 and H.173; Paris, BN, MS It. 1684 (all discussed on pp. 62–3 above with accompanying notes; see also Appendix IIc below); and Rome, BAV, MS Barb. Lat. 6461, fols. 148–9, with another copy on fols. 157–8. Two drafts for the letter are preserved in Rome, BANL, Archivio Linceo 2, fols. 57–60, 62.

75. Cesi in Acquasparta to Cassiano in Rome, 2 December 1624, in: Gabrieli 1959, no. 799, p. 967.

76. Gabrieli 1959 (p. 1188, n. 2) rightly noted that this *argento ramoso* (or 'branchy silver') is not in fact described by Imperato. It is, however (as Ian Rolfe has pointed out to the present author), the naturally occurring form of dendritic silver.

77. Colonna in Naples to Stelluti in Rome, 10 November 1628, in: Gabrieli 1959, no. 982, pp. 1187–8.

78. Colonna in Naples to Cesi in Rome, 16 December 1628, in: Gabrieli 1959, no. 984, p. 1190.

79. Colonna confessed his puzzlement over a whole variety of phenomena, such as the opuntia or Indian fig, various kinds of Balanides, types of petrified corals, mineralized plants, and apparent combinations of plant with bone or rock, all phenomena illustrated, in one form or another, in the volumes of fossils drawings in Windsor and elsewhere in the Cesian corpus, especially in the Institut de France. Many are also referred to in Stelluti 1637, pp. 11–12. Colonna in Naples wrote to Cesi in Rome, 16 December 1628 (in: Gabrieli 1959, no. 984, pp. 1190–91): '*Potriasi, mentre se pigliasse per pietro l'istesso osso del Testaceo, così chiamarsi quel Zoophyto che* saxis adhaeret, *et da un principio et animale, a modo chè fa la Fico de India Opuntia, germina pij et altri Zoophyti poco minori, servendo il primo per tronco; ma non mi sodisfaccio.* Balani 2ª species *ne dipinge molti Aldrovandi. L'altro animale è soluto et camina, et pigliandosi per pietra la sua testa se potria dir esser animale et pianta, perchè dall'istessa testa o scorza vegetante, vi è una pelle herbacea come la lichene, et in molte parti fimbriata et pij fogliosa, cioè nelle extremità delli vecchi orificij, et nell'ultima bocca, chè cresce insieme l'herba et l'osso dell'animale.*

Scrissi al Signor Stelluti chè, se alcuna specie di quelle Pore ramose come coralli o altra cosa simile havesse senso, se potria così dire; ma io desidero imparare da V. Ecc.'.

80. Colonna knew something about accurate illustration of the world of nature, having been the first ever to have etched such subjects in the famous *Phytobasanos* of 1592, then in the *Ekphrasis* of 1606, and the *Purpura* and the second *Ekphrasis*, both of 1616.

81. *Thesaurus*, pp. 905-50. See the edition and brief commentary by Pirotta 1904.

82. Cf., for example, Gabrieli 1959, p. 1188, n. 1, and Gould 1998b.

83. All this in *Thesaurus*, p. 905.

84. Faber's commentary on Mexican animals was actually written between 1625 and 1628.

85. *Thesaurus*, pp. 573–4.

86. Ibid., pp. 502–3.

87. Montpellier, BEM, MS H.267, fol. 34.

88. *'Due calamari di legno-fossile' and 'Un tavolino di legno fossile rotto senza piedi, e semplice senza profili'* in Rome, BANL, MS Archivio Linceo 32, fol. 87; also available in Nicolò/Solinas 1986, p. 208.

89. Stelluti 1630.

90. See comments and discussion in Freedberg 1998, as well as in Freedberg forthcoming.

91. The illustration and discussion of the magnified bee is one of the great incunabula of microscopic illustration. See Freedberg 1998 with earlier literature.

92. Stelluti 1630, p. 21; and this is also the source for the quotations that follow.

93. Ibid.; cf. p. 170 as well.

94. Stelluti 1630, pp. 169–70.

95. *'Ma pare che la natura invidiasse il gran sapere di questo Signore'*; Stelluti 1637, p. 12.

96. Stelluti to Cioli, 2 June 1635, cited in Nicolò 1986, pp. 176–7; Florence, BNC, MS Gal. 102, fol. 18.

97. Montpellier, BEM, MS H.170, fols. 6–7v; MS H.173, fols. 5–6.

98. Montpellier, BEM, MS H.271, fol. 146; Lhote/Joyal 1989, pp. 172–3.

99. Although there are many other references to La Ferrière in Peiresc's correspondence (e.g. in the letter preserved in Carpentras, Bibliothèque Inguimbertine, MS 1821, fol. 221 –221v, or that from Peiresc to Jean Bourdelot in Rome, 21 March 1635, with instructions to La Ferrière on the proposed trip to Sicily mentioned above), we know from this 20 March 1635 letter of Peiresc's to Cassiano that La Ferrière was 'a gentleman most competent and curious in the secrets of nature and of human art, a doctor, but more

noble and more generous'. In addition, 'he has discovered very great marvels, and I am sure that your lordship will be hugely pleased by his accounts of these, and by his experiments' (Montpellier, BEM, MS H.271, fol. 151; also in Lhote/Joyal 1989, pp. 175–6).

100. Montpellier, BEM, MS H.271, fol. 151; in Lhote/Joyal 1989, pp. 175–6, with useful supplementary notes on Peiresc's interest in the fossil remains of Sicily.

101. See the remarkable letter of 1 February 1635 from Peiresc to Menestrier, in which Peiresc discusses at length the fossil and other curiosities of Sicily (Tamizey de Larroque *Correspondants*, V, pp. 756–9); cf. also another letter, preserved in Carpentras, Bibliothèque Inguimbertine, MS 1821, fol. 140–140v.

102. Montpellier, BEM, MS H.271, fol. 151v; Lhote/Joyal 1989, p. 177, with further important supplementary notes.

103. Cf. Lhote/Joyal 1989, p. 177, n. 27.

104. Montpellier, BEM, MS H.271, fol. 152v; also in Lhote/Joyal 1989, p. 180.

105. Rome, BANL, Archivio dal Pozzo, V, fol. 228; cited in Nicolò 1982, pp. 94–5 and again in Nicolò 1986, p. 176. Cf. Stelluti 1637, p. 9, describing the sample illustrated in Stelluti 1637, pls. 7–8.

106. Stelluti 1637, sig. A2.

107. Montpellier, BEM, MS H.271, fol. 157v; Lhote/Joyal 1989, p. 185.

108. Tamizey de Larroque *Lettres*, IV, p. 502, and V, pp. 784–5; cited and commented upon by Alessandrini 1978, p. 117.

109. Cf. Montpellier, BEM, MS H.271, fols. 161 and 164–164v.; transcribed in Lhote/Joyal 1989, pp. 189–99.

110. Montpellier, BEM, MS H.271, fol. 169v; Lhote/Joyal 1989, p. 203.

111. Montpellier, BEM, MS H.271, fol. 172v; Lhote/Joyal 1989, p. 208.

112. Tamizey de Larroque *Correspondants*, XIII (Naudé, *Lettres inédites*), p. 43.

113. Montpellier, BEM, MS H.170, fols. 11–12; this letter (La Ferrière to Peiresc, November 1635) is accurately transcribed in Alessandrini 1978, pp. 214–17. See Appendix IIIb below for the whole letter.

114. Peiresc expressed his views on fossils, for example, in the letter of 2 August 1635 (Lhote/Joyal 1989, pp. 198–201).

115. The views of Cesi and Stelluti on fossils were set out, especially, in Stelluti 1637; discussed above, pp. 56–62.

116. Despite the wide circulation of Stelluti's *trattatello*, no-one seems seriously to have discussed the classificatory problems at stake—perhaps because these do not seem to have been much expounded in the *trattatello*.

117. Cf. the letter written by Stelluti to Cioli and cited in n. 96, above.

118. Stelluti to Galileo, 3 November 1635, in Galileo *Opere*, XVI, p. 337.

119. Montpellier, BEM, MS H.271, fol. 194r; Lhote/Joyal 1989, p. 237.

120. Gabrieli 1989, p 938; slightly expanded in Nicolò 1986, p. 177.

121. 'The hidden parts of nature'—*Gli occulti parti della natura* (Stelluti 1637, sig.A2)—was a phrase that occurred over and over again at the time, and especially in the early correspondence of the Linceans. Generally used in regard to that which was hidden below the surface of things, to the secrets and mysteries of nature—the *arcana naturae*—the phrase here referred literally to things hidden beneath the earth.

122. Stelluti 1637, pp. 3–4.

123. As also observed by Gould 1998b.

124. Stelluti 1637, p. 5, also referring to Pliny, *Natural History*, Book 13, Cap.15.

125. With its indications of the sites where *metallophita et succensiones* were to be found, it is probable that the map was supplied by Cesi to Stelluti; cf. Gabrieli 1989, pp. 940–41. On the general location shown, see Boncambi 1961, pp. 67–76 and Biondi 1984, p. 16, citing the important work of R. Almagià, *Documenti cartografici dello Stato Pontificio*, Vatican City 1960. See p. 79 below for the geological significance of this map.

126. Stelluti 1637, p. 6.

127. Cf. the discussion on pp. 38–42 above.

128. Stelluti 1637, p. 6; cf. also p. 56 above.

129. Stelluti 1637, p. 6.

130. Stelluti 1637, p. 8. On 'jokes of nature', see Findlen 1990, Freedberg 1992b and Freedberg/Baldini 1997.

131. Stelluti 1637, p. 8.

132. See **7**, **8**, **11**, **19**, **76**, **93–95**, **97**, **104**, **112**, **117** and **197–199**.

133. Cf. p. 79, below.

134. Stelluti's *Trattato* is here described as 'enervated' because the work as a whole is so very thin in comparison with what Cesi is said to have promised.

135. Stelluti 1637, Pls. 3–6; there is in fact a fifth plate of the same kind (Pl. 9).

136. Stelluti 1637, p. 7.

137. See Gould 1998b.

138. On the 'pregnant stone', cf. nn. 21 and 70, above.

139. Several of these unusual stones are mentioned by Colonna in his letter of 16 December 1628, cited in n. 78 above, as well as Cesi's Naples manuscript, Naples, BN, MS XII.E.4, fol. 15r–v. According to *Thesaurus*, p. 536, lyncurium—the *Linceo*'s stone—was a kind of woody concretion on which mushrooms grew, and was produced by the *Linceo*'s urine.

140. Especially in Paris, IDF, MS 978.

141. B. IX (forthcoming) of the present series.

142. Montpellier, BEM, MS H.170, fols. 11r–12r, and headed '*Copie de la lettre escrite à Monsieur de Peyresc sur les bois fossiles d'Aqua Sparta*'. Translated *in toto* in Appendix III b.

143. Paris, BN, Res. S.309, drawn to the author's attention by Francesco Solinas.

144. Cf. Morello 1986, p. 58, no. 42. This manuscript, in a vellum binding with Cardinal Francesco Barberini's coat of arms on it, also has a slightly different title from that of the final *Trattato*: it is here entitled *Breve trattato della natura e qualità del legno fossile minerale ondato*.

145. The slightly more distant variants in Montpellier, BEM, MS H.173 suggest that it should be dated just a little before Paris, BN, MS It. 1684.

146. Montpellier, BEM, MS H.173, fol. 3v and Paris, BN, MS It. 1684, fol. 3v.

147. Stelluti 1637, p. 9.

148. Montpellier, BEM, MS H.173, fol. 4r; Paris, BN, MS It. 1684, fol. 3v.

149. Stelluti, 1637, p. 8.

150. Cf. p. 63 above.

151. On the use of this particular metaphor (of the cobbler sticking to his last) in the context of Galileo's interpretation of scripture (and his opponents' interpretations of data about the natural world), see Freedberg forthcoming.

152. Cf. p. 47 and n. 77, above.

153. Colonna in Naples to Stelluti in Rome, 10 November 1628, in: Gabrieli 1959, no. 982, pp. 1187–8.

154. Ibid., p. 1188.

155. Ibid.

156. In order to demonstrate that fire was not transparent, Galileo made a rather forced distinction between 'fire' and 'flame' (cf. Reeves 1991, pp. 3–4).

157. The situation was complicated, and touched on other critical Galilean issues as well—such as whether any star could be seen through the brilliant light of flames. On all this, see, for example, Galileo *Opere*, VI, pp. 174–95 *passim*, as well as pp. 361–5 and 495–7. In his initial use of this example of transparency of flame, Grassi happened to use the example of semi-burnt wood as seen through the flames of a fire (Galileo *Opere*, VI, p. 174). Cf. also the important discussion in Reeves 1991.

158. Montpellier, BEM, MS H.271, fol. 167; Lhote/Joyal 1989, pp. 199–200. The next extract is from the same source.

159. Montpellier, BEM, MS H.271, fol. 167v; Lhote/Joyal 1989, p. 200. Cf. also Reeves 1991 on the topic of those more practical people who can reliably penetrate to the mysteries that would otherwise be obscured (with the metaphor of the mysterious writing on the wall at Belshazzar's Feast, in Daniel, 5).

160. Tamizey de Larroque *Correspondants*, XIII (*Lettres de Naudé*), pp. 42–3.

161. '*Mais pour bien esclaircir cette matiere, il en faudroit bien faire un plus grand raisonnement, ce que ne pouvant pour le present a cause de la haste en laquelle je vous escrips*' in Ibid., p. 43.

162. Alessandrini 1978, pp. 128–30, 215–17; cf. also Appendix IIIb below.

163. For a brief account of the significance of Naudé's relationships with Mazarin and Queen Christina, as well as more on his criticisms of the Cesian and Stellutian position on fossils, see Alessandrini 1978, pp. 128–35.

164. Tamizey de Larroque *Correspondants*, XIII (*Lettres de Naudé*), p. 52.

165. Naudé 1650, pp. 667–8.

166. Alessandrini 1978, p. 133.

GEOLOGICAL BACKGROUND

Andrew C. Scott

THE DRAWINGS preserved in Volumes XIV, XV, XVI and XVII of the Royal Library's series of albums, Natural History: Fossils, represent what remains of an important geological and palaeontological investigation by the *Accademia dei Lincei* undertaken by its founder Federico Cesi and a fellow member, Francesco Stelluti, to elucidate the origin of fossils. This material is unique[1] for such an early date, since it contains field drawings of the sites from which rocks and fossils were collected, shows some of the objects drawn *in situ* before their removal from the rock and includes descriptions of the specimens, often drawn from a number of angles.[2] Some of the annotations on the drawings state the location, and indications given elsewhere show that the material came from the area around Acquasparta.

Geology of the Acquasparta District

Plate 1 of Stelluti's *Trattato del Legno Fossile Minerale*[3] [Fig. 5]— a map of the district around Acquasparta—not only provides evidence of the Linceans' systematic, documentary approach but allows many of the drawings to be placed in their geological context. Acquasparta is situated about 90 km due north of Rome in the foothills of the Apennines, the range of hills that forms the backbone of Italy. This part of Umbria has a variety of rock types, both sedimentary and volcanic,[4] ranging in age from the Jurassic through the Cretaceous to the Tertiary and up to the Recent [Fig. 6]. The illustration of both Jurassic and Tertiary fossils in the drawings is therefore hardly surprising. Stelluti's perspective map is like an oblique aerial photograph, with north to the left, and shows hills, rivers, towns and villages.[5]

Although the view is diagrammatic, the scale (apparently of Roman miles) shows it was intended to be accurate. It depicts several features of geological significance. The background shows a range of steep-sided hills behind Acquasparta, while the land in the foreground is much lower and more undulating, with a few small hills. By giving the modern geological map the same orientation as Stelluti's, many of the features become clear. Prominent is the geological fault which runs north-south (left to right across the geological map) along the line of the via Flaminia (as seen on Stelluti's map). To the east of the fault lie harder Jurassic and Cretaceous rocks, elevated and forming the hills behind the town of Acquasparta. The lower lying ground between Acquasparta and Todi is composed of younger, softer sediments.[6] Comparison with the modern map shows the position of rivers and settlements to be remarkably accurate, and many of the localities shown on Stelluti's map can be

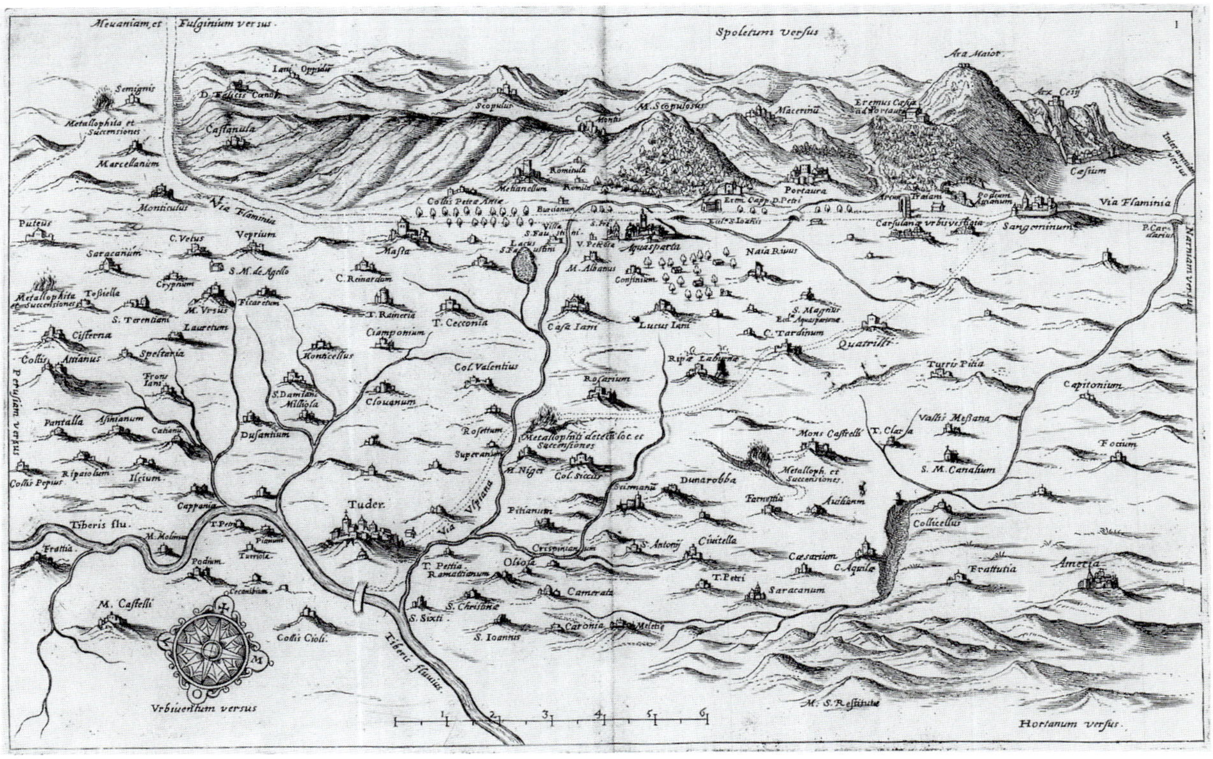

Fig. 5. Stelluti 1637, Pl. 1: map of the area around Acquasparta

Fig. 6. Modern geological map of the area around Acquasparta

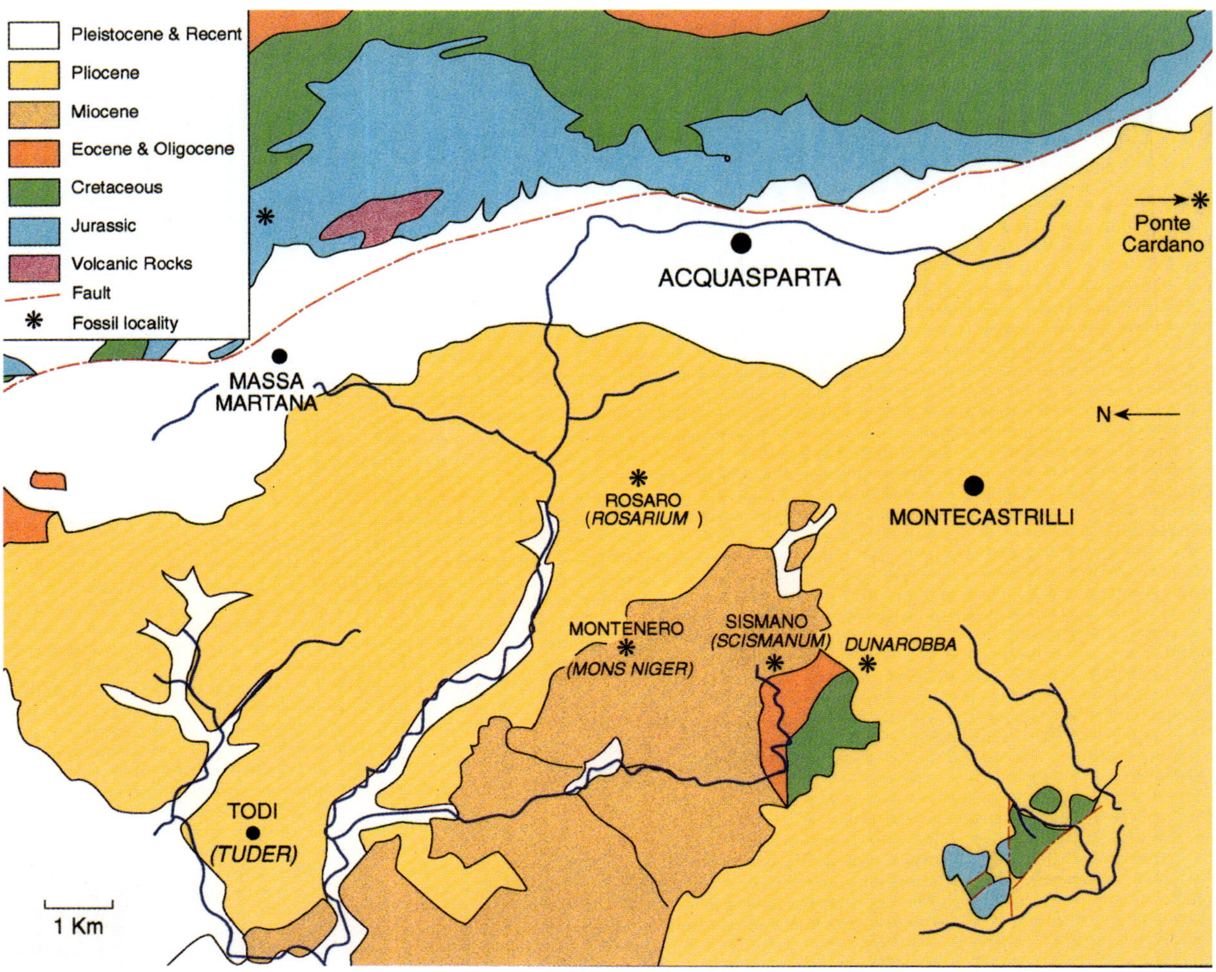

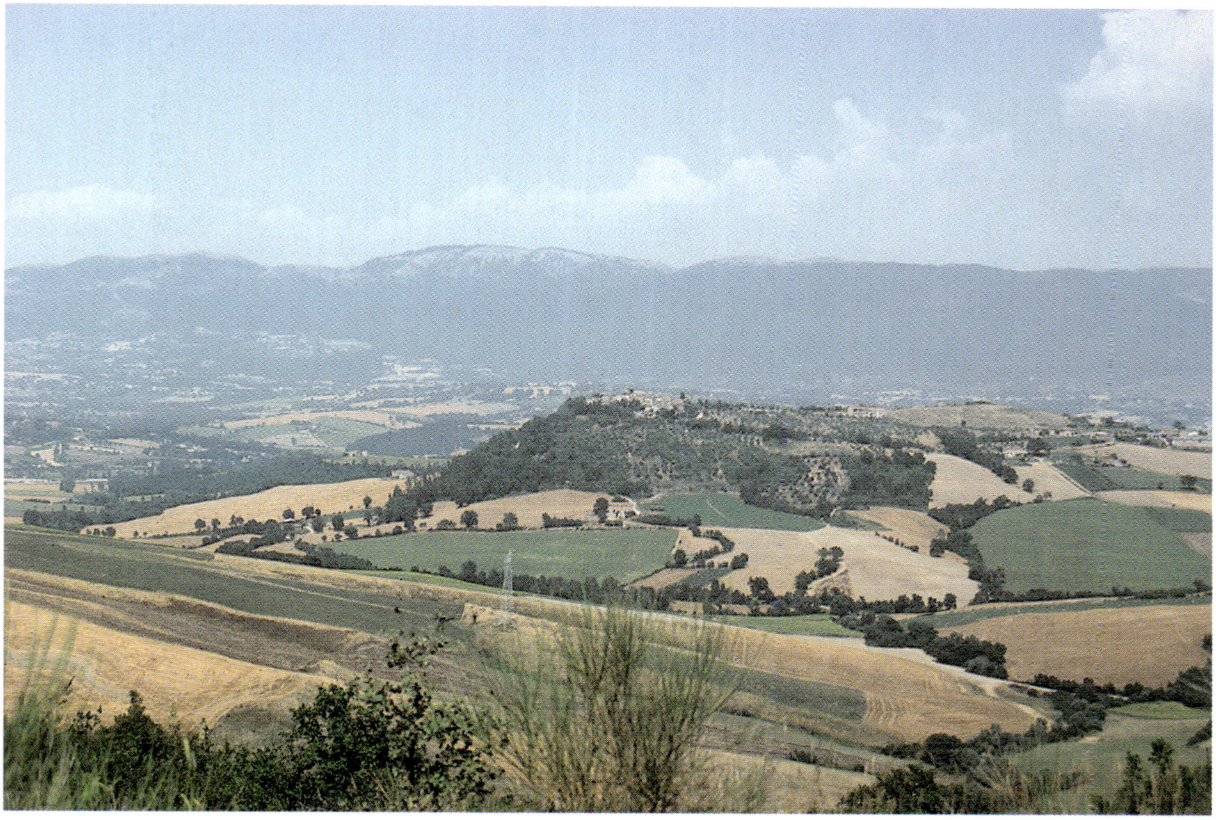

Fig. 7. View across the Acquasparta district

identified clearly. A good impression of the general topography of the area is given in Fig. 7. The majority of the rocks and fossils, and all of the sites illustrated, are from the low-lying rocks in the foreground of Stelluti's view while the ammonites probably come from the older rocks in the hills behind Acquasparta.

Jurassic Rocks and Fossils

The Jurassic rocks of the Apennines have been extensively studied.[7] A typical sequence of Lower Jurassic age comprises massive white crystalline limestones with gastropods, overlain with limestones full of ammonites, followed by red marls and limestones.[8] These later deposits contain abundant ammonites and are known as the *Ammonitico Rosso*. The Jurassic System is divided into a sequence of stages, each with its typical assemblage of fossils. Ammonites from Lower Jurassic rocks are common in central Italy.[9] It is no surprise, therefore, that such fossils, striking in their appearance, should be among those illustrated in Cesi's corpus. In contrast to the

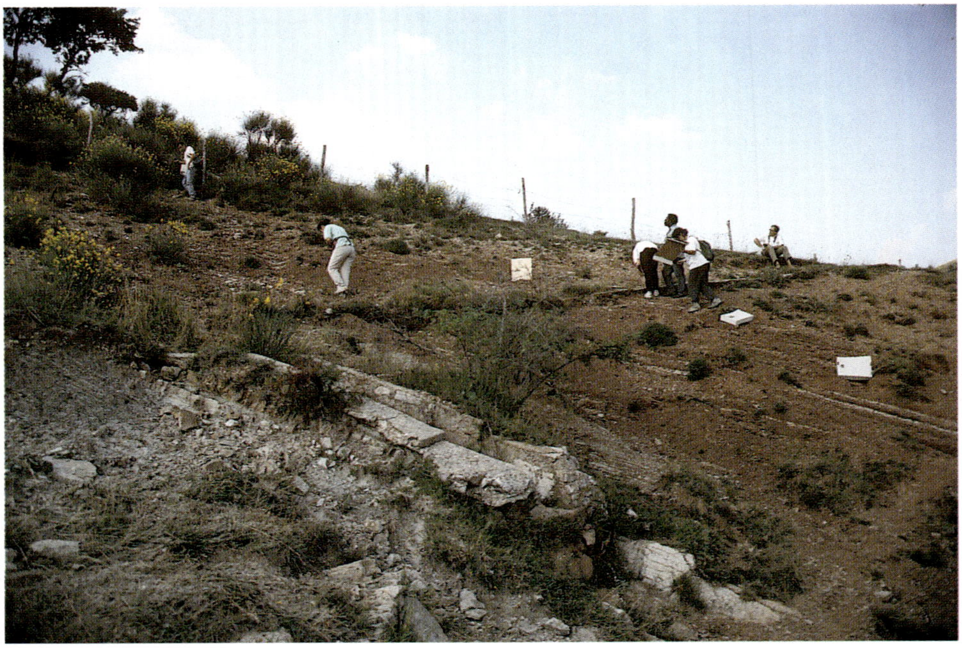

Fig. 8. The ammonite locality, Fosso di Pozzale

fossil woods, no indication of place of origin is given for any of the ammonites, making exact identification difficult.

However, one locality yielding ammonites was apparently known in Cesi's time, material having been collected there by Michele Mercati (1514–1593), the papal physician and naturalist.[10] This is near Massa Martana where the rocks of early Jurassic age (Toarcian) yield numerous ammonites [Fig. 8].[11] While definitive evidence is lacking for the provenance of the ammonites, the closeness of this site to Acquasparta, and the similarity of Cesi's ammonite drawings to specimens collected there, suggest this was the most likely source.

Cretaceous rocks also occur in the region in a small outcrop around Acquasparta but there are no fossils typical of the Cretaceous period among those illustrated in the collection of drawings.

Tertiary Rocks

The rocks of the Tertiary comprise several distinct lithological units. The detail of the geology of the region can be found on the *Carta Geologica d'Italia* (1:100 000 sheets F 131, F 130, F 137 and F 138, and adjacent sheets) indicating a succession of sediments. The oldest rocks are Eocene marine sands and marls. It would appear that no fossils from this stratum are recorded in the drawings. Oligocene and Miocene rocks are also present and these are predominantly marine limestones and sandstones. The youngest rocks are of Pliocene and early Pleistocene age and comprise lacustrine muds and silts with freshwater snails together with clays and

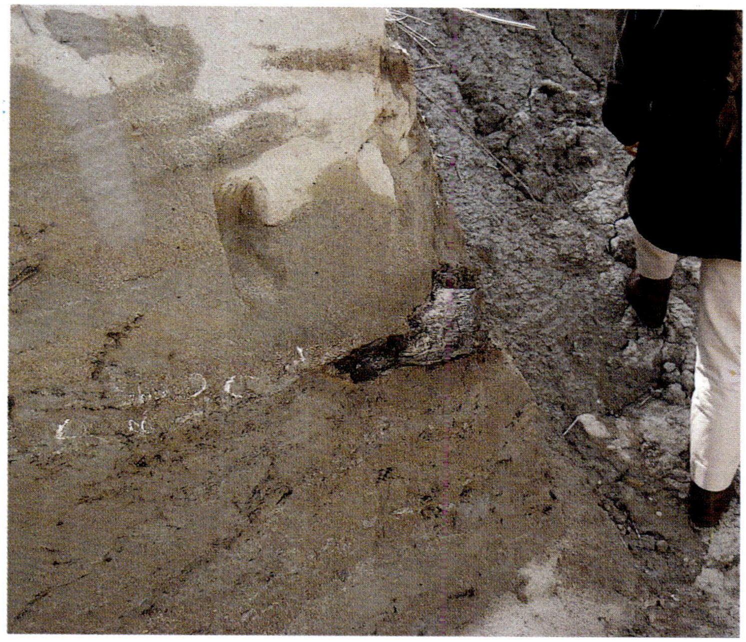

Fig. 9. Fossil wood and shells from the Santa Maria di Ciciliano Formation

lignites.[12] This rock unit was referred to by Bromehead[13] as the *Argille Lignifère*, but it is not shown as a separately defined stratum on this particular geological map. The lignite from this level may have been burned as a fuel in local iron furnaces and forges[14] but also some of this material is on record as having caught fire and burnt underground.

It is plant fossils from this Tertiary level in the Acquasparta district which appear to be the most commonly illustrated in the drawings. The rocks exposed in the area around Rosaro, Sismano and Dunarobba were deposited in the Tiberino Basin,[15] whose lake has waxed and waned over the last three million years. Recent work has redefined the rock sequence in the area and several new names for the different rock formations have been proposed.[16] These sediments were deposited in lakes and on swampy flood plains by rivers. Snails and clams lived in the lakes and the surrounding swamp was rich with trees and inhabited by animals.[17] A typical rock sequence at Dunarobba includes sands, silts and clays, together with plant-rich sediment. Associated with the fossil forest now exposed at Dunarobba are greenish-grey clays which in addition to plant fossils also yield freshwater snails, sideritic (iron) nodules or concretions and lignites.

While there have been several recent studies of the rocks around Dunarobba,[18] less is known of the rocks around Sismano and Rosaro. The lignite-bearing Fosso Bianco Formation outcrops extensively between Sismano and Rosaro. At Sismano itself, deep marine Miocene sediments are exposed which have not yielded plant fossils. Overlying the Fosso Bianco Formation are non-marine sediments of the Santa Maria di Ciciliano Formation of Pliocene/Pleistocene age and of the Acquasparta Formation of Pleistocene age, which have yielded some fossil woods [Fig. 9].

Italian Tertiary Fossil Woods

Fossil woods are well known from Tertiary deposits throughout Europe.[19] The study of plant fossils from these deposits indicates that the climate was very different from that of today and many plant genera had then a much wider geographical range. In the Eocene, the climate of Europe was tropical; and even though cooling took place, it was at least sub-tropical and wet during the Miocene and Pliocene (15–1.8 million years bp).[20] During this wet phase many peat deposits were formed which, through burial, turned into lignite or brown coal.[21] It is clear from studies of the continental Pliocene deposits of central and northern Italy that the climate was warm and wet.[22] Fossil woods are known from several areas.[23] In addition, studies of fossil fruits and seeds have helped in palaeo-ecological and palaeo-climatological interpretation of the vegetation.[24]

Of particular relevance is the widespread occurrence of conifers belonging to the family Taxodiaceae. Today these are represented by several genera known from relict populations world-wide. Best known is the swamp cypress or *Taxodium*. The large Giant Redwood, *Sequoiadendron*, and the Dawn Redwood, *Sequoia*, are now confined to North America; and the Chinese Swamp Cypress, *Glyptostrobus*, is found only in Southern China and North Vietnam.[25] Unfortunately it is difficult to distinguish between pollen of these genera and so the distribution of particular genera cannot be stated with certainty. In addition, their wood is known now from relatively small populations and may have had a greater range of anatomical features in the past. The use of modern specimens for precise identification of fossil species is thus questionable without supporting data, particularly as the genera we distinguish now may have overlapped then. For this reason, the wood of some of these genera known from fossils is given the general fossil name *Taxodioxylon*. Other genera of fossil wood include *Glyptostroboxylon*, but it appears that wood of this type is much less common in the fossil record than leafy branches and cones which can be identified securely as *Glyptostrobus*.[26] It is therefore important to examine cones, seeds and leaves from the sites alongside the fossil wood in order to establish the identify of the plant concerned.

It is not possible to identify woods from their external appearance alone, but it may be possible, by studying such features as branching, arrangements of knots and other external features, at least to guess the general group of plants to which the woods belonged, for example conifers or flowering plants (angiosperms). Such data, together with information on the flora known from modern excavations, may help us make an intelligent guess as to the identity of the material illustrated in the drawings.

Fig. 10. Dunarobba fossil forest ➛

Fig. 11. Roots from the Dunarobba fossil forest ➛

Fossil Plants of the Acquasparta District

Continental Pliocene deposits yielding plant fossils are widespread in Italy.[27] Of particular interest are the deposits exposed in the area west of Acquasparta.[28] A geological map of the region shows the fossil sites, and these are probably mainly Pliocene but also possibly Pleistocene in age (see Fig. 6). Many place-names are mentioned on the drawings, and several can be located on Stelluti's map. The most significant sites for the fossil woods and lignites are Dunarobba, Sismano (*Sismanum*), Rosaro (*Rosarium*), and Montenero (*Mons Niger*). In recent years, little work has been undertaken at any of these sites, apart from Dunarobba, which has attracted considerable interest.[29]

It is evident that at Dunarobba, and at a number of the other sites near Acquasparta, there were once several genera of conifers, none of which are found in the region today. Among the drawings, several fossil woods from Dunarobba are illustrated as well as views of the site itself. The large quantity of wood exposed at the site, as seen in these views, is particularly notable. In 1980, during excavation of clays for brick-making, a fossil forest was discovered there.[30] Subsequent excavation indicated a significant number of upright tree trunks *in situ* measuring from two to eight metres in circumference and up to ten metres in height,[31] the roots of which were often well exposed [Figs 10 and 11]. The palaeobotanists who carried out the research noted that the woods were not permineralized (see p. 129 and Fig. 31) and that the trees were all slightly inclined to the north-east despite being *in situ*. It is clear, however, that this is because of the dip of the beds. This may account for some of the crumpling of cellular structure in some of the woods, although most show excellent internal anatomy. After detailed analysis of some of the wood,[32] it was concluded that the trees belong to a fossil species *Taxodioxylon gypsaceum* which, it has been claimed, may be closely related to the living species *Sequoia sempervirens* or Coast Redwood, a member of the conifer family Taxodiaceae [Fig. 12]. Interestingly, this species has also been recorded from the Miocene at Monte Castellaro near Pesaro.[33] Palynological samples yielding similar pollen were taken from the Pietrafitta deposit (to the north), which was thought to be Upper Pliocene in age but is now considered Pleistocene.[34]

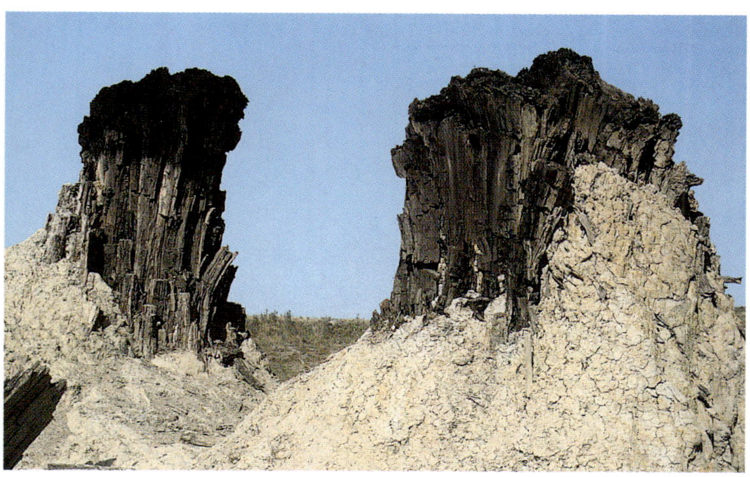

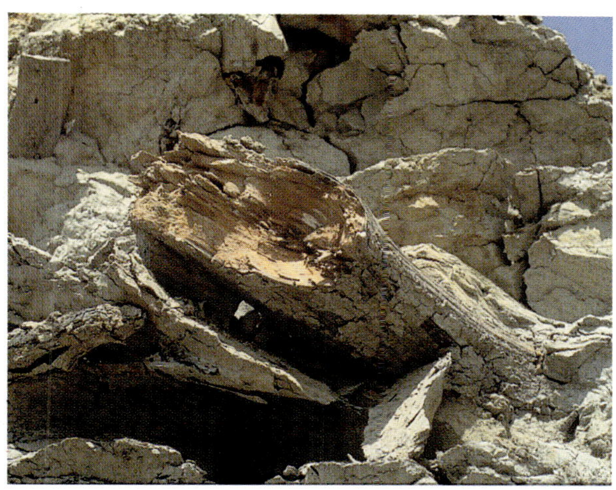

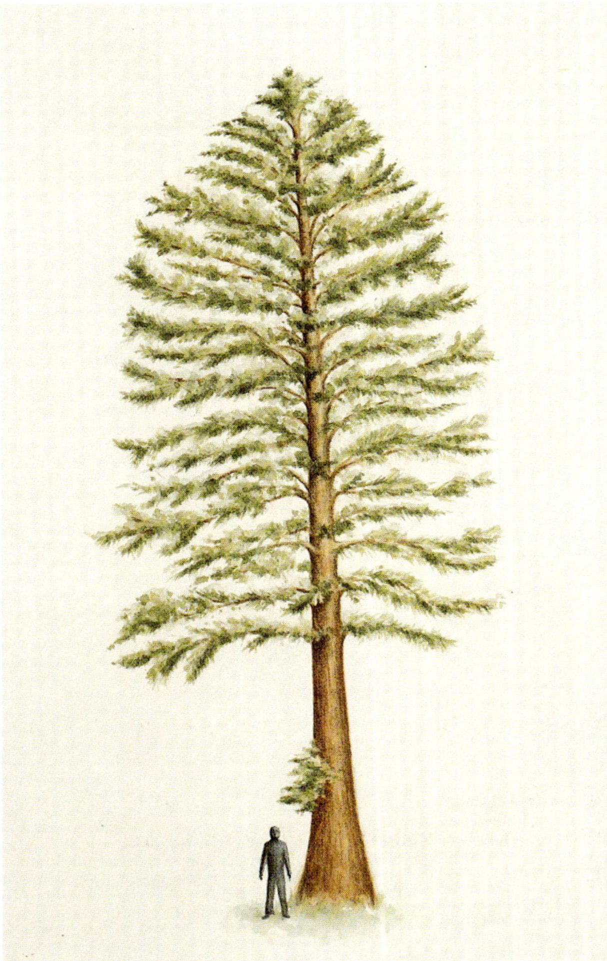

Fig. 12. Modern *Sequoia sempervirens*

Fig. 13. Foliage of modern *Sequoia sempervirens*

Fig. 14. Foliage of modern *Glyptostrobus*

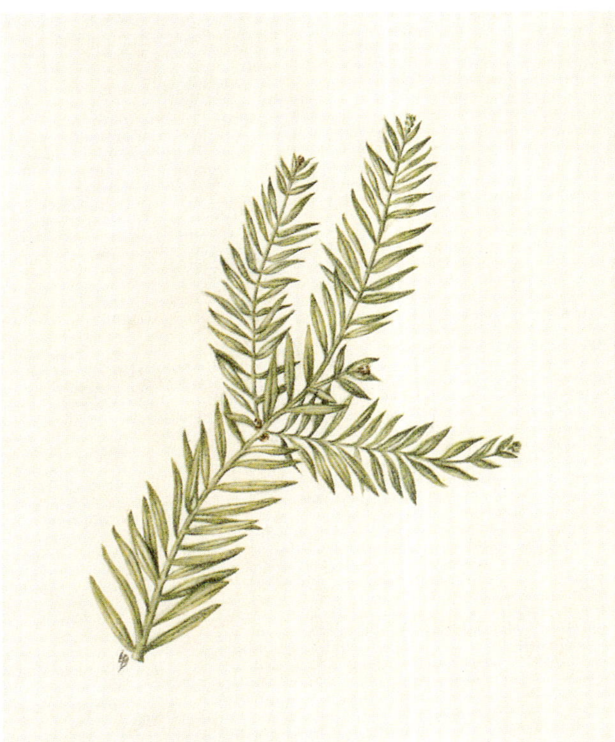

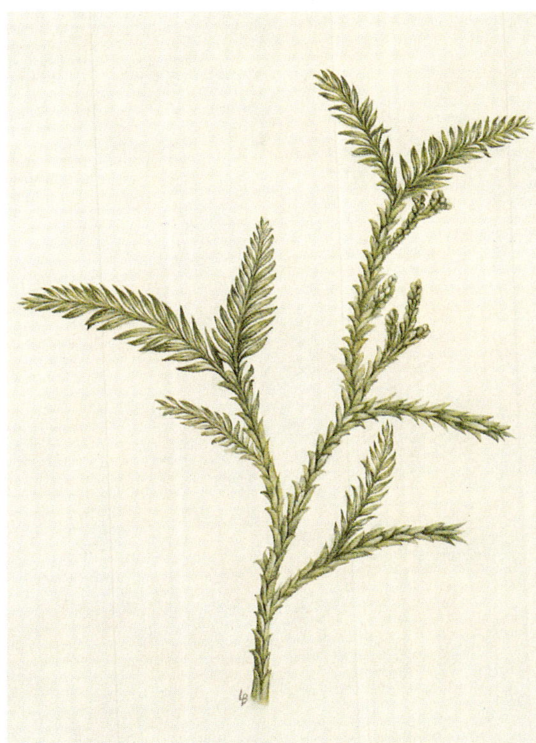

Fruits and seeds recovered from Dunarobba and other north Italian Pliocene deposits have indicated that other Taxodiaceous taxa such as *Glyptostrobus europaeus* occur in some of the associated muds and organic-rich clays.[35] It is possible that *Glyptostrobus* leaves, cones and seeds were produced by the trees whose trunks have been named *Taxodioxylon*. If this is correct then only two million years ago *Glyptostrobus* lived in Italy—whereas it is now only found in Southern China, where it is thought to grow to a height of at least 50 m (165 ft). If the trunks really represent *Sequoia* then this too is interesting: today the genus only survives in its natural state in the western United States, where specimens have been measured over 110 m (362 ft) high. Several authors have shown that both *Sequoia* and *Glyptostrobus* are found together at a number of European Pliocene sites.[36] In addition, palaeo-ecological studies indicate that these species form distinctive plant communities.[37] Whereas the wood of *Sequoia* and *Glyptostrobus* is rather similar,[38] their leafy shoots are quite different [Figs 13 and 14], as are their cones.[39] Material collected by the present author from the Dunarobba site contains both cones and leafy shoots; and their appearance seems to support the hypothesis that *Glyptostrobus* was present at the site [Fig. 15]. Material from this conifer occurs both below and above the fossil forest level.[40] Recently Ravazzi and Van der Burgh[41] have reported woods of *Glyptostrobus* from the Dunarobba site; Van der Burgh[42] considers that the large upright trunks also belong to this genus.

The fruit and seed flora that occur in sediments both above and below the trunk level at Dunarobba indicate the presence of wetland plants such as the Cyperaceae *Cladium, Scirpus* and *Carex.* This is in contrast to the presumed drier habitat of the modern *Sequoia sempervirens.*[43] In other samples from Dunarobba, a high dominance

Fig. 15. Fossil *Glyptostrobus* foliage from Dunarobba

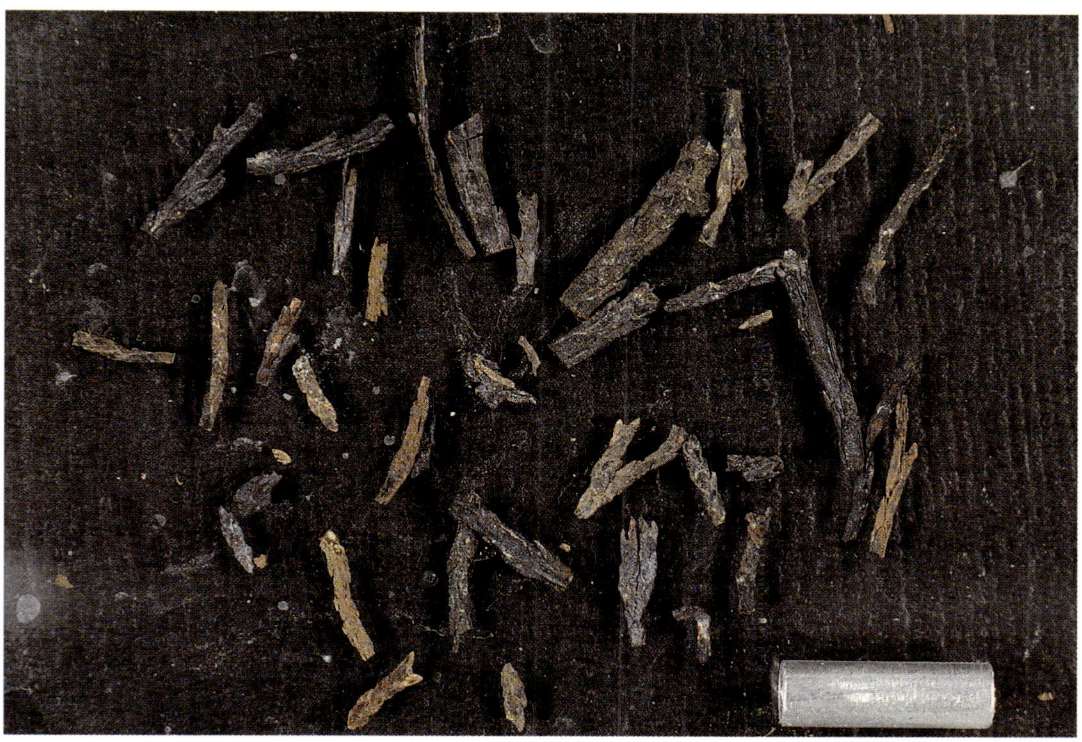

of *Scirpus* and *Cladium* is recorded; in yet another sample a dominance of *Swida*.[44] Most of these taxa favour wetland or swampy conditions, as does *Glyptostrobus*. As the living *Glyptostrobus* is rare, little is known about its timber, shape and habit, and comparative material has been hard to obtain.

The significance of Stelluti's work for the study of fossil woods has recently been documented.[45] The stratigraphic sequence at Dunarobba has also been studied,[46] showing there is a thin lignite layer overlying the fossil forest horizon. The fossil forest currently exposed reveals at least forty closely spaced, upright trunks.[47] While there are several drawings of fossil woods from Dunarobba, the engravings included in Stelluti's *Trattato* appear to reproduce specimens only from Sismano. Evidence of fossil woods in the area (particularly near Rosaro) is recorded in a letter from Jacques de la Ferrière to Nicolas Fabri de Peiresc in 1635 (see Appendix III b). He wrote:

> ... the main road from Narni to San Gemini and from there to Todi passes northwards beneath the said Castle of Rosaro, leaving it to the right, and the terrain or land where this wood is found an equal distance to the left, it being no more than a musket-shot from Rosaro to where the fossils are found ... The discovery of these fossils was made on account of the torrents of water at the time of the heavy rains which have eroded the earth on the slope of the hill over a period of time, this aided by constant ploughing which has made the earth more liable to be washed away.

It is interesting to note that no upright trunks had been found at that time at any of the sites: 'none of these trunks has ever been found standing or even inclined but always lying on the ground'. La Ferrière was told that 'Prince Cesi had once harnessed eight pairs of oxen to a trunk to uproot or exhume it, but that was because of its bulk, as none have been found that were longer than ten or twelve feet at the most'. However, he mentioned that Cesi had had a table made of a single piece of wood nine *palmi* wide and ten long.[48] He qualified another report—that a trunk 40–50 feet (12–15 m) long had been discovered by Cassiano—with the comment that he could not find anyone to confirm it. La Ferrière said the wood had been set on fire and had burnt for seven years, an observation made by Cesi to which we will return (see below, p. 292). It is quite possible that some of the tree trunks were large, from the evidence of the modern excavation at Dunarobba[49] [see Fig. 10] and from those found between Sismano and Rosaro [Fig. 16].

A study of spores and pollen from the deposit at Montesano, near Todi, showed evidence of a forest of *Taxodium* and *Nyssa*, with *Dacrydium, Podocarpus, Sciadopitys, Sequoia, Cryptomeria, Keteleeria* and *Englehartia*, among others.[50]

At Dunarobba, fossil snails have also been found.[51] A number of them bear a striking similarity to some of those illustrated in the drawings (see **147–152**).

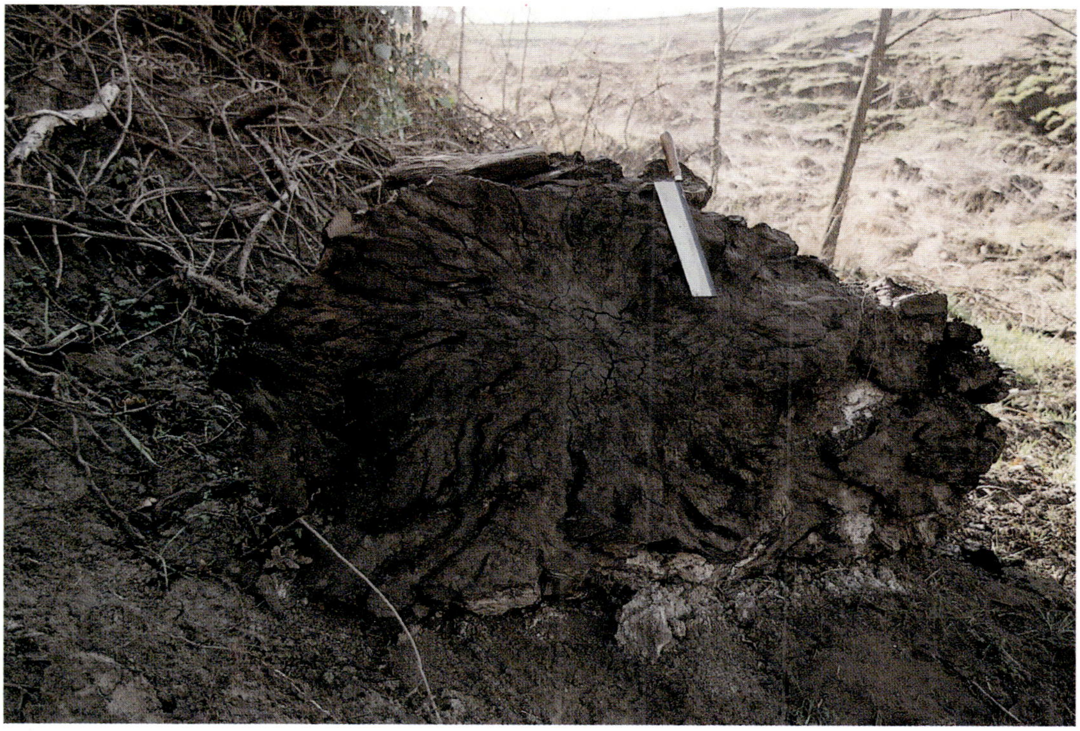

Fig. 16. Large fossil trunk from between Rosaro and Sismano

At first glance, the juxtaposition in the albums of drawings of woods, shells, ironstone nodules and baked clays may seem odd. It is clear, however, following recent field studies by the present author, that the material belongs together: it forms a corpus of material found in the area of Prince Cesi's estate and constituting part of a single investigation.

NOTES

1 Making field drawings of fossil sites did not become common practice until the end of the 18th and the beginning of the 19th centuries (see Rudwick 1976).

2 The earliest truly scientific, detailed palaeobotanical studies had been thought to be those from the 18th century. The work of the *Accademia dei Lincei* has not previously been recognized in volumes on the history of palaeobotany. See, e.g., Andrews 1980, although Gould casts this work into sharp relief (Gould 1998a).

3 Stelluti 1637; see above, pp. 31, 42, 56.

4 Desio 1973, pp. 285, 669; Venturi 1996, p. 13.

5 See above, p. 58.

6 Ambrosetti 1995, p. 465; Venturi 1996, p. 42.

7 Arkell 1956, p. 199; Desio 1973, p. 285; Venturi 1982 and 1986.

8 Arkell 1956, p. 214.

9 Ibid., p. 217; Venturi 1982 and 1986.

10 Venturi: personal communication to the author.

11 Venturi 1982 and 1986; Venturi 1996, pp. 20–23.

12 Ambrosetti 1995, p. 465. As with many geological maps it is not clear if the term 'lignite' refers to true lignite beds (i.e. fossil peat) or just to accumulations of fossil wood.

13 Bromehead 1947a, p. 73.

14 Bromehead 1947a, p. 73. The identification by Bromehead of *Metallophyti et Succensiones* (marked on Stelluti's map) as being furnaces and forges should be set aside. *Metallophyta,* a word of Greek origin, translates as 'metal-like plants' and certainly refers to the pyritized woods (and other iron-impregnated woods). *Succensiones* is derived from Latin *succendo,* relating to 'fires kindled from below' and, while this could be used of furnaces, the context clearly points to the 'sites of underground fires' or fumaroles, as depicted on Stelluti's map. Drawings of specimens from two sites in the Rosaro – Dunarobba area are in the corpus but there is no mention of the other two localities near Semignis and Tossiella (upper left on Stelluti's map). These last two areas expose extensive lignites and have been mined in recent years (G. Basilici: personal communication, 1999).

15 Basilici 1997, p. 73.

16 Ambrosetti 1995, p. 471; Basilici 1997, p. 77.

17 Ambrosetti 1995, p. 466.

18 Cerquaglia 1992, pp. 1–90; Martinetto 1994, p. 190; Ambrosetti 1995, p. 466; Ambrosetti/Cerquaglia 1996, pp. 1–131; Basilici 1997, p. 85.

19 Knobloch 1993, p. 357.

20 Ambrosetti 1995, p. 465.

21 Hower 1996, p. 630.

22 Desio 1973, p. 669; Ambrosetti 1995, pp. 465–6.

23 Biondi 1982, p. 227.

24 Martinetto 1994, p. 190.

25 Page 1990, p. 353.

26 Knobloch 1993, p. 357.

27 Ravazzi/Van der Burgh 1994, p. 597.

28 Ambrosetti 1995, p. 470.

29 Work at Dunarobba is summarized in Ambrosetti 1995; Ambrosetti/Cerquaglia 1996, pp. 1–131.

30 Cerquaglia 1992, pp. 9–18.

31 Biondi/Brugiapaglia 1991, p. 111.

32 Ibid.

33 Biondi 1992, pp. 74–82.

34 Follieri 1977, p. 925; Venturi 1996.

35 Martinetto 1994, p. 190.

36 Van der Burgh 1987, p. 299; Roiron 1981, p. 27.

37 Knobloch 1993, p. 357.

38 Kräusel 1947, p. 83; Mathiesen 1970, p. 1.

39 Page 1990: pp. 353–61: 'In *Sequoia*, an evergreen tree, leaves are variable, short, scale-like all around the shoot or long, linear to slightly falcate and +2 ranked. Female cones are broadly ovoid-oblong, scales narrow wedge-shaped'. 'In contrast *Glyptostrobus*, a deciduous tree, the leaves are variable and heterotaxic; linear-oblong, set spirally and spreading all around most leading shoots. Female cones are ellipsoid-ovoid, elongated with inclined scales.'

40 Martinetto 1994, p. 189; personal observation by the author.

41 Ravazzi/Van der Burgh 1994, p. 604.

42 Van der Burgh: personal communication.

43 Martinetto 1994, p. 190.

44 Ambrosetti/Cerquaglia 1996, p. 91.

45 Biondi 1984, p. 11; Biondi 1992, pp. 74–6; see also Gould 1998a and 1998b.

46 Basilici/Gentili 1992, pp. 47–73.

47 Cerquaglia 1992, pp. 9–18.

48 For more on Cesi's table (and other objects made from fossil wood) see p. 44.

49 Biondi 1992, pl. xx, p. 74.

50 Follieri 1977, p. 925.

51 Basilici/Gentili 1992, p. 63; Ambrosetti 1995, p. 466. The snails are illustrated in Girotti 1992, p. 30 and Ambrosetti 1995, p. 499.

CATALOGUE

The drawings are divided into categories, each one preceded by a short introduction. The fossil woods are further ordered into branch specimens, unbranched woods, knot ends and various fragments. The ammonites are kept separate as they are not part of the material collected from the fossil-wood sites.

NOTES FOR THE READER

Key terms, headings and typographical conventions used in the Catalogue

CATALOGUE NUMBER: Each Catalogue entry has a unique number composed of the SERIES letter (A = Antiquities and Architecture, B = Natural History); the PART number in roman numerals, and the ENTRY number, for example, A.II.32. Cross-references to entries within a Part are indicated by the **bold** entry number only. References to entries in other Parts of the Catalogue Raisonné are to the full Catalogue number. The present volume is B.III. A list of all planned Parts in the Series appears on p. 6.

PROVENANCE: The provenance for all the drawings in this volume is: Cesi, dal Pozzo, Albani, George III and by descent to Her Majesty The Queen. The abbreviation RL signifies the Royal Library.

DRAWING SHEET: Listed in this order are: media; size, in millimetres (height followed by width); condition and watermark. For discussion and types of watermarks see pp. 349–353 Iron-gall ink has produced holes in the paper; in some of these cases, the drawing was retouched in ink after being placed on the mount sheet (as in **1**) but this has not been noted for specific entries.

NUMBERING: This refers to the arabic numerals inscribed on the drawing sheets in the seventeenth century.

ANNOTATION: In brackets are given position, medium and hand, followed by transcription (in italics) of any annotation. Most of the annotations are in the usual scribal hand (see p. 33) although a few are in another unidentified hand (as seen in **6**, **76** and **84**). Cesi's partially erased annotations in Latin can be seen on many sheets. The annotations are transcribed in clockwise order, starting top left. Because the annotations are abbreviated and in poor Latin, and the precise meaning often debatable, full translations have not been provided.

Frame-lines—ruled in graphite or pen and ink, or both—and crosses can be seen on a certain number of sheets (see p. 71).

LITERATURE: Full details of titles cited in abbreviated form in the Catalogue will be found in the Bibliography.

REFERENCES: These are references to discussion or mention of the drawings in (i) Stelluti's *Trattato* (see Appendix II a for transcription and II b for translation) and (ii) the 17th-century Index (see Appendix I for transcription). The page references given in (ii) are to the present volume.

OTHER DRAWINGS: This refers to other drawings of the same specimen in this corpus.

Italian Measurements

Of the 199 drawings, 141 either have the measurements of the specimen written on them or are marked as being actual size. Those drawings of specimens without an indication of their size appear to have been drawn actual size. In other cases, actual size is specified as follows:

di misura na[tura]le

magnitudo (often abbreviated to *mag.*) *na[tura]lis*

na[tura]li (often abbreviated to *nali / e*)

The following measurements (in order, from smallest to largest) are used:

dit.[*o*] or *dig.*[*itale*] or *d.*	width of a finger (approx. 2 cm)
unc. (abbreviated from *unciale / uncio*)	inch
palmo / i (often abbreviated to *pal.*)	usual unit of length (approx. 23 cm)
palm.1. per ogni verso	1 palm each side
Pique	length of a pike (approx. 1.8 m)
cann.[*a*]	length of a cane or rod (approx. 2 m, with regional variations)
canna di Roma	10 *palmi* (approx. 2.4 m)

Other terms used are:

alt. (abbreviated from *altezza*)	height
crass.[*o*]	thickness
cir. (abbreviated from ?*circonferenza*)	circumference
lat. (abbreviated from *lato*)	width
long. (abbreviated from *lunghezza / longitudine*)	length

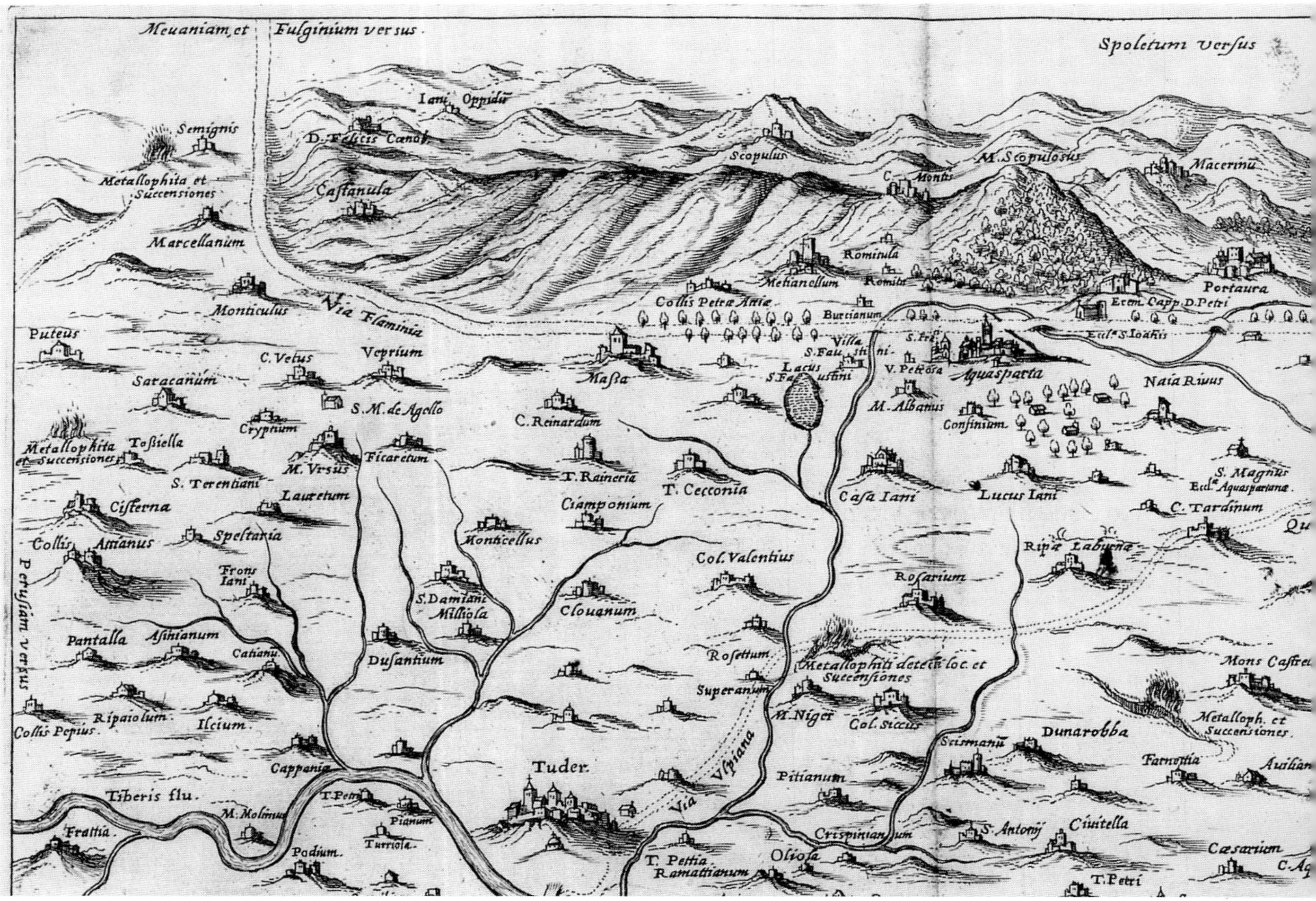

Fig. 17. Detail of Stelluti 1637, Pl. 1, showing fossil sites near Rosano and Dunarobba

I. FIELD DRAWINGS [1–17]

Field illustrations (showing rocks in the field) are essential for modern geological and palaeontological studies. Until recently they had been thought to be an innovation of late eighteenth- and nineteenth-century geological investigations.[1] The presence of seventeen such drawings in Cesi's collection represents a unique attempt at this period, and indeed the only one for the next 150 years, in that it represents a close and detailed record of field localities, and documents fossils *in situ* in the rocks, sometimes from several angles.

A number of these field drawings are annotated with the name of the site. Pl. 1 of Stelluti's *Trattato* [Fig. 17], for which there is no corresponding drawing, shows four sites marked *Metallophyti et Succensiones* (metallophytes and underground fires). Stelluti states in his accompanying text: 'the first shows the territory of Todi and that of Acquasparta, with its boundaries, in which four places in particular are marked where this type of wood grows, as is seen in the said illustration, marked there with the name of metallophytes'. These four places lie (i) in the Rosaro– Sismano–Montenero triangle; (ii) between Dunarobba and Montecastrilli; (iii) near Tossiella and (iv) near a place tellingly called Semignis. Pl. 2 of the *Trattato*—for which, likewise, there is no corresponding drawing—shows an engraving of one of these sites [Figs 23, 55]. Although the locality is not stated, Stelluti notes: 'The second illustration is of a particular place not far distant from the Castle of Rosaro, where a large quantity of the said wood of various sizes has been found, as is indicated in the drawing; and those fumes which are seen to issue from the earth have been observed by someone of the castle for ten consecutive years and have never ceased; nevertheless, these fumes did not always issue from the same spot, but kept on changing their site, depending, perhaps, on whether or not the fire found material to burn'. The four fires shown in **14**, a view of a locality between Sismano and Rosaro, may be compared with those in this plate [Fig. 23]. They represent the site of the 'everlasting fires' (see below, p. 292).

The topography may have been modified over the past 400 years and identification of actual sites is difficult. Localities mentioned on the drawings are described below. Many of the drawings of the field sites are from the area between Rosaro, Sismano and Montenero. These are all within the Fosso Bianco Formation. The rocks in this formation include clays with wood, thin lignite bands, gravels and shelly carbonates. Fieldwork in 1998 and 1999 has demonstrated that there are at least six sites with fossil woods in this area alone. At each of the localities both branches and trunks have been found in a variety of states of preservation. In the area between Rosaro and Montenero several outcrops were found which included gravels, as shown in the field drawings. Areas with burnt clays were also found at four of the fossil sites.

Dunarobba

Several of the fossil woods are reported to have come from Dunarobba, and the field drawings of this site are of outstanding quality. The recent discovery of a fossil forest here has aroused particular interest, and it has been presumed that this was the locality mentioned by Stelluti.[2] During fieldwork in 1995 by the present author, the locality was identified on the hillside above the site of the fossil forest. The similarity between the modern site and its depiction in the drawings is striking (see **1** and Comp. fig. 1). Here, fossil woods occur in abundance and have

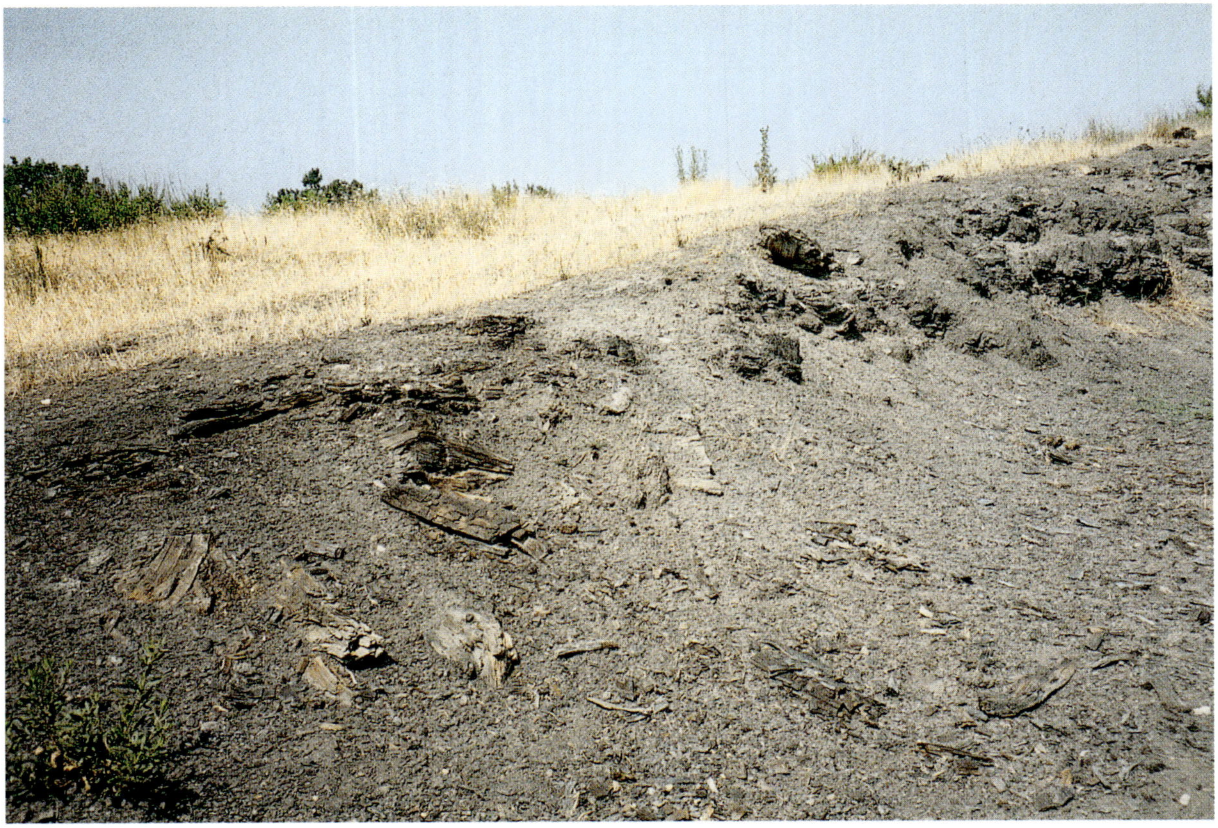

Fig. 18. Detail of fossil wood site at Dunarobba, showing permineralized wood

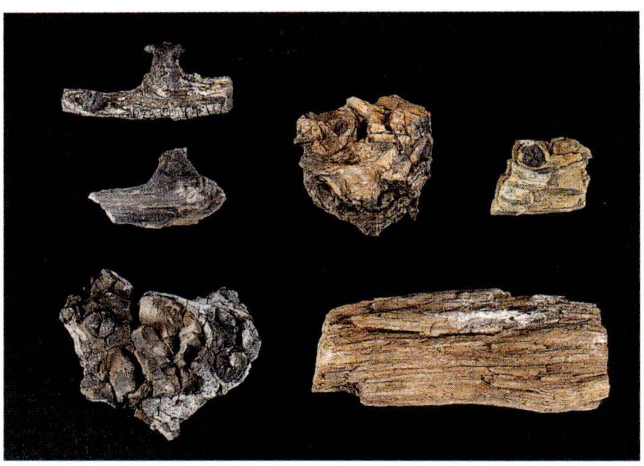

Fig. 19. Woods from Dunarobba showing range of shapes and preservation states

Fig. 20. Permineralized wood from Dunarobba

Fig. 21. Log with iron permineralization in centre

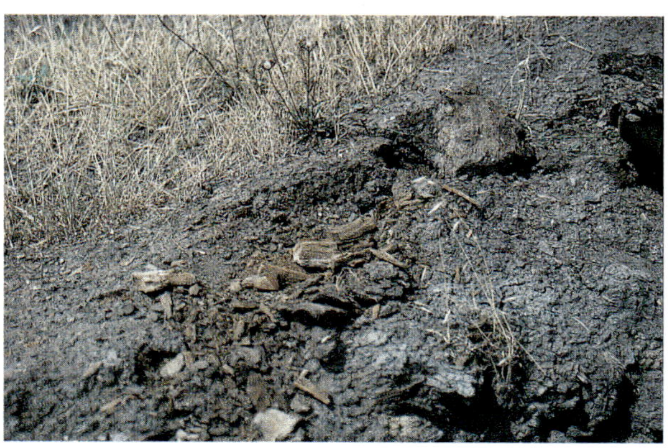

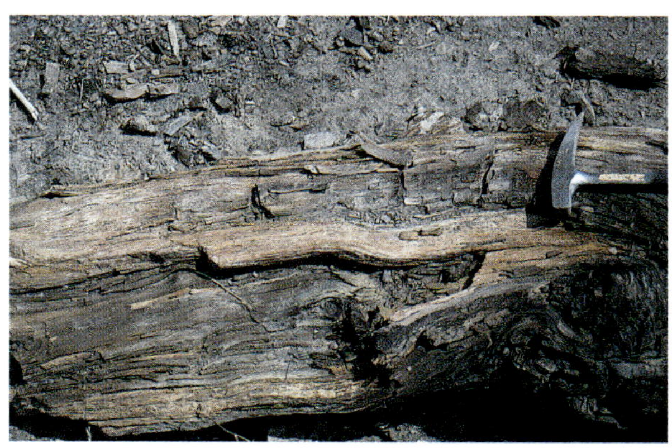

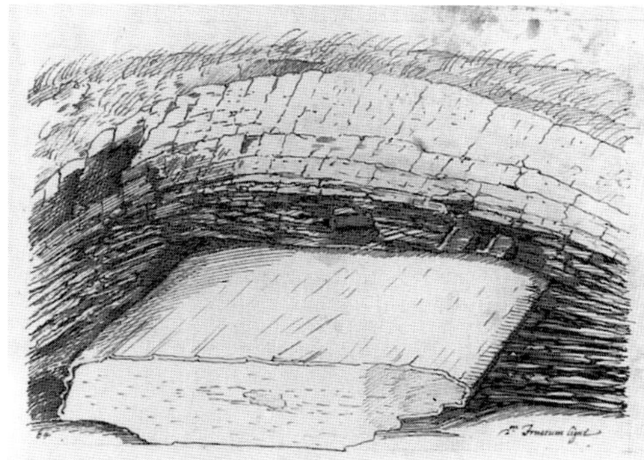

7

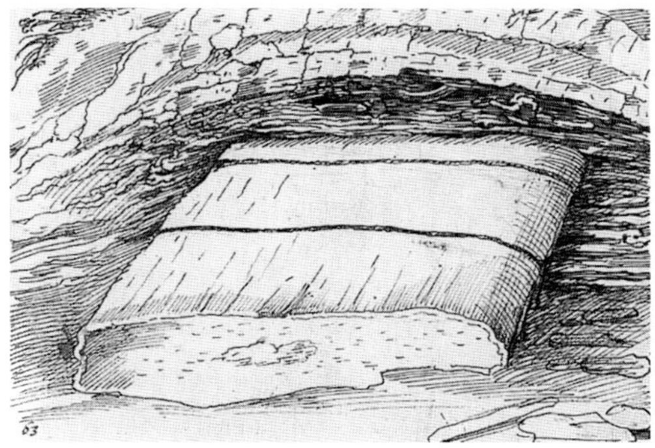

8

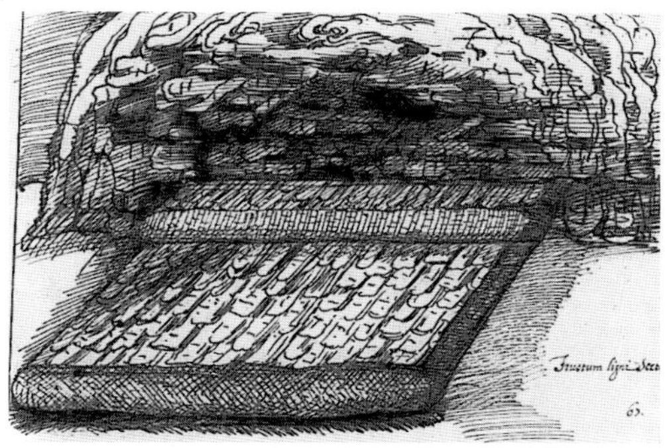

9

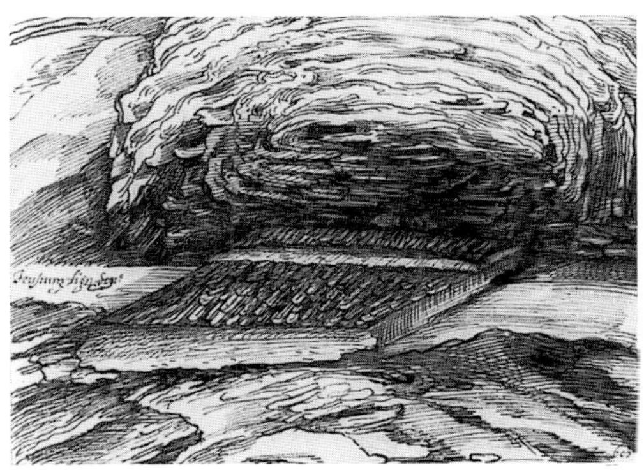

10

11

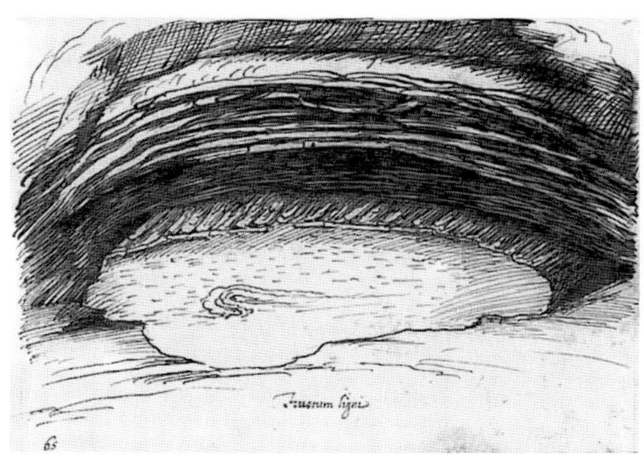

Fig. 22. Sequence (cat. nos **7–11**) showing excavation of a fossil log near Sismano

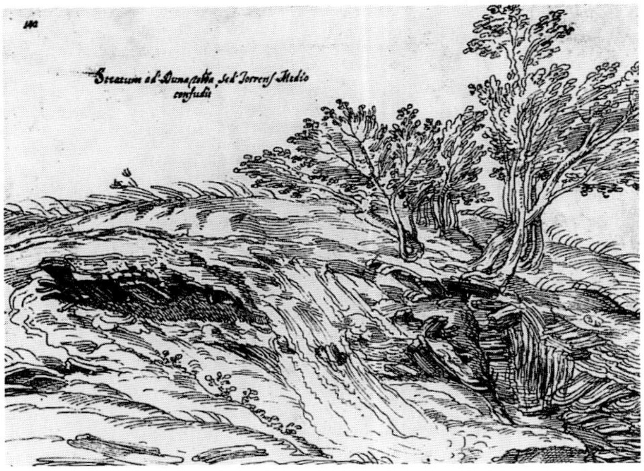

1

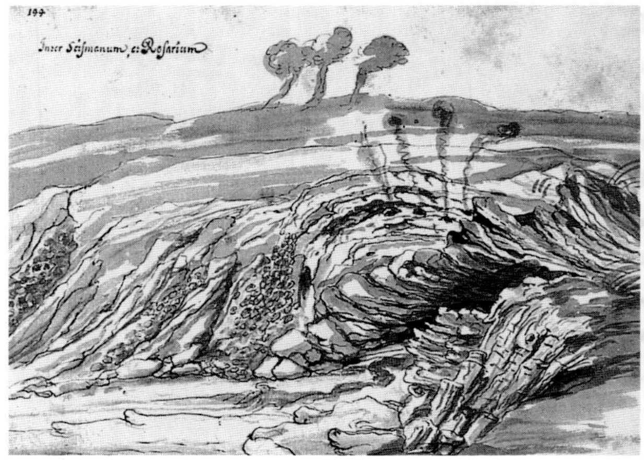

14

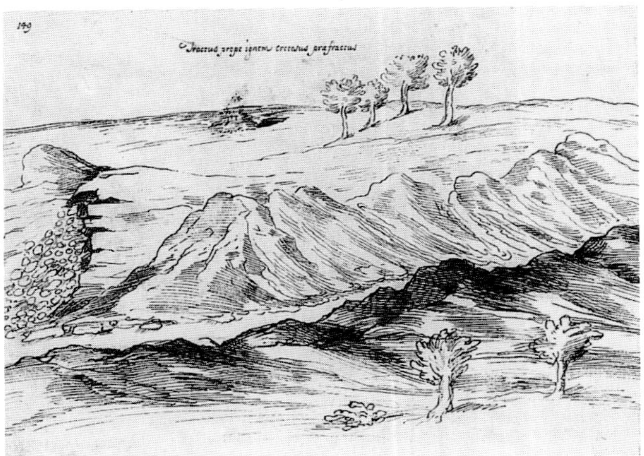

13

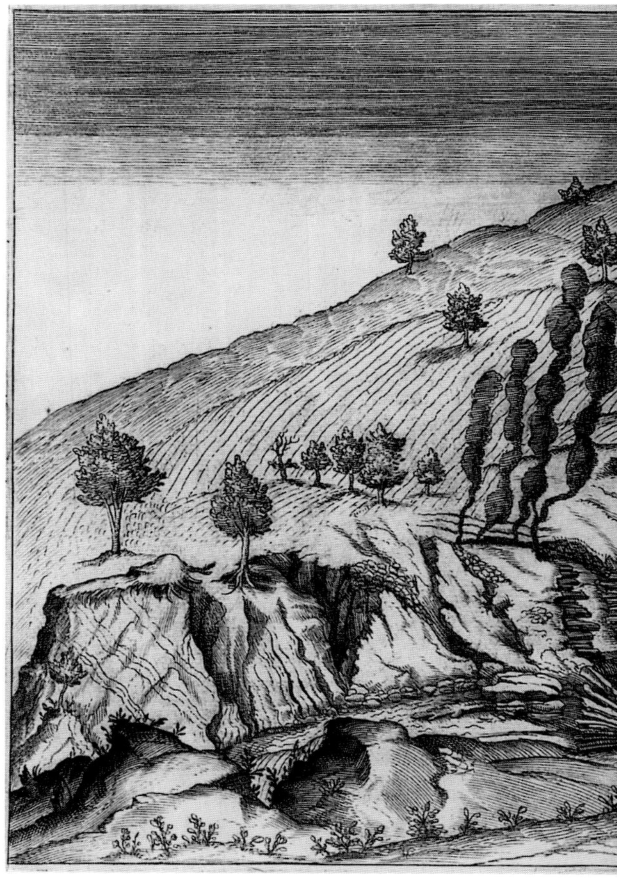

Stelluti Pl. 2

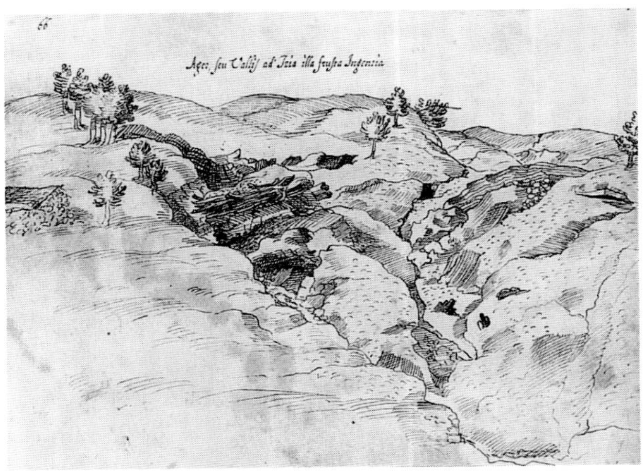

6

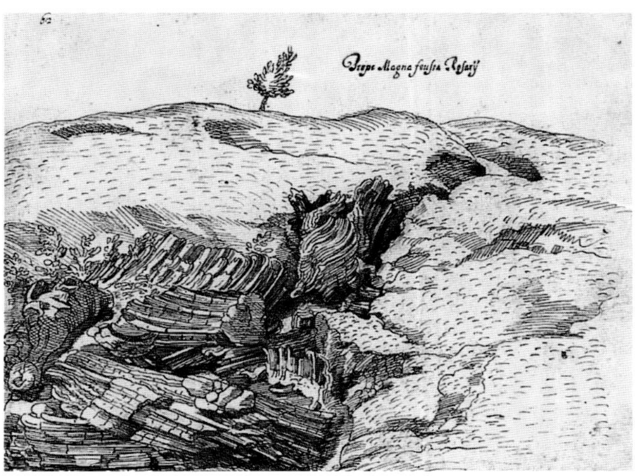

15

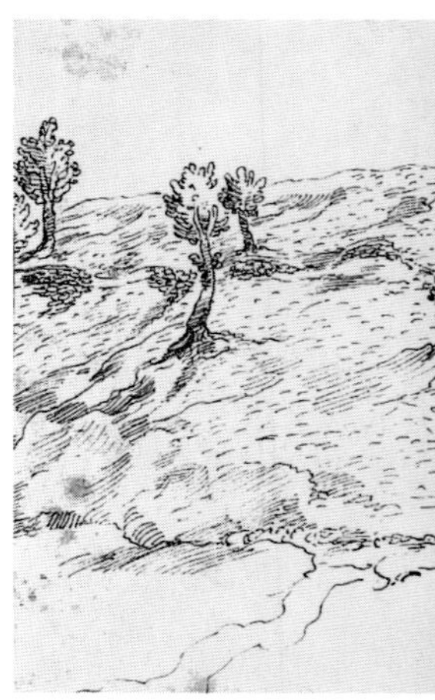

17

Fig. 23. The landscape views (Stelluti 1637, Pl. 2 and cat. nos **1, 4–6, 12–17)**

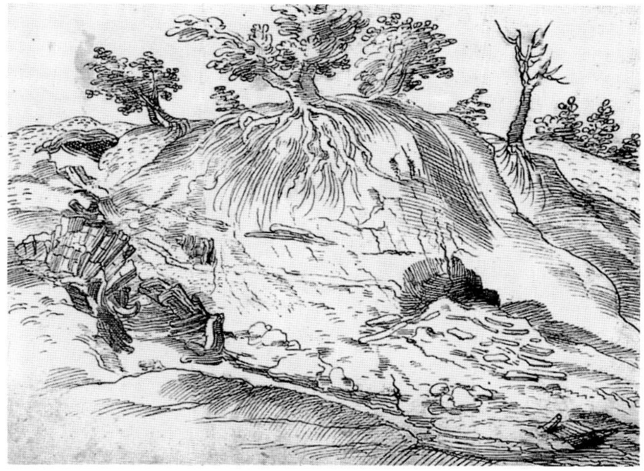

16

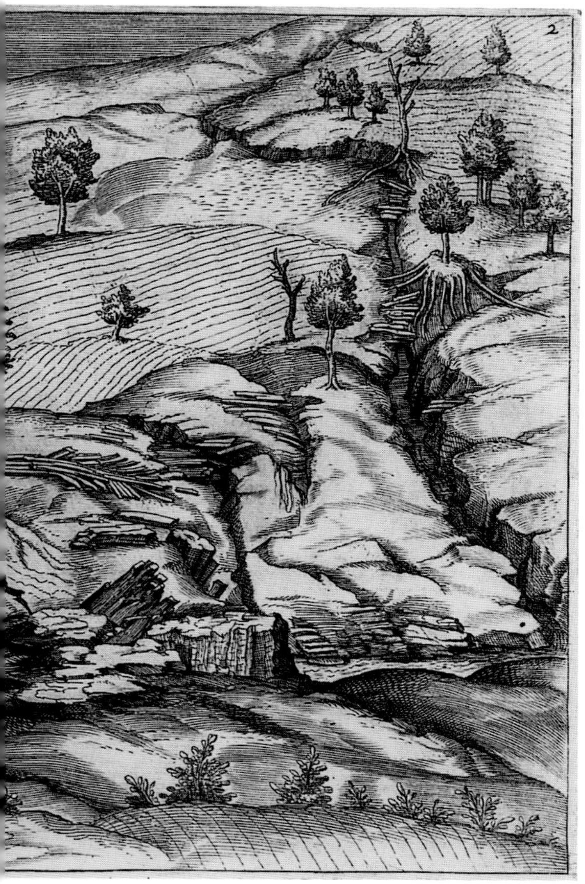

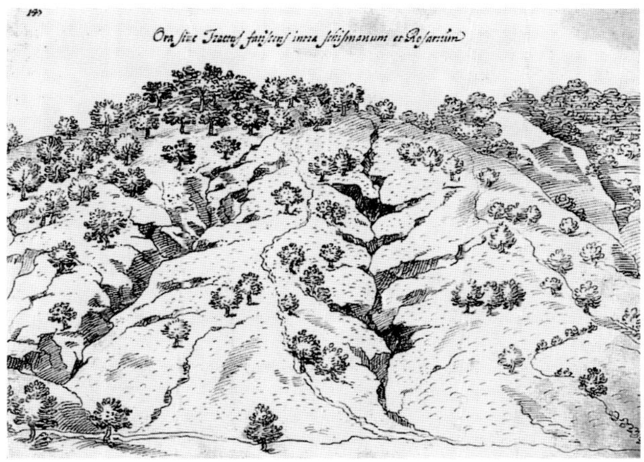

12

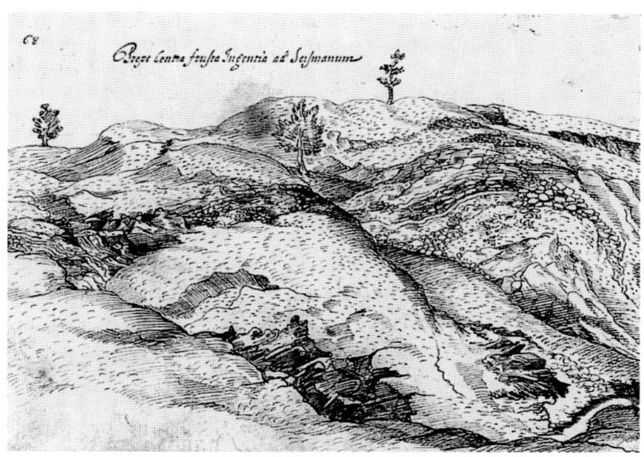

5

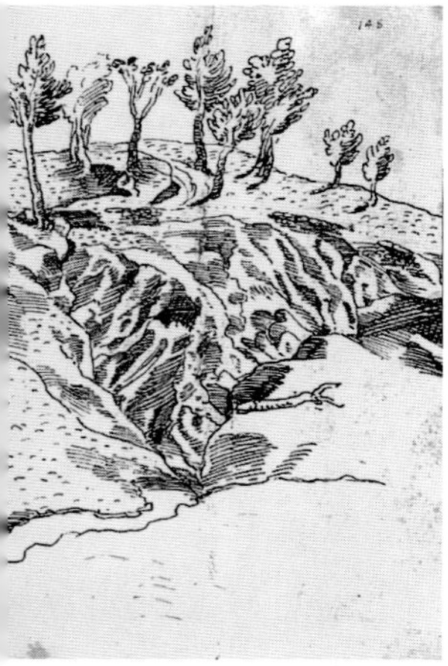

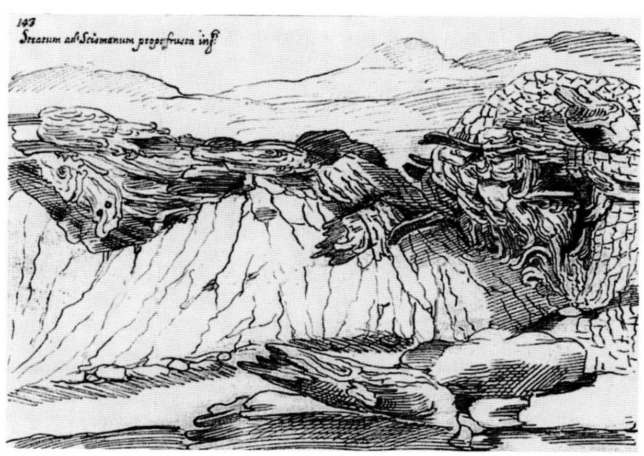

4

been preserved in a variety of ways (Figs 18–21). The rocks are late Pliocene in age (about 2.5 million years old).

Several fossil woods from this site are illustrated, and these include branches, roots and trunk wood. All are consistent with their belonging to conifers. Indeed the commonest plant from this site is the conifer *Glyptostrobus*.[3]

Most of the drawings of woods from this site carry the comment that they are 'petrified' or stone-like. Samples recently collected from this locality showed a very wide range of states of preservation. In the upper part of the outcrop several examples of permineralized woods[4] were collected [Fig. 18] and some of them bore a remarkable similarity to those illustrated in the drawings [Fig. 20]. Other woods were hardly altered and might even be confused with modern wood [Fig. 19]. Still others had been infiltrated with iron, especially iron carbonate and were variously coloured from brown to yellow to red, as described on some of the drawings [Fig. 21]. No pyritized wood specimens were found at this locality. Cesi's 'metallophytes' were probably fossil forms infiltrated with an iron mineral which in some cases could well have been iron pyrite, but in other cases might have been another iron mineral such as iron carbonate (in the form, for example, of siderite) or various iron oxides such as goethite and limonite which range in colour from brown to red and yellow.

Rosaro

The second most important area where the fossil woods in this corpus were found was the area between Rosaro and Sismano. The 1998 fieldwork succeeded in locating several fossil sites in this area. Whilst exact identification with the field drawings has not yet been achieved, it is clear that this area was the source of most of the material used in Cesi's study.

Rosaro is represented in three drawings, two of which show the area between Rosaro and Sismano (**12**, **14**) and one the locality itself (**15**). La Ferrière made many important observations about this site in his letter to Peiresc (see below, Appendix III b). Those concerning the burning of wood, and the origin and significance of the everlasting fires, will be considered later (see below, p. 292). One drawing (**14**) probably shows burning wood. It may also show the site of some of the burnt, multicoloured clays (see below, p. 292). The rocks are Pliocene.[5]

Fossil wood from this locality is certainly shown in **23, 58** and **59**. It probably includes both gymnosperm and angiosperm wood.

Sismano

This locality is shown in three drawings (**3**, **4**, **5**). Fossil wood *in situ* is shown in **7–8**, and probably also **9–11**. The fossils depicted almost certainly did not come from Sismano itself, which is built upon deepwater marine Miocene sediments, but from an area to the east or south where rocks of the Fosso Bianco Formation are exposed.

Material from this site includes a number of woods (**3**, **4**, **7–12**, **14** and **82**), some of which were illustrated by Stelluti in the *Trattato*, among them possibly both conifer and angiosperm. Woods from this site include some placed by Cesi in the category of *metallophyta*; they may have been pyritized, but no pyritized woods have been found near here in recent years.

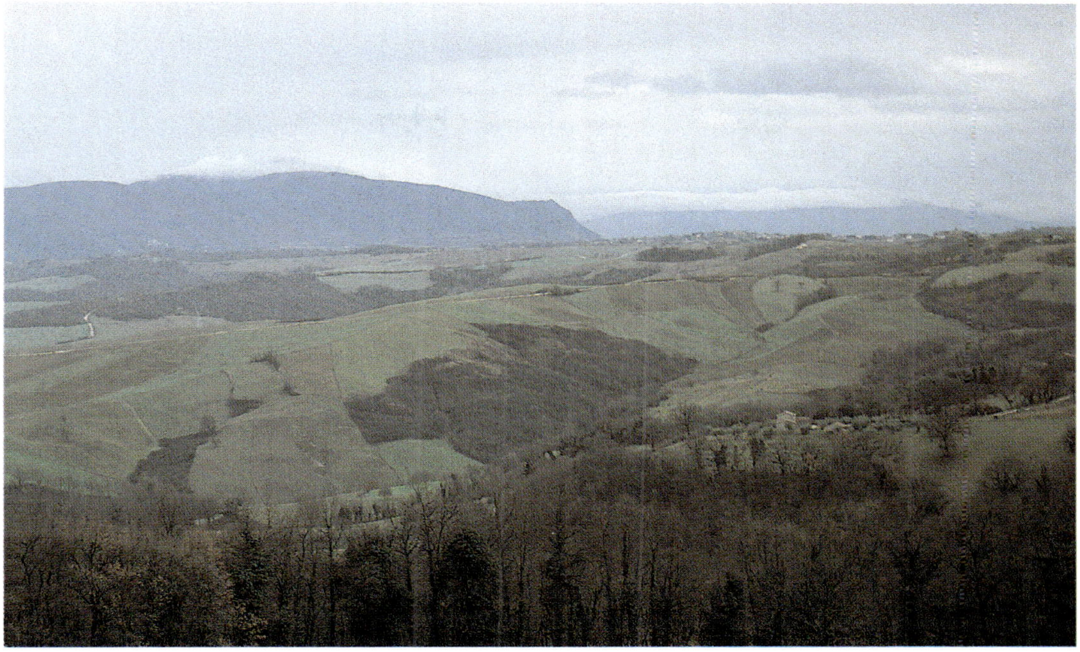

Fig. 24. Area between Rosaro and Sismano where fossil sites have been found

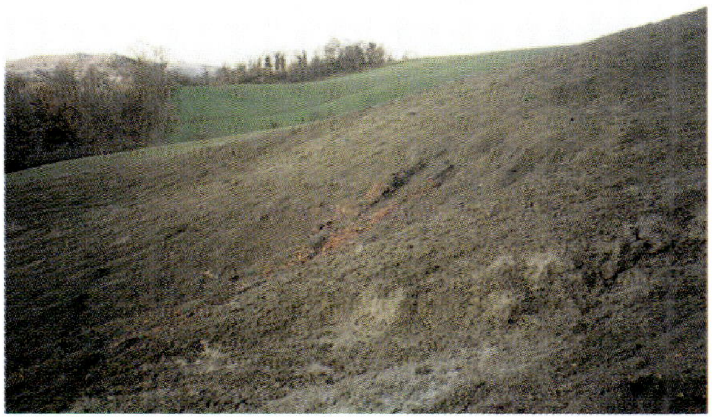

Fig. 25. Landslipped area near Rosaro with baked clays

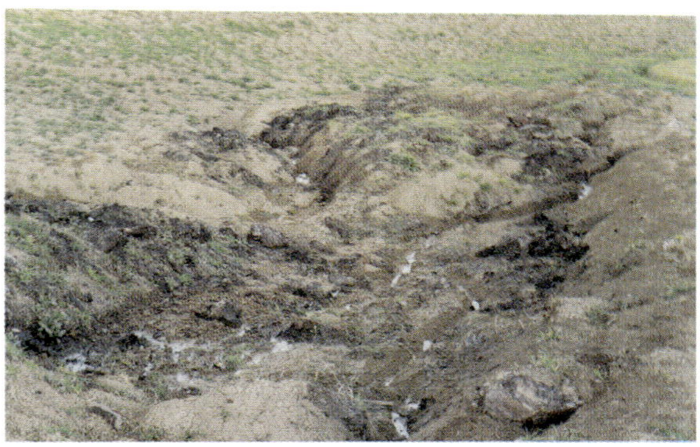

Fig. 26. Landslipped area between Rosaro and Sismano with large pieces of fossil wood

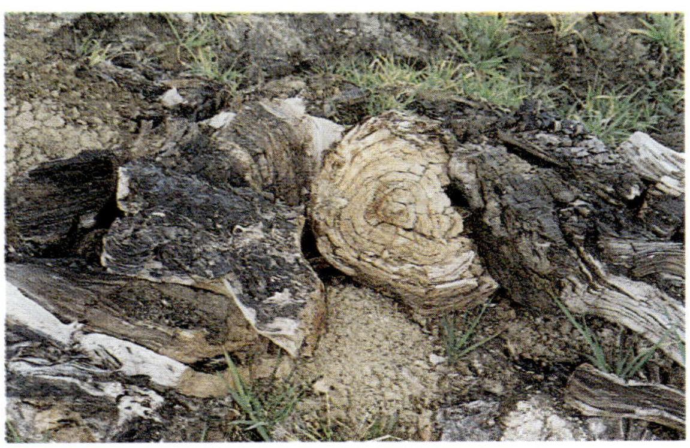

Fig. 27. Large pieces of fossil wood from site illustrated in Fig. 26

Montenero

No view of this locality is supplied. The rocks belong to the Fosso Bianco Formation and are Pliocene in age. Wood in clay is illustrated from this locality (**37**).

'Picchia'

The exact location of this site is uncertain. No view of 'Picchia' is given, but the black wood illustrated in **48** is supposed to have come from a place with this name. Could it be the same place as the Turris Pitia indicated on Stelluti's map, about 5 km (3 ml) south of Montecastrilli? Black coalified wood was collected by the present author here (its modern name is Torre Piccium) in 1998. Even if the wood in **48** was not from this particular location, it is probable that the wood depicted comes from sands of the Santa Maria di Ciciliano Formation as this is typically black [Comp. fig. 117]. It is not known whether the black colour indicates that the wood was from an older and thus probably more coalified sequence, or if it had been partially charred, either before or after fossilization. No information is given by Cesi to help resolve this point.

Ponte Caldaro

Fossil woods from this site (*Pons Caldarius*) are shown in several drawings (**34-36**, **62**). It is 15 km (9 ml) south of Dunarobba; there is no overall view. The rocks are Pliocene in age. Some woods from this site are described as being black like 'fossil ebony'. Another specimen (**63**), also described as being 'fossil ebony', may be from this site. Fossil woods from this locality occur in the gravel bars under the old Roman bridge [Fig. 28]. In 1999, gravels below the bridge yielded a wide variety of woods to the present author. These included brown wood, black wood and iron-impregnated woods, probably derived from both the Fosso Bianco and the Santa Maria di Ciciliano Formations, transported by the river and redeposited in the river gravels [Comp. fig. 34].[6]

Capitoni

This locality is mentioned only on one drawing (**130**). It is shown on the Stelluti map as *Capitonium* near Amelia. The area exposes rocks of the Santa Maria di Ciciliano Formation.

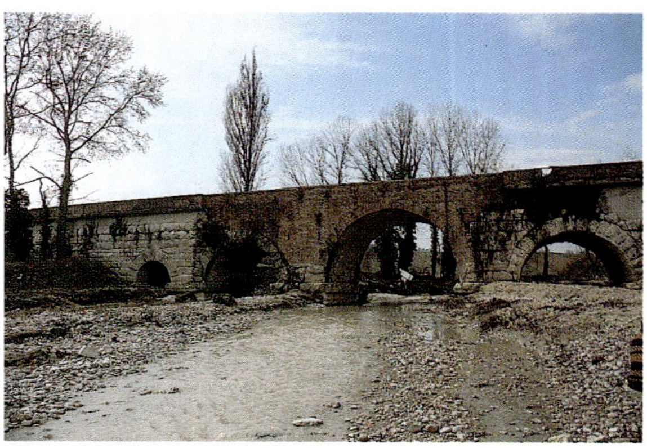

Fig. 28. Ponte Caldaro: fossil woods are found in the gravels beneath the bridge

NOTES

1. Rudwick 1976. The present author conducted fieldwork in the Acquasparta district in 1995 with Pierluigi Ambrosetti, and in 1998 and 1999 with Giorgio Basilici. Studies on the preservation of the fossil plants collected are in progress.
2. Cerquaglia 1992, pp. 14 and 69–71; Ambrosetti/Cerquaglia 1996, pp. 14 and 74–6.
3. Martinetto 1994, p. 189; Ravazzi/Van der Burgh 1994, p. 604; Ambrosetti/Cerquaglia 1996, p. 91.
4. See p. 129 for discussion of fossil plant preservation.
5. Ambrosetti 1995, p. 465; Basilici 1997, p. 75.
6. Woods with a range of preservation from brown to black were found by the author here in 1999.

1. Field view of Dunarobba

Windsor, RL 25681, *Fossils*, vol. XVI, fol. 42
Pen and ink, over traces of black chalk
186 × 259 mm. *Watermark*: Bird 41B
NUMBERING: [ink, top left corner] *142*
ANNOTATION: [ink] *Stratum ad Dunarobba, sed Torrens Medio/ confudit*

A field view of the fossil site at Dunarobba with, according to the inscription, a 'torrent pouring down the middle'. The locality shows clays at the base with a few pieces of fossil wood overlain by a wood horizon, possibly lignite in part.

Eight of the specimens of woods illustrated in the corpus are annotated as coming from this locality.

Some indication of scale is suggested by a comparison with the modern photograph (Comp. fig. 1).

REFERENCES: 17th-cent. Index, p. 369

OTHER DRAWINGS: **2**

Comp. fig. 1. Exposure of fossil-wood site at Dunarobba

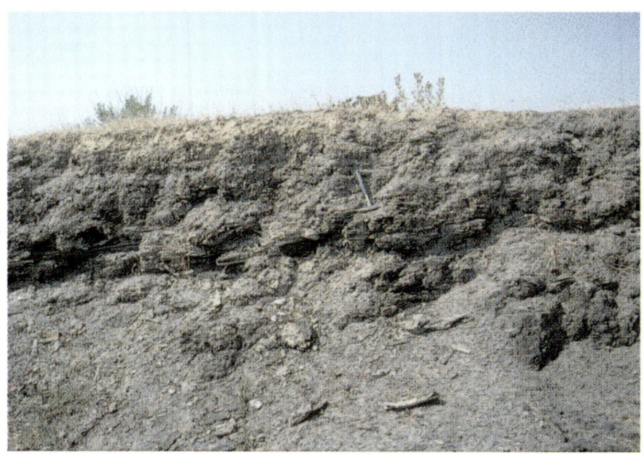

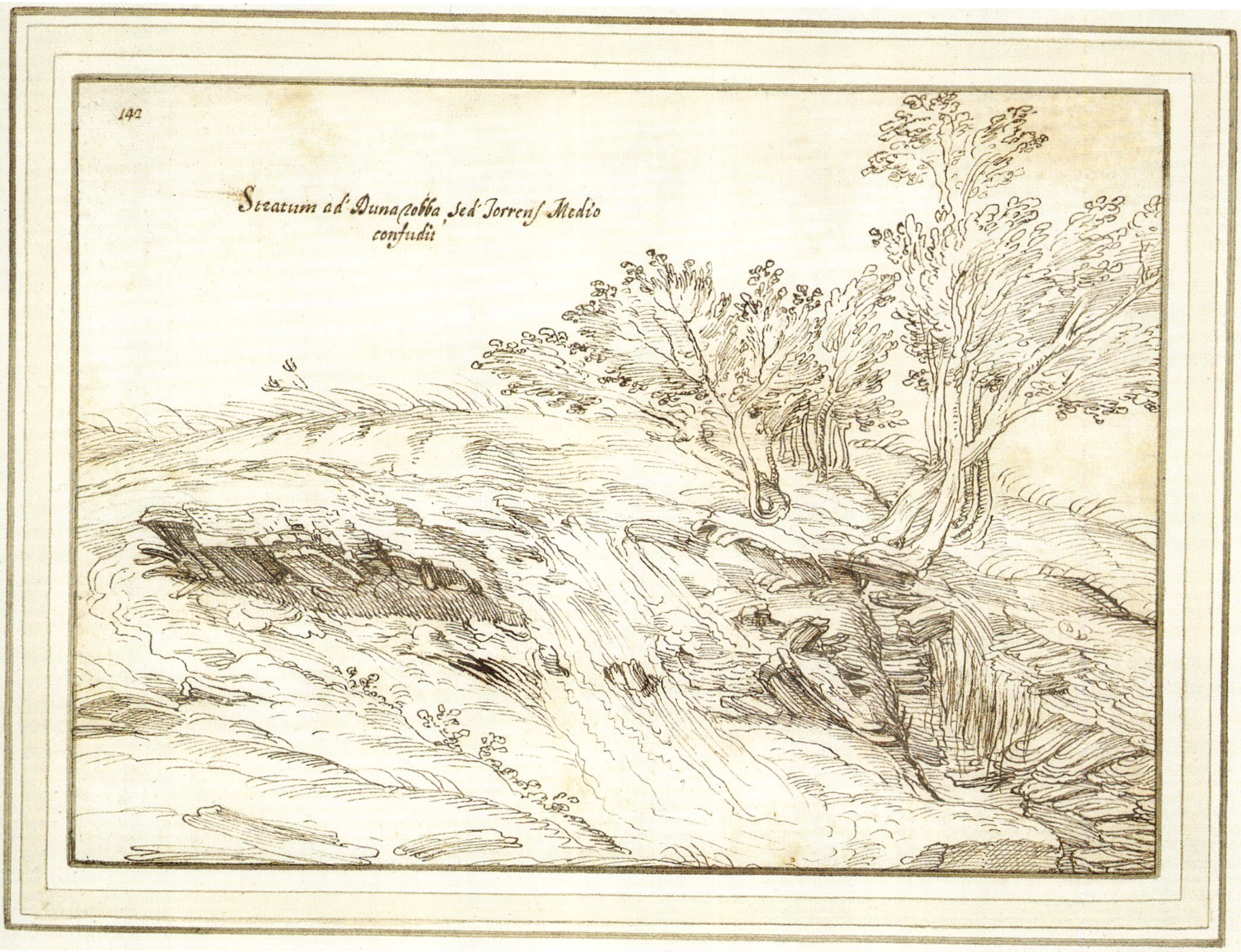

1

2. Field view of Dunarobba

Windsor, RL 25680, *Fossils*, vol. XVI, fol. 41

Pen and ink, over traces of black chalk; retouched in ink on mount sheet; ink frame line

178 × 244 mm; small holes caused by ink

NUMBERING: [ink, top left corner] *141*

ANNOTATION: [ink] *Stratum ad Dunarobba/ Supra cretam*

This field drawing shows the horizontally layered rock sequence at Dunarobba. The lowermost rocks are probably clays and the uppermost unit the wood-bearing horizon.

The inscription states that this is a 'stratum above the clay'. Several specimens of fossil wood from Dunarobba are elsewhere in the corpus.

REFERENCES: 17th-cent. Index, p. 369

OTHER DRAWINGS: **1**

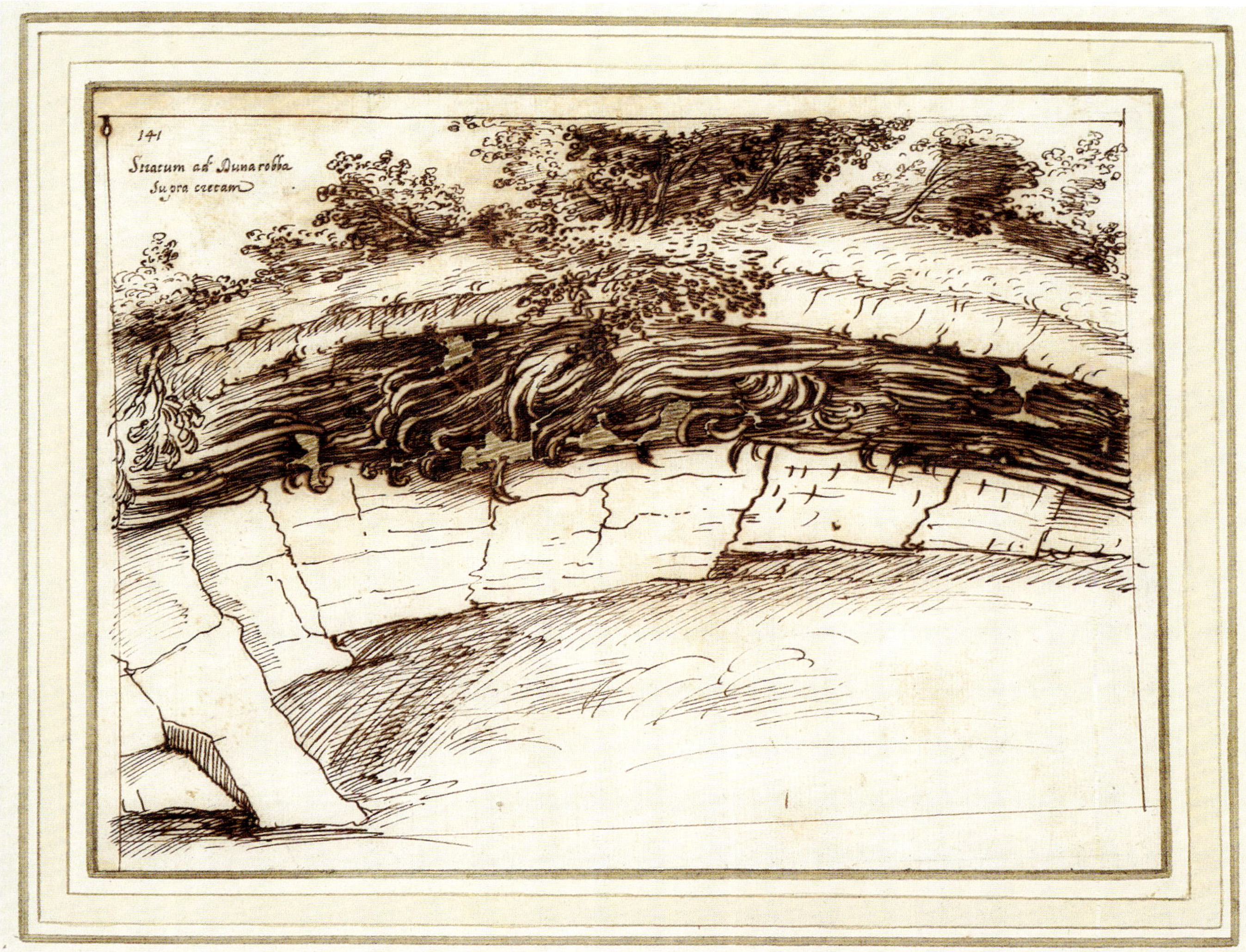

2

3. View of the deposit at Sismano

Windsor, RL 25684, *Fossils*, vol. XVI, fol. 45

Pen and ink, over traces of black chalk

190 × 259 mm

NUMBERING: [ink, top left corner] *145*

ANNOTATION: [ink] *Stratum ligneum crassum superigentia frusta ad Scismanum*

According to the inscription, the drawing shows the 'thick, woody stratum above the large pieces at Sismano'. This site is mentioned both in Stelluti's *Trattato* and on some of the drawings (for example, **7**). A number of large trunks are shown. They display the typical cracking that results from drying caused by exposure to the air.

REFERENCES: 17th-cent. Index, p. 369

OTHER DRAWINGS: **4**, **5**

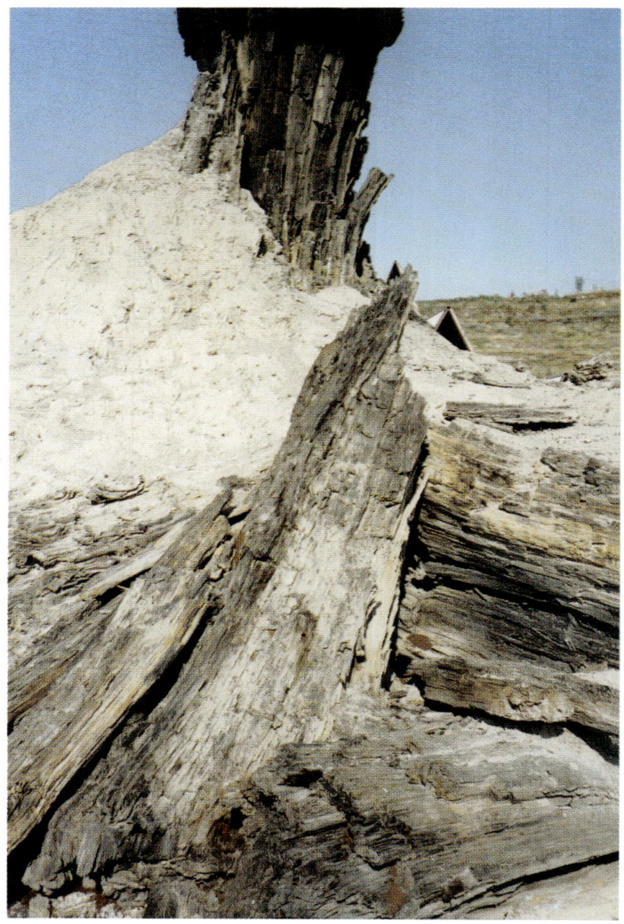

Comp. fig. 3. Upright trunks from the Dunarobba fossil forest

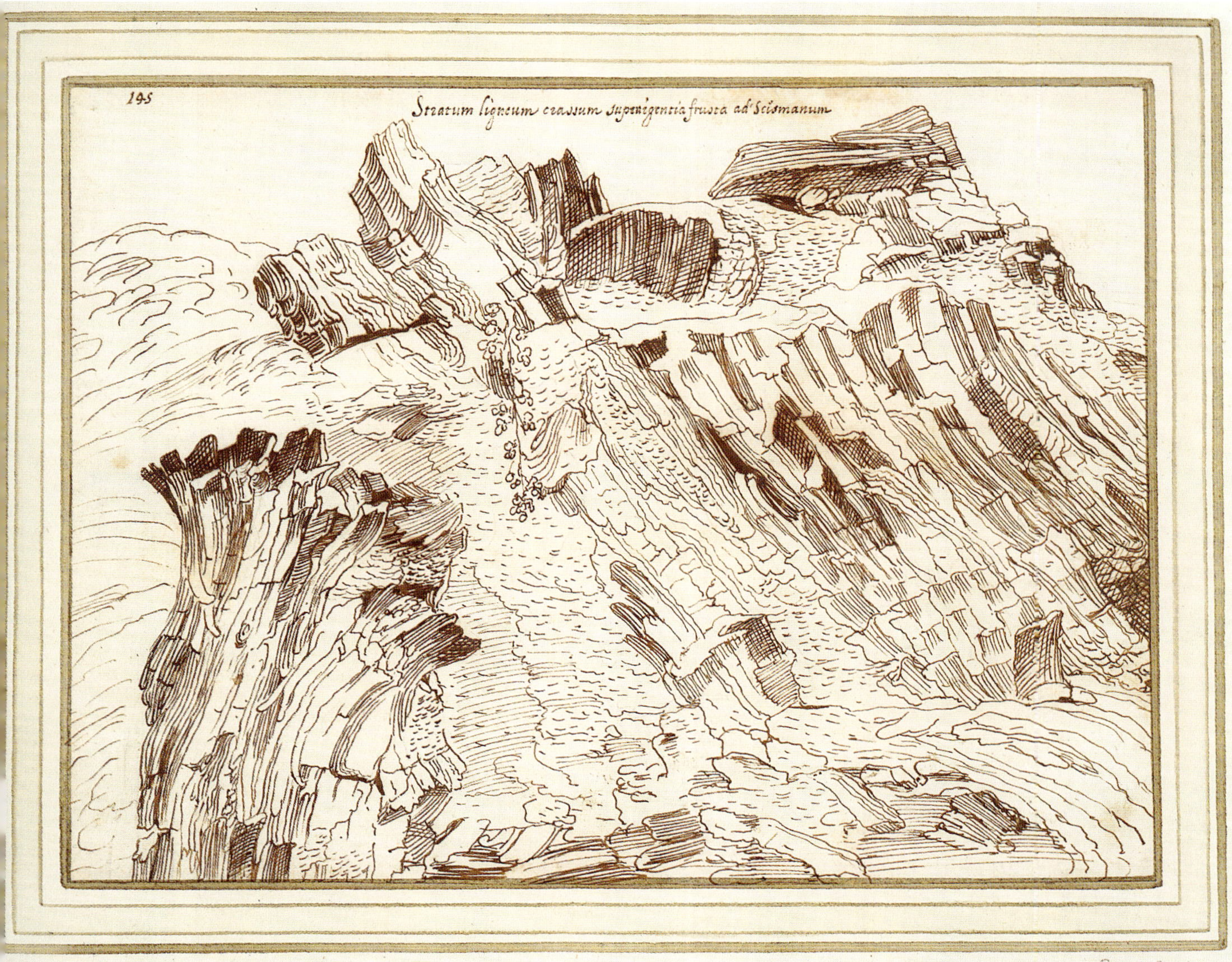
145
Stratum ligneum crassum superigentia frusta ad Scismanum

3

4. Field view of Sismano

Windsor, RL 25682, *Fossils*, vol. XVI, fol. 43

Pen and ink, over black chalk; measurements not given; ink frame line

177 × 240 mm; small holes caused by ink

NUMBERING: [ink, top left corner] *143*

ANNOTATION: [ink] *Stratum ad Scismanum prope frusta ing:ti*

Field view of a fossil plant-bearing stratum at Sismano. The inscription states that this is 'near the huge piece [of wood]'. This site is referred to by the inscriptions on some of the drawings (for example, **5**, **8** etc.). Large trunks of fossil wood are shown lying horizontally above a clayey layer.

REFERENCES: 17th-cent. Index, p. 369

OTHER DRAWINGS: **3**, **5**

4

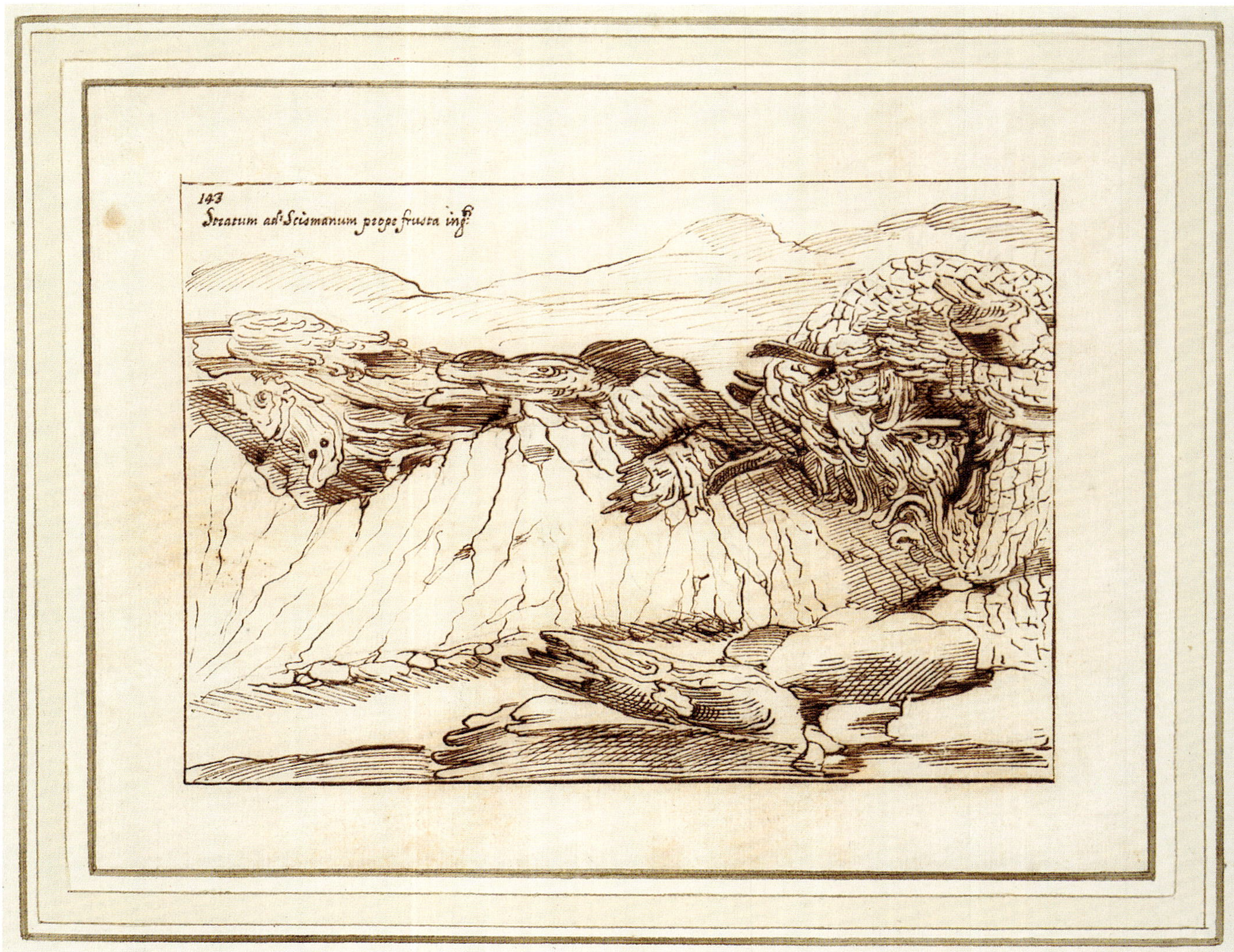

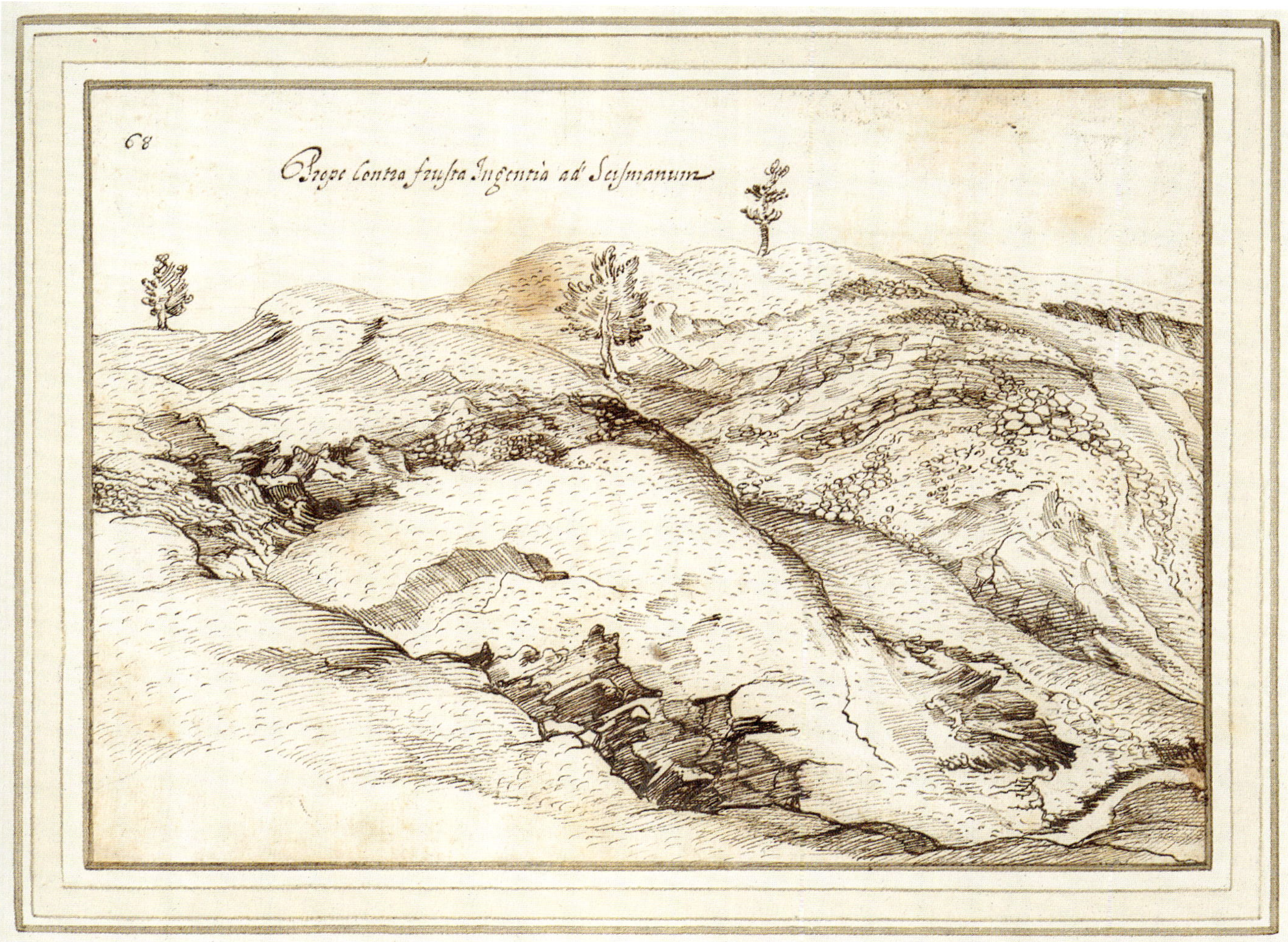

5

5. View of the fossil wood deposit at Sismano

Windsor, RL 25607, *Fossils*, vol. XV, fol. 16

Pen and ink, over traces of black chalk

182 × 261 mm

NUMBERING: [ink, top left corner] *68*

ANNOTATION: [ink] *Prope Contra frusta Ingentia ad Scismanum*

According to the inscription, this field sketch is 'nearly opposite the huge pieces at Sismano'. They may show an area near that depicted in **6**. The sketch clearly shows interbedded sediments outcropping across the hillsides and comprising layers of conglomerate and fossil plant-bearing deposits, particularly in the centre left of the drawing.

REFERENCES: 17th-cent. Index, p. 364

OTHER DRAWINGS: **3**, **4**

6. View of a fossil wood deposit between Sismano and Rosaro

Windsor, RL 25604, *Fossils*, vol. XV, fol. 13

Pen and ink, over traces of black chalk

188 × 258 mm

NUMBERING: [ink, top left corner] *66*

ANNOTATION: [ink] *Ager, seu Vallis ad' Tria illa frusta Ingentia*

This locality drawing depicts a hillside with two small gullies, the left-hand one clearly showing some large fossil logs. The scale is indicated by small trees. The inscription records that this is 'The field or valley at [near] those three huge pieces of wood'. One may assume that this is the location, close to Sismano (cf. **4**), from which the series of embedded fragments depicted in **7–11** came.

REFERENCES: 17th-cent. Index, p. 364

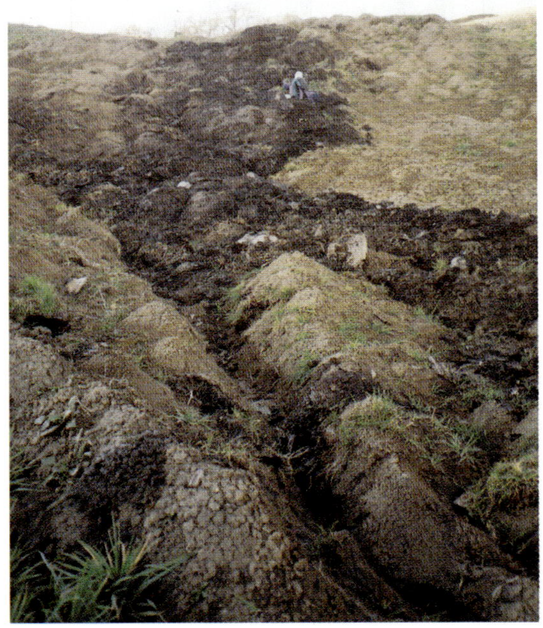

Comp. fig. 6. Landslipped area between Rosaro and Sismano, with abundant fossil woods

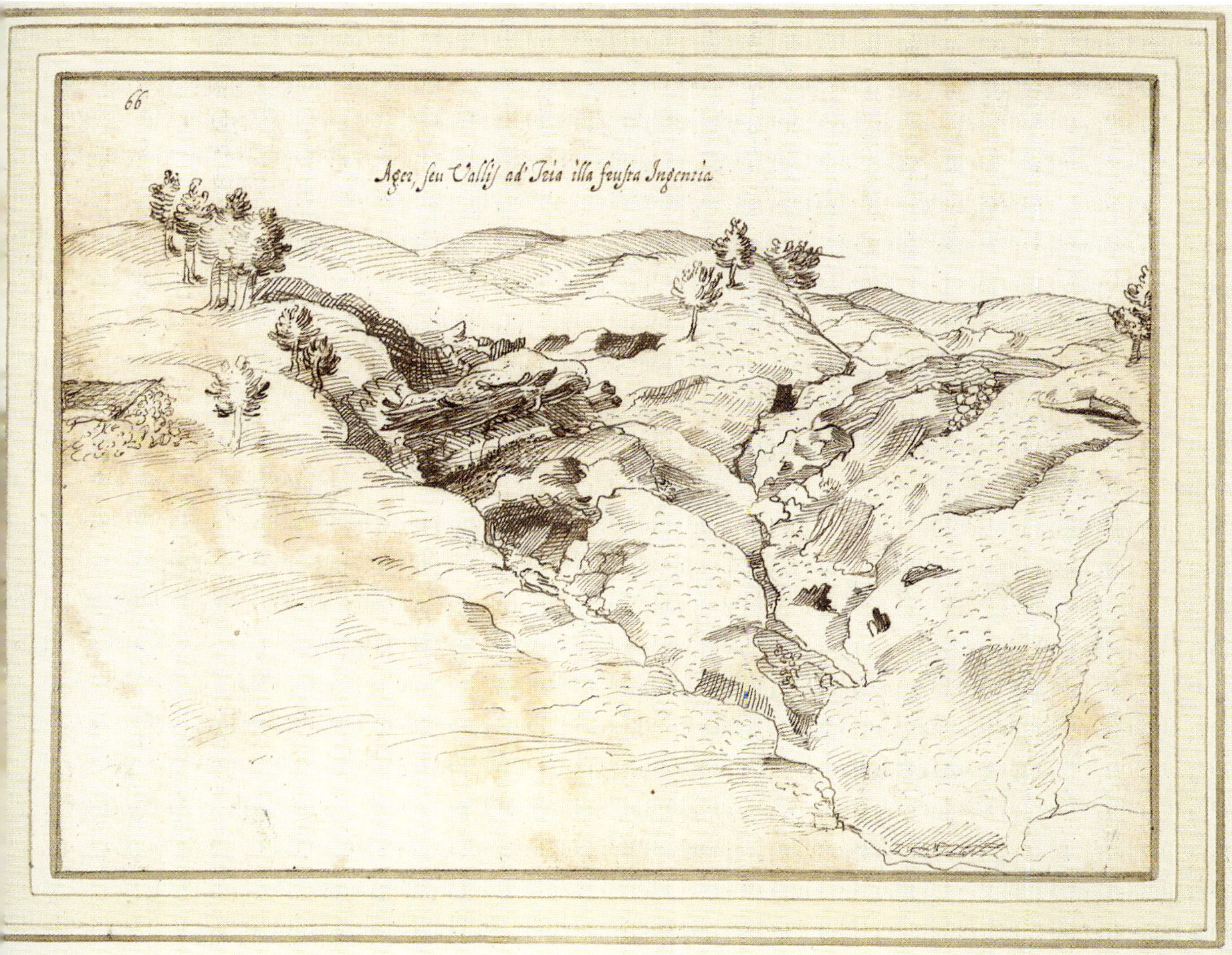

6

7. Piece of fossil wood from Sismano, *in situ*

Windsor, RL 25602, *Fossils*, vol. XV, fol. 11

Pen and ink, over traces of black chalk; measurements not given; black chalk frame line

190 × 257 mm; small holes caused by ink

NUMBERING: [ink, lower left corner] *64*

ANNOTATION: [ink] *2m. Frustum ligni*

Part of the series of drawings showing the excavation of a fossil log from Sismano. The drawing was engraved as the upper part of Stelluti's plate 8 (cf. **11** for the lower part of this plate). In the accompanying text Stelluti wrote: 'The two pieces of the following figure, number *8*, found just as they are illustrated here, also appeared sawn off at the front, and were somewhat thicker than the aforementioned piece, but shorter, and having bark like the trunks of the other trees, still having some sign of their inner core, Nature here straining to make these her imperfect parts resemble as closely as possible those which are perfect. In that part where the wood appears sawn off, a resin was found, white in colour, like gum, and white incense, the wood here sweating just as those walls do which have been freshly plastered in damp places. The other large piece, mentioned above, was also found with this resin, both on the front and in each of its divisions, these being nothing other than cracks in the earth while the clay had not yet lignified, as has been observed in other pieces in several places. Then above, all around these two large pieces were others of varying size, as is seen in the same illustration, some of which were not wholly turned into wood, some were all wood, and some like the bark of trees and black as coal.'

LITERATURE: Solinas 1989c, pp. 86–9, figs 65, 66

REFERENCES: Stelluti 1637, p. 9; 17th-cent. Index, p. 364

ENGRAVINGS: Stelluti 1637, Pl. 8, upper part; proof engravings in Montpellier, BEM, MS H.173, fol. 17, and Paris, BN, MS It. 1684, fol. 15 (see p. 403)

OTHER DRAWINGS: **8–11** (whole sequence shown at Fig. 22, p. 97); for a related drawing in BAV, Vat. Lat. MS 8258, see p. 403

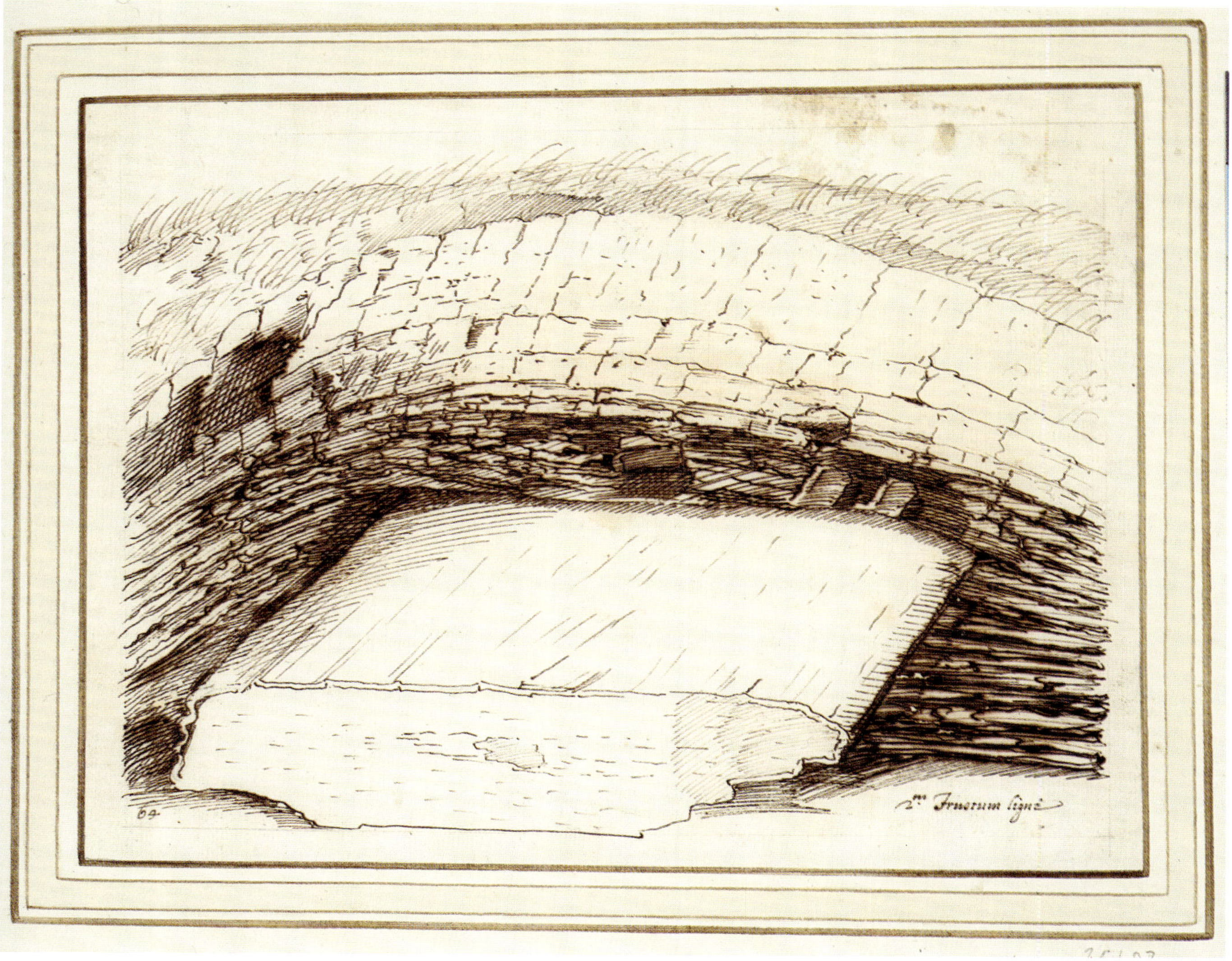

7

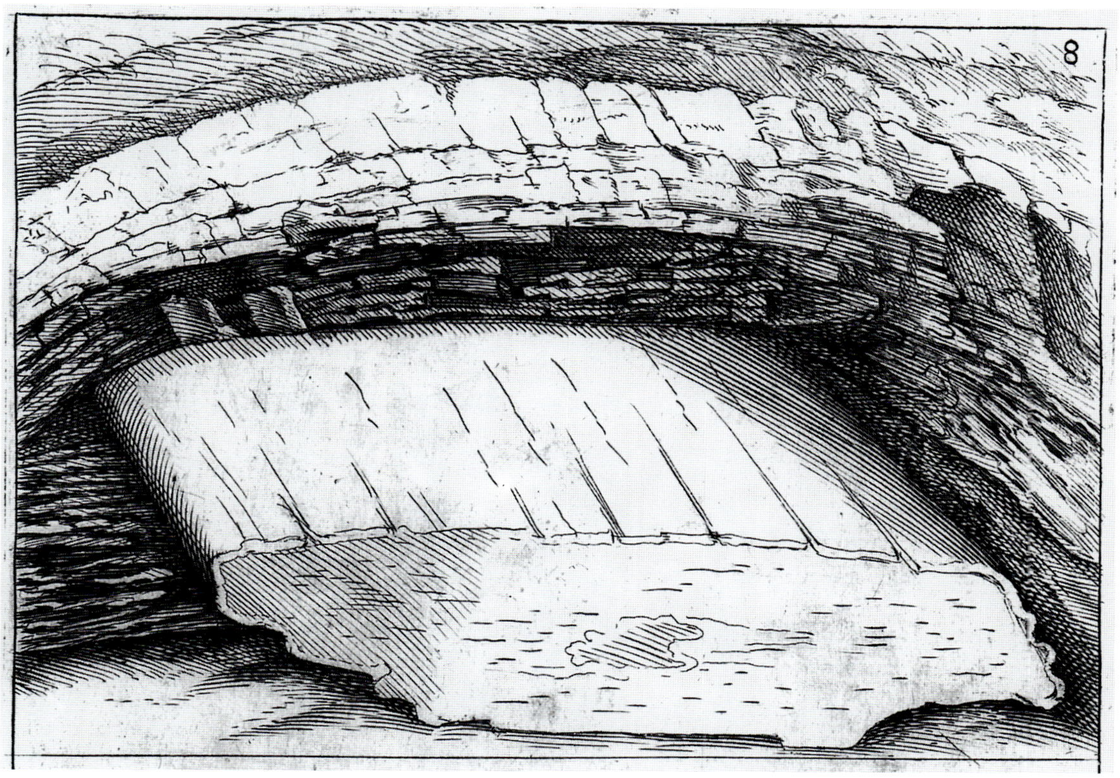

Comp. fig. 7. Stelluti 1637, Pl. 8, upper part

8. Piece of fossil wood from Sismano, *in situ*

Windsor, RL 25601, *Fossils*, vol. XV, fol. 10

Pen and ink, over traces of black chalk; ink framing line over black chalk around upper image only

184 × 229 mm

NUMBERING: [ink, lower left] *63*

ANNOTATION: [ink] *Tria ligni frusta cum Sectionibus ad Scismanū*

The drawing is one of the series (**7–11**) showing a piece of fossil wood within the outcrop of clay at Sismano. Part of the specimen has been extracted from the clay and two saw cuts are shown across it. Whilst the inscription states that the drawing represents 'Three pieces of wood in sections at Sismano', it is highly probable that, in fact, they represent the same piece of wood shown at three different stages of its emergence from the clay.

The drawing was engraved as plate 7 of the *Trattato*, where measurements were added. In his accompanying text Stelluti notes: 'The illustration of page number *7* shows the piece of wood of vast proportions mentioned above, discovered not as it is shown engraved, with its side uppermost, but with its belly, if I may so term it, on the ground, and its crest turned towards the sky, as are the other two pieces on the following sheet [Plate 8, as drawn in **7** and **11**]; and for this reason we have noted the dimensions of its size at the foot of the illustration, where it can be seen: adding that the part which appears to have been sawn off, bearing the letters *A, B, C, D,* was the first to be discovered; and the earth which covered it having been removed as far as point *F*, it was here that the severed piece was found, as if it had been sawn off; and more earth having been removed, another juncture was found at the part indicated by the letter *G*, as illustrated in the said figure; and in continuing to dig, to make a large hole in order to uncover the entire piece, and to ascertain its length, after so much had been removed that an arm's length of this wood was discovered, the earth fell over it and it was covered once again; nor were more attempts made to finish uncovering it, because the owner of the land did not want his field to be ruined by further holes; so that the remainder was left covered by the earth; but now, while writing this account, I have heard that the waters have once more uncovered it'.

LITERATURE: Solinas 1989c, pp. 86–7, figs 63 and 64

REFERENCES: Stelluti 1637, p. 9; 17th-cent. Index, p. 364

ENGRAVINGS: Stelluti 1637, Pl. 7; proof engravings in Montpellier, BEM, MS H.173, fol. 16, and Paris, BN, MS It. 1684, fol. 14 (see p. 403)

OTHER DRAWINGS: **7**, **9–11** (whole sequence shown at Fig. 22, p. 97)

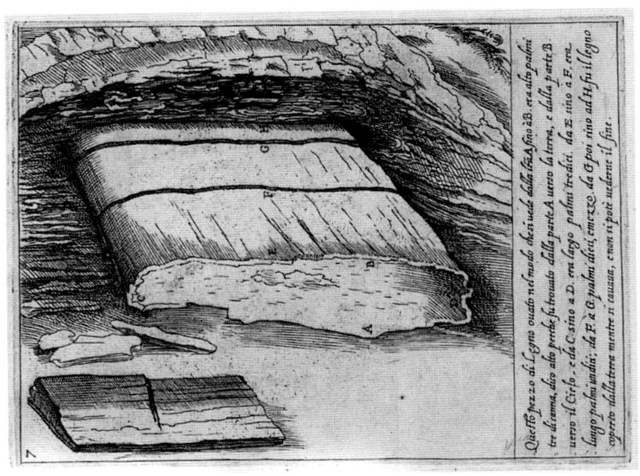

Comp. fig. 8. Stelluti 1637, Pl. 7

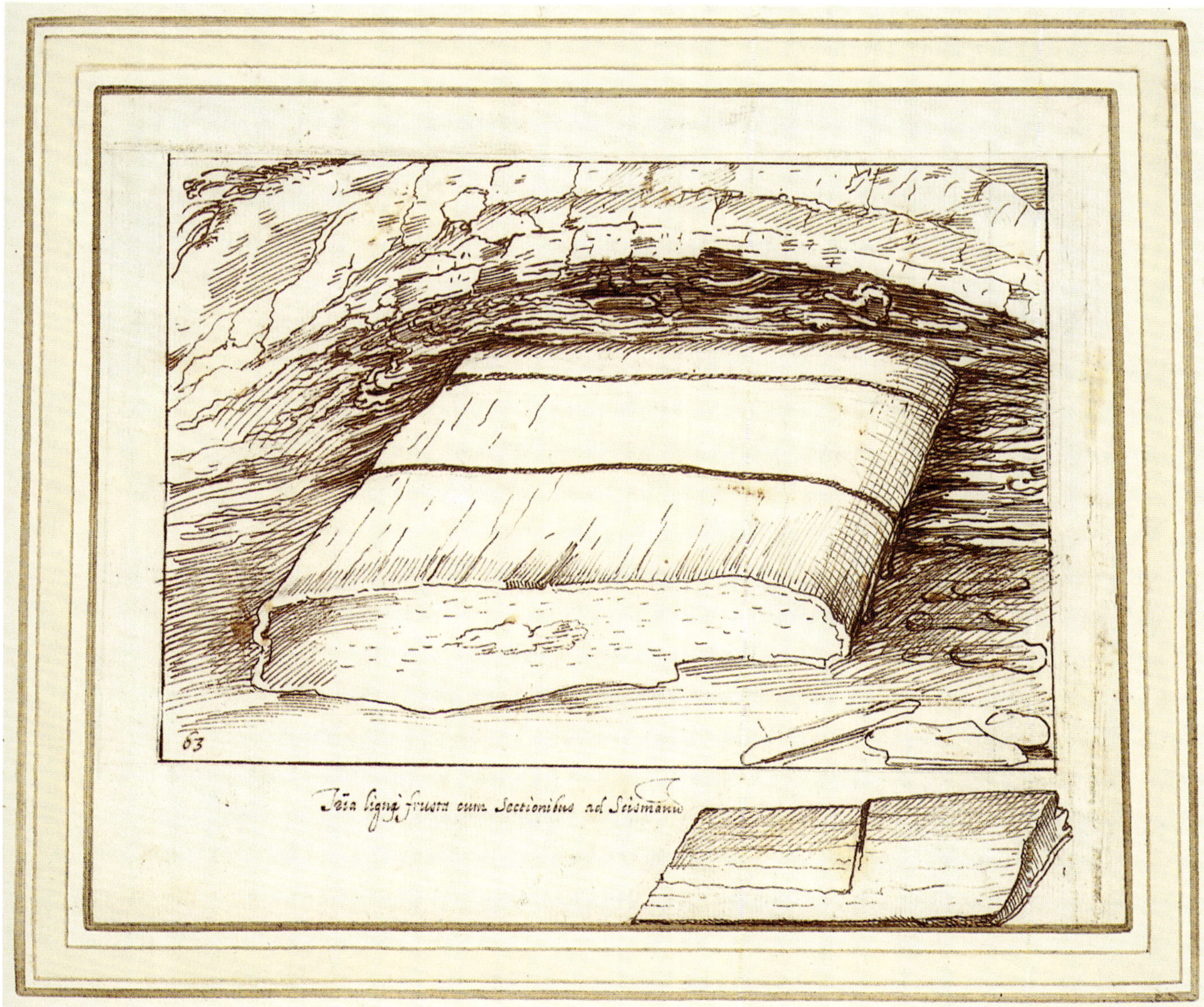

8

9. Piece of fossil wood *in situ*, probably from Sismano

Windsor, RL 25605, *Fossils*, vol. XV, fol. 14

Pen and ink, over traces of black chalk; retouched in ink on mount sheet; measurements not given; frame line in ink

176 × 240 mm; small holes caused by ink

NUMBERING: [ink, lower right corner] *67*

ANNOTATION: [ink, cut] *Frustum ligni Sect.*

This drawing, one of the series showing the excavation of fossil wood, illustrates the specimen with a slice or layer removed from its upper surface.

REFERENCES: 17th-cent. Index, p. 364

It is described as a 'section of a piece of wood'. Neither this drawing nor that following (**10**) was reproduced by Stelluti.

OTHER DRAWINGS: **7**, **8**, **10**, **11** (whole sequence shown in Fig. 22, p. 97)

10. Another view of a piece of fossil wood from Sismano, *in situ*

Windsor, RL 25606, *Fossils*, vol. XV, fol. 15

Pen and ink, over traces of black chalk; retouched in ink on mount sheet; measurements not given; frame line in ink

171 × 228 mm; small holes caused by ink

NUMBERING: [ink, lower right corner] *68*

ANNOTATION: [ink] *Frustum lign. Sect.*

This drawing shows the specimen with the slice removed as in **9**. It is viewed from slightly further away and from a different angle, thus giving a clearer view of its two-layered side section.

REFERENCES: 17th-cent. Index, p. 364

OTHER DRAWINGS: **7-9**, **11** (whole sequence shown in Fig. 22, p. 97)

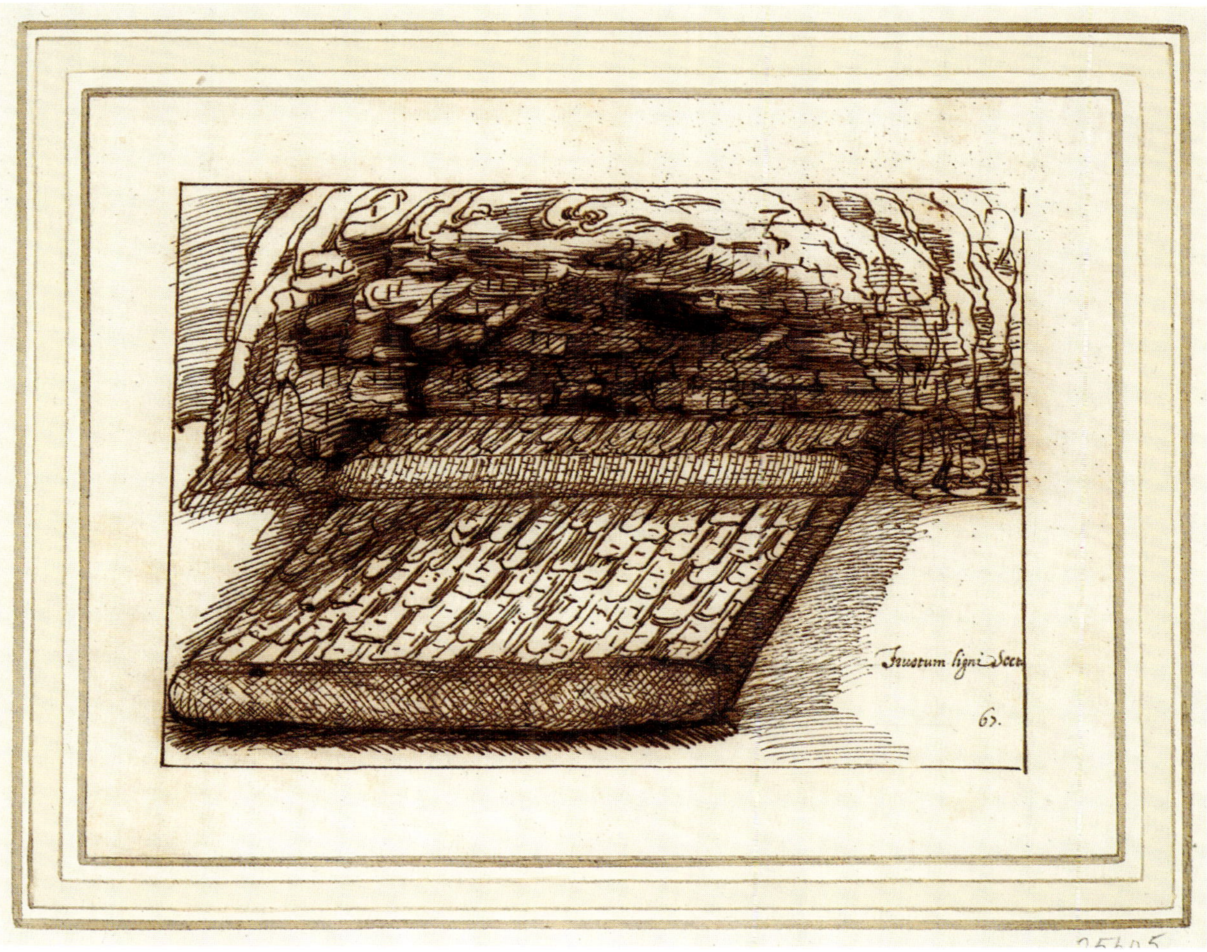

9

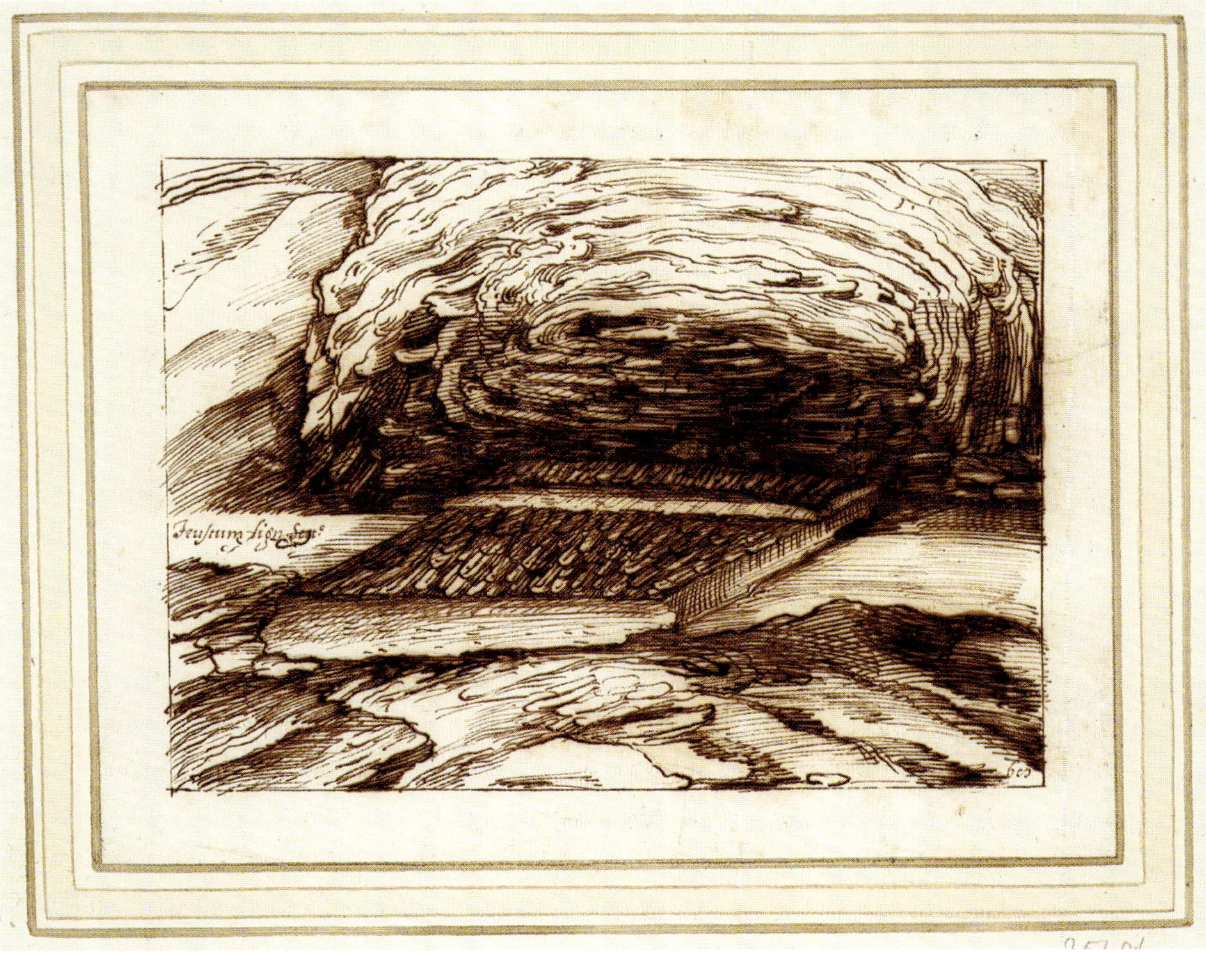

10

11. Piece of fossil wood from Sismano, *in situ*

Windsor, RL 25603, *Fossils*, vol. XV, fol. 12

Pen and ink, over traces of black chalk; retouched in ink on mount sheet; measurements not given; frame line in black chalk

175 × 235 mm; small holes caused by ink

NUMBERING: [ink, lower left corner] 65

ANNOTATION: [ink] *Frustum ligni*

The final drawing in the series showing the uncovering of a fossil wood specimen from Sismano. The drawing was engraved as the lower part of Stelluti's plate 8 (for the upper image, see **7**). For the text that accompanied this plate, see under **7**.

LITERATURE: Solinas 1989c, pp. 86–9, figs 65, 67

REFERENCES: Stelluti 1637, p. 9; 17th-cent. Index, p. 364

ENGRAVINGS: Stelluti 1637, Pl. 8, lower part

OTHER DRAWINGS: **7–10** (whole sequence shown in Fig. 22, p. 97); for a related drawing in BAV, MS Vat. Lat. 8258, see p. 403

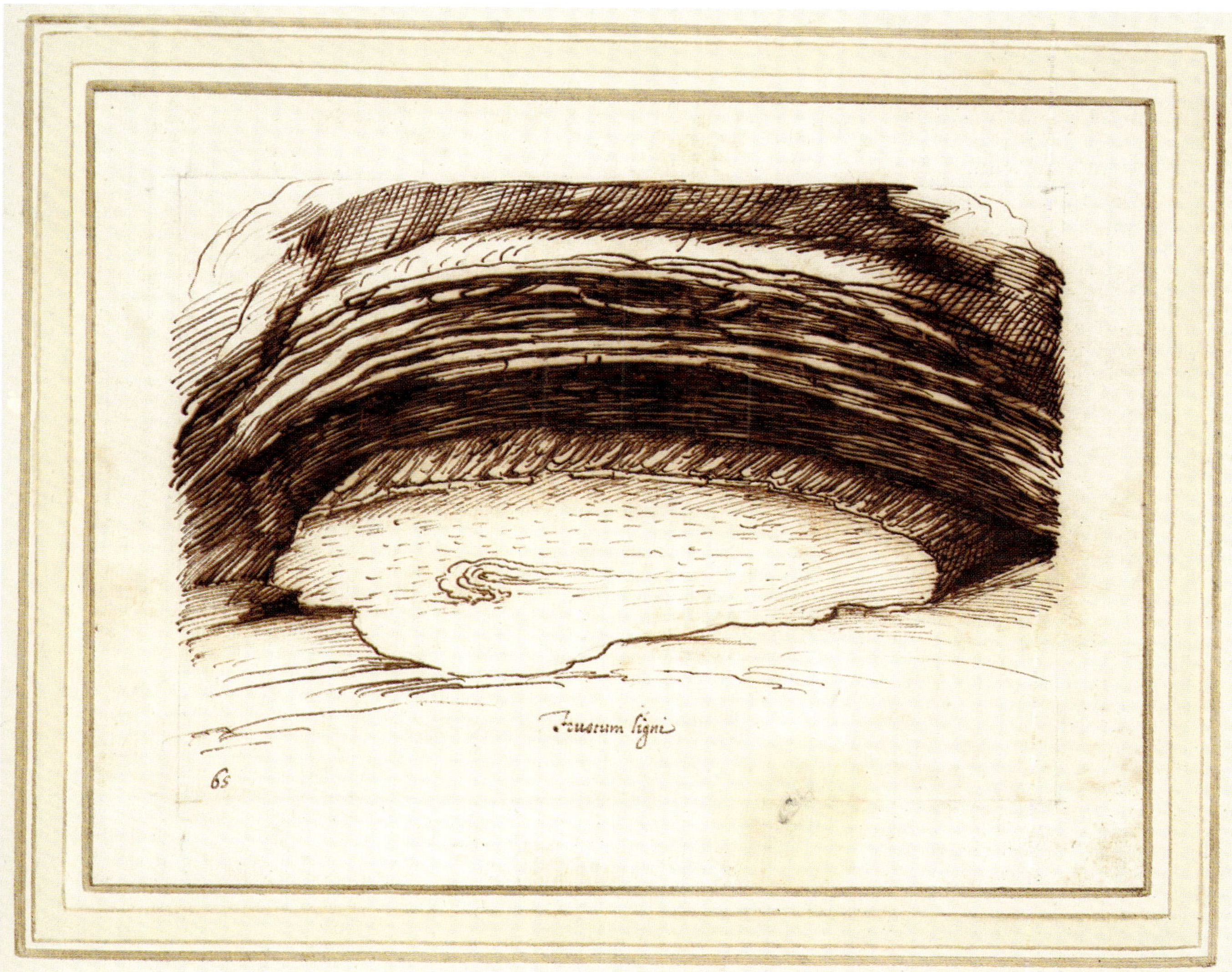

11

12. Field view between Sismano and Rosaro

Windsor, RL 25686, *Fossils*, vol. XVI, fol. 47

Pen and ink

183 × 255 mm

NUMBERING: [ink, top left corner] *147*

ANNOTATION: [ink] *Ora sive Tractus fatiscens intra Schismanum et Rosarium*

View of a large part of the area between Sismano and Rosaro bearing fossil wood. (For further details see especially **4** and **5**.) Two prominent gullies are seen (the inscription notes the 'cracking bank' or 'tract'), with further gullies in the background and tree-covered slopes. No fossil trees are visible.

REFERENCES: 17th-cent. Index, p. 370

12

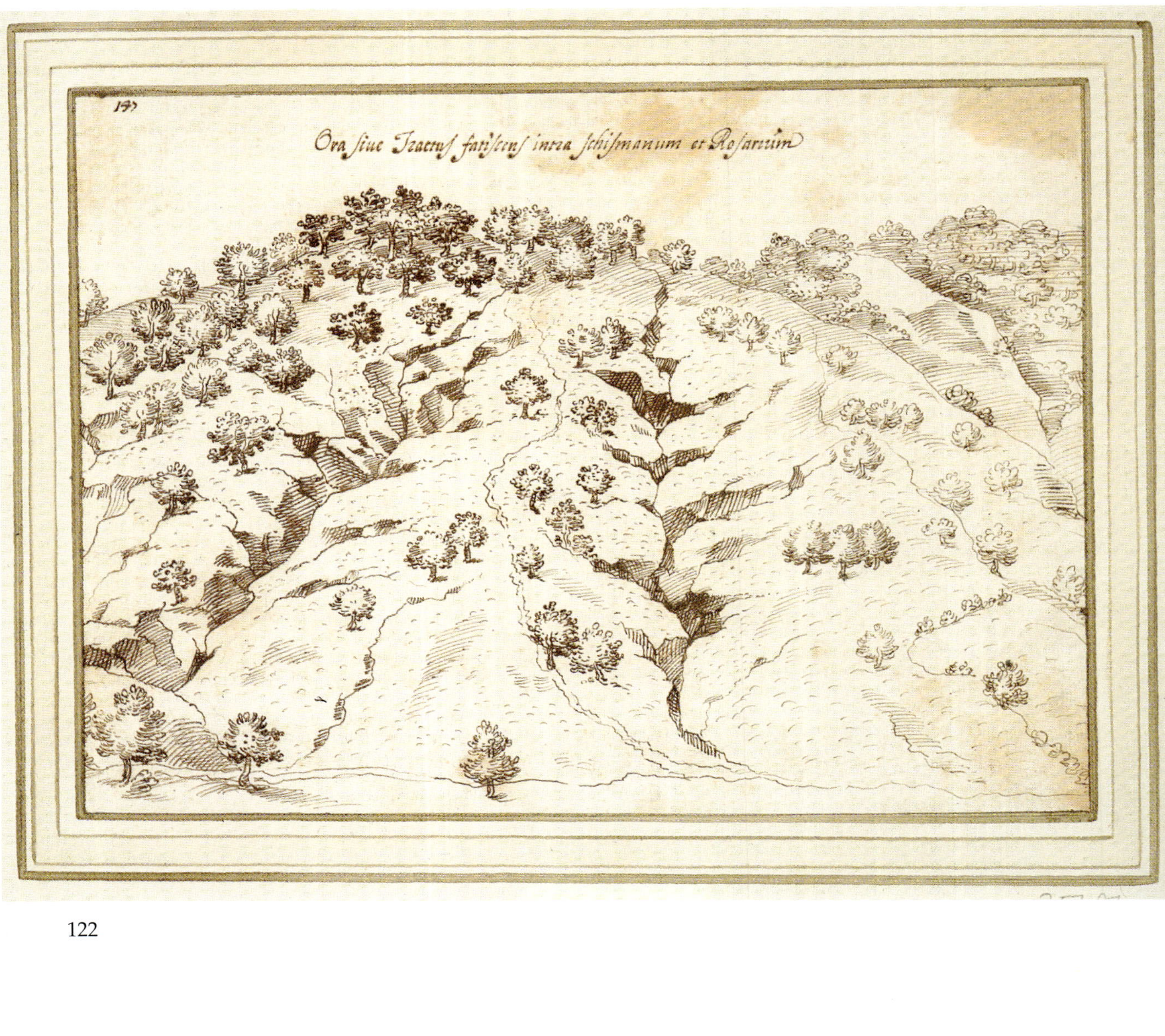

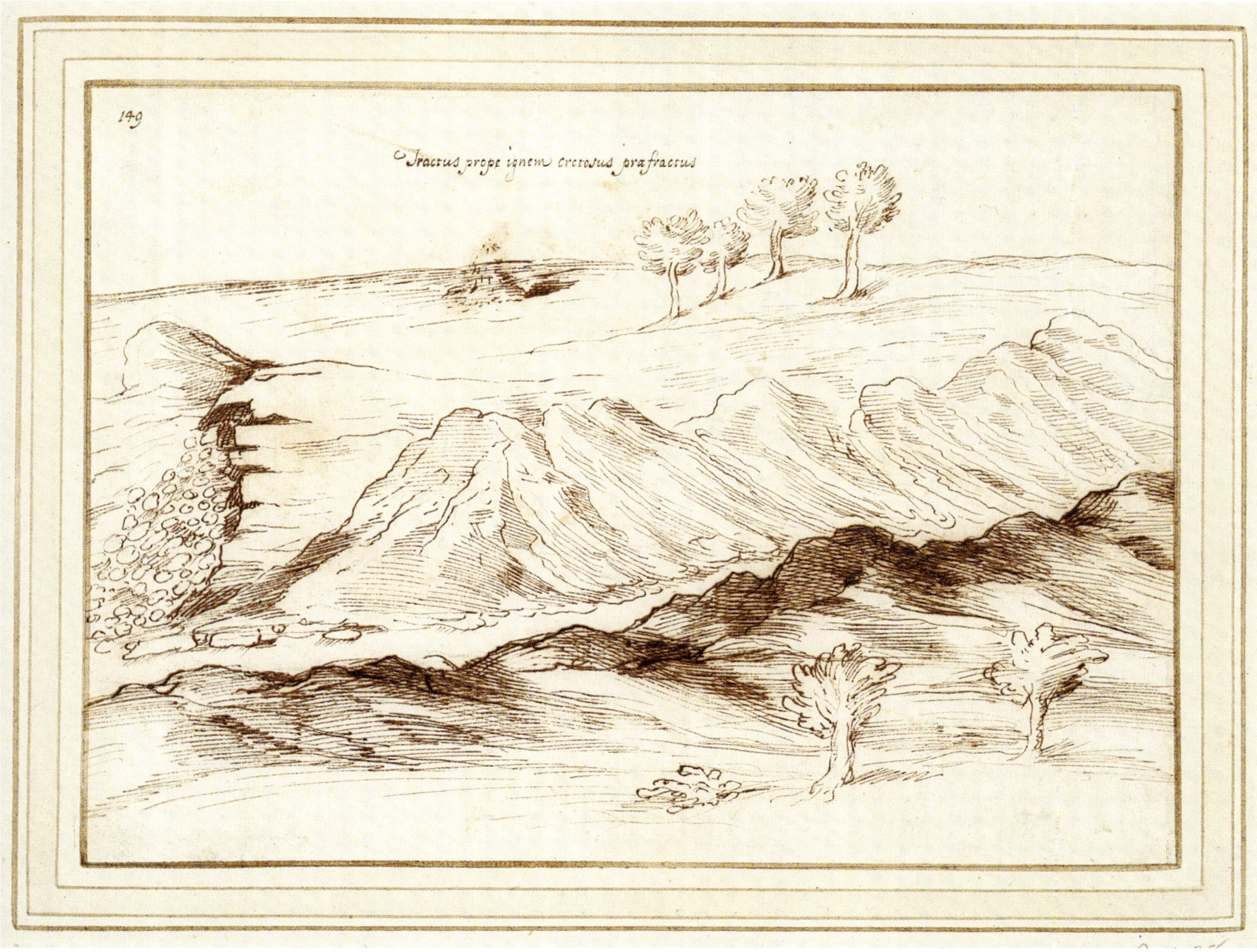

13

13. Site of everlasting fires

Windsor, RL 25688, *Fossils*, vol. XVI, fol. 49

Pen and ink, over traces of black chalk

180 × 255 mm

NUMBERING: [ink, top left corner] *149*

ANNOTATION: [ink] *Tractus prope ignem cretosus præfractus*

The inscription states that the view is of a 'broken-away clayey tract near the fire'. If it is a site of the everlasting fires (for discussion of these see below, p. 292, it is probably near Rosaro and can be compared to Stelluti, Pl. 2, from the relative position of the scree, gully, trees and fire.

REFERENCES: 17th-cent. Index, p. 370

14. Field view of a locality between Sismano and Rosaro

Windsor, RL 25683, *Fossils*, vol. XVI, fol. 44

Watercolour and pen and ink, over black chalk

182 × 257 mm

NUMBERING: [ink, top left corner] *144*

ANNOTATION: [ink] *Inter Scismanum, et Rosarium*

This is the only coloured field sketch. The drawing is notable for its depiction of four rising smoke plumes. These are the everlasting fires mentioned by Stelluti as 'those fumes which are seen to issue from the earth' (see also p. 292 below). A gully with a stream, together with screes of pebbles, indicates that this may be the site shown in Stelluti's Pl. 2 (See Fig. 23, p. 98). Some trunks of fossil wood are shown in the foreground.

REFERENCES: Stelluti 1637, p. 5; 17th-cent. Index, p. 369

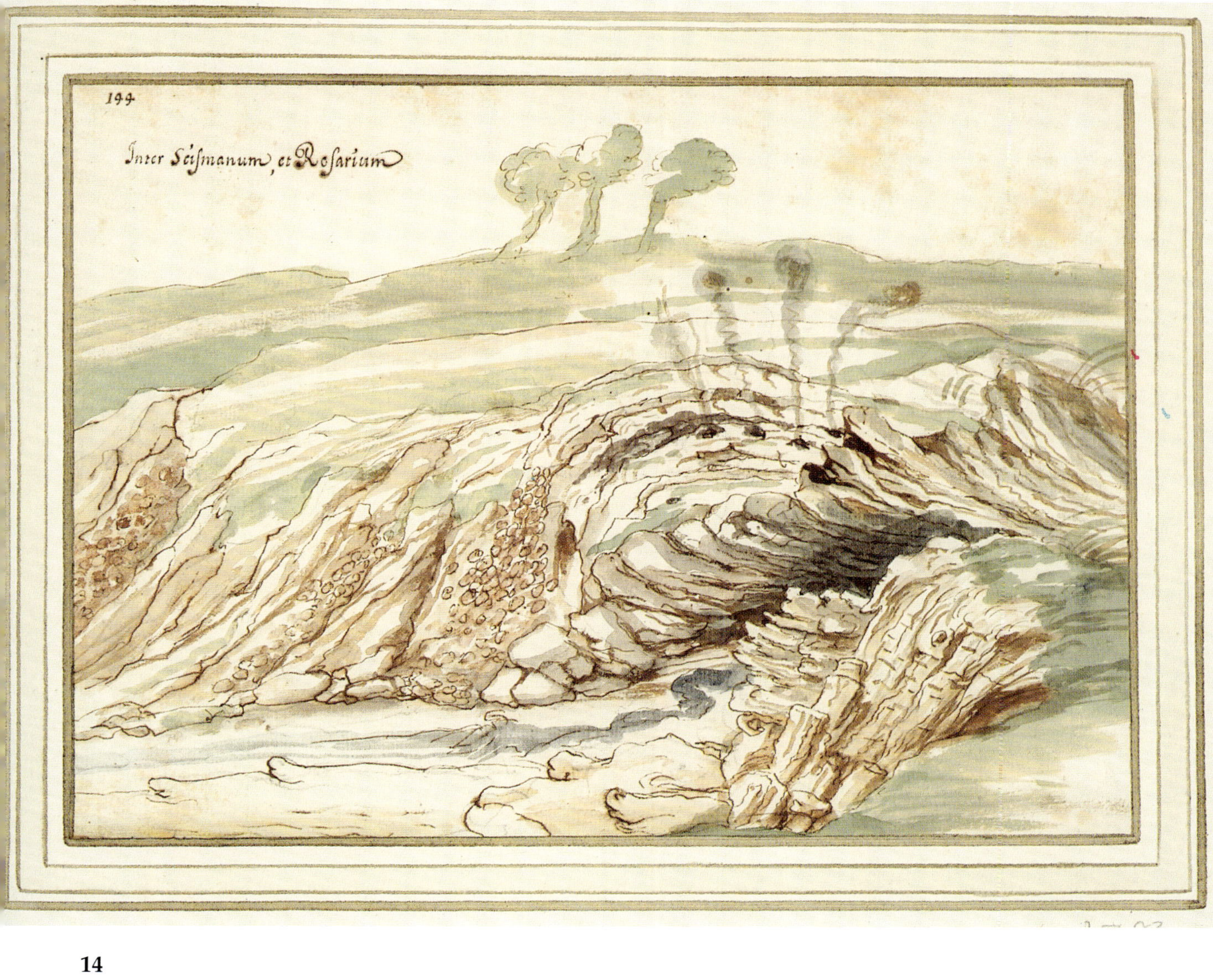

14

15

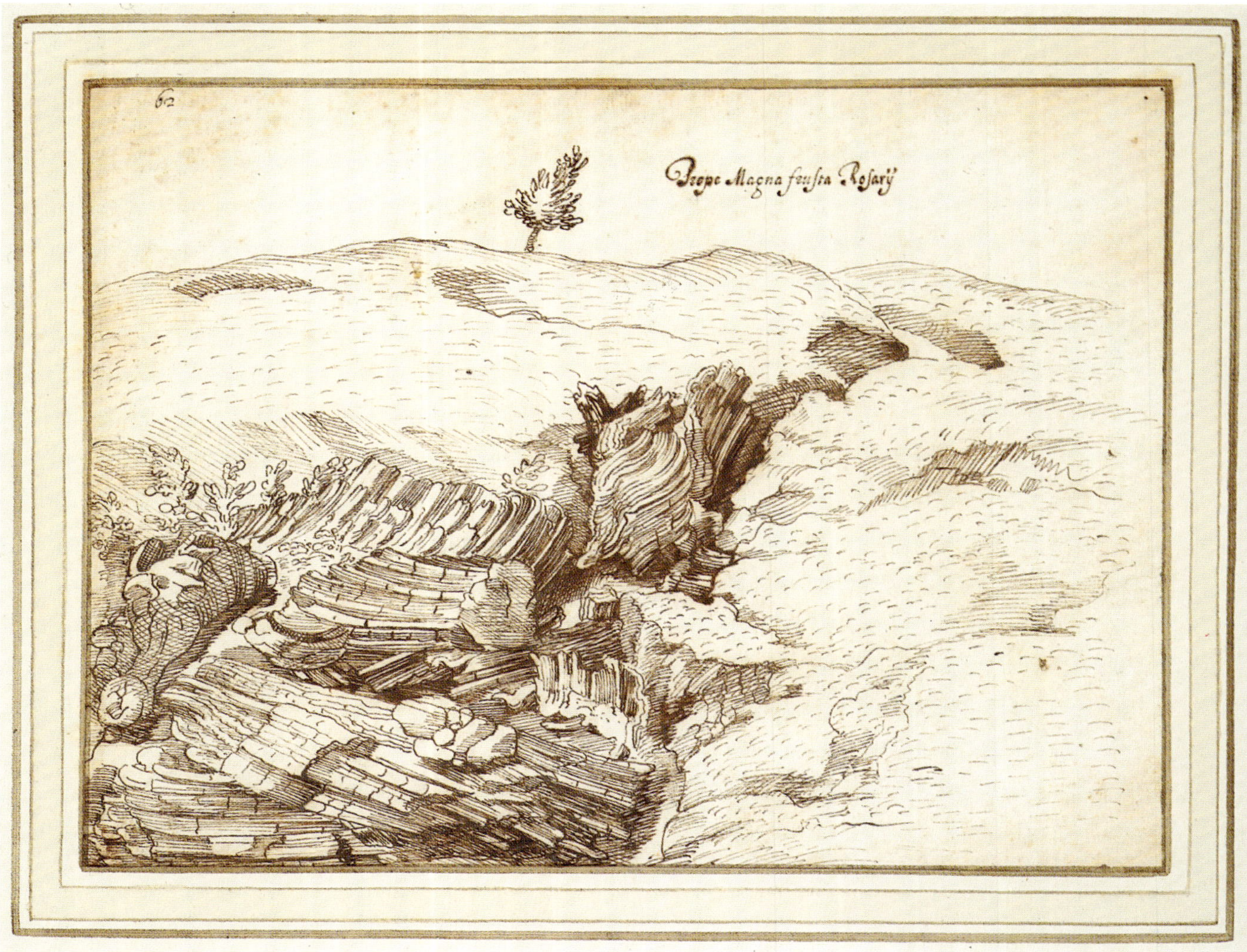

15. View of the fossil wood deposits in the vicinity of Rosaro

Windsor, RL 25600, *Fossils*, vol. XV, fol. 9

Pen and ink

188 × 260 mm

NUMBERING: [ink, top left corner] *62*

ANNOTATION: [ink] *Prope Magna frusta Rosarij*

This field sketch of the site near Rosaro shows numerous large logs weathering out of the clay in a small gully. It appears that none of the trunks is upright, *in situ*. The inscription states that the sketch is 'near the big pieces at Rosaro'. La Ferrière reported seeing logs in this locality which, already sawn, measured eight to ten feet (2.5–3.0 m) in length; and a further tree said by Cassiano to be 40 or 50 feet (12–15 m) long (letter to Peiresc, November 1635, see Appendix III b).

REFERENCES: 17th-cent. Index, p. 364

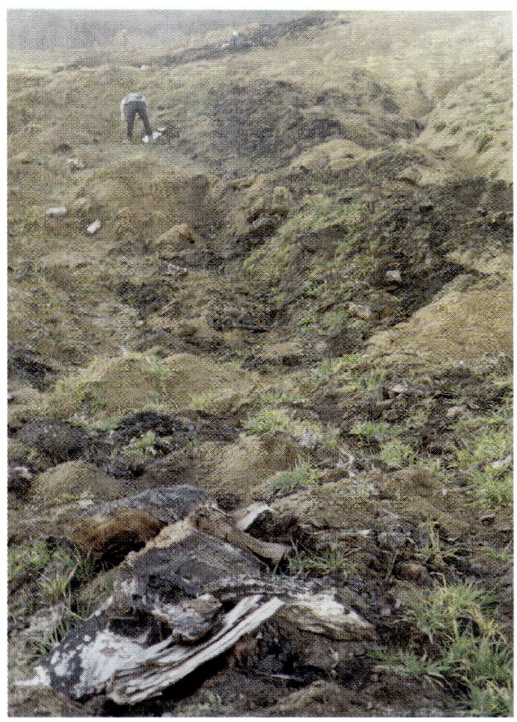

Comp. fig. 15. Landslipped area between Rosaro and Sismano with large trunks of fossil wood

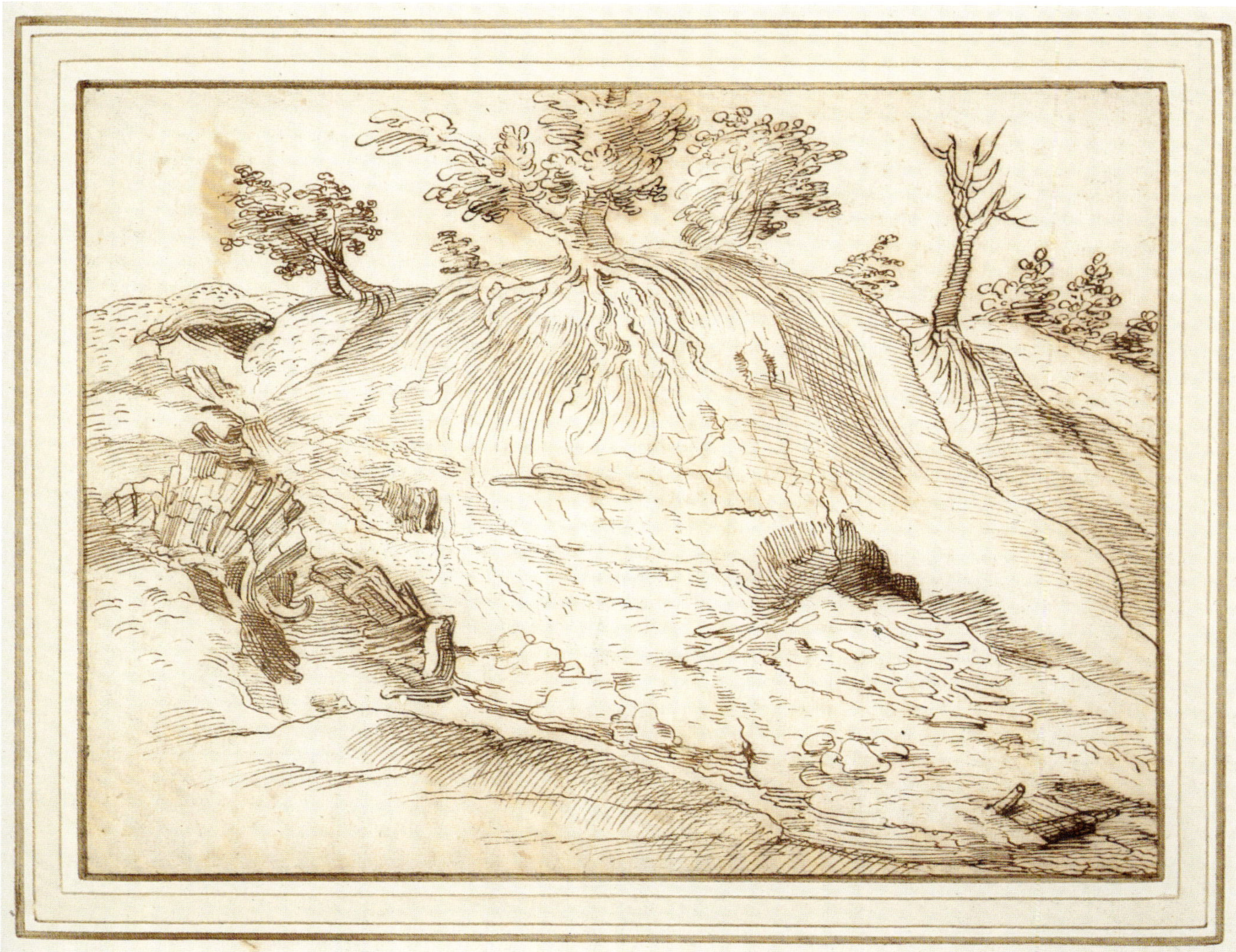

16

16. Sketch of one of the fossil wood sites, locality uncertain

Windsor, RL 25538, *Fossils*, vol. XIV, fol. 9

Pen and ink

199 × 270 mm

The view shows fragments of fossil wood being eroded from the natural hillside. The roots of the living trees are shown penetrating the rocks. On the left-hand side of the drawing, a large trunk of wood is shown still in place. Although the site is not identified, it is probably in the Rosaro–Sismano–Dunarobba area.

LITERATURE: Bromehead 1947a, p. 81

REFERENCES: missing from 17th-cent. Index

17. Field view of Perrosciolo

Windsor, RL 25687, *Fossils*, vol. XVI, fol. 48

Pen and ink

219 × 391 mm; old vertical fold at centre; repaired vertical tear on right.
Watermark: Bird 41C

NUMBERING: [ink, top right corner] *148*

ANNOTATION: [ink] *Tractus Perroscioli Strato Secante*

Field sketch of a site identified as Perrosciolo with, according to the inscription, 'a stratum splitting apart'. Gullies, streams and trees are shown, as are boulder screes near the top of the sketch. These boulder screes could represent an outcrop of a conglomerate horizon. Perrosciolo has not yet been located.

REFERENCES: 17th-cent. Index, p. 370

17

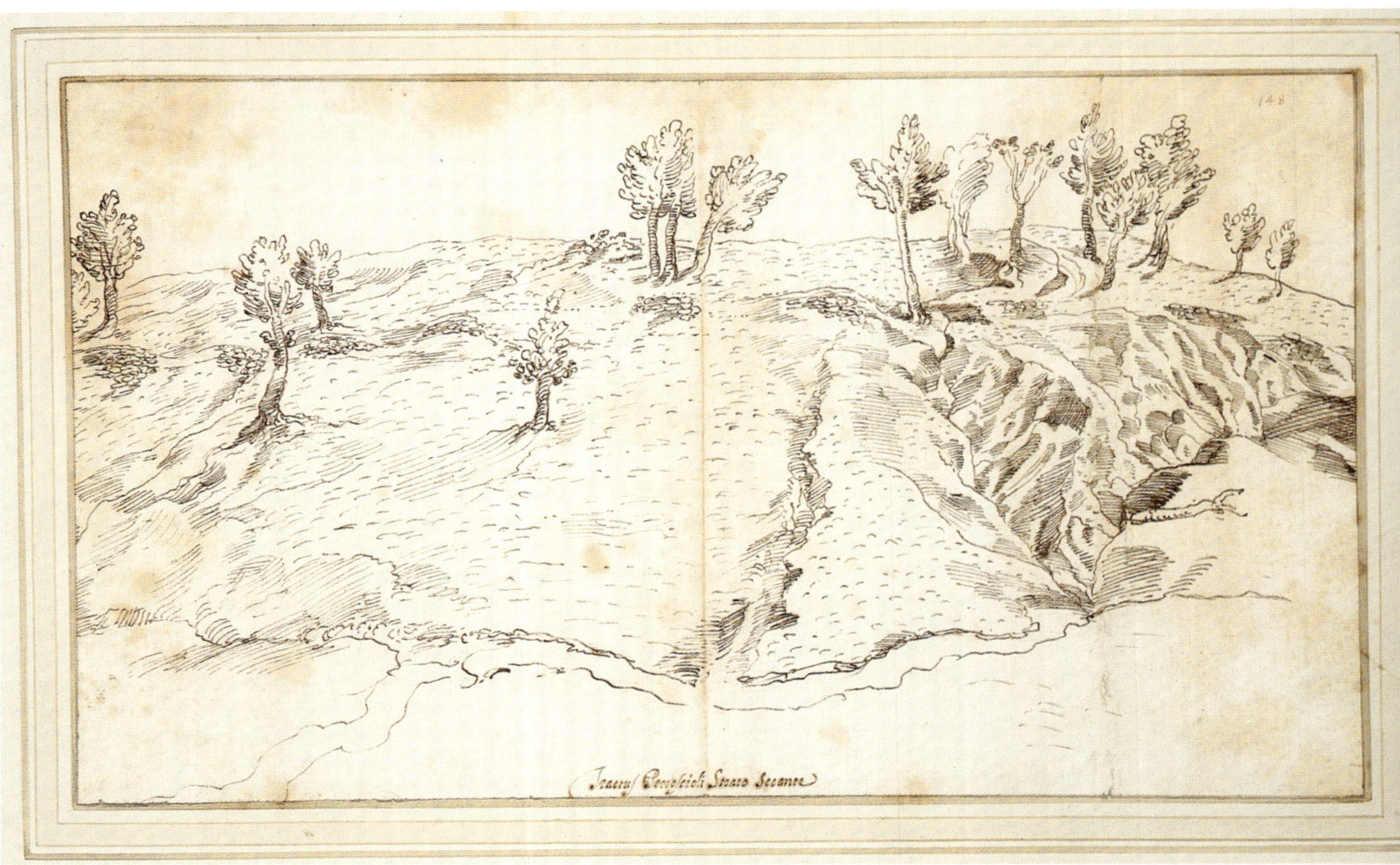

II. FOSSIL WOODS [18–133]

Only recently has a reasonable understanding been attained of the wide variety of ways in which plants are fossilized.[1] Cesi's and Stelluti's study of fossil woods is remarkable in a number of ways. It was the first to record detailed field evidence of the occurrence of fossil woods and to document the extraction process. The specimens were described with their size, colour and weight, as well as the locality of discovery, all recorded. They were drawn, often from several angles, just as in a modern study. Finally, observations and a few experiments were undertaken to note the changes that occurred to the fossils. This helped Cesi and his colleagues to come to a judgement as to the nature of the fossils and allowed them to draw far-reaching conclusions based on the evidence and on comparative material. While the conclusions published by Stelluti and, as he claimed, by Cesi also, are often incorrect, it is possible to argue that they were based on sound observation and argument. In order fully to appreciate the problems which Cesi faced it will be necessary to review our current understanding of how woods may be fossilized.

The preservation of fossil woods

Fossil wood, while often appearing much like recent wood, may change some of its characteristics during the fossilization process. The unusual nature of some preservation states can, even today, lead an observer to an incorrect conclusion regarding the nature and affinity of the fossils. In view of the fact that the meaning of many terms (such as 'petrifaction') used in the description of fossil wood has changed considerably over the years, it is necessary to describe in some detail the current thinking on the preservation of plant fossils.

Wood is comprised of a variety of organic molecules which are variously decay-resistant. Most wood is dead cellular tissue used to carry water in the plant. It is broadly composed of two materials: cellulose, which constitutes 70 per cent of the cell walls, and lignin, which adds structural strength and resistance to decay to the plant tissues and accounts for the remaining 30 per cent. Wood, like other plant tissues, will decay and break down in air, and when subject to fungal and bacterial activity. Restricting the availability of oxygen by, for example, waterlogging or burial in peat or mud, may slow or stop its decay. The partial decay of the wood releases both humins and humic acids which, when mixed, cause the precipitation of an ulmin gel within the cell spaces of the wood. This gellification is, therefore, a biochemical process and is part of peatification. Such gellified wood is much harder than unaltered wood. The buried wood will also be subjected to forces of compaction as well as to an increase in temperature through continued burial. This increase in temperature, in particular, promotes changes to the organic composition of the wood driving off water, as well as methane and carbon dioxide gases. This process, known as coalification [Fig. 29], affects all organic material including both wood and peat. In the case of peat—defined as an accumulation of plant material with 50 per cent organic matter—the change brought about by coalification causes a loss of water and volatile matter, as well as an increase in the carbon content of the material. The peat changes in both colour and texture, reflecting these chemical or rank changes. Peat is progressively changed to lignite or brown coal, to black sub-bituminous, then to bituminous coal and finally to black shiny anthracite [Fig. 29].[2] In the same way wood, which generally

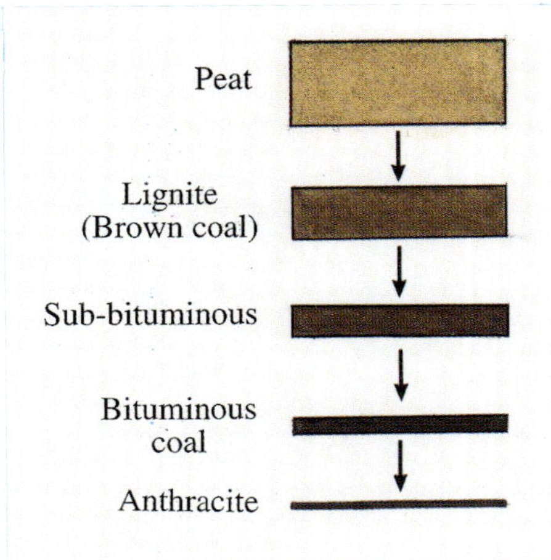

Fig. 29. Diagram showing changes in colour and volume during coalification

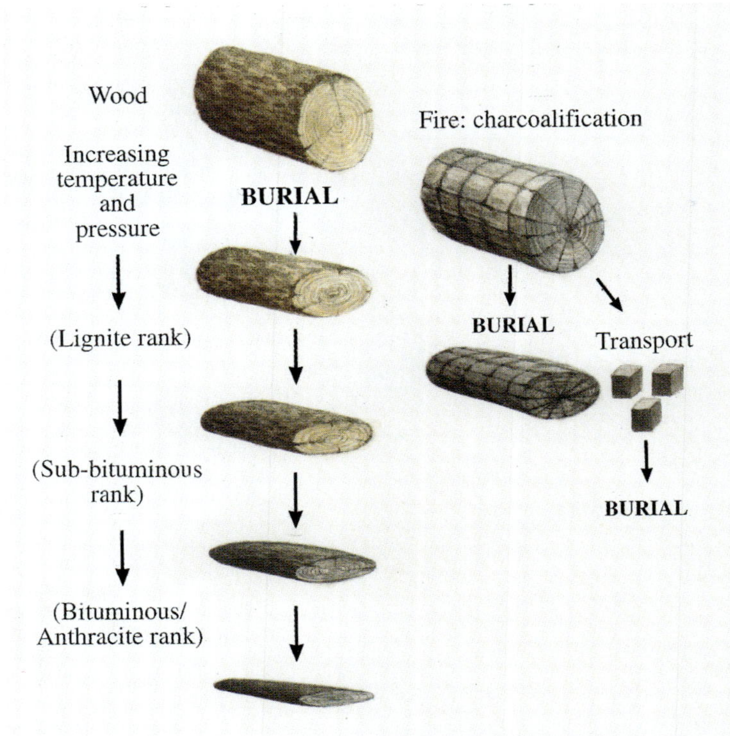

Fig. 30. Diagram showing alteration of wood during burial

Fig. 31. Diagram showing alternative ways in which wood may be fossilized

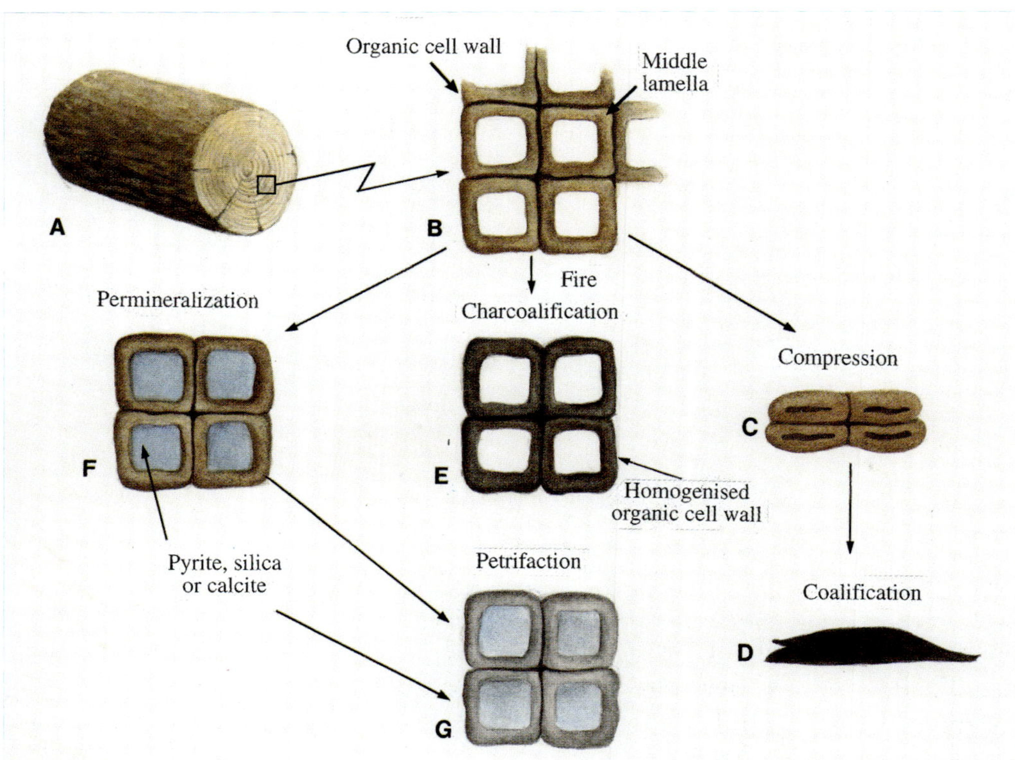

starts light brown in colour, will eventually darken until it becomes black [see Fig. 29]. It is important to remember, however, that the material is still organic and remains a slightly altered wood, though its cellular structure may be obscured during compaction. In many cases, woods which have been buried and partly compressed and have reached only a low level of alteration will appear much like modern wood, apart from being slightly darker and more earthy in texture.

The organic material in wood is not, however, always preserved. In some cases, the cellular spaces within the wood may be filled by a mineral [Fig. 31B] which precipitates from water percolating though the sediment, generally after burial [Fig. 31F]. This process is called 'permineralization'.[3] Common minerals found in fossil wood are: iron pyrites (gold or brown in colour), calcite (white) or silica (which may be white, colourless or even reddish-brown). If the wood is buried in organically rich muds, pyritization is common, but if associated with volcanic activity then the permineralizing fluids may carry calcite or silica. In the past, and still today, some geologists use the term 'petrifaction' for these fossils. The density of the wood is increased but its organic material, as well as its anatomical structure, remains. When, following permineralization, the remaining wood decays, further mineral precipitation may occur and the wood is gradually replaced by minerals; this is what today is known as petrifaction [see Fig. 31G].

The final method of preservation of fossil wood is through burning. This converts the wood into charcoal [Fig. 31E].[4] Such burning would have occurred before the wood was buried in the sediment. Black pieces of wood, which readily mark the hand, are easily recognized—fossil charcoal is known as 'fusain'. Although such material is known from the Dunarobba site,[5] none can be identified with certainty among the drawings in the corpus.

Some woods are described as ebony (**36**) indicating that they were black. Also some were described as gagate (**62**)—a term used for jet. However, neither category can be identified for certain, ebony being a species of wood and jet being a black coalified coniferous wood that can be worked into jewellery (see **62** for further discussion of gagate).

Observations on fossil woods from Dunarobba, and comments on the observations of Cesi and Stelluti

During reinvestigation of the Dunarobba site by the present author in 1995, 1998 and 1999 a variety of fossil woods was collected [Fig. 19]. These fossil woods commonly occurred within the clays of the small exposure shown in Fig. 18 and were remarkably similar to the specimens in the drawings [Fig. 32]. Scattered woods, both branches and small trunks as well as leaves, were also found[6] and included numerous specimens of the conifer *Glyptostrobus*. Much of the timber looked like wood from modern trees, and some specimens were almost identical to those illustrated in the drawings [Fig. 32]. Towards the top of the outcrop, plant material was concentrated to form an organically rich or lignite layer. In some cases clays and woods were compacted together.

Of particular interest was the range of preservation of the woods found at the site. Many specimens at the base of the outcrop had been impregnated with iron and were 'sideratized' or 'limonatized', showing a variety of colours from browns through yellows to red [Fig. 21]. At the top of the outcrop, several specimens of wood permineralized and petrified by iron and

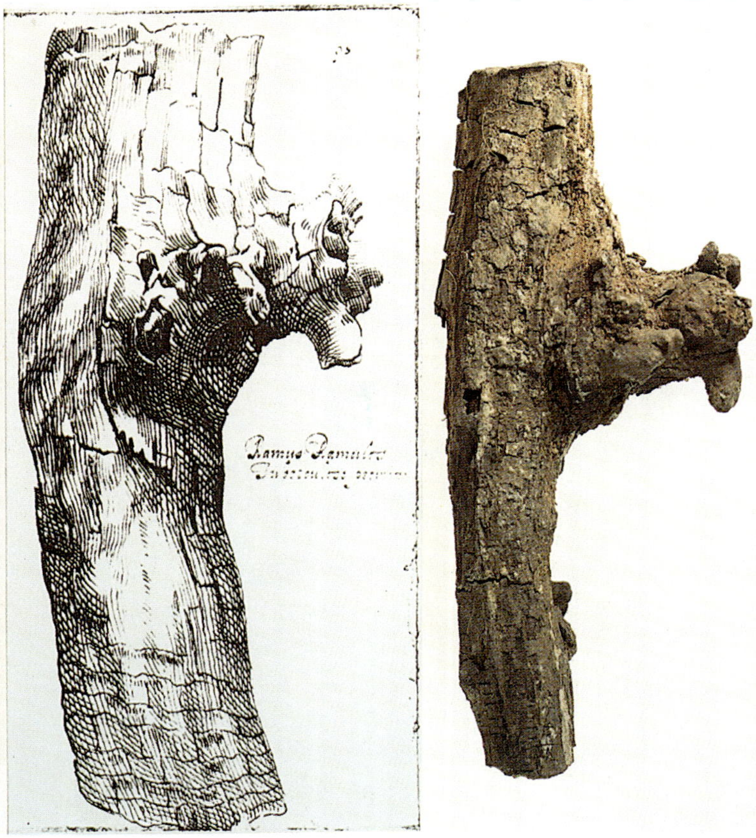

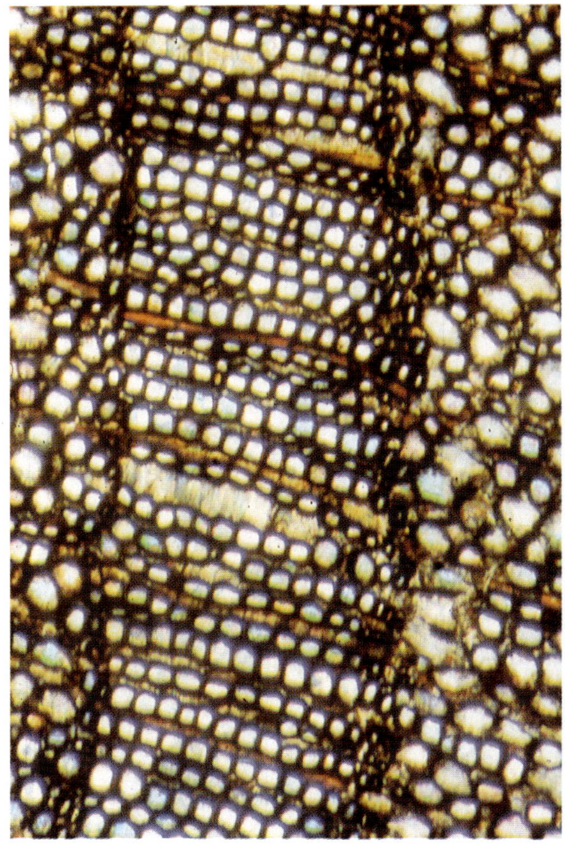

Fig. 32 (*left*). Drawing of fossil wood with branch (**21**) with specimen from Dunarobba

Fig. 33 (*above*). Thin section of permineralized fossil wood from Dunarobba

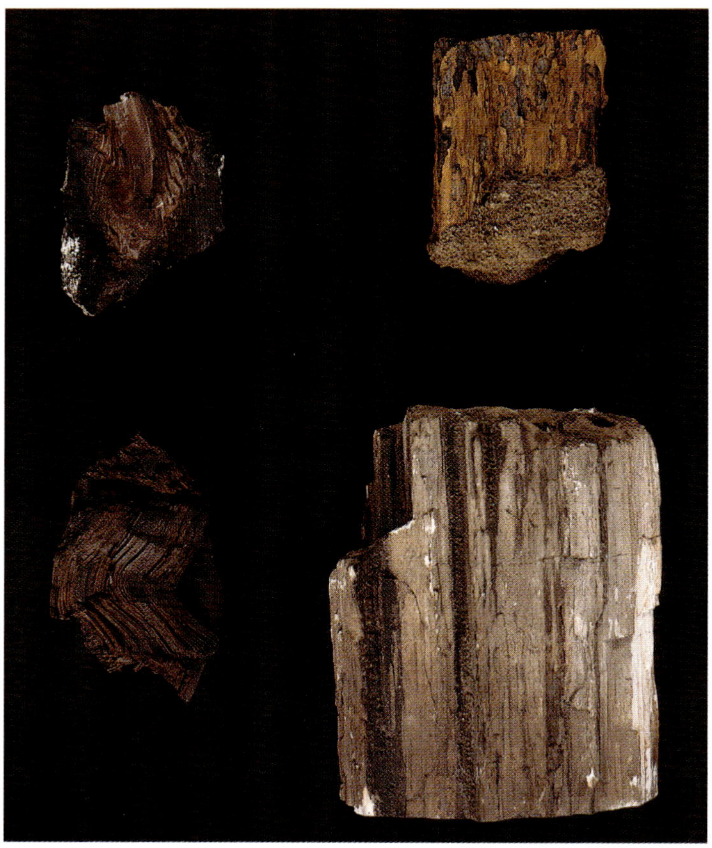

Fig. 34. Preservation of fossil woods, from compressed woods to petrifaction

calcium carbonate were found, as well as some specimens which were petrified at one end and not at the other [Fig. 20]. This range of preservation of fossil woods is unusual for a single locality and is difficult even for modern palaeobotanists to explain. Fieldwork undertaken in 1998 and 1999 by the present author and Giorgio Basilici identified several other fossil wood deposits in the triangle between Rosaro, Sismano and Montenero.

In his classification of fossil wood, Cesi used two major categories: *cretilignum* and *metallophyta*. It is probable, both from the drawings and the annotations on them, that the term *cretilignum* was used to describe a range of materials from single pieces of fossil wood with adhering clay, either on the outside or in hollows (that is, washed in), to complete admixtures of clay with wood fragments sometimes described as '*cretilignum* with bark-like scales of wood' (see, for example, **113**).

The intimate association of wood and clay was important to Cesi and Stelluti. In many ways it was their key to the interpretation of the fossils. Some of their contemporaries believed that fossil woods were from once-living trees no longer found in the area, which had been turned in part to stone or other minerals (see below, pp. 391–2). But Cesi—and after him Stelluti—took a different view:

> We have now only to pass to an explanation of the illustrations which follow. But first we will say that, when there is no other evidence or proof of the origins of this wood, let this suffice: that a quantity of damp earth, which was around a piece of this wood, having been removed and placed in a room in the Palace of Acquasparta belonging to the Lord Duke Cesi, it was found after some months to be wholly converted into wood, to the great wonder of the aforementioned lord, and of all the others who saw it: whence he has no doubt whatsoever that the earth itself is the seed and mother of this wood, the earth of these parts being most suitable for its generation.

It is quite possible, however, that what they had witnessed was a feature of the drying-out of fossil material: a piece of fossil wood surrounded by wet clay was collected, and then, when it dried, the clay shrank and the wood expanded, breaking the clay open to reveal the wood. Could it be that Cesi actually knew better, but that the ongoing pressure on Galileo—and thus the few remaining Lynxes too—made him reluctant to advance any more radical explanation for the origin of fossils?

In some cases iron pyrite was evident in the material, and one specimen is actually described as a '*cretilignum* with grains of iron and copper' (**133**). The occurrence of the mineral iron pyrite was recognized as a separate substance—both in its own right, 'pyritic clay from our well'—and as incorporated into the wood. In such cases the wood was sometimes described as 'metallized'. During the recent excavations, no pyritized woods were found at Dunarobba and indeed none of the specimens noted by Cesi to have come from that locality were classified by him as belonging to the category he called *metallophyta*. However, *metallophyta* were said to have come from Rosaro and Sismano. La Ferrière, in his letter written in 1635 to Peiresc (see below, Appendix III b) recalled that the site at Rosaro was described by Cesi as 'sulphurous'. A feature of pyritized wood is that when exposed to humid conditions the pyrite decays, undergoing a significant chemical reaction. Sulphuric acid is then released and the pyrite oozes liquid, and decays to a white powder [see Fig. 36]. It is clear that Cesi and Stelluti had observed this too. Not only did they mention a sulphurous smell, a feature easily recognized, but also

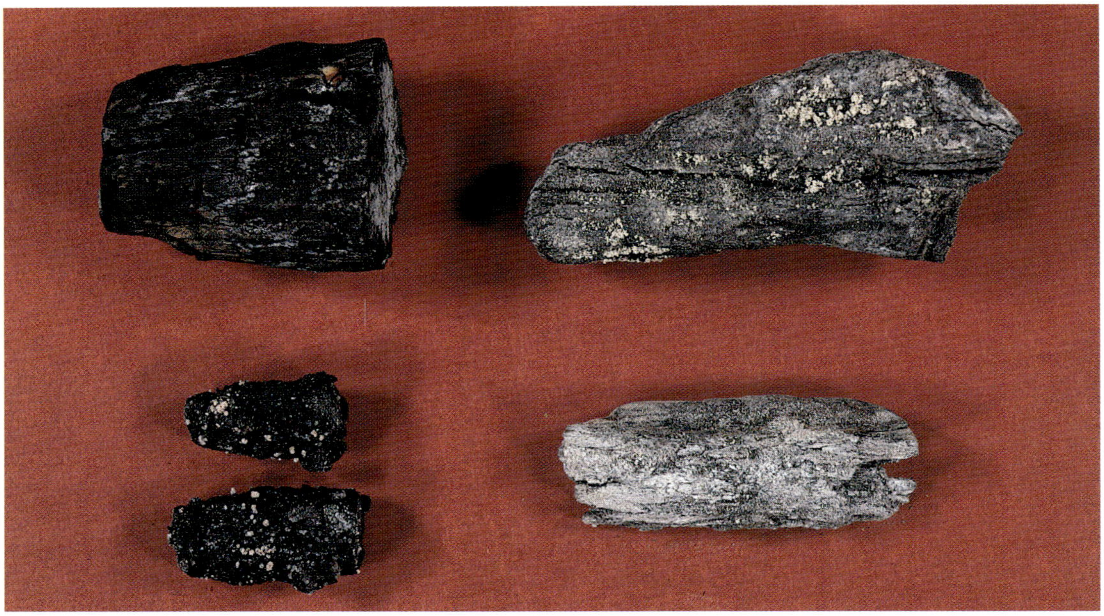

Fig. 35. Pyritized wood showing decay when exposed to moist air

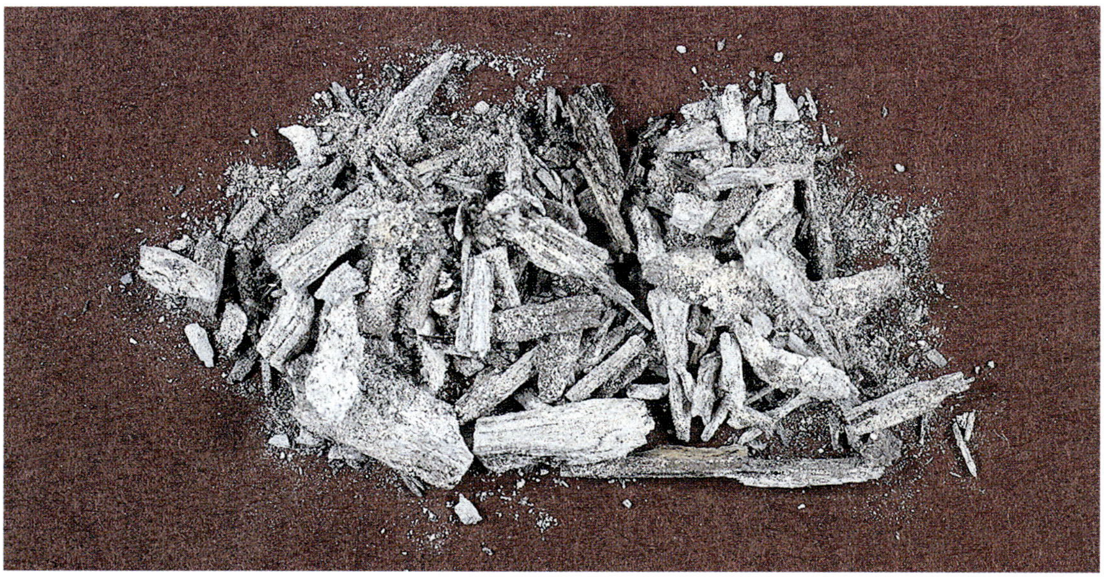

Fig. 36. Decay of pyritized wood, showing white powdery residue

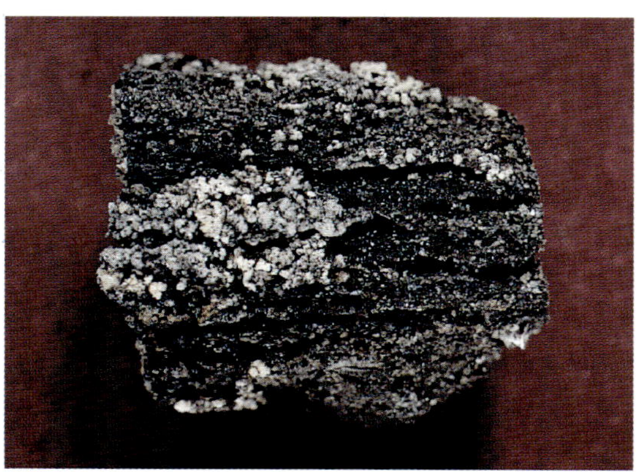

Fig. 37. Pyritized wood decaying in a moist atmosphere, showing 'glue-like' decay products described by Stelluti

recorded many of the tell-tale signs of pyrite decay. They noted the occurrence of small whitish flecks on some of the specimens and Stelluti wrote:

> In that part where the wood appears sawn off, a resin was found, white in colour, like gum, and white incense, the wood here sweating just as those walls do which have been freshly plastered in damp places' [Fig. 37].

In the woods from Rosaro small amounts of sulphur may be present. Some of this may be due to the action of recent bacteria. If the wood is burnt (either in a fire or by underground fires) then sulphur dioxide is released which gives a distinctive smell even though the concentration may be small.

It is not surprising that both Cesi and Stelluti had difficulty in interpreting what they saw. Using density, colour, shape and mineral associations to classify the wood may seem strange to us, but it should be remembered that the microscope had only just been invented. It was to be another fifty years before the microscopic structure of fossil wood, so necessary for the modern classification of woods, was to be observed[6] and many years before the now accepted forms of botanical classification were established.

NOTES

1. Scott 1990.
2. Scott 1991, p. 25.
3. Scott 1990, p. 264.
4. Scott 1990, p. 264.
5. Scott, personal observation from field trip 1995.
6. Hooke 1665; see Rudwick 1976, p. 55.

18. Two views of a branch of fossil wood

Windsor, RL 25677, *Fossils*, vol. XVI, fol. 38

Pen and ink, over black chalk

183 × 272 mm

NUMBERING: [ink, top left corner] *138*

ANNOTATION: [ink, top] *Ramus ligneus multi nodosus ab utraq: parte* [bottom] *long: pal: 3. lat. d: 3. alt: d: 3*

The inscription describes the specimen as 'very knotted on both sides'. The drawing indicates spirally-arranged branch scars, suggesting that the plant was a conifer.

The lower view is not dissimilar to that shown in **19**.

REFERENCES: 17th-cent. Index, p. 369

OTHER DRAWINGS: For similar specimen, see **19**

19. Piece of fossil wood with groups of branch scars

Windsor, RL 25616, *Fossils*, vol. XV, fol. 25

Pen and ink, over traces of black chalk; retouched in ink on mount sheet

187 × 267 mm; small holes caused by ink

NUMBERING: [ink, upper right corner] *77*

ANNOTATION: [ink, top] *Intus lapid: rubens, extra veste lignea, nigra / Rugosa* [bottom] *long. pal: 6: lat: p: 1. Crass. d. 4*

This specimen of fossil wood shows well-preserved groups of branch scars. It is described as 'inwardly stony and ruddy, outwardly having a black, woody covering, wrinkled'. The drawing was engraved as figure D in Stelluti's Plate 10.

In the accompanying text he writes: 'Trunk *D*, in between the two capillaries mentioned above, also excited great wonder: because its inner part was already totally petrified, and reddish in colour, and the outside covering of a bark-like wood, and knotty, exactly as it has been drawn. The length of the entire trunk was six palms, and its width approximately one palm in diameter. Other similarly large trunks have been found there, that is with their centre petrified but their outside covered with wood, of which drawings have already been made'. It is interesting to note the comment that only the centre of the wood was petrified.

REFERENCES: Stelluti 1637, p. 10; 17th-cent. Index, p. 364

ENGRAVINGS: Stelluti 1637, Pl. 10 (Fig. D); proof engraving in Montpellier, BEM, MS H.173, fol. 19 (see p. 403)

OTHER DRAWINGS: For a similar specimen, see **18**

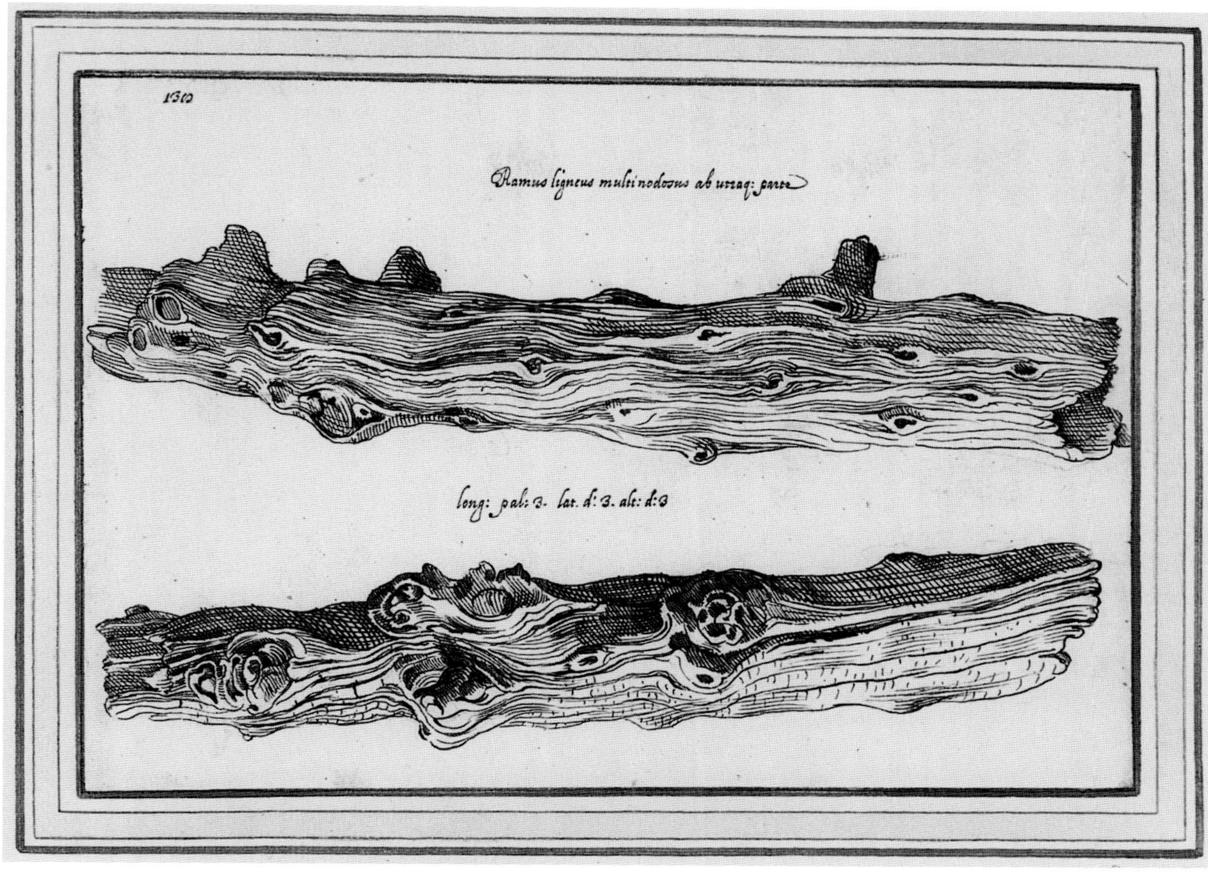
1310
Ramus ligneus multinodosus ab utraq: parte
long: pal: 3. lat. d: 3. alt: d:3

18

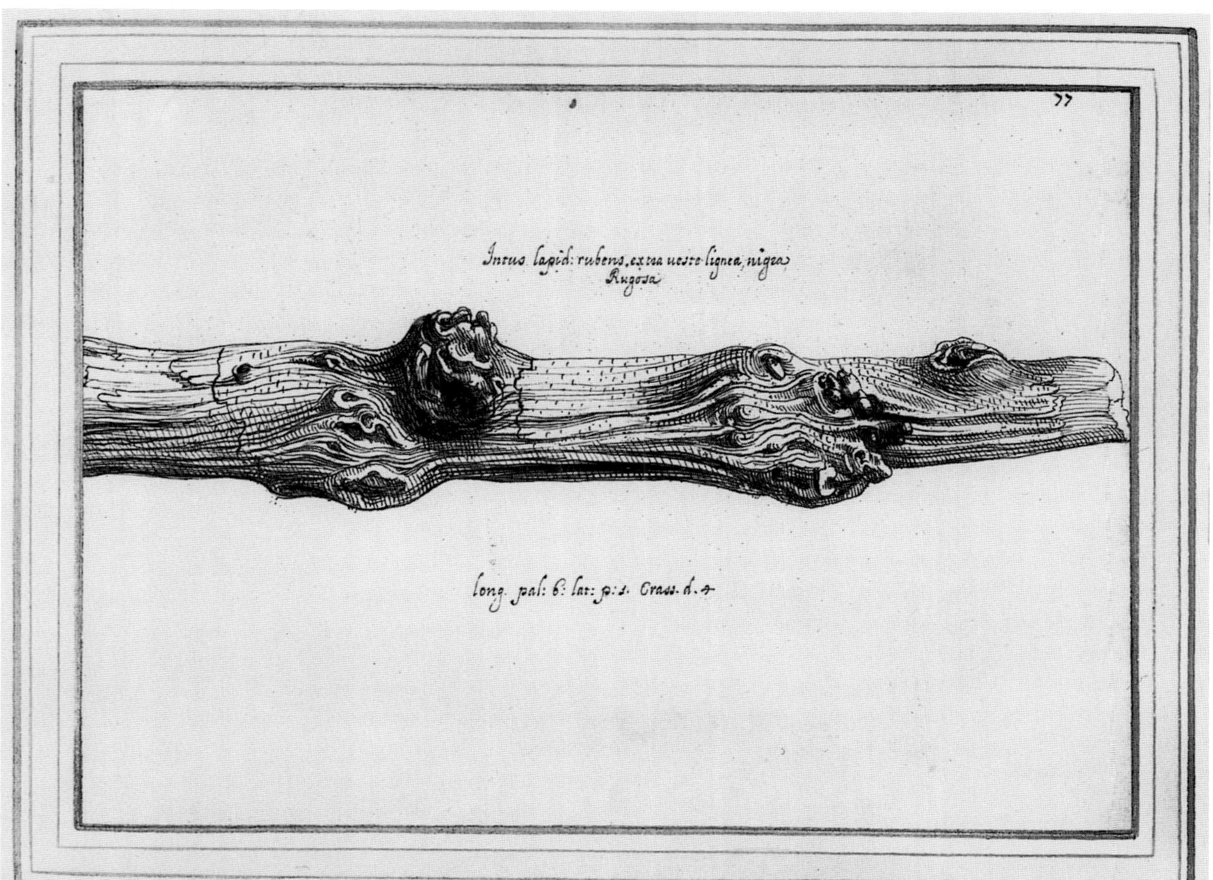
77
Intus lapid: rubens, extra ueste lignea, nigra
Rugosa
long. pal: 6: lat: p: 1. Crass. d. 4

19

20 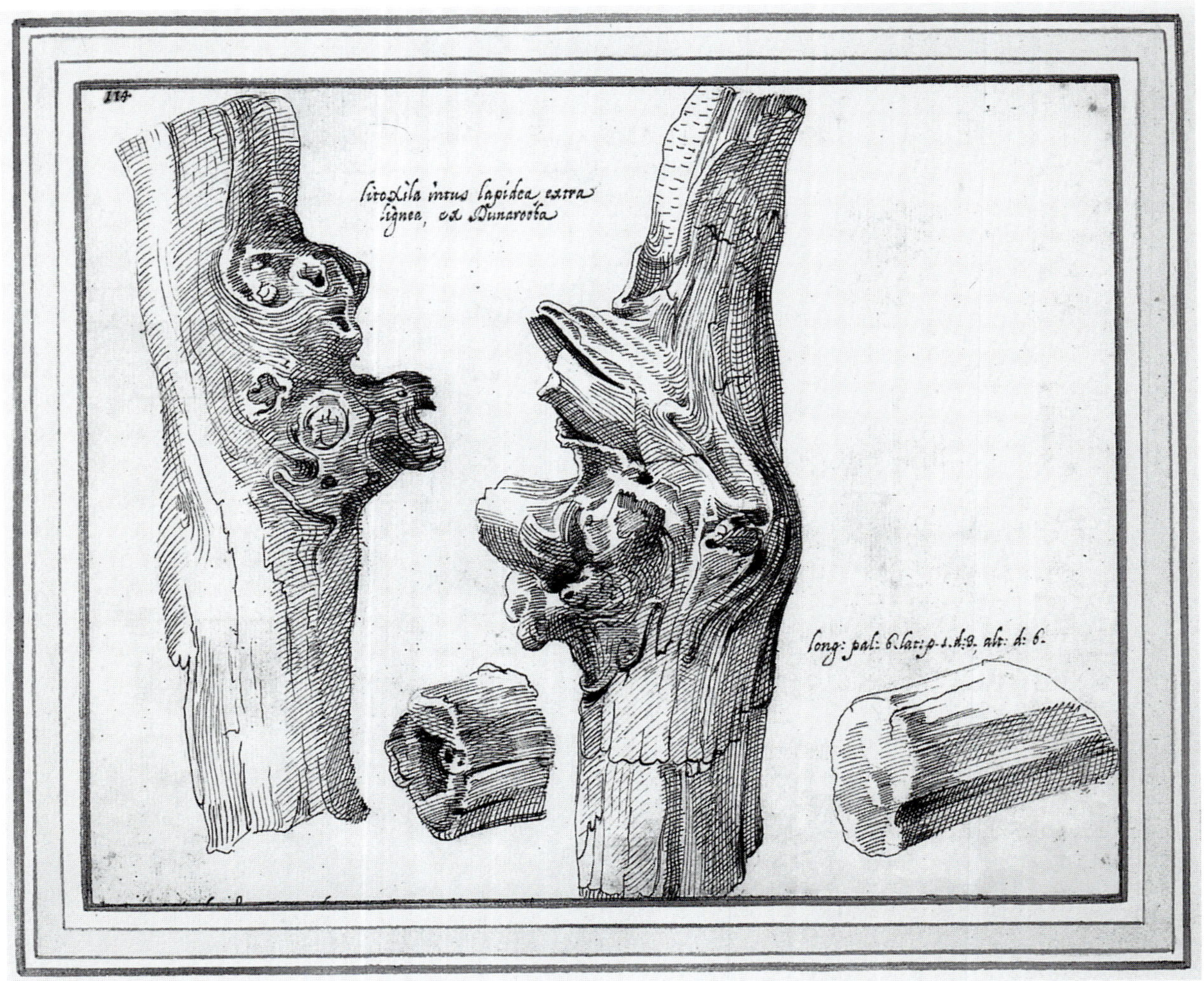

20. Two views of a piece of fossil wood from Dunarobba, with branch clusters, and two end views

Windsor, RL 25653, *Fossils*, vol. XVI, fol. 14

Pen and ink, over black chalk

199 × 249 mm

NUMBERING: [ink, top left corner] *114*

ANNOTATION: [ink, top] *Litoxila intus lapidea extra / lignea, ex Dunarobba*
[bottom] *long: pal: 6. lat: p. 1. d: 3. alt: d: 6.*
[traces of Cesi's inscriptions, bottom right]

The sheet depicts two views of a single piece of fossil wood, together with separate views of either end. The specimen, from Dunarobba, shows a typical branch cluster, possibly from the conifer *Glyptostrobus*, common at this locality (see above, p. 87). The description of 'stone wood, the interior stony and the outside woody' indicates that it must have been partially petrified.

REFERENCES: 17th-cent. Index, p. 367

OBJECT DRAWN: from Dunarobba

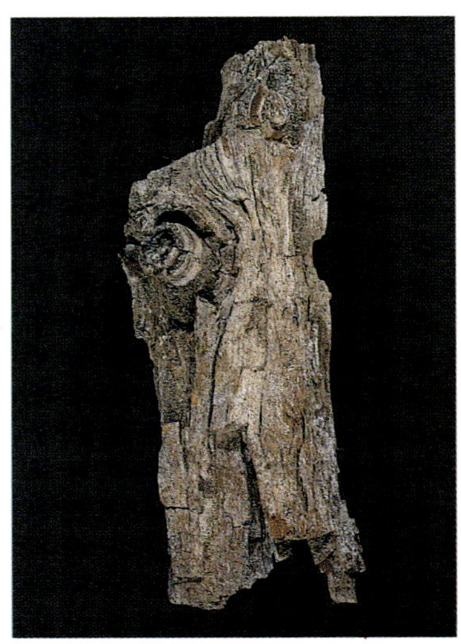

Comp. fig. 20. Branch with branch scars, Dunarobba

21. Two pieces of fossil wood, right-hand specimen with branches

Windsor, RL 25634, *Fossils*, vol. XV, fol. 43

Pen and ink, over black chalk; retouched in ink on mount sheet;
measurements given of the left-hand specimen only; sheet marked with a cross

204 × 257 mm; small holes caused by ink; stained at edge

NUMBERING: [ink, top right corner] *95*

ANNOTATION: [ink, lower centre] *lignum corticosum suberaceum cris-/ pum levius*
[bottom left] *long: pal. 1. lat: p. 1. d: 2 alt: d. 2½*
[centre right] *Ramus Ramulos / Tubercolose prominen*

Two pieces of fossil wood. The left-hand specimen is described as 'a lighter piece of cork-like barky and curly wood'. The right-hand specimen, with its prominent knotty excrescence noted by the inscription, shows well-preserved knotty branches. An almost identical specimen, identified as the conifer *Glyptostrobus*, was found at Dunarobba on a field trip in 1995 (Comp. fig. 21).

REFERENCES: 17th-cent. Index, p. 366

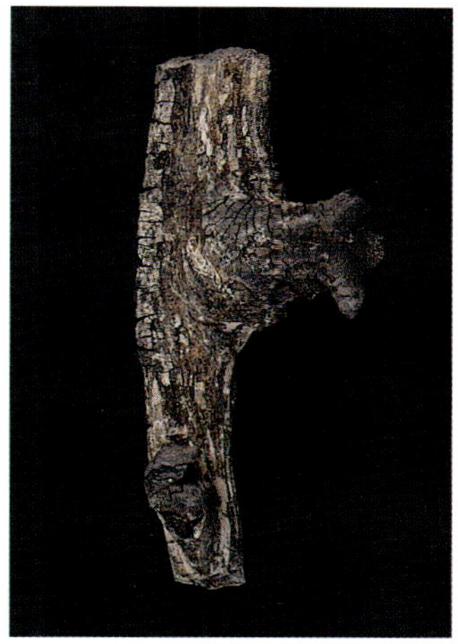

Comp. fig. 21. Fossil wood from Dunarobba

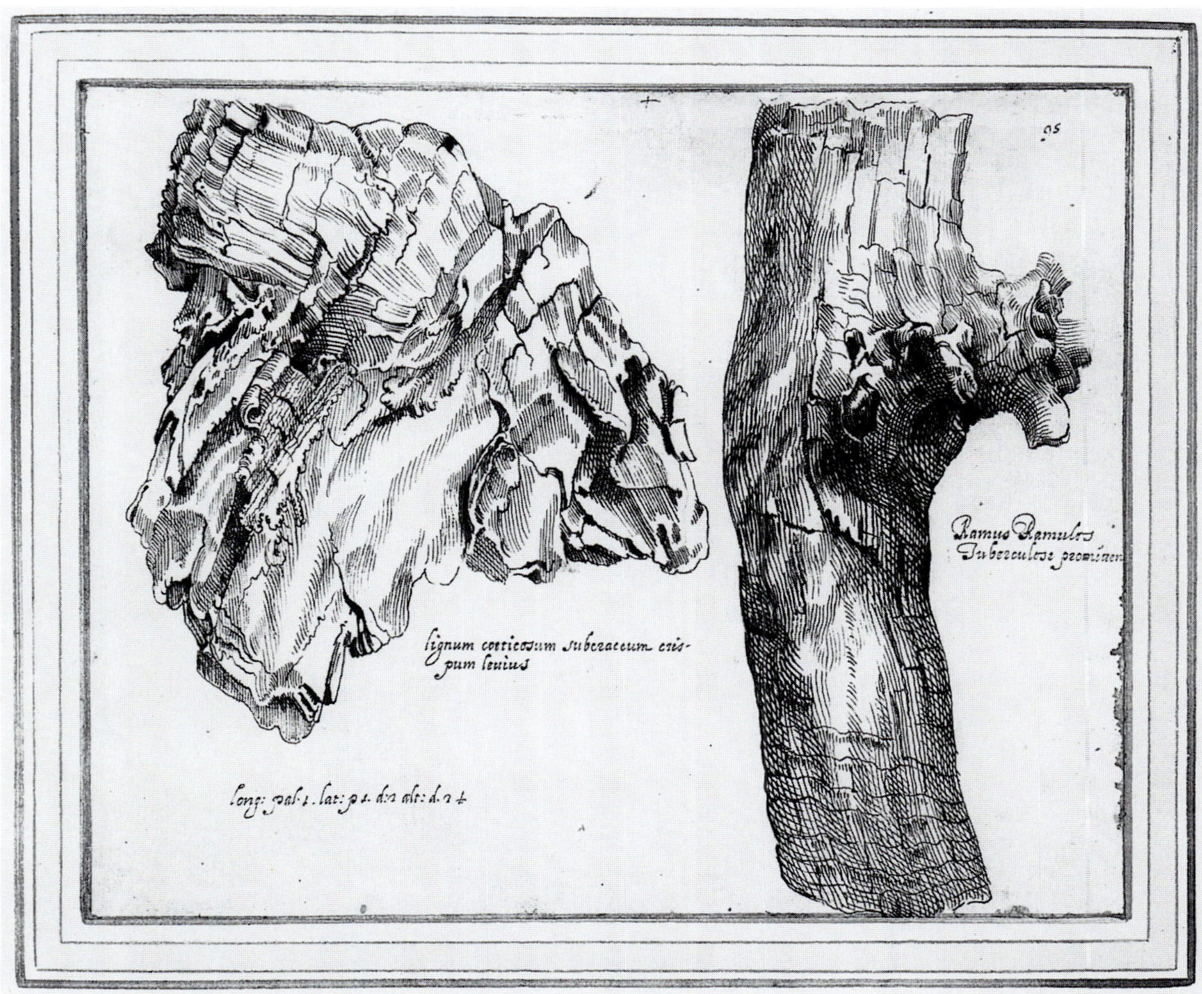

21

22. Two pieces of fossil wood

Windsor, RL 25678, *Fossils*, vol. XVI, fol. 39

Pen and ink, over traces of black chalk; left-hand specimen is marked with a cross

190 × 258 mm

NUMBERING: [ink, top left corner] *139*

ANNOTATION: [ink, top centre] *Ramus Ramulos Tuberculose promittens. B*
[centre] *lignum corticosum suberaceum / multipliciter crispum leve. A*
[bottom left] *long. pal: 1. lat. d. 4 alt. d. 4.*
[bottom right] *long: pal: 1. d: 3. lat: d. 3. alt. d 1½.*
[Cesi's inscriptions, right, cut]

The left-hand specimen is described as 'light, barky, cork-like wood, curled in many parts', while the right-hand specimen is described as a 'tuberous (that is, knotty) branch putting forth branchlets'.

An almost identical view of the right-hand specimen is shown the other way up in **23**.

REFERENCES: 17th-cent. Index, p. 369

OTHER DRAWINGS: **23** (right-hand specimen)

23. Three specimens of fossil wood, one from Rosaro

Windsor, RL 25679, *Fossils*, vol. XVI, fol. 40

Pen and ink, over black chalk; measurements given for left and right-hand specimens only

209 × 268 mm

NUMBERING: [ink, top left corner] *140*

ANNOTATION: [ink, top centre] *Ramulus Ramulos Tuber-/ culose prominens*
[top right] *Terri lignum / Tribus nodis ab alia / Parte / long: pal: 6½ /lat. p. 1. d. 4. alt / d: 7.*
[lower left] *Ex Visceribus crassissime / molis ad Rosarium Secta / superficies, transversis rectisq. / multis confertis nodis com / prehensis, ex Ramulis / long: pal: 7. d. 4. lat. pa. 2. alt / d: 2.*
[in Cesi's hand, bottom left] [ex] *vicerib.*[us] *crassissime molis / Rosarium secta superficies /* [tra]*nsversis rectisq multis con /* [fe]*rtis nodis comprehensis / ex ramuli*
[also fragments top and lower-right edge]

The left-hand specimen is a piece of wood from the centre of a trunk from Rosaro, with many small knots or branch scars. According to the inscription, the specimen was cut from the trunk perhaps to demonstrate the 'knots from the little branches'. It may have been from such a specimen that the cross-sections illustrated in Stelluti's Plate 9 were cut.

The central specimen with the prominent branchlets is the same as that in **22.**

The right-hand specimen is a piece of fossil wood with—as the inscription indicates—three knots on the other side (that is, the side not represented in the drawing).

LITERATURE: Cassiano cat. 1993, no. 149

REFERENCES: 17th-cent. Index, p. 369

OTHER DRAWINGS: **22** (centre specimen)

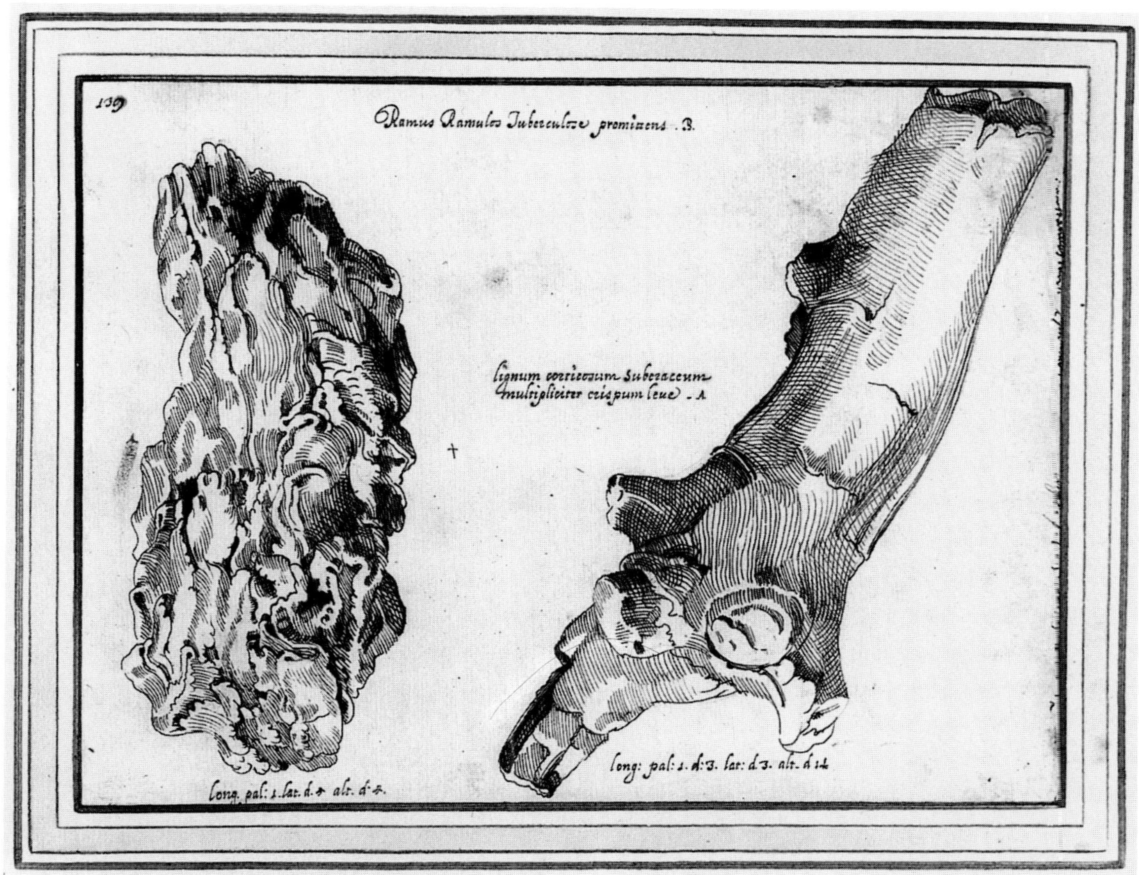
139
Ramus Ramulos Tuberculose promitens . B.
lignum corticosum Suberaceum
multipliciter crispum leue . A

22

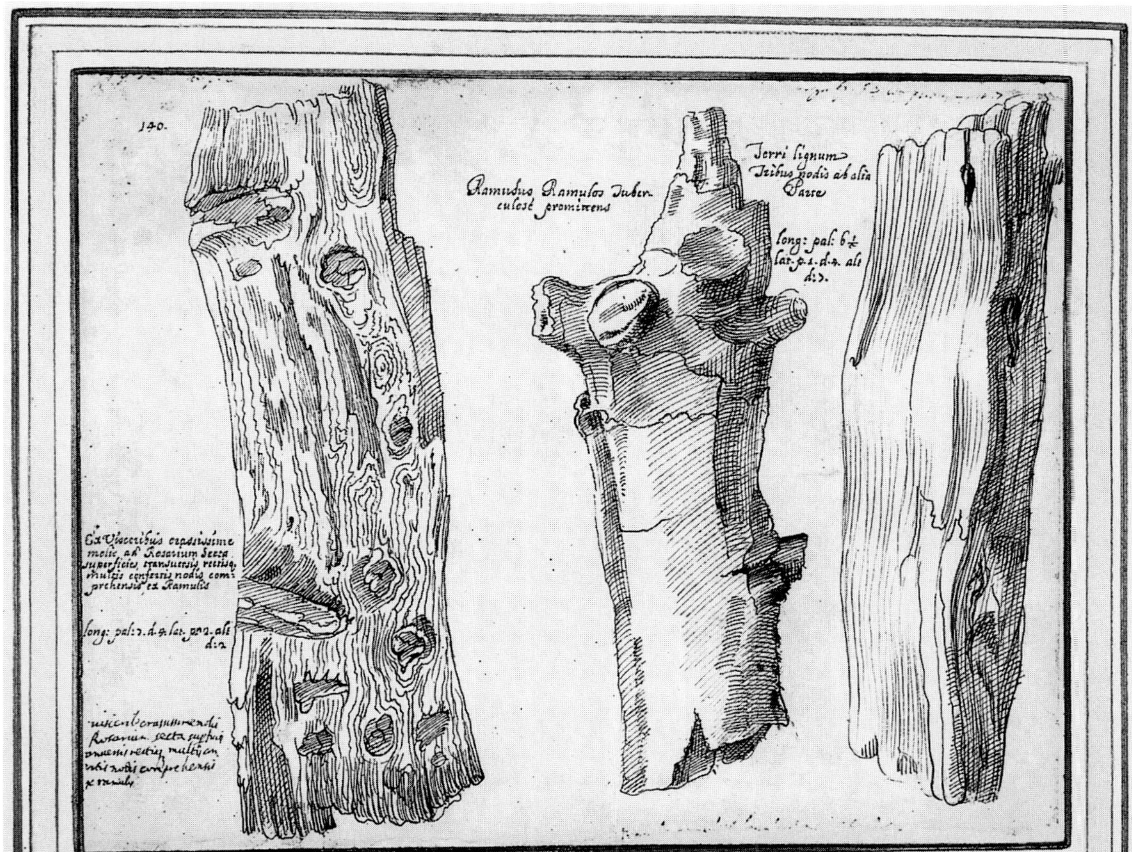
140.
Ramulus Ramulos Tuber
culose promitens
Terri lignum
Tribus nodis ab alia
Parte

23

24. Three pieces of fossil wood from Dunarobba

Windsor, RL 25654, *Fossils*, vol. XVI, fol. 15

Pen and ink, over black chalk; pen-and-ink drawing of a cross-section, possibly of one of the three pieces, is visible on the verso

194 × 256 mm

NUMBERING: [ink, top left corner] *115*

ANNOTATION: [ink, top left] *litoxilum ex Duna-/Robba*
[lower left] *long: pal: 4: d 4. lat / p: 2. alt. pal: 1.*
[centre] *Frustum ligneum nodo in / medio extuberante / long: pal. 5. lat: p 1½ /d: 4. alt: d: 5.*
[top right] *litoxilum ex / Dunarobba*
[right] *long. pal. 3. d. 4 / lat: pal. 1: alt: d. 4*

Three specimens of fossil wood from Dunarobba. The left- and right-hand specimens are described as *litoxilum* ('stone wood') and may therefore be petrified. The centre specimen is described as being a 'woody piece, having a knot projecting from the middle'.

REFERENCES: 17th-cent. Index, p. 367

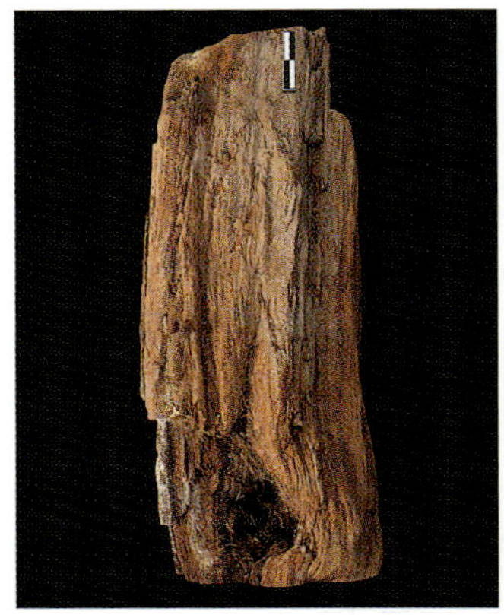

Comp. fig. 24. Permineralized wood from Dunarobba

25. Two pieces of petrified fossil wood, one from Dunarobba

Windsor, RL 25674, *Fossils*, vol. XVI, fol. 35

Pen and ink, over black chalk

199 × 255 mm; torn, lower edge

NUMBERING: [ink, top right corner] *135*

ANNOTATION: [ink, upper left] *Nigricans extra lapideum, intus lapideum et Varium/ ex Dunarobba*
[lower centre] *long: pal. 2½ lat: pal: 1½, alt: d: x*
[upper right] *Petri lignum extra, et pro maiori / parte lapideum, intus alicubi / ligneum*
[lower right] *long: pal: 5. lat. p 2. alt. pal: 1½.*
[Cesi's inscriptions at lower edge, cut]

The left-hand specimen, from Dunarobba, is described as 'blackish and woody on the outside and stony on the inside', while the right-hand specimen is 'petrified wood, stony on the outside and woody on the inside'.

REFERENCES: 17th-cent. Index, pp. 368, 369

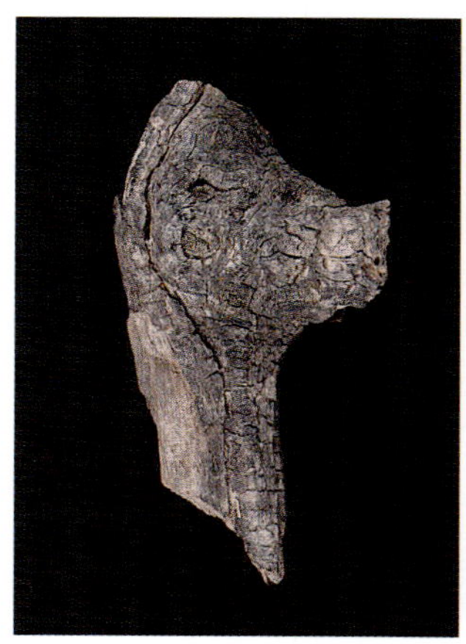

Comp. fig. 25. Wood from Dunarobba

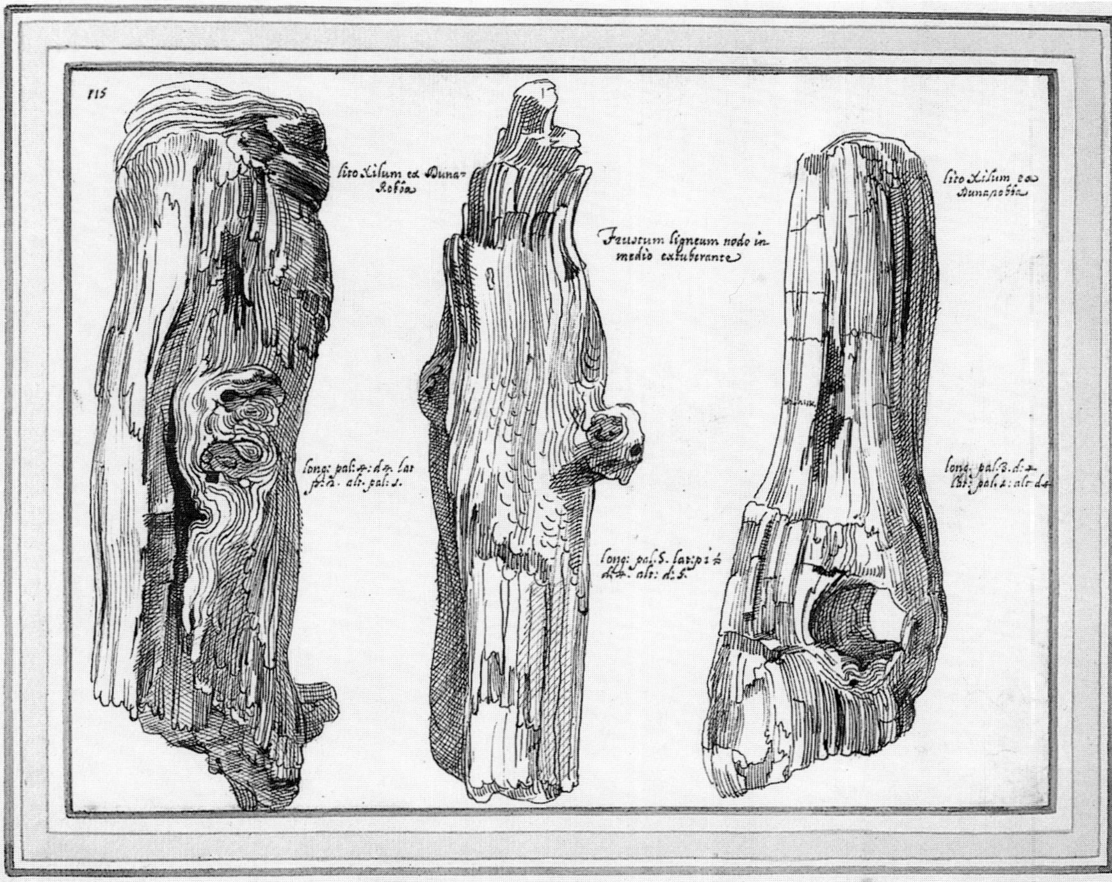
115
lito Xilum ex Duna-Robba
Frustum ligneum nodo in medio exuberante
lito Xilum ex Dunarobba

24

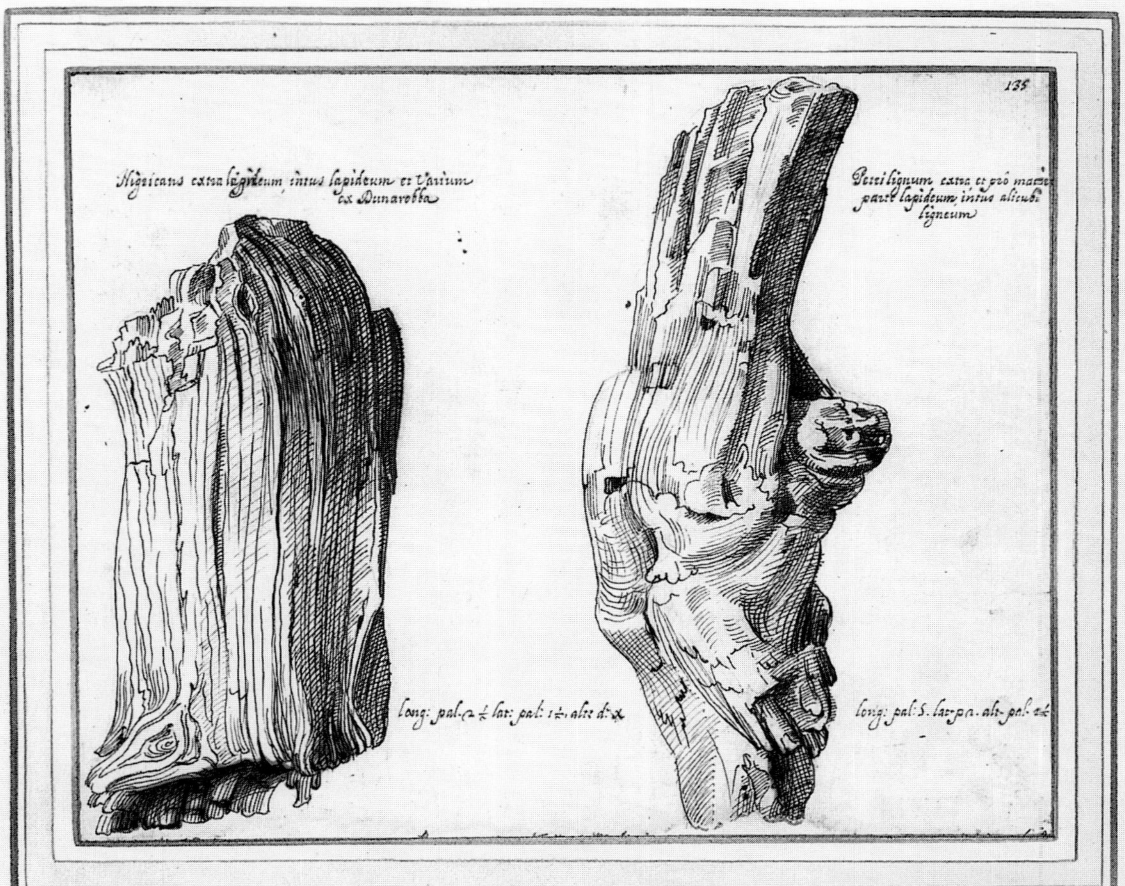
135
Nigricans extra ligneum, intus lapideum et varium ex Dunarobba
ligneum

25

26. Section through a piece of fossil wood, showing growth rings

Windsor, RL 25613, *Fossils*, vol. XV, fol. 22
Pen and ink, over black chalk
186 × 258 mm
NUMBERING: [ink, top left corner] *74*
ANNOTATION: [ink] *lat: palm: 1. alt. palm. med.*

This is a cross-sectional view of a branch or log. Well-defined growth rings are shown. There is no inscription to indicate how Cesi might have descibed or classified it.

REFERENCES: missing from 17th-cent. Index

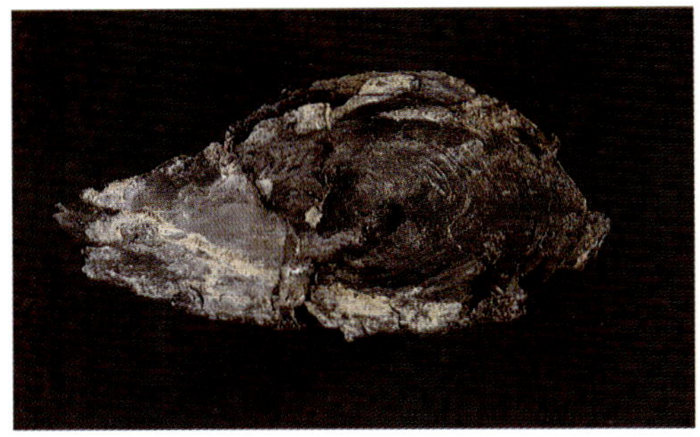

Comp. fig. 26. Black coalified wood showing growth rings, found between Rosaro and Sismano

27. Fossil wood with branches

Windsor, RL 25689, *Fossils*, vol. XVI, fol. 50
Pen and ink, over traces of black chalk
185 × 248 mm
NUMBERING: [ink, top right corner] *150*
ANNOTATION: [ink, top] *Tuberatio lign.*
[bottom] *di Misura nāle*

Fossil wood with several branches. It is described in the inscription as a *tuberatio lign*[ea], suggesting a woody knot or fibrous excrescence of some further unspecified kind. It is drawn actual size.

REFERENCES: 17th-cent. Index, p. 370

Comp. fig. 27. Fossil wood from Dunarobba, showing concentration of knots

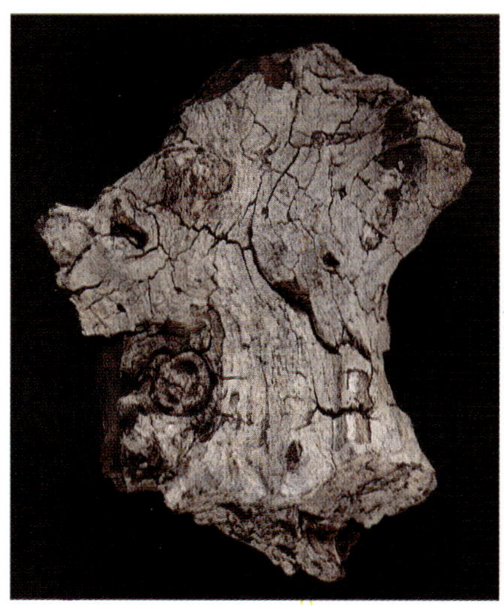

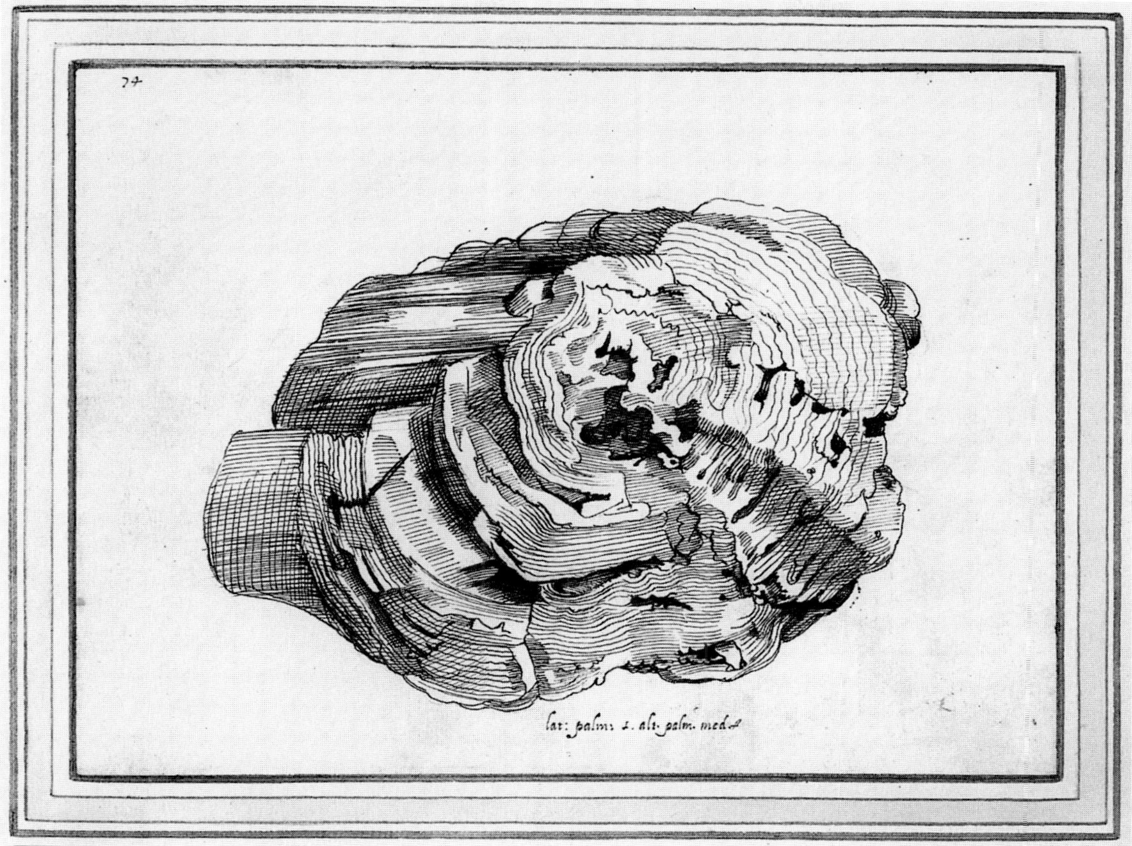

26

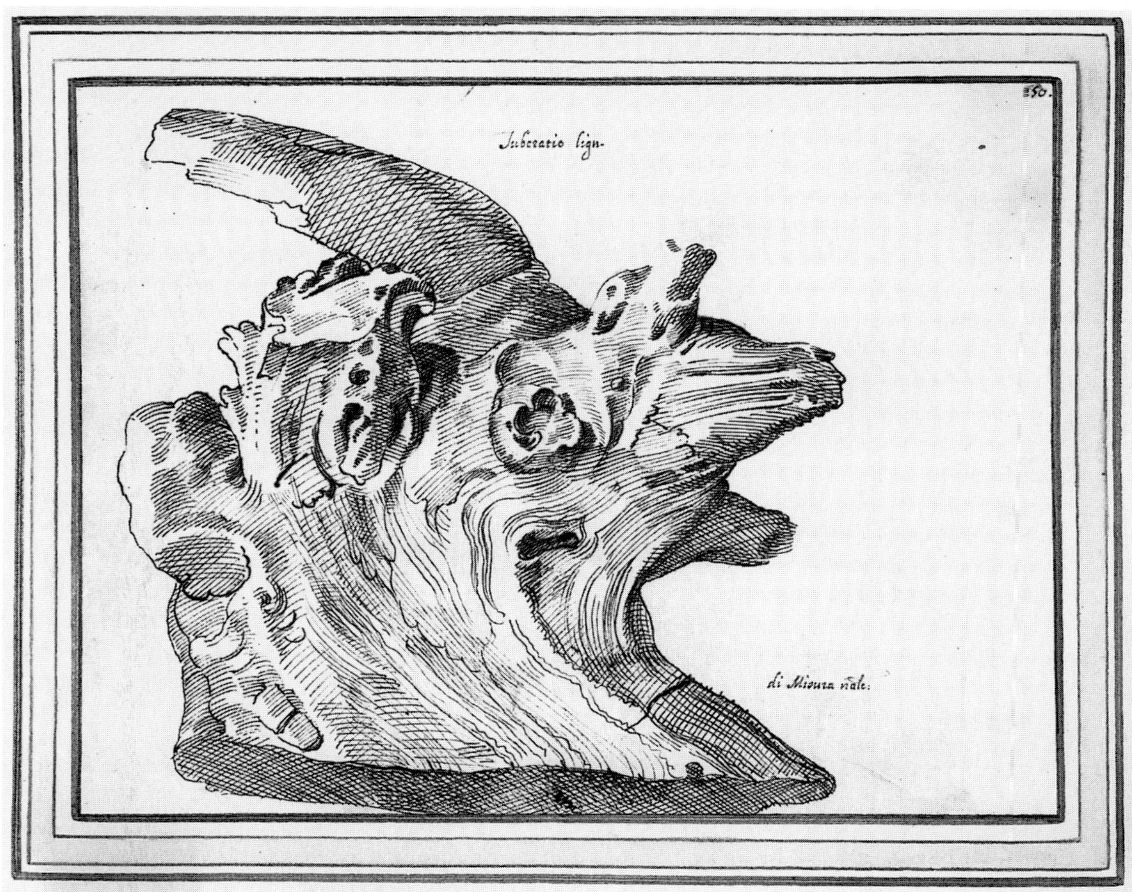

27

28. Fossil wood

Windsor, RL 25617, *Fossils*, vol. XV, fol. 26

Pen and ink, over black chalk

193 × 255 mm

NUMBERING: [ink, upper right corner] *78*

ANNOTATION: [ink, top left] *lapideum veste lignea nigra* [bottom centre] *long: palm. 2½ lat: p. 2: alt: p. 1*

This specimen of fossil wood is described as being 'stony, with a black woody covering' and may, therefore, be partly petrified.

REFERENCES: 17th-cent. Index, p. 365

28

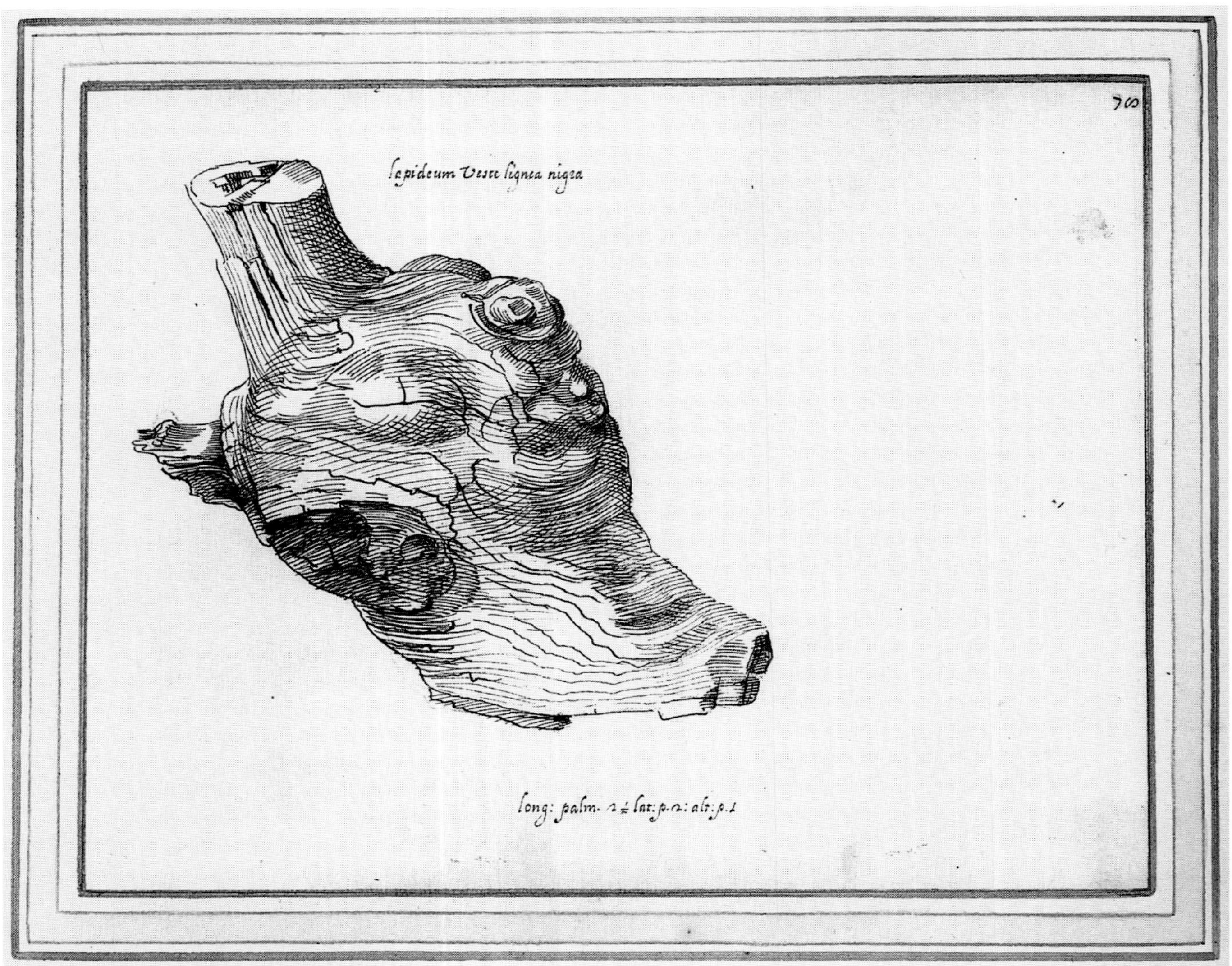

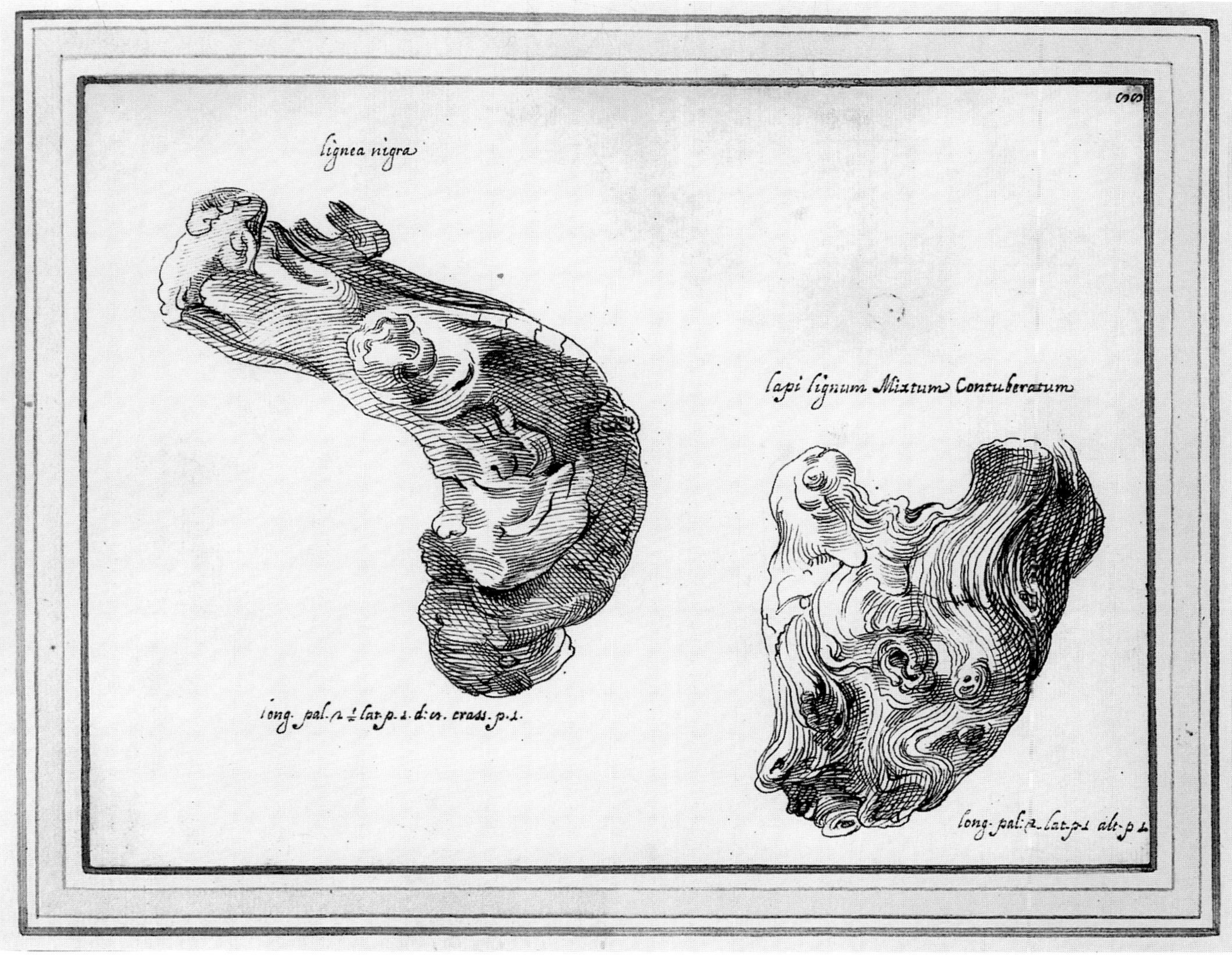

29

29. Two pieces of fossil wood

Windsor, RL 25627, *Fossils*, vol. XV, fol. 36

Pen and ink, over black chalk

190 × 257 mm

NUMBERING: [ink, top right corner] *88*

ANNOTATION: [ink, top left] *lignea nigra*
[lower left] *long. pal. 2½ lat. p. 1. d: 8. crass. p. 1.*
[centre right] *lapi lignum Mixtum Contuberatum*
[bottom-right corner] *long. pal: 2. lat. p. 1 alt. p. 1*

The left-hand specimen of wood is described as being black and woody, while the right-hand one is described as 'stone wood with knots'.

REFERENCES: 17th-cent. Index, p. 365

30. Piece of fossil wood

Windsor, RL 25656, *Fossils*, vol. XVI, fol. 17

Pen and ink, over traces of black chalk

193 × 246 mm. *Watermark*: Bird 41 B

NUMBERING: [ink, top left] *117* over *119*

ANNOTATION: [ink, top] *Metallophyti lignei squama pressa et volutata reflexo simplici anno/ 1611 a nobis invent et Romam allat.*
[bottom] *Mag:o Nālis. crass. dig:* ⅓

A piece of contorted, possibly pyritized wood with a knot scar; according to the inscription, it was found by Cesi in 1611 and taken to Rome. The annotation also states that the specimen is drawn actual size and is a third of a finger width (*c.* 7 mm) thick. The arrangement on the page would suggest that another view of the same specimen was to have been added to the sheet.

REFERENCES: 17th-cent. Index, p. 367

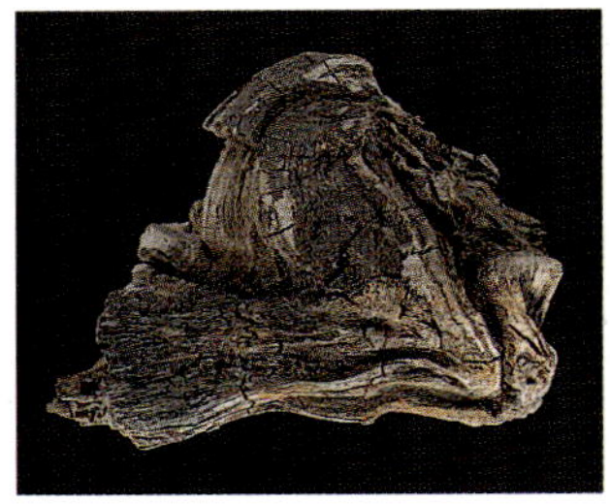

Comp. fig. 30. Contorted wood from Dunarobba

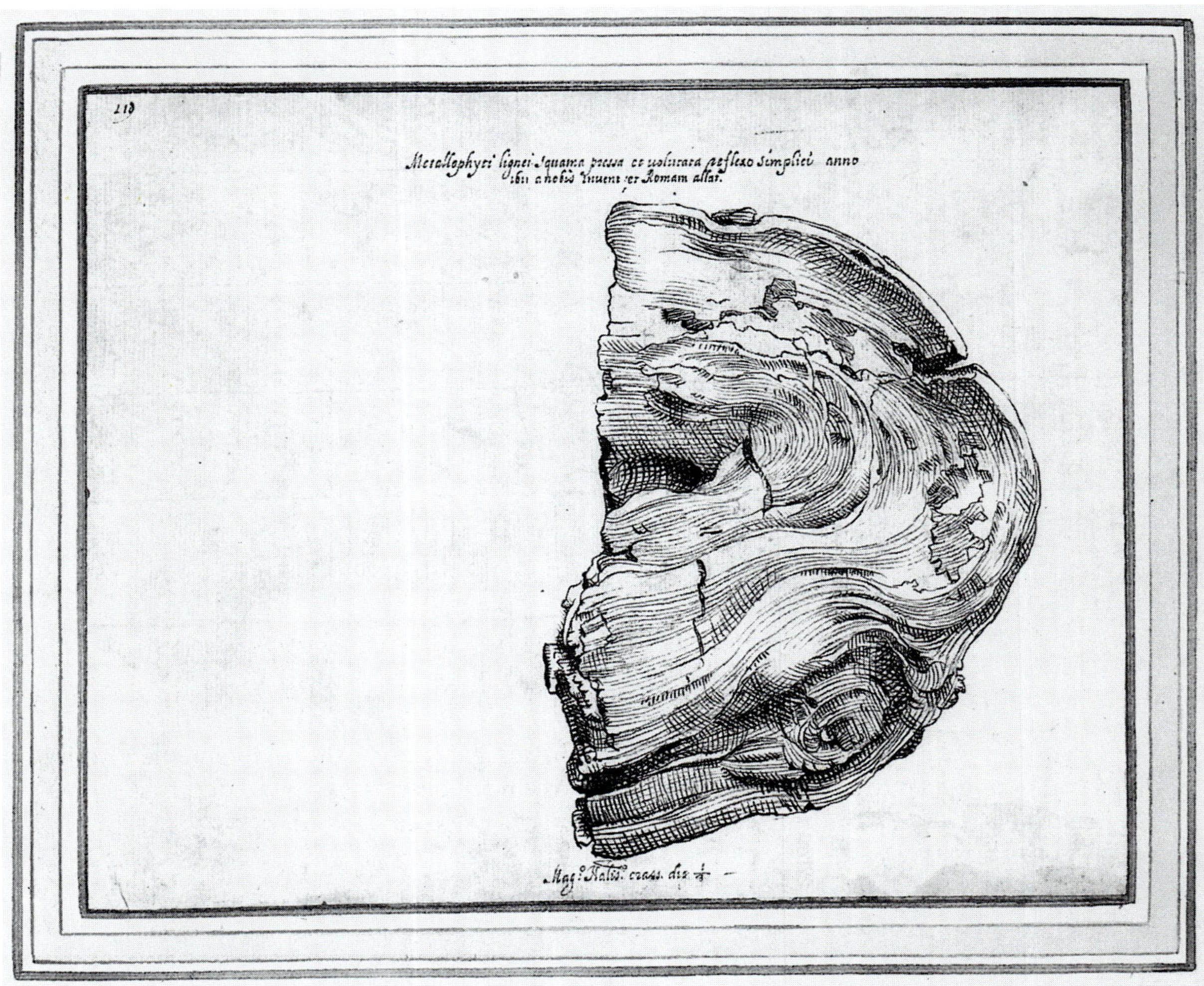

30

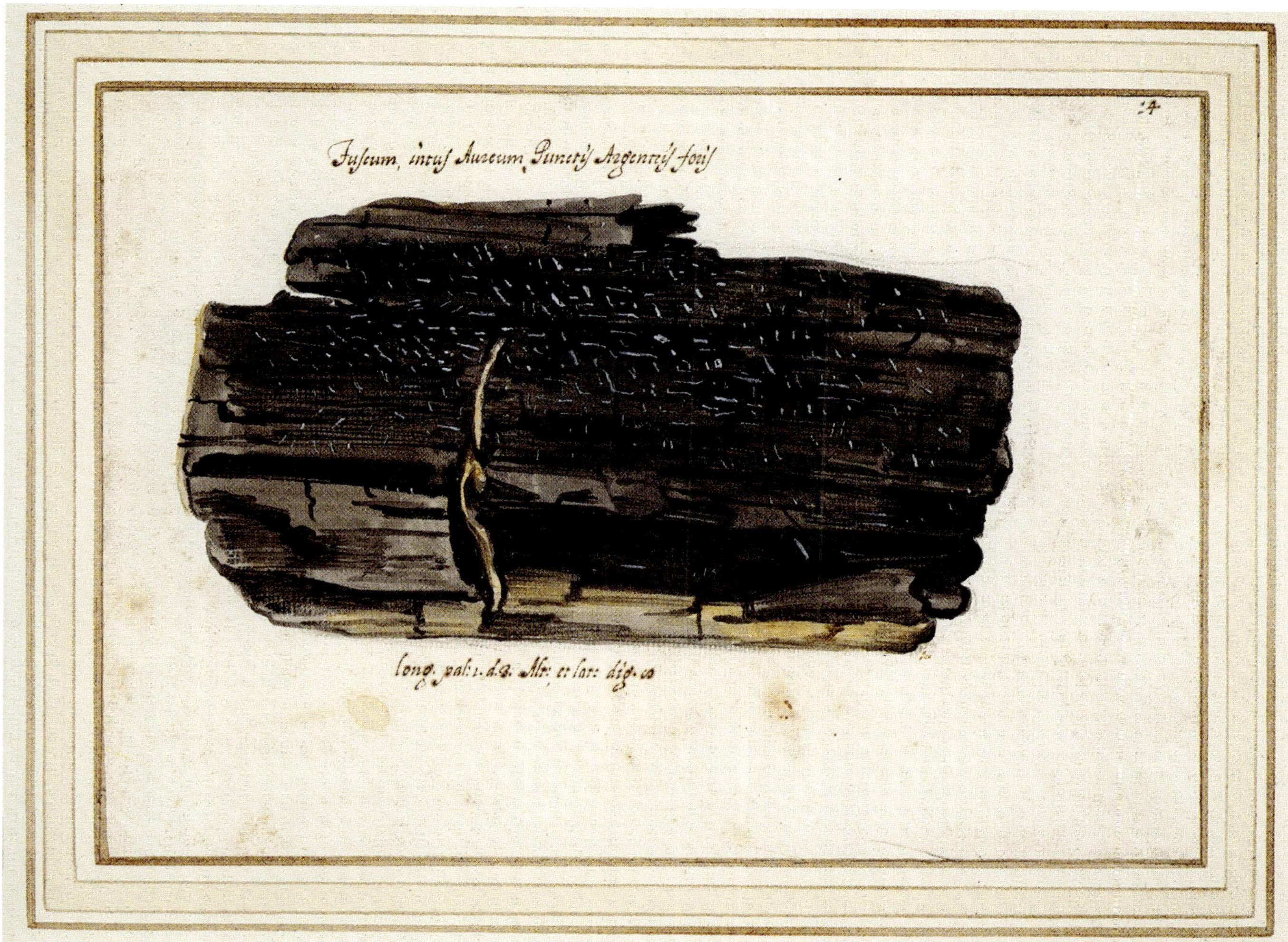

31

31. Fossil wood

Windsor, RL 25711, *Fossils*, vol. XVII, fol. 17

Watercolour and bodycolour and pen and ink, over traces of black chalk; retouched in ink on mount sheet

180 × 259 mm; small holes caused by ink

NUMBERING: [ink, top right corner] *15*

ANNOTATION: [ink, top] *Fuscum in Medio lateritium*
[bottom] *long: pal: 2. lat. pal: 1. d 2. alt: pal: 4.*

A piece of fossil wood infiltrated by various iron minerals concentrated in the centre of the specimen. As is noted in the inscription, it is a dark fragment, baked in the middle (hence *lateritium*). Alternatively, the inscription may have been intended to describe the whole fragment as a *lateritium*, but darker in the centre.

REFERENCES: 17th-cent. Index, p. 371

32. End view of broken fossil branch

Windsor, RL 25667, *Fossils*, vol. XVI, fol. 28

Pen and ink, over traces of black chalk; retouched in ink on mount sheet

177 × 240 mm; small holes caused by ink; ink frame line

NUMBERING: [ink, top right] *128*

ANNOTATION: [ink, top] *Nodos: lign: medio Terros.* [bottom] *Long: pal: 1. lat. p. 1. d: 3: alt: d: 8.*

The specimen shows the end of a branch of knotty fossil wood with a typical radiating pattern caused by shrinkage. The annotation describes the centre as being 'earthy'.

REFERENCES: 17th-cent. Index, p. 368

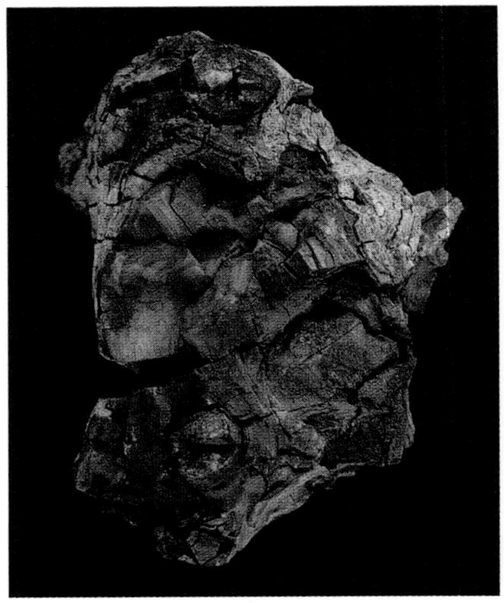

Comp. fig. 32. Fossil wood from Dunarobba showing broken end of branch

33. End view of a broken fossil branch

Windsor, RL 25664, *Fossils*, vol. XVI, fol. 25

Pen and ink, over traces of black chalk; retouched in ink on mount sheet

180 × 251 mm; small holes caused by ink; ink frame line

NUMBERING: [ink, top right corner] *125*

ANNOTATION: [ink, top] *Nodosa ligna* [bottom] *Palm. 1. per ogni verso —*

A similar specimen to that shown in **32**. It is described as measuring 'one palm for each side'.

REFERENCES: 17th-cent. Index, p. 368

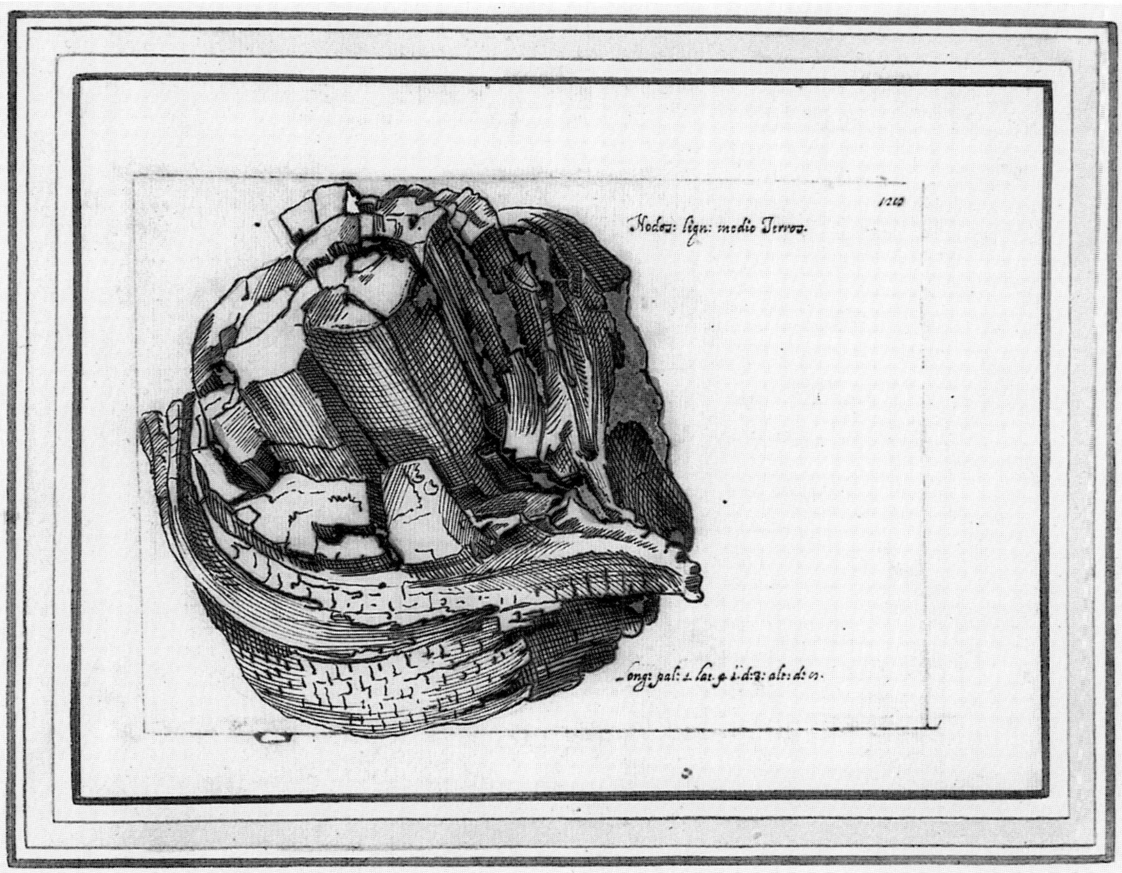
Nodos: lign: medio Terros.

32

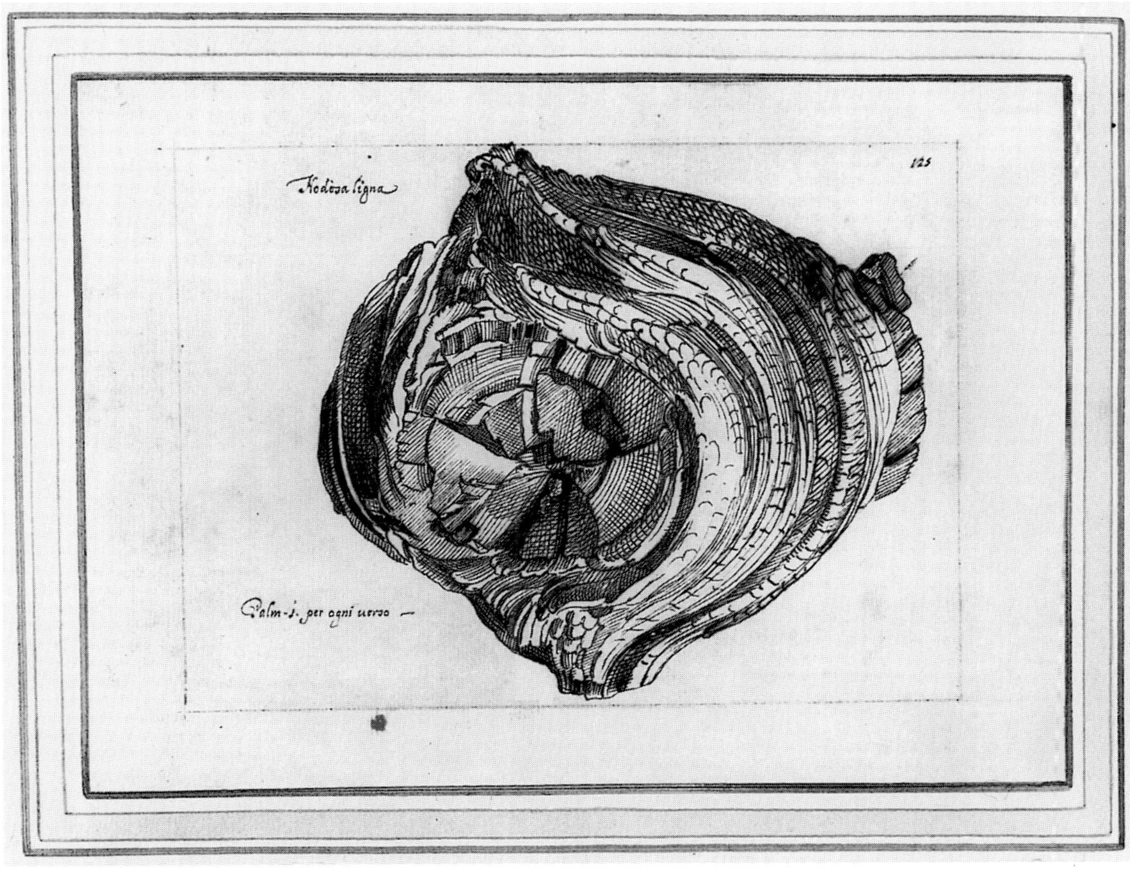
125
Nodosa ligna
Palm. 1. per ogni uerso

33

34. Fossil wood from Ponte Caldaro

Windsor, RL 25578, *Fossils*, vol. XIV, fol. 49

Pen and ink, over traces of black chalk; retouched in ink on mount sheet

196 × 248 mm (trimmed through left edge); small holes caused by ink

NUMBERING: none; formerly *39* (see Appendix I)

ANNOTATION: [ink, top left] *Ex Ponte Cardano*
[lower centre] *Long. pal. 1. dig. 8 lat. d. 7. alt: d. 3.*

No descriptive details are given for this. The specimen is reported to be from Ponte Caldaro. With **35**, **36** and **62**, it is one of a group of specimens, some but not all of which are inscribed as coming from the same location.

REFERENCES: 17th-cent. Index, p. 363

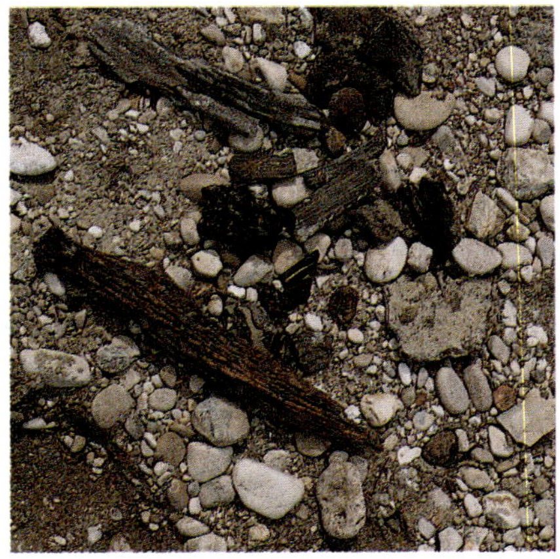

Comp. fig. 34. Range of preservation in fossil woods, found in gravels beneath Ponte Caldaro

34

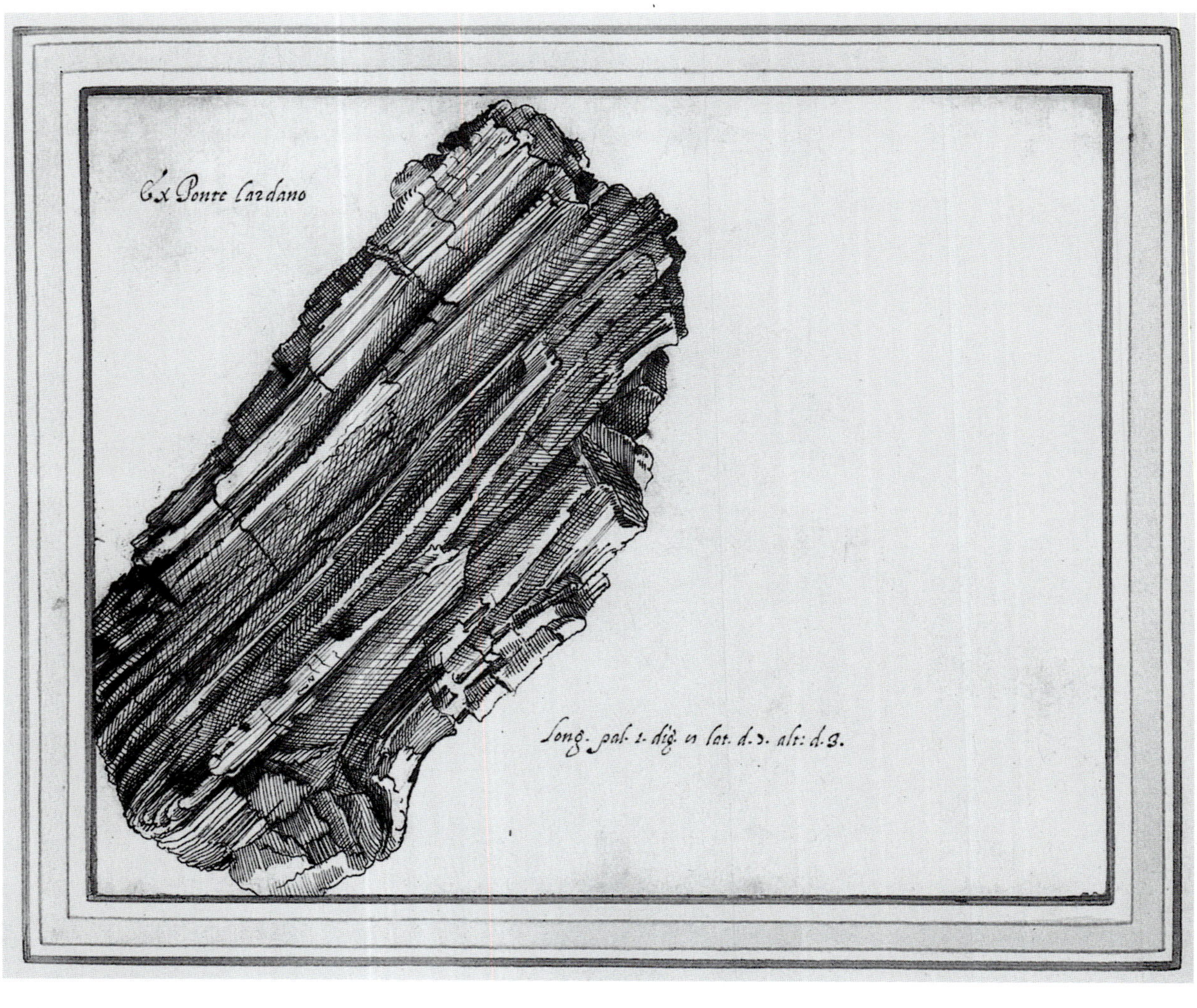

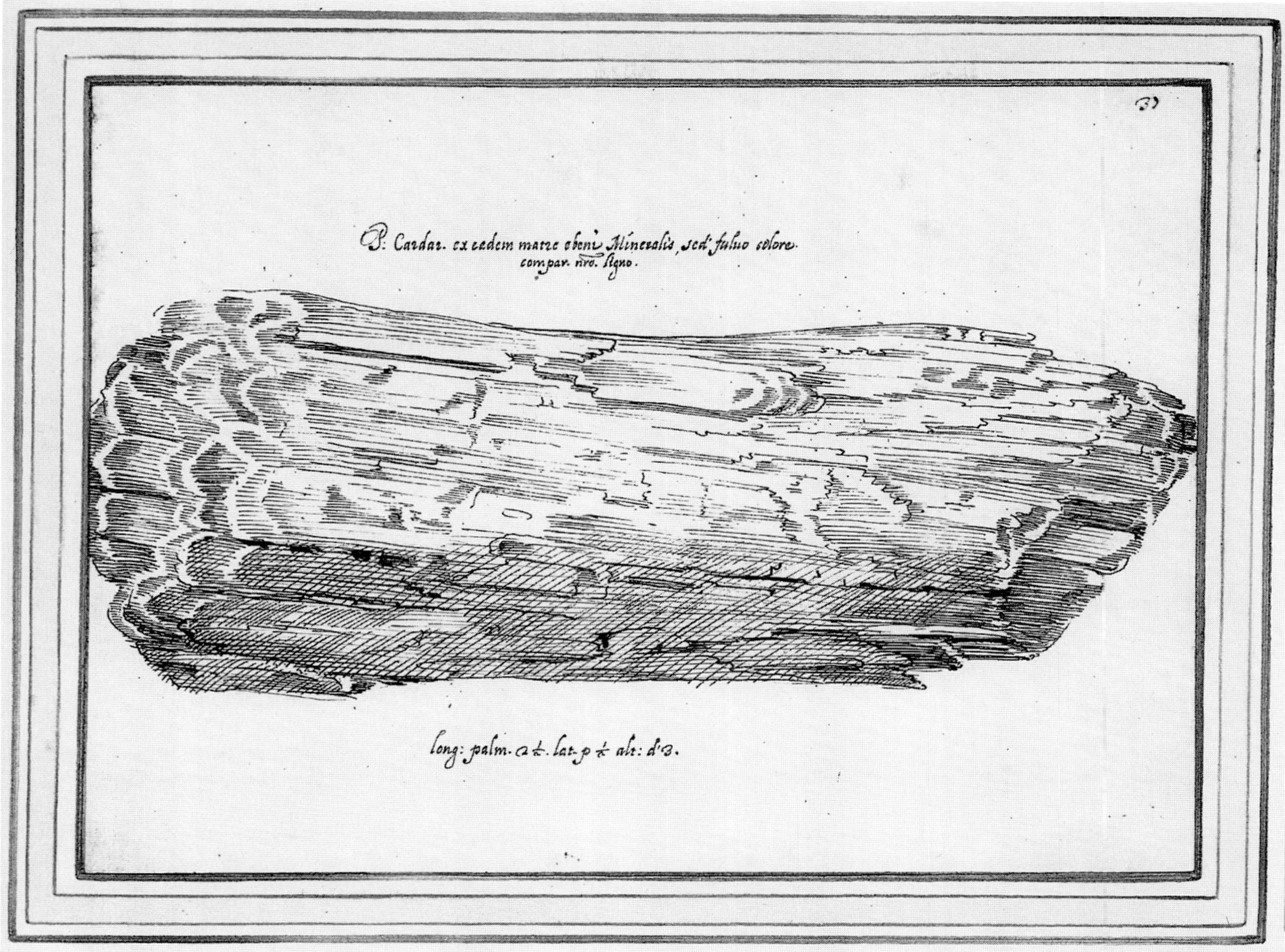

35

35. Fossil wood from Ponte Caldaro

Windsor, RL 25575, *Fossils*, vol. XIV, fol. 46

Pen and ink, over black chalk

185 × 261 mm

NUMBERING: [ink, upper right] *37*

ANNOTATION: [ink, top] *P: Cardar. ex eadem matre ebeni Mineralis, sed' fulvo colore / compar. nr̄o. ligno.*
[bottom] *long: palm. 2½. lat. p ½ alt: d. 3.*

Another piece of fossil wood, of a yellow colour, but also described as being 'mineral ebony'. The light colour may be a result of the petrifaction (hence the specimen is described as mineral) and it was thought that this was a variation on the predominantly black woods also found in Ponte Caldaro.

REFERENCES: 17th-cent. Index, p. 363

36. Piece of fossil wood and cross-section from Ponte Caldaro

Windsor, RL 25576, *Fossils*, vol. XIV, fol. 47

Pen and ink, over black chalk; cross, in pen and ink, above left-hand specimen

186 × 254 mm

NUMBERING: [ink, upper right] *38*

ANNOTATION: [ink, upper right] *Ebenus fossil. ex P. Card: cum Virgulis/ Metallicis interceptis Niger medio / Fulvus—*
[bottom right] *long. pal. 2. lat. p: 1. alt: pal. ½*

The description that this piece of 'fossil ebony' had 'metallic nodules' probably implied that it was partially pyritized, a probability reinforced by the 'yellowish black' colour. Such a combination of colours is typical of decaying pyrite.

The horizontal line across the specimen appears to represent a partial cut. The right-hand drawing is a view of the end of the specimen.

REFERENCES: 17th-cent. Index, p. 363

36

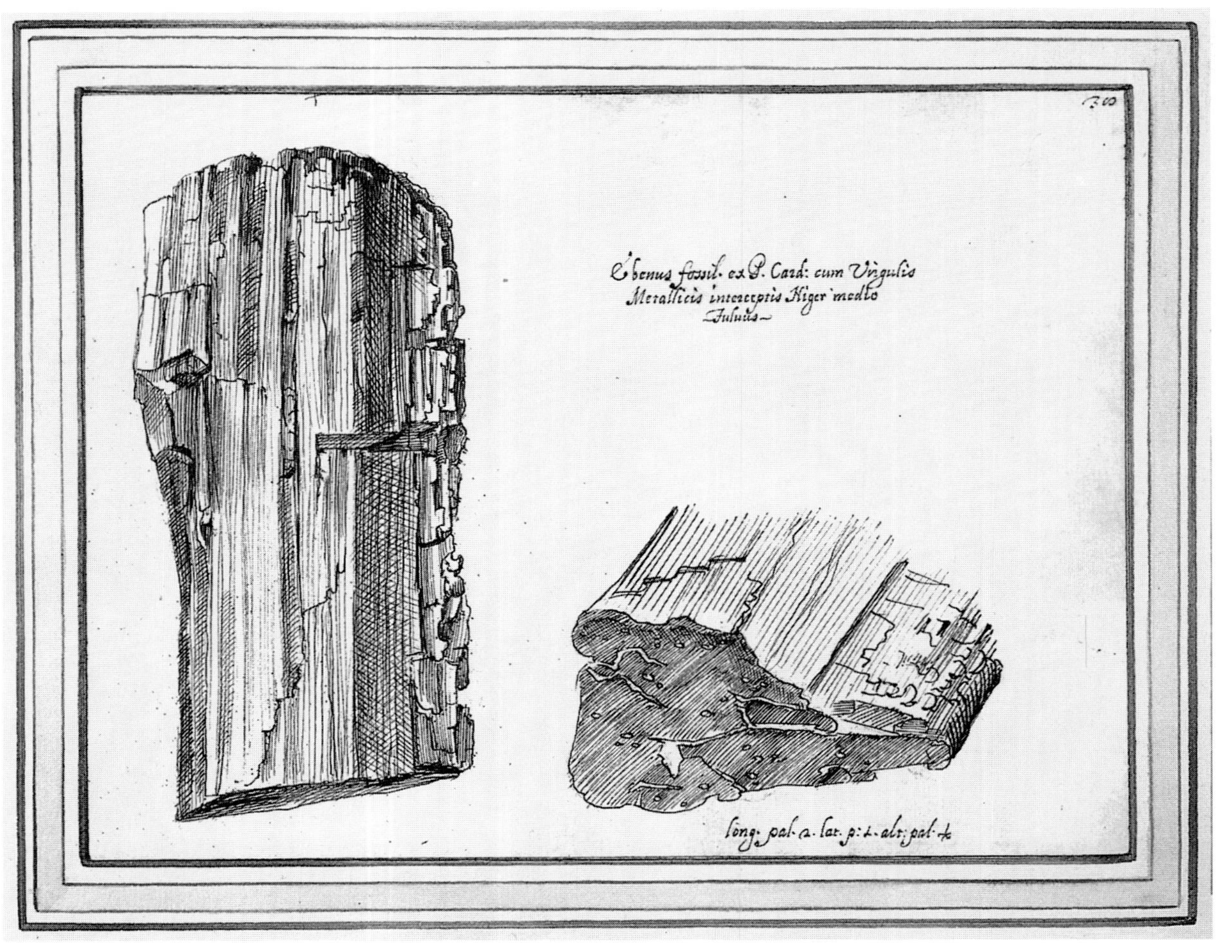

37

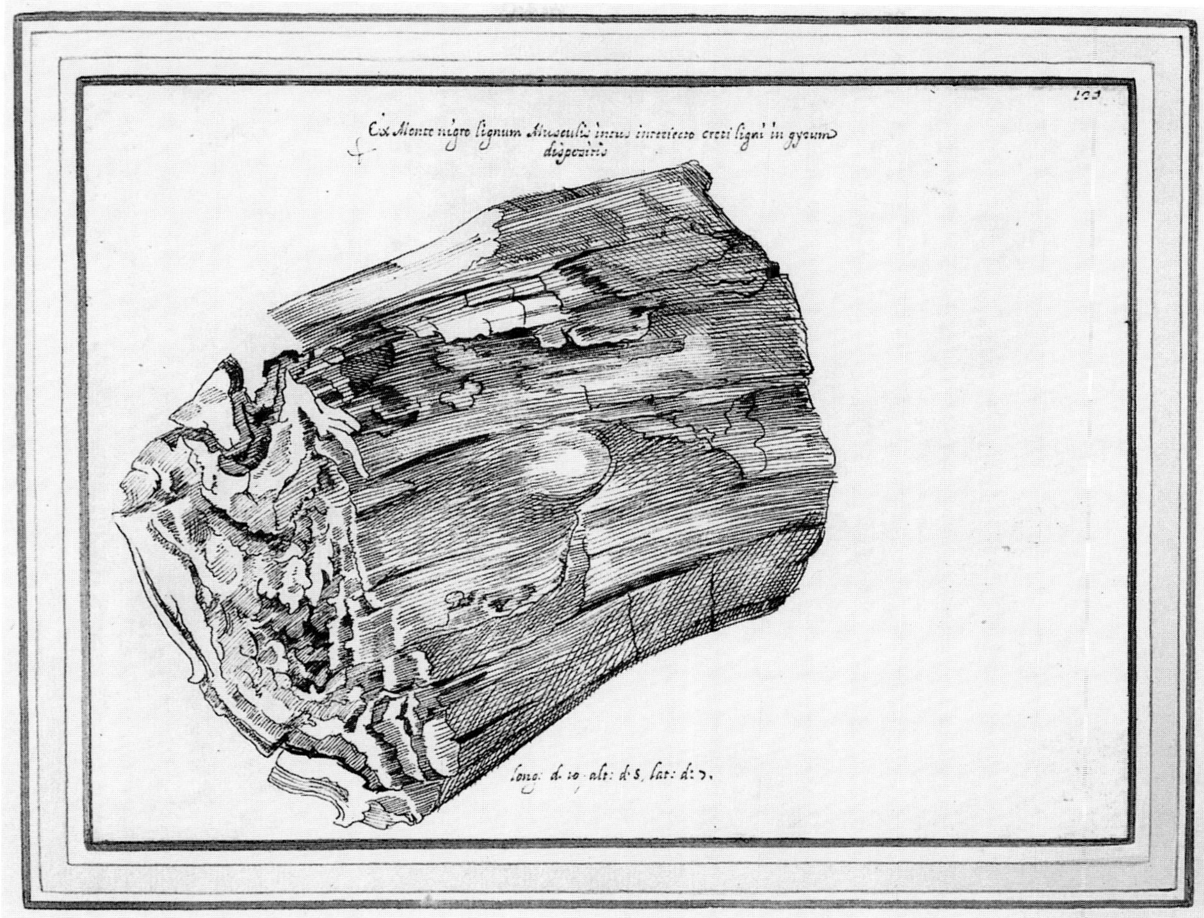

37. Fossil wood in clay from Montenero

Windsor, RL 25659, *Fossils*, vol. XVI, fol. 20

Pen and ink, over traces of black chalk; marked with a cross

185 × 254 mm

NUMBERING: [ink, top right corner] *120*

ANNOTATION: [ink, top] *Ex Monte nigro lignum Musculis intus interiecto creti ligni in gyrum / dispositis*
[bottom] *long: d. 10, alt: d. 5, lat: d: 7.*

A piece of fossil wood with clay interspersed in the cracks. It is said to be 'a piece of fossil wood from Montenero with sinews around it, interspersed with clay in the inside'.

A small area of the drawing appears incomplete.

REFERENCES: 17th-cent. Index, p. 367

38. Two pieces of petrified fossil wood with detail

Windsor, RL 25619, *Fossils*, vol. XV, fol. 28

Pen and ink, over traces of black chalk

194 × 268 mm

NUMBERING: [ink, top left corner] *80*

ANNOTATION: [ink, top] *lapila teri lignum homo geneum sub. ochreus tingens* [centre] *long: palm: 2½ lat: d. 4. alt: d: 2* [bottom] *long: palm. 1. alt: d: 2*

The larger, uppermost piece of fossil wood is described as a mixture of stone and earth (the scribe's ignorance of Latin here makes any more precise translation impossible) being 'stone-like earth-wood with a slight ochre tinge'. This is a typical colour of some of the petrified and permineralized woods from the area. The lower specimen is not described but the tilted end of the specimen is shown.

REFERENCES: 17th-cent. Index, p. 365

OTHER DRAWINGS: **55**

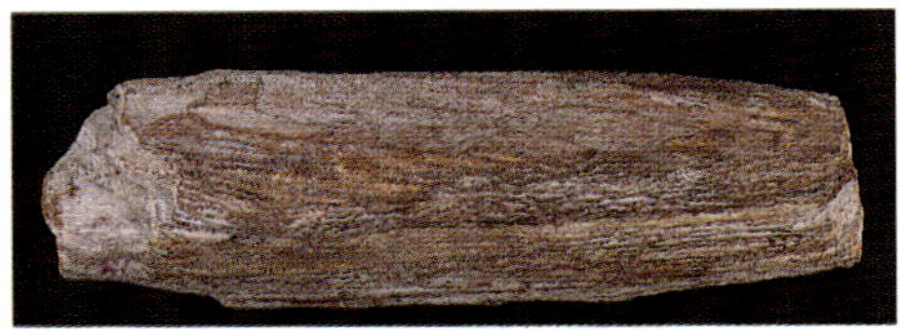

Comp. fig. 38. Permineralized wood from Podere Casanova

38

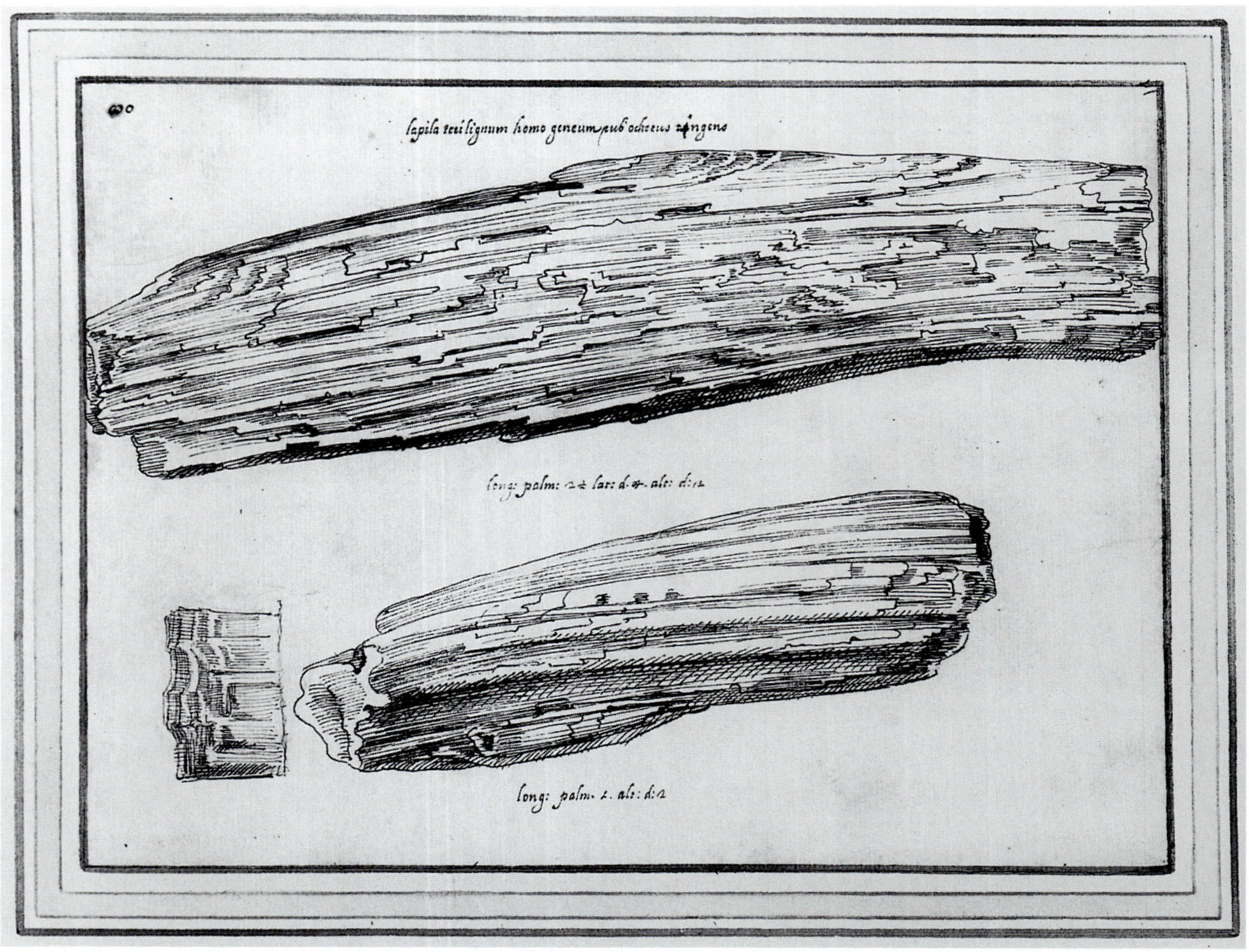

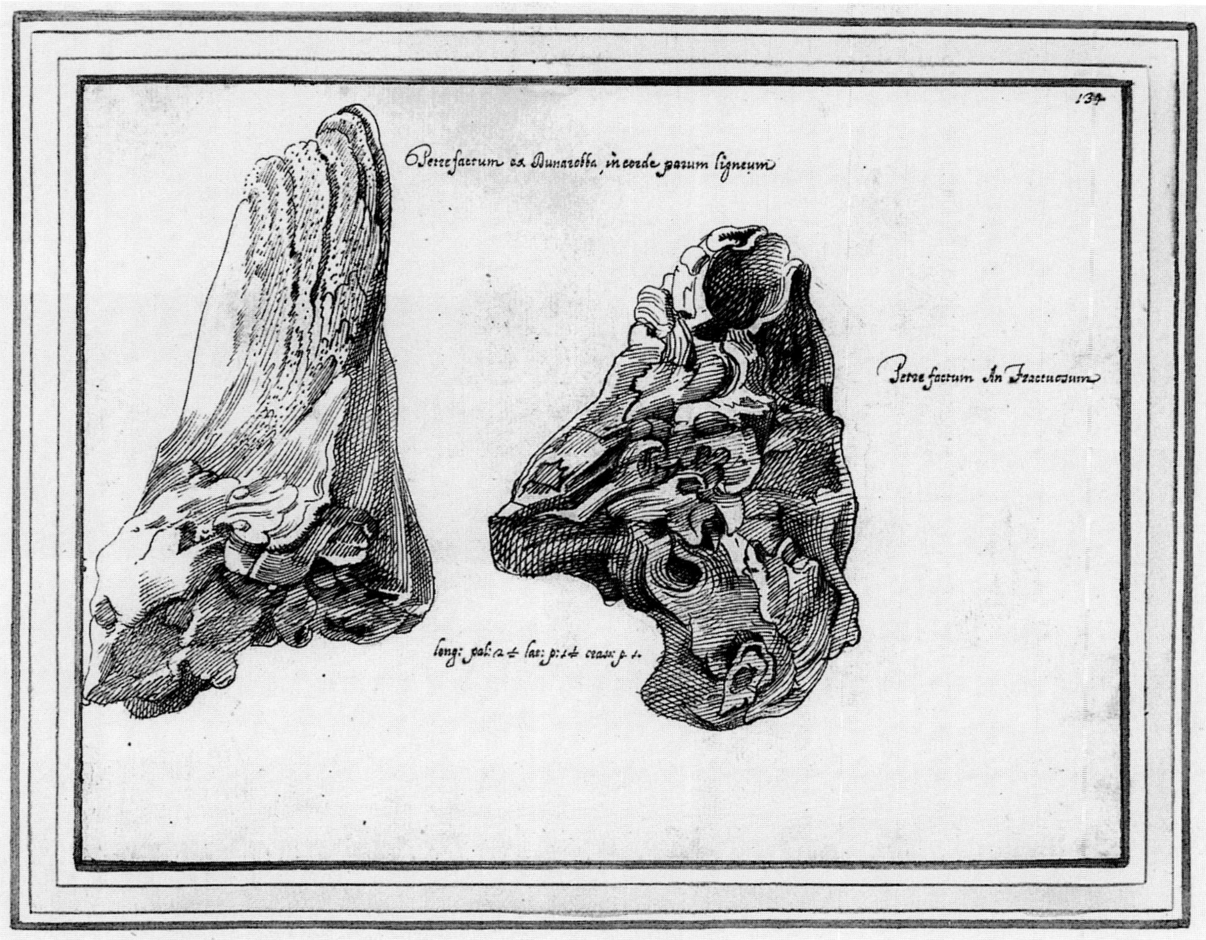

39

39. Two pieces of petrified wood, one from Dunarobba

Windsor, RL 25673, *Fossils*, vol. XVI, fol. 34

Pen and ink, over traces of black chalk; measurements given for right-hand specimen only

191 × 257 mm

NUMBERING: [ink, top right corner] *134*

ANNOTATION: [ink, top] *Petrefactum ex Dunarobba, in corde parum ligneum* [right] *Petrefactum An Fractuosum* [bottom] *long: pal: 2½ lat: p: 1½ crass: p. 1.*

The left-hand specimen is a piece of petrified wood from Dunarobba annotated: 'its centre a little woody'. The right-hand specimen is described as 'a twisted piece of petrified wood'.

REFERENCES: 17th-cent. Index, p. 369

40. Two pieces of fossil wood

Windsor, RL 25615, *Fossils*, vol. XV, fol. 24

Pen and ink, over black chalk

191 × 260 mm

NUMBERING: [ink, upper right corner] *76*

ANNOTATION: [ink, top centre] *lapid: folijs compactum sub rufum*
[top right] *long: pal: 1½. lat: d: 5. alt: d: 3.*
[centre] *ligniterr. Nervosum fulvo fuscum*
[bottom] *long: pal 1½ alt: d: 3. crass: d: 2.*

The upper piece of fossil wood is described as a 'reddish, stony concretion', indicating that the specimen was probably petrified. The lower specimen is called simply a 'dark yellow, fibrous piece of woody earth'.

REFERENCES: 17th-cent. Index, pp. 364, 367

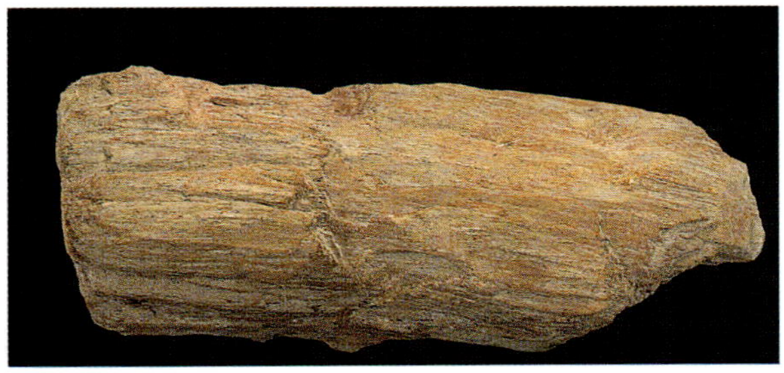

Comp. fig. 40. Permineralized wood from Podere Casanova

40

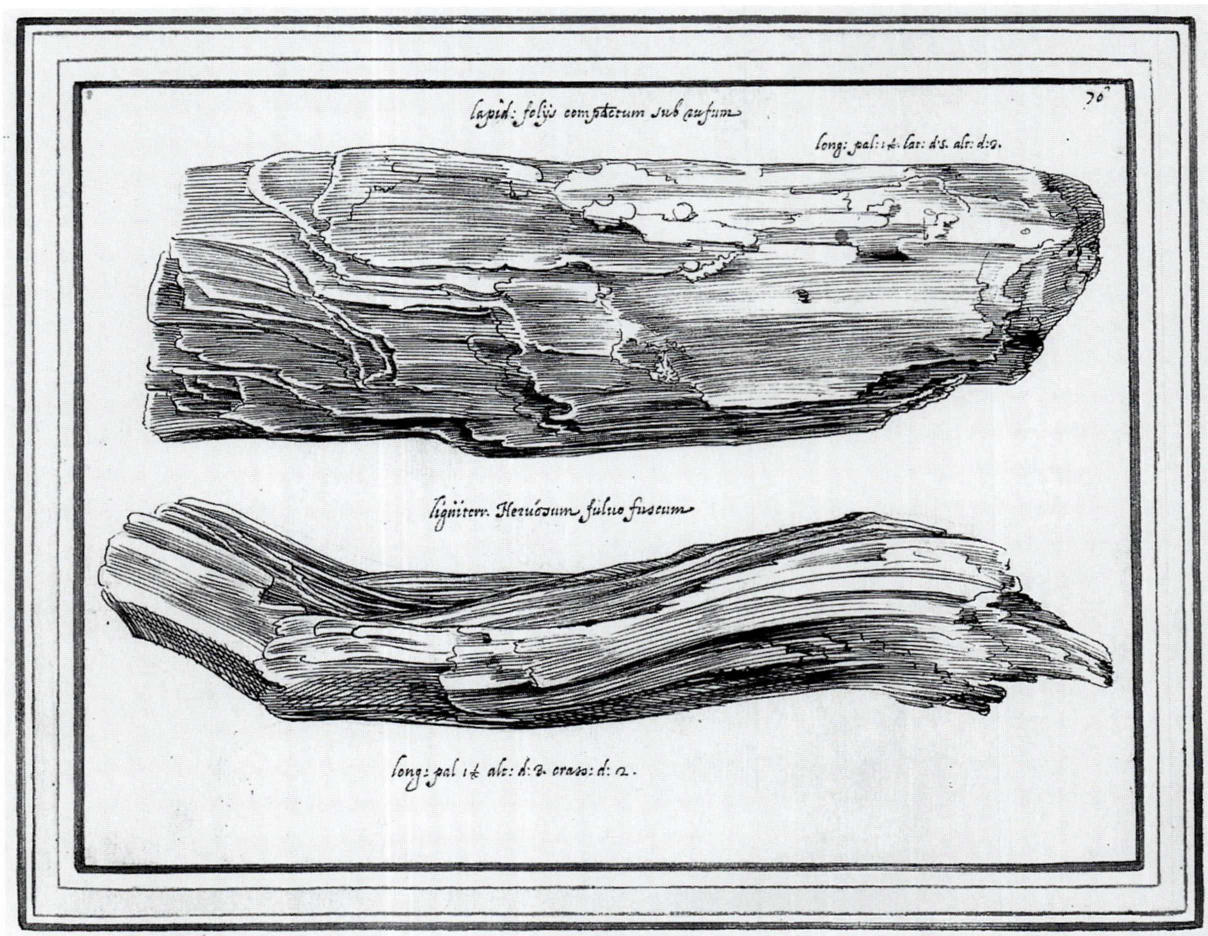

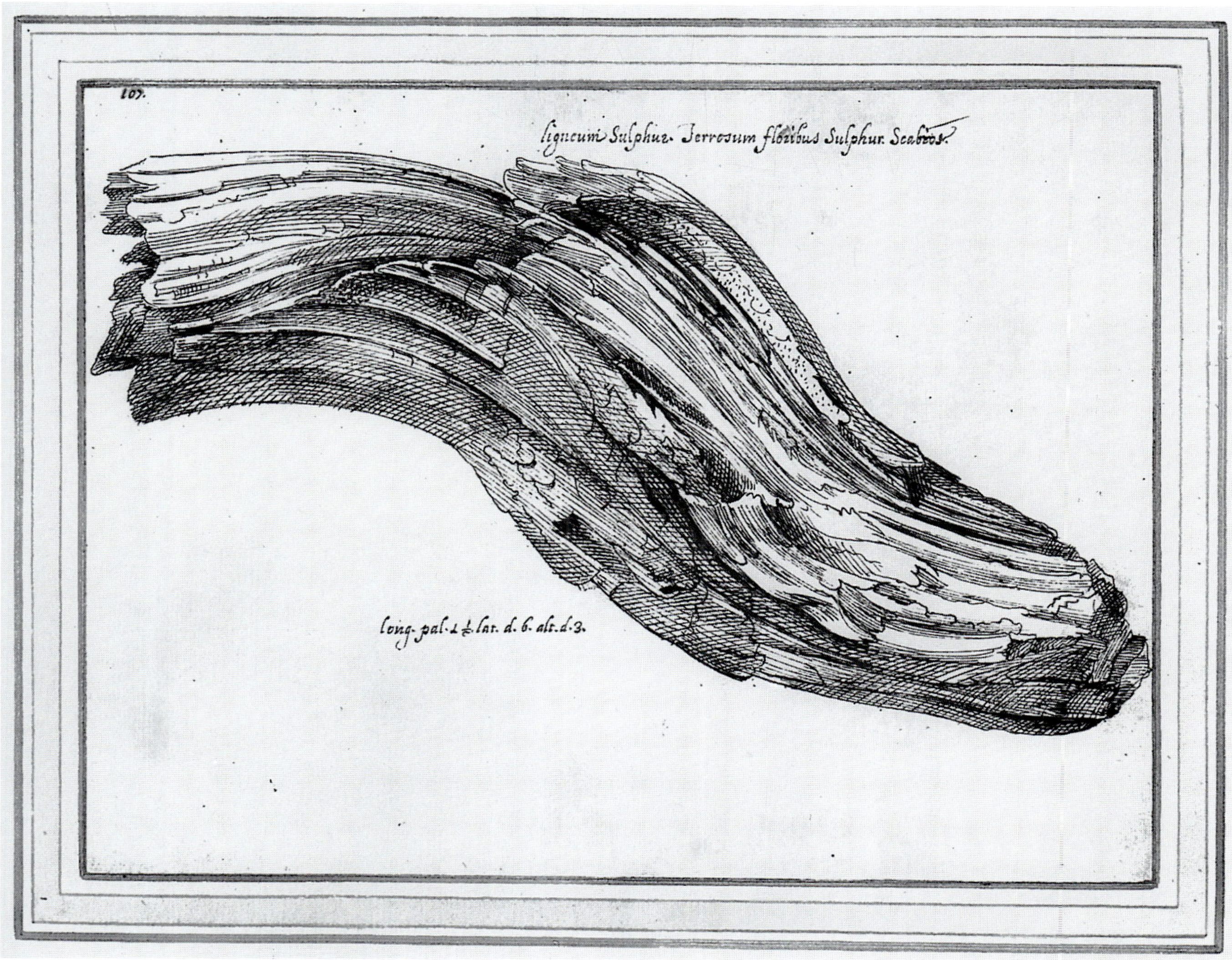

41. Fossil, possibly pyritized wood

Windsor, RL 25646, *Fossils*, vol. XVI, fol. 7

Pen and ink, over traces of black chalk

190 × 256 mm

NUMBERING: [ink, top left corner] *107*

ANNOTATION: [ink, top] *Ligneum Sulphur. Terrosum floribus Sulphur. Scabros.*
[bottom] *long. pal. 1½ lat. d 6. alt. d. 3*

Fossil wood described as being 'sulphurous earthy wood, flaky with sulphurous florets'. The presence of sulphur may indicate pyrite decay, or the resulting bacterial decay.

REFERENCES: 17th-cent. Index, p. 366

Comp. fig. 41. Wood showing sulphur precipitate from decay, from Rosaro

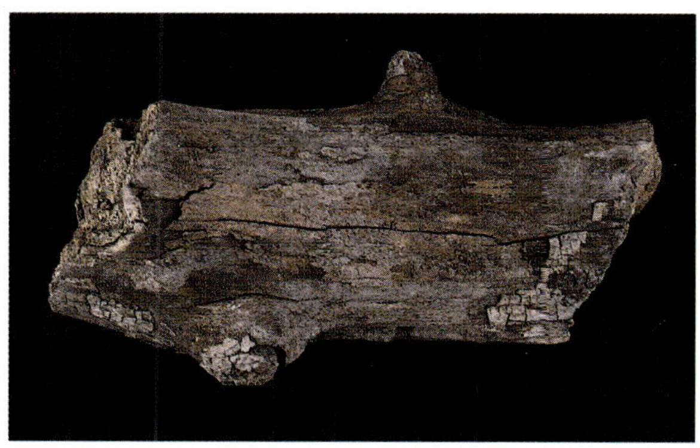

42. Fossil wood branch with clustering of branches

Windsor, RL 25647, *Fossils*, vol. XVI, fol. 8

Pen and ink, over traces of black chalk

186 × 261 mm; small holes caused by ink

NUMBERING: [ink, top left corner] *108*

ANNOTATION: [ink, top] *lign: Terr. Nigr.* [bottom] *long: palm. 3: d: 3. lat: d. 8: alt: d: 4.*

Specimen of fossil wood described as 'black, earthy wood' with a branch cluster.

REFERENCES: 17th-cent. Index, p. 367

42

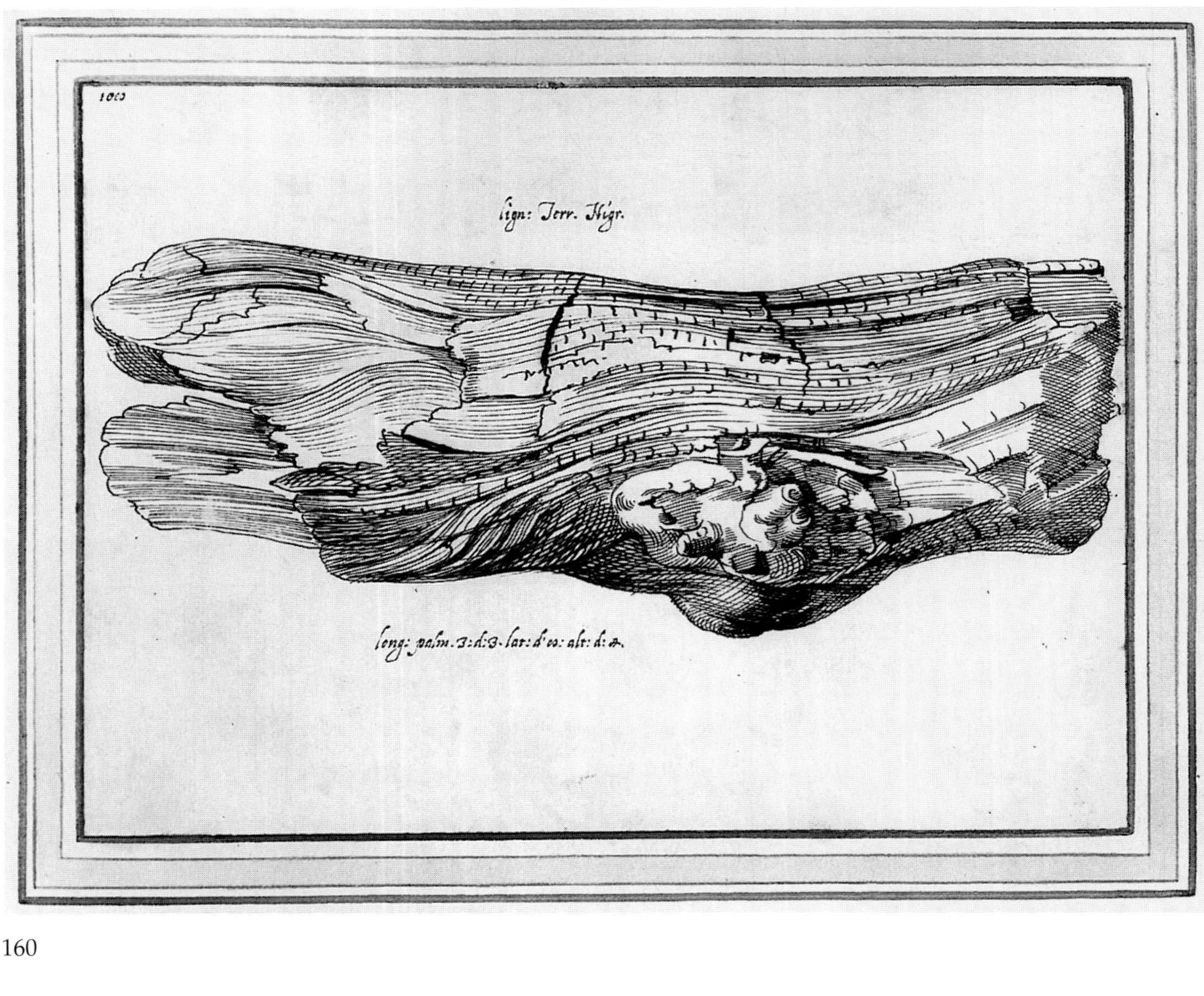

43

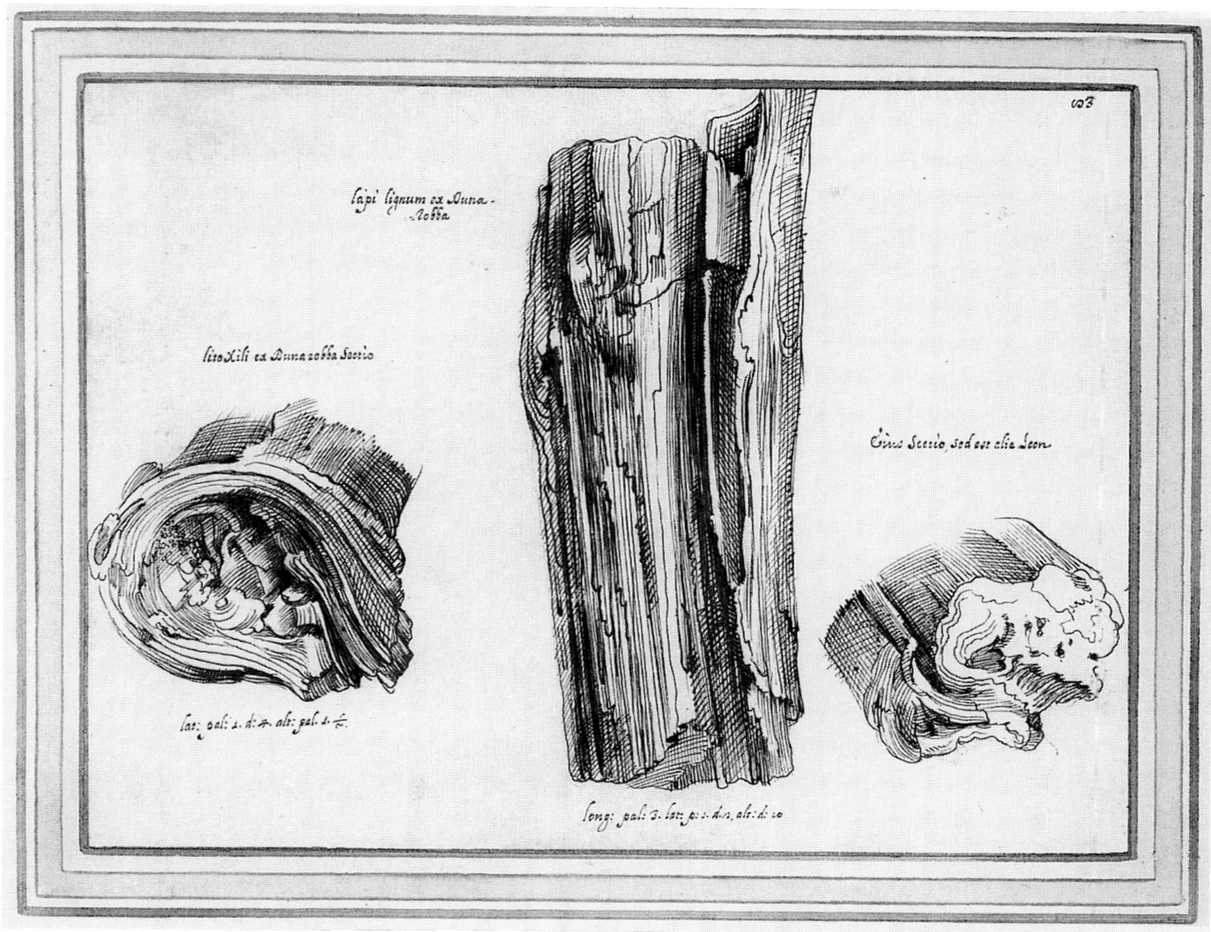

43. Two pieces of petrified fossil wood from Dunarobba, and section detail

Windsor, RL 25622, *Fossils*, vol. XV, fol. 31

Pen and ink, over black chalk

192 × 262 mm

NUMBERING: [ink, upper right corner] *83*

ANNOTATION: [ink, top left] *lapi lignum ex Duna-/Robba*
[centre left] *litoxili ex Dunarobba Sectio*
[lower left] *lat: pal: 1. d: 4. alt: pal. 1. ½.*
[bottom centre] *long: pal: 3. lat: p: 1. d. 2, alt: d: 10.*
[centre right] *Eius Sectio, sed est alia scon*

These specimens from Dunarobba are called, respectively, *lapilignum* and *litoxilum*. Both words are translatable as 'stone-wood' although a distinction was clearly intended.

REFERENCES: 17th-cent. Index, pp. 365, 367

44. Two pieces of fossil wood

Windsor, RL 25649, *Fossils*, vol. XVI, fol. 10

Pen and ink, over traces of black chalk; retouched in ink on mount sheet

204 × 255 mm; small holes caused by ink; measurements given
for right-hand specimen.

NUMBERING: [ink, top right corner] *110*

ANNOTATION: [ink, lower left] *In Secta ligni crassi parte media Sectio / Naturale*
[upper right] *lignum Undatum*
[lower right] *long. pal. 1½. lat. d. 10. alt. d. 3*

Two pieces of fossil wood. The left-hand specimen, illustrated life-size (43 cm long) is described as a 'section in the cut middle part of thick wood'. The right-hand specimen has evidently been compacted: hence the 'waviness' noted in the inscription.

REFERENCES: 17th-cent. Index, p. 367

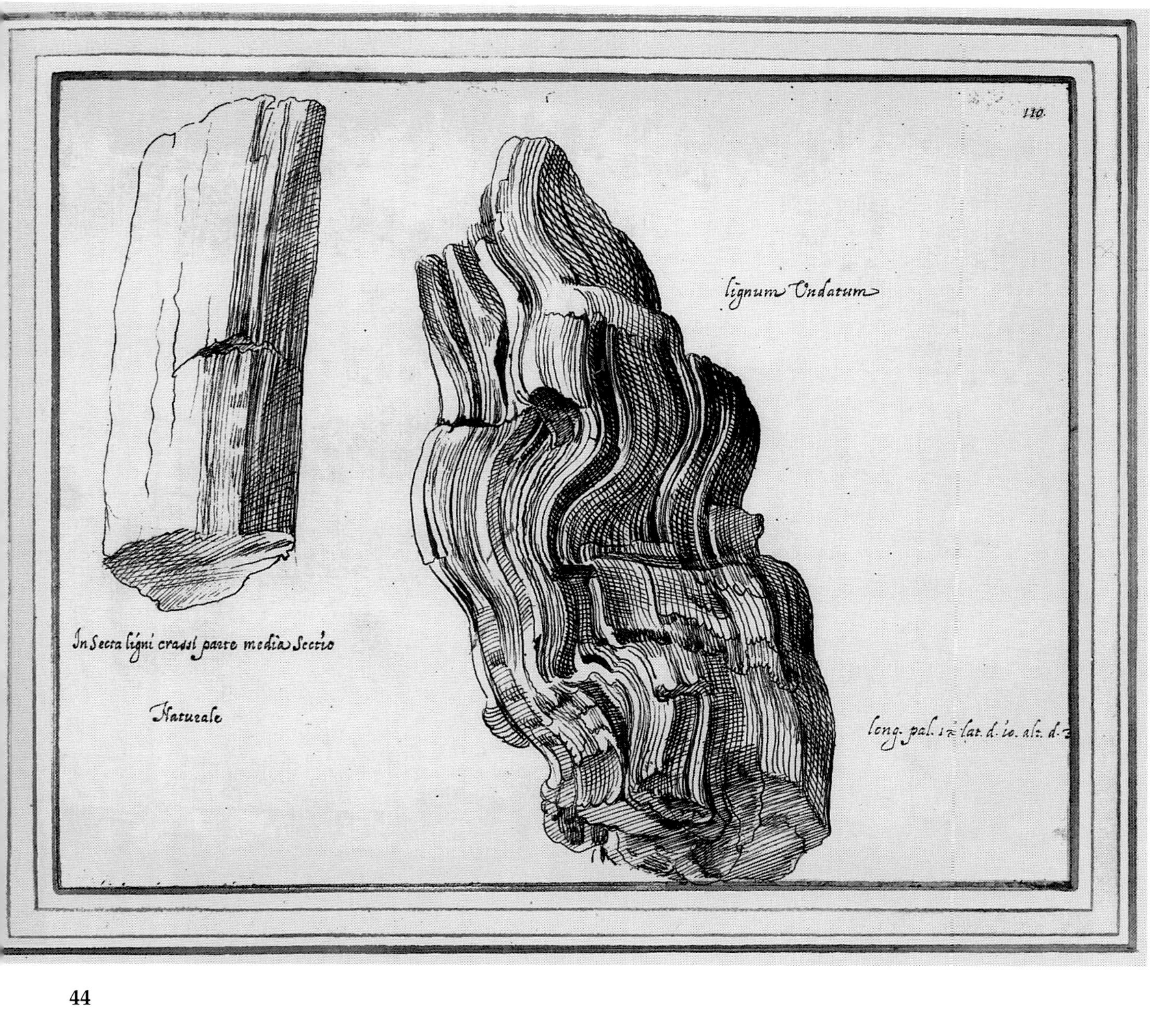
110.
lignum Undatum
In Secta ligni crassi parte media Sectio
Naturale
long. pal. 1½ lat. d. 10. alt. d. 3

44

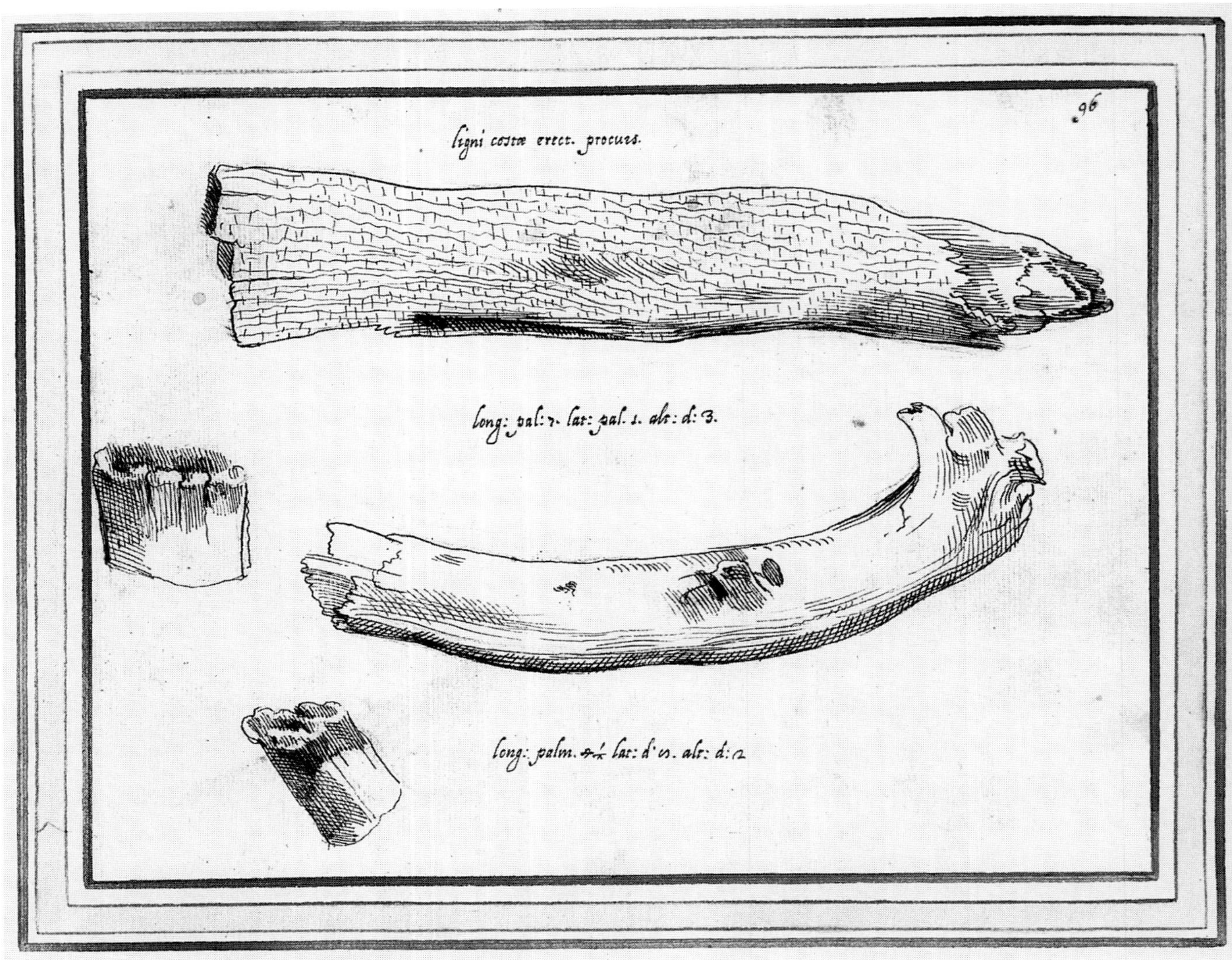

45

45. Two pieces of fossil wood, with end views

Windsor, RL 25635, *Fossils*, vol. XV, fol. 44

Pen and ink, over black chalk

187 × 253 mm. *Watermark*: Bird 41B

NUMBERING: [ink, top right corner] *96*

ANNOTATION: [ink, top] *ligni costae erect. Procurs.*
[centre] *long: pal: 7. lat: pal. 1. alt: d: 3.*
[bottom] *long: palm. 4½ lat: d. 8. alt: d: 2*

Two specimens of fossil wood, together with details of their ends. The specimens are described as 'ribs of wood'.

REFERENCES: 17th-cent. Index, p. 366

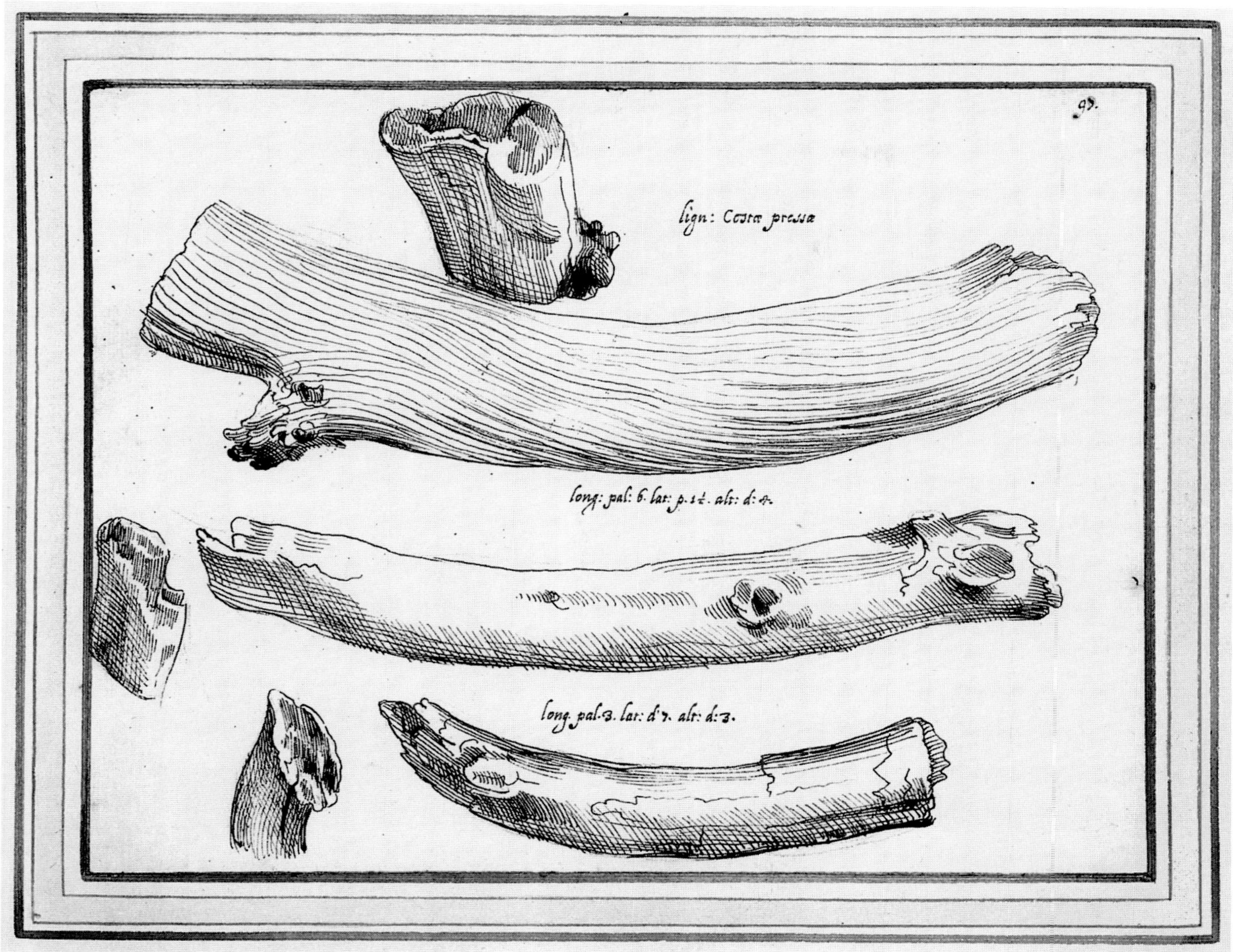

46

46. Three pieces of fossil wood with end views

Windsor, RL 25636, *Fossils*, vol. XV, fol. 45

Pen and ink, over black chalk

192 × 258 mm

NUMBERING: [ink, top right corner] *97*

ANNOTATION: [ink, upper right] *lign: Costae pressae*
[centre] *long: pal: 6. lat: p. 1½. alt: d: 4.*
[bottom] *long. pal. 3. lat: d. 7. alt: d: 3.*

The specimens are similar in appearance to those in **45** and are likewise described as 'ribs of wood', with the addition of the adjective 'compacted'. End views are shown for each of the three specimens.

REFERENCES: 17th-cent. Index, p. 366

47. Piece of fossil wood, possibly coniferous with branch scars

Windsor, RL 25608, *Fossils*, vol. XV, fol. 17

Pen and ink, over traces of black chalk; measurements not given; black chalk sketch of three ends of fossil wood specimen on verso

186 × 247 mm. *Watermark*: Bird 41C

NUMBERING: [ink, top left corner] *69*

ANNOTATION: [ink] *Frustum ligneum nodo in medio extuberante*

This piece of fossil wood shows a branch scar, a feature also noted in the annotation, 'a woody piece with a knot protruding from the centre'. There is clearly space left for another drawing beneath.

REFERENCES: 17th-cent. Index, p. 364

47

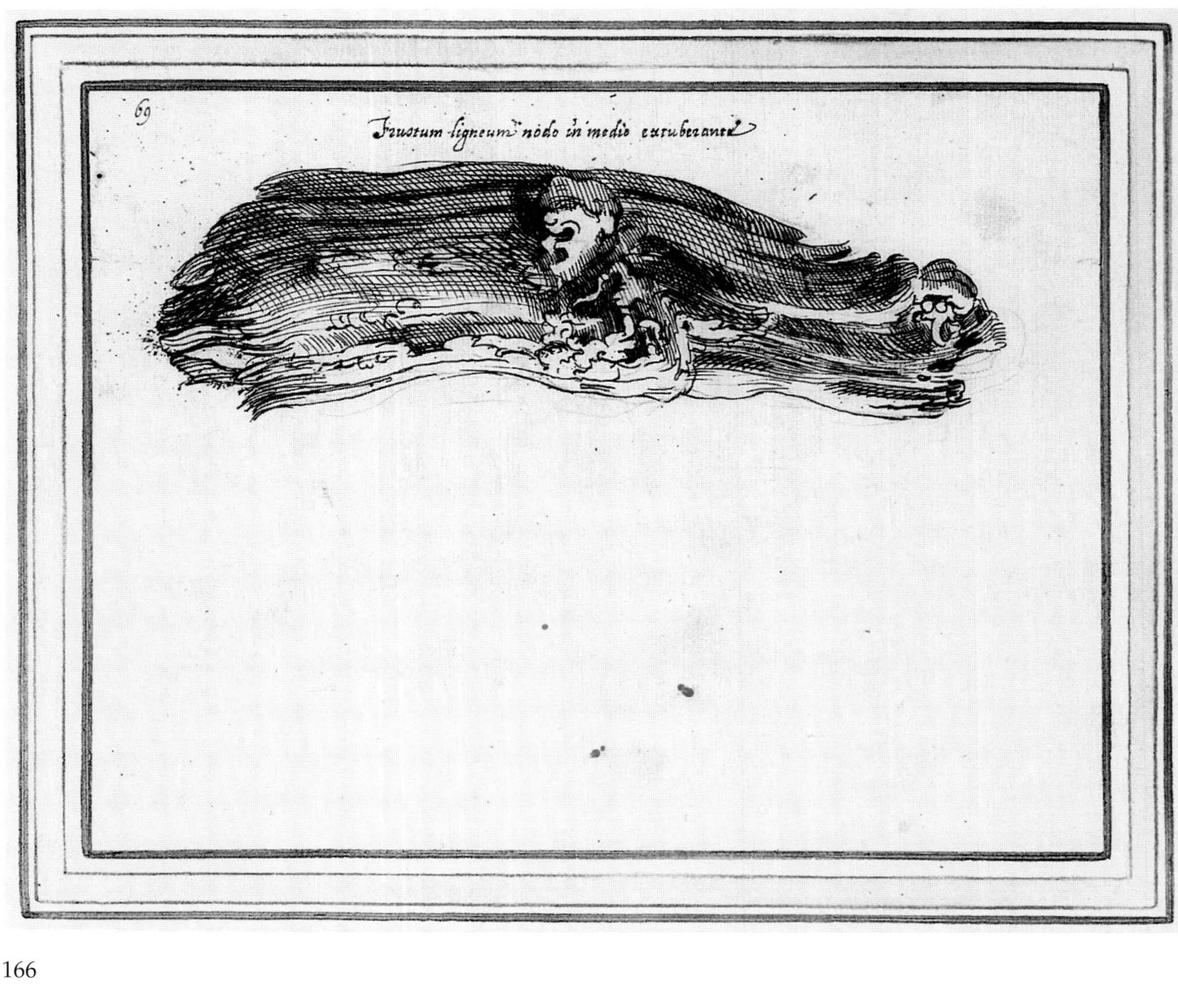

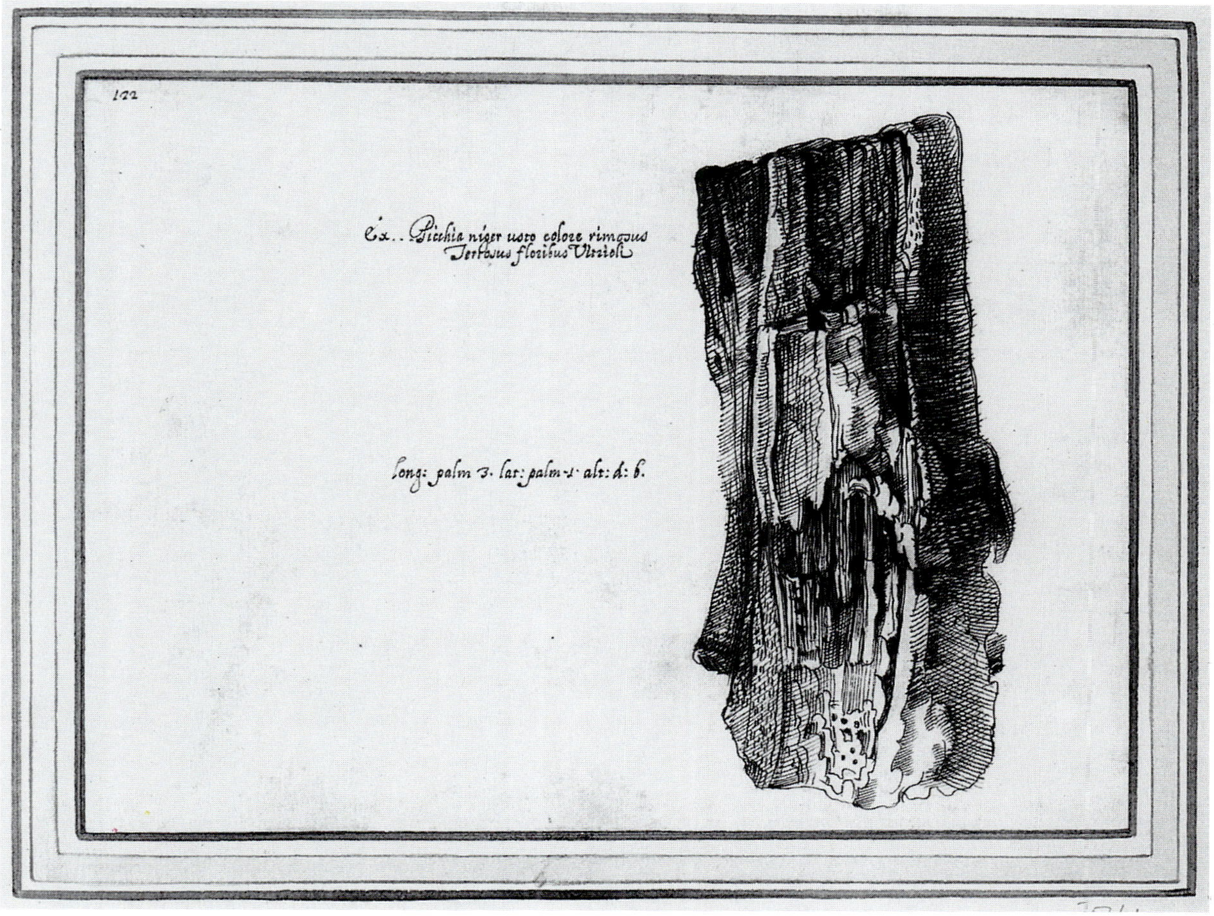

48

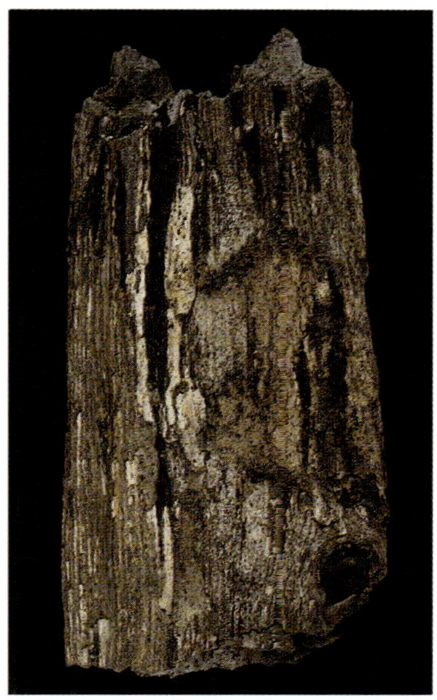

Comp. fig. 48. Partly coalified and partly decayed wood from between Rosaro and Sismano

48. Pyritized wood

Windsor, RL 25661, *Fossils*, vol. XVI, fol. 22

Pen and ink, over traces of black chalk

184 × 256 mm; small holes caused by ink

NUMBERING: [ink, top left corner] *122*

ANNOTATION: [ink, top] *Ex ... Picchia niger usto colore rimosus/ Terrosus floribus Vitrioli* [bottom] *long: palm 3. lat: palm. 1. alt: d: 6.*

A piece of decaying pyritized wood showing what the inscription refers to as flowers of vitriol. This could mean that it is a decaying pyrite. The name *Picchia* may refer to a place, possibly Turris Pitia (see above, p. 102).

REFERENCES: 17th-cent. Index, p. 368

49. Petrified wood

Windsor, RL 25672, *Fossils*, vol. XVI, fol. 33

Pen and ink, over black chalk

188 × 270 mm

NUMBERING: [ink, top right corner] *133*

ANNOTATION: [ink, top] *Petrefactum cavum voluti ad corticem, Musculus obrutundus albus.*
[bottom] *long: pal: 1. alt. p: 1. lat. rotund.*

A specimen of petrified wood with branch scar—described as being an 'almost round, white muscle' with a 'petrified cavity'. The white substance may be a mineral precipitate. See Comp. fig. **24**.

REFERENCES: 17th-cent. Index, p. 368

50. Fossil wood, possibly permineralized

Windsor, RL 25637, *Fossils*, vol. XV, fol. 46

Pen and ink, over traces of black chalk; frame line in black chalk

185 × 256 mm

NUMBERING: [ink, top right corner] *98*

ANNOTATION: [ink, top] *ligni lapiferreum*
[right] *long: palm: 2. lat. pal: ½ alt: d: 2*

Specimen of fossil wood described as 'ferrous stone wood' indicating iron impregnation (see also **38**).

REFERENCES: 17th-cent. Index, p. 366

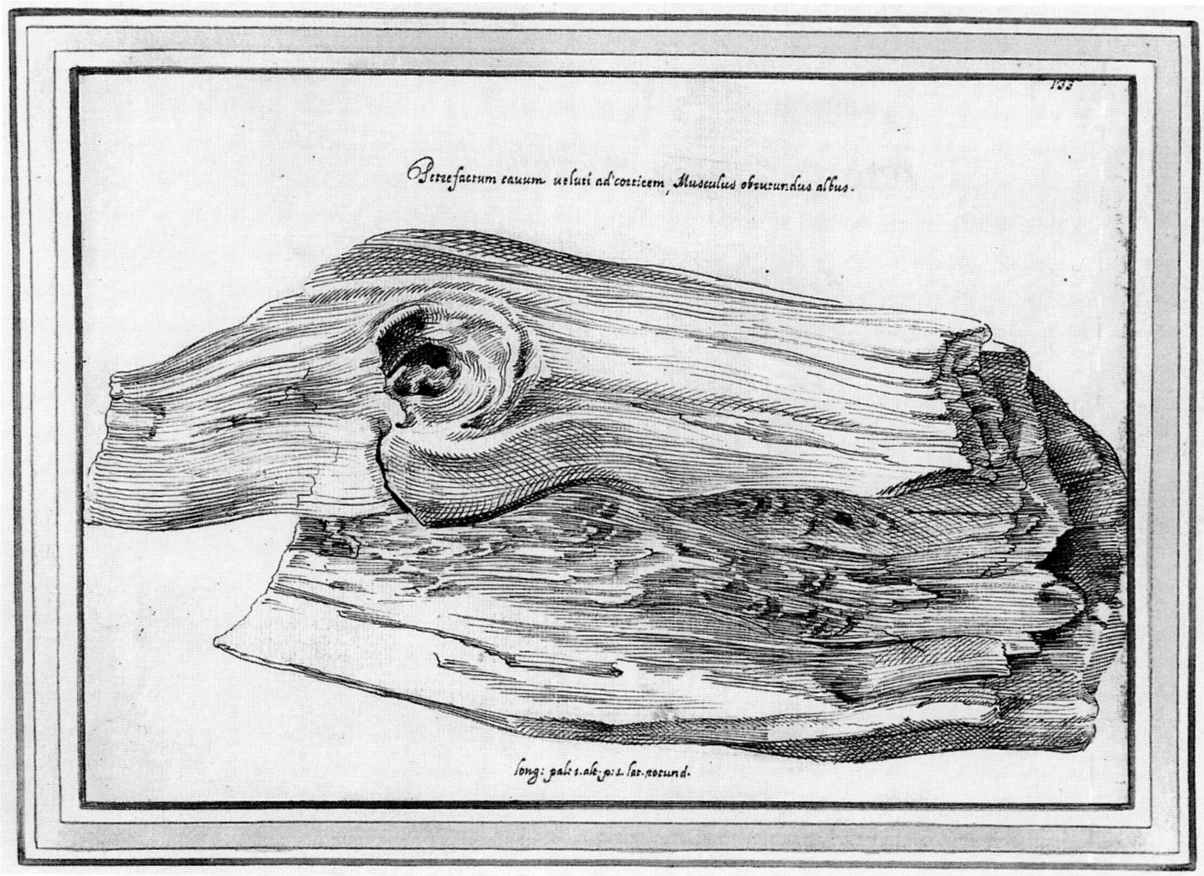
Petrefactum cauum ueluti ad corticem, Musculus obturundus albus.
long: pal: 1. alt: p: 1. lat. rotund.
49

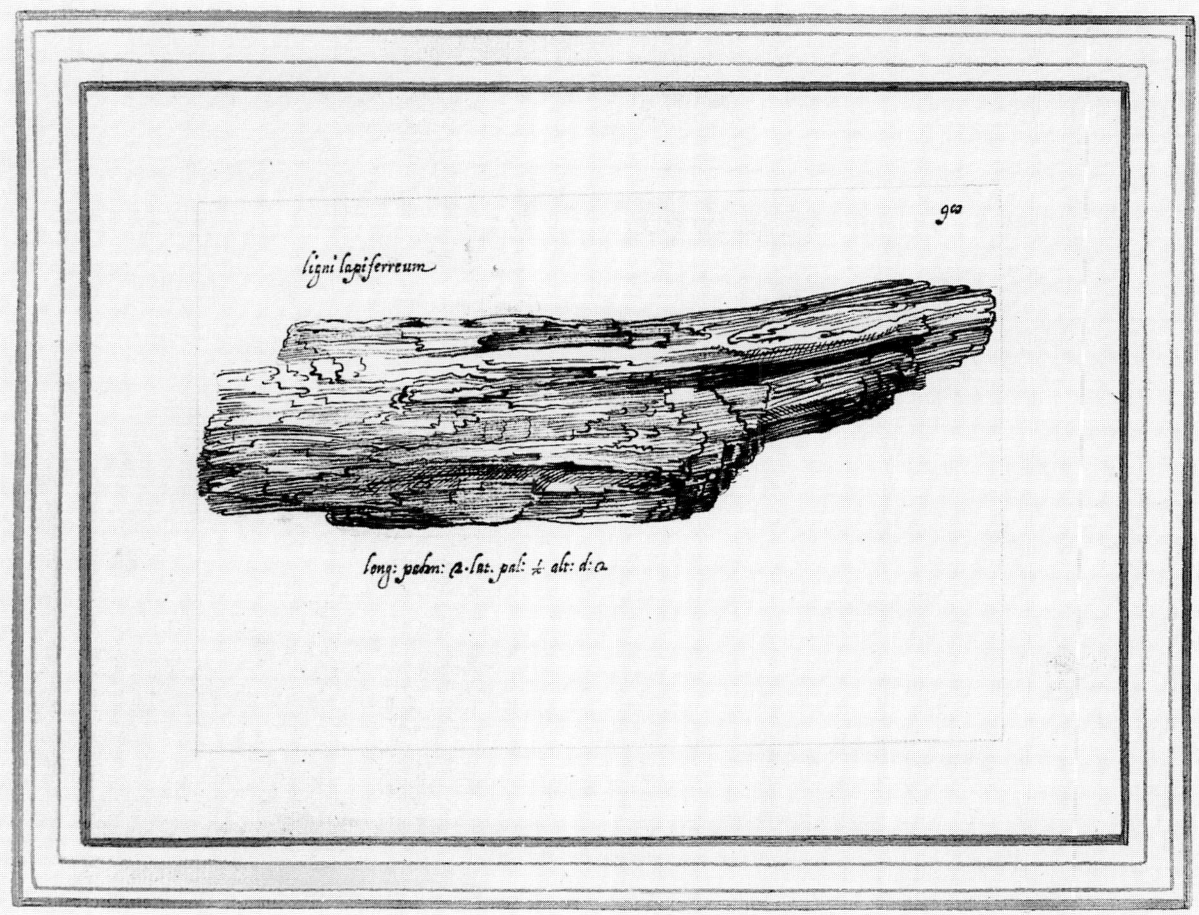
ligni lapiferreum
long: palm: 2. lat. pal: ½ alt: d: 2
50

51. Piece of fossil wood; branch with scars

Windsor, RL 25610, *Fossils*, vol. XV, fol. 19
Pen and ink, over traces of black chalk
193 × 259 mm. *Watermark*: Bird 41B
NUMBERING: [ink, top left corner] *71*
ANNOTATION: [ink] *long: p: 6½ lat: p.1. d. 4. alt: d: 3.*

This specimen of fossil wood shows well-preserved branch scars. No description is inscribed on the sheet, although further space—perhaps for another drawing or two—is left on the right-hand side of the sheet.

REFERENCES: missing from 17th-cent. Index

Comp. fig. 51a. A modern drawing showing recent *Sequoia* with branch

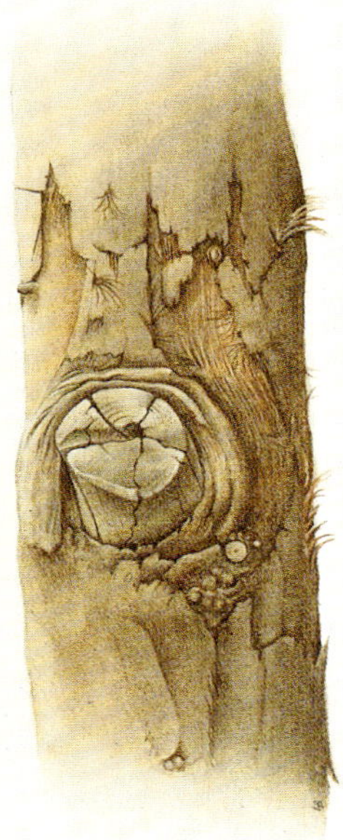

Comp. fig. 51b. Fossil wood with branch scars from Dunarobba

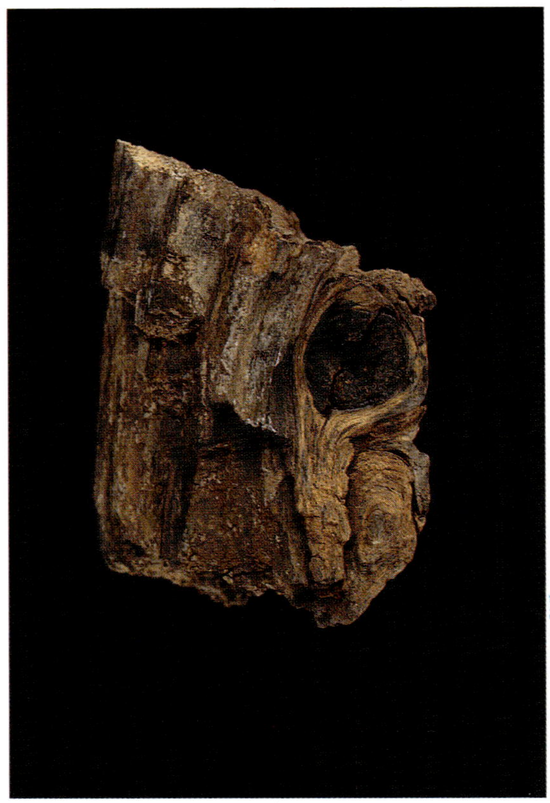

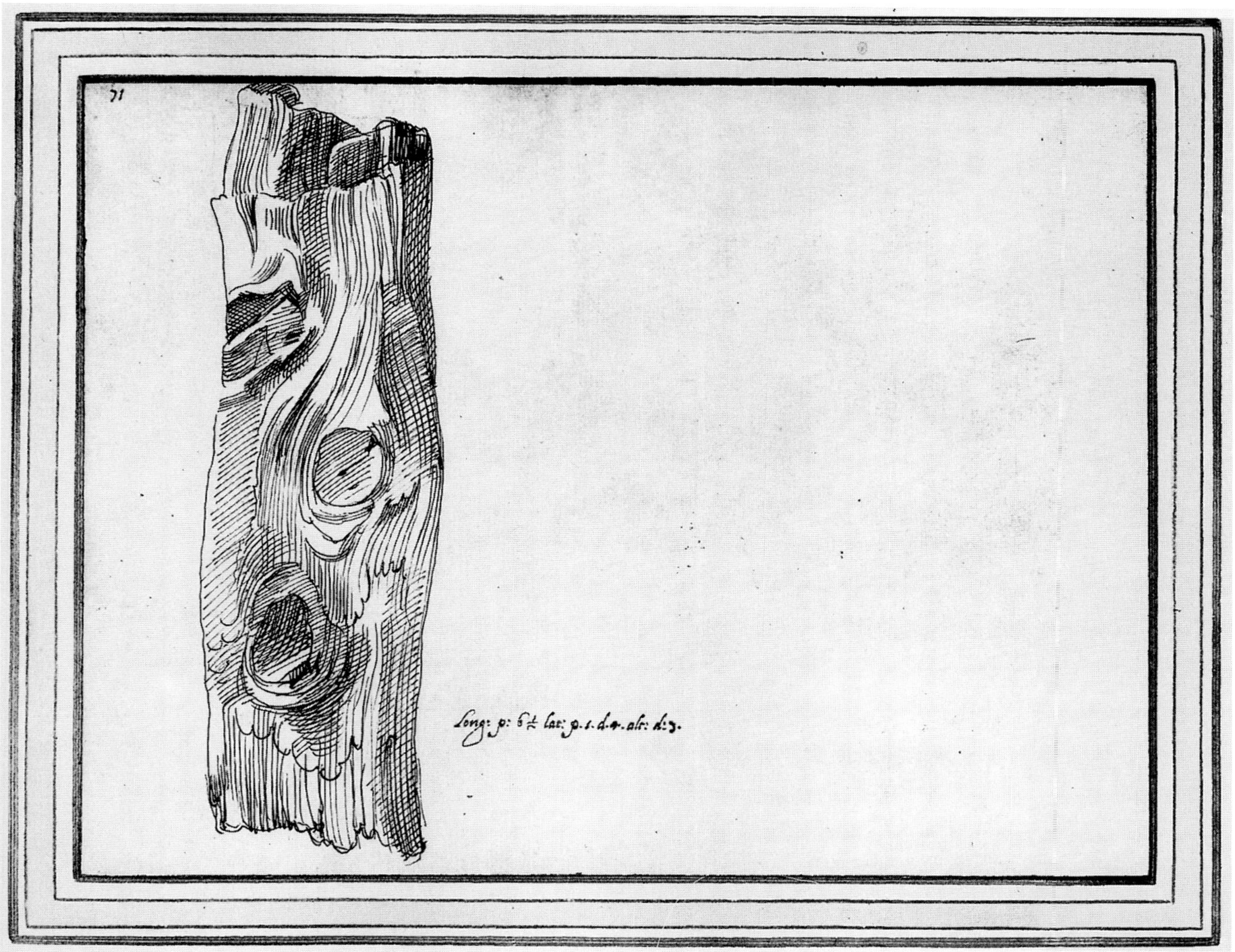

51

52. Two pieces of fossil wood, twisted with branch scars

Windsor, RL 25609, *Fossils*, vol. XV, fol. 18

Pen and ink, over traces of black chalk; retouched in ink on mount sheet

181 × 250 mm; small holes caused by ink

NUMBERING: [ink, top left corner] *70*

ANNOTATION: [ink, top centre] *Frustum ligneum Tribus volutis, nodo, et plexu Varium*
[centre] *long: palm. 7. lat p. 1½ alt: pal: 1.*
[bottom] *long: pal. 6. d: 2. lat. pal: 1½, alt: d. 9*

These two pieces of fossil wood are typical branch fragments. The top specimen is noted as having 'three whorls, a knot and a fold'.

REFERENCES: 17th-cent. Index, p. 364

53. Fossil wood

Windsor, RL 25546, *Fossils*, vol. XIV, fol. 17

Pen and ink, over traces of black chalk

168 × 234 mm; small holes caused by ink

NUMBERING: [ink, upper left] *10*

ANNOTATION: [ink, top] *Brachium Cretæ non dum lignificatæ in ramum cuntis*
[centre] *long: pal: 1. d. 1. lat: d: 2. alt: d: 1.*
[bottom] *long: pal: 1. d. 2. lat: d. 3. alt: d. 2*

Both specimens are branches (possibly coniferous). The lower specimen shows pieces of clay still attached, while the right-hand side of the upper specimen reveals the outer layers of the branch to be finely cracked. The observation on the upper specimen that the 'branch of clay' is 'not yet lignified turning into a branch' supports Stelluti's claim that Cesi believed the fossil wood to be formed out of the clay; 'whence he has no doubt whatsoever that the earth itself is the seed and mother of this wood, the earth of these parts being most suitable for its generation'.

REFERENCES: 17th-cent. Index, p. 361

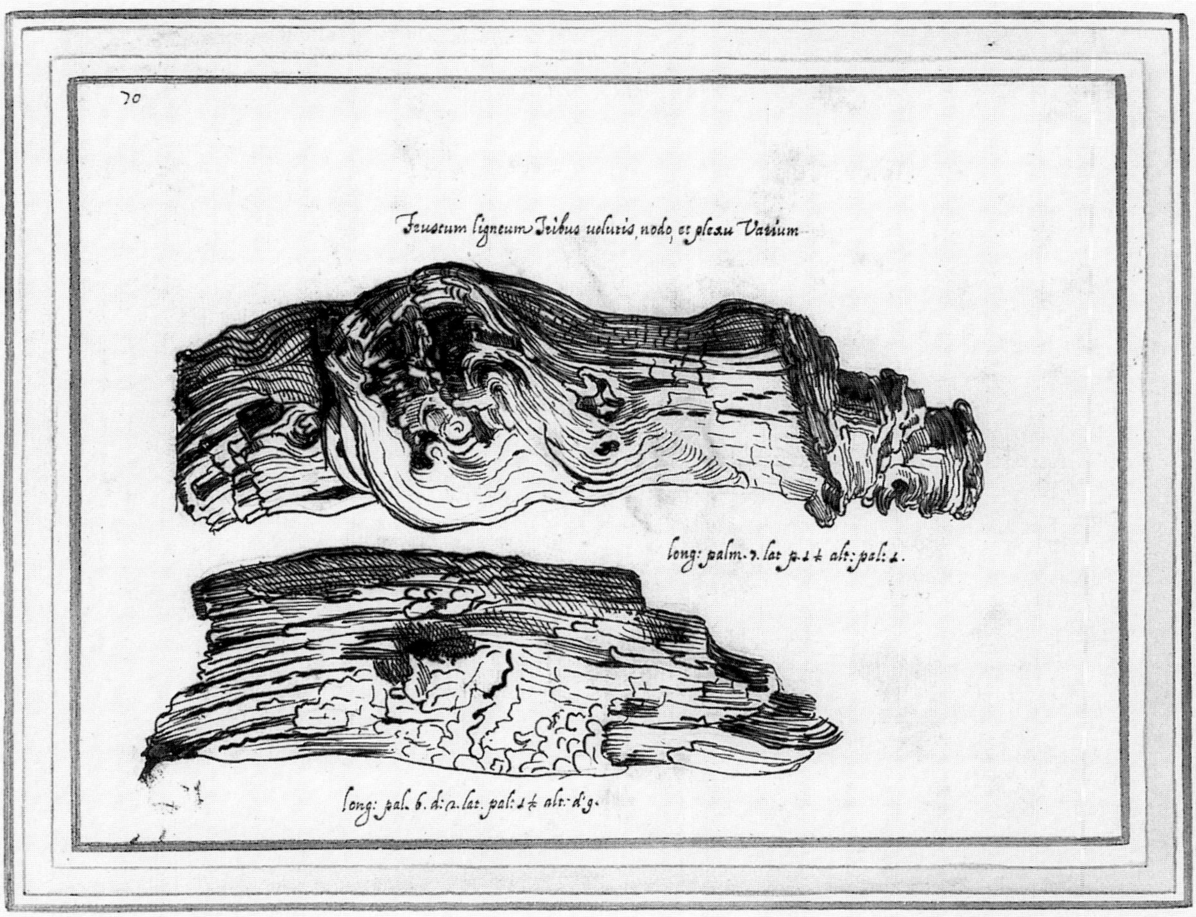
70
Frustum ligneum Tribus uelutis, nodo, et plexu Varium
long: palm. 7. lat. p. 1 ½ alt: pal: 1.
long: pal. 6. d: 2. lat. pal: 1 ½ alt: d: 9.

52

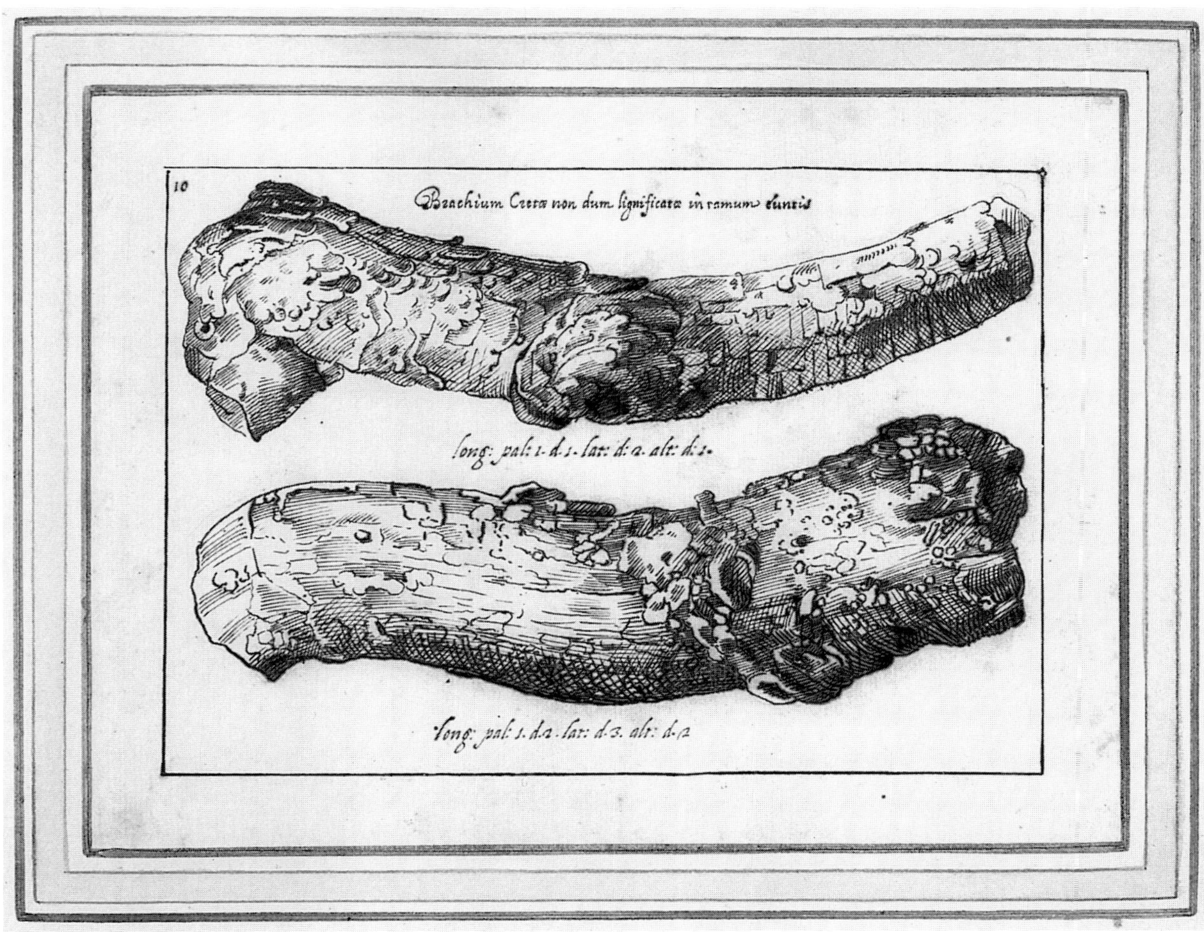
10
Brachium Cretæ non dum lignificatæ in ramum euntis
long: pal: 1. d: 1. lat: d: 2. alt: d: 1.
long: pal: 1. d. 2. lat: d: 3. alt: d. 2

53

54. Piece of fossil wood with clay-filled cracks

Windsor, RL 25633, *Fossils*, vol. XV, fol. 42

Pen and ink, over black chalk

188 × 254 mm. *Watermark:* Bird 41B

NUMBERING: [ink, top right corner] *94*

ANNOTATION: [ink, upper right] *ligneum Caput cum Intruso ramo in fractione/et Creta mixta in Cavitate fractionis* [bottom right] *long: pal: 1. lat: p. med. alt ½.*

According to the inscription, this specimen of fossil wood contained clay in the cracks between the wood: 'a branch intruded into the break, and clay mixed in the cavity of the break'.

It is possible that the remaining space was intended for a further drawing on the right.

REFERENCES: 17th-cent. Index, p. 366

54

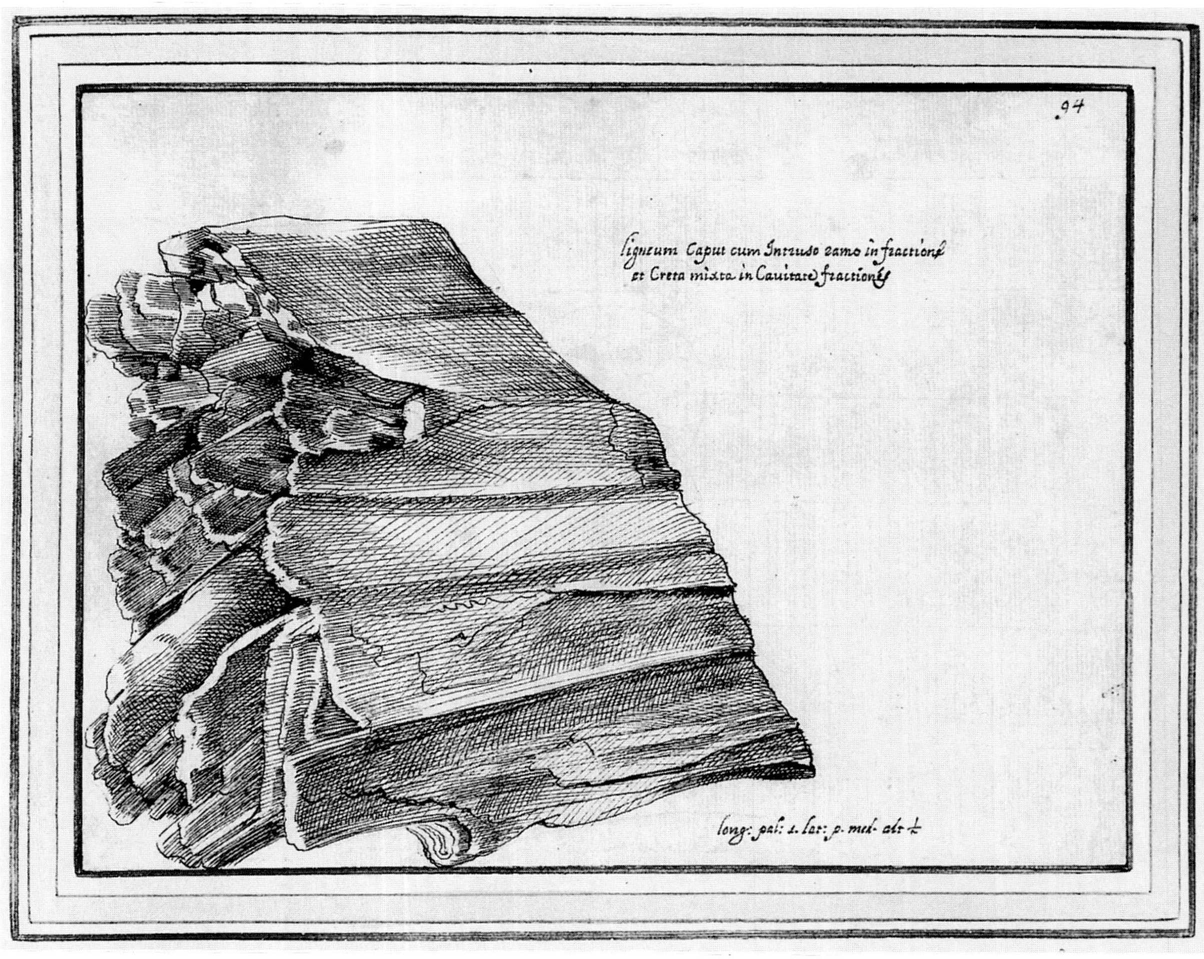

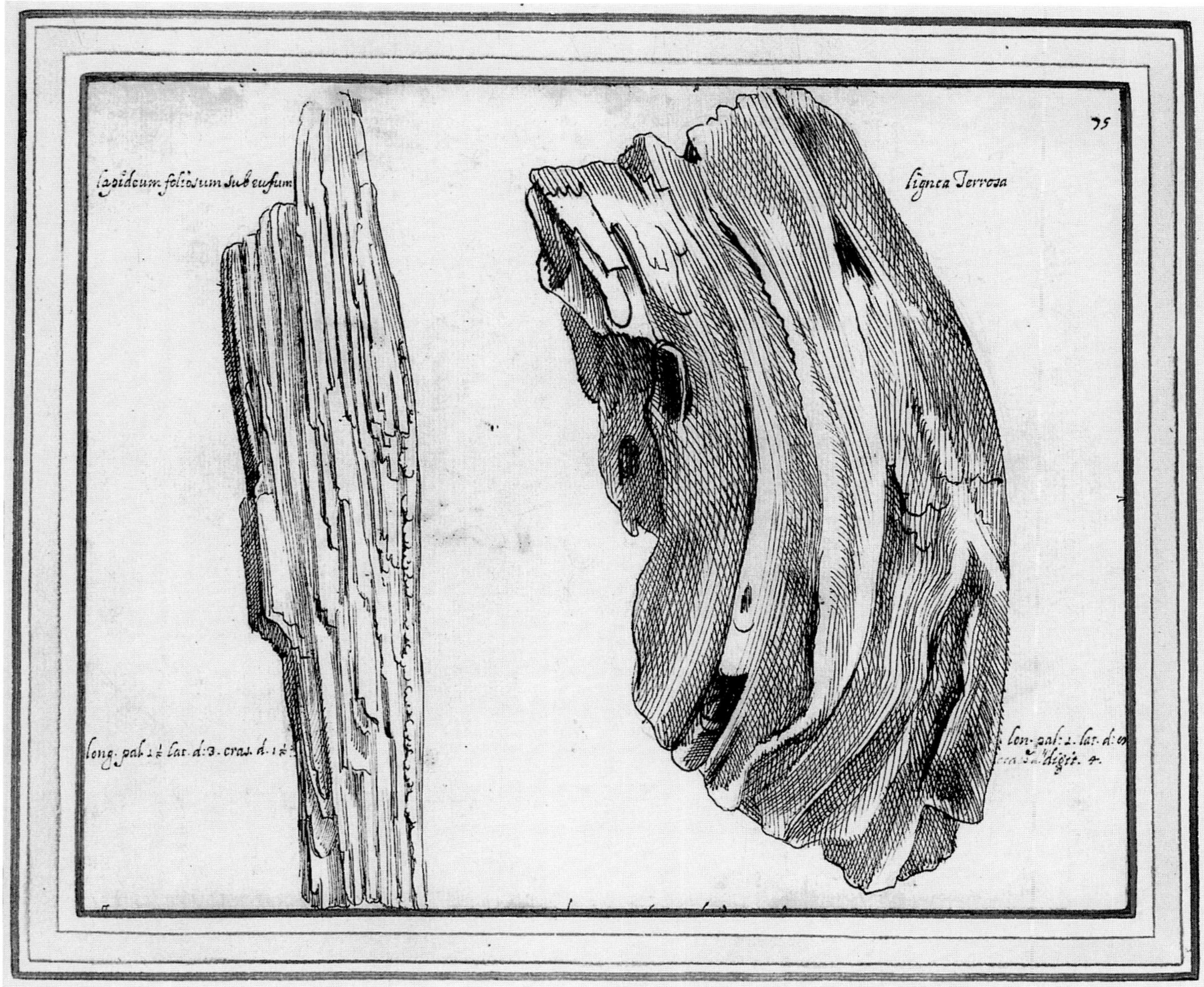

55

55. Two pieces of fossil wood

Windsor, RL 25614, *Fossils*, vol. XV, fol. 23

Pen and ink, over black chalk

202 × 255 mm

NUMBERING: [ink, upper right corner] *75*

ANNOTATION: [ink, top left] *lapideum foliosum sub rufum*
[lower left] *long. pal. 1½ lat. d: 3. cras. d. 1½*
[top right] *lignea Terrosa*
[lower right] *long. pal. 1. lat. d. 8*/[faint: *crass. d. 4*] *digit. 4*
[Cesi's inscriptions, cut, lower and right edges]

The left-hand specimen of fossil wood is described as 'stony, leafy, reddish' and may have been petrified or permineralized by siderite. The right-hand specimen is described as 'woody' and 'earthy' and probably represents unaltered wood.

REFERENCES: 17th-cent. Index, pp. 364, 367

OTHER DRAWINGS: **38**

56. Piece of fossil wood with branch bases

Windsor, RL 25631, *Fossils*, vol. XV, fol. 40

Pen and ink, over traces of black chalk

193 × 264 mm

NUMBERING: [ink, top right corner] *92*

ANNOTATION: [ink, top left] *lignea*
[lower left] *long: pal: 2½. lat: d: 6. alt: d: 1*

The drawing has been closely trimmed through the left edge. The specimen is simply described as 'woody'. The smooth outer surface may represent the part from which the bark has been stripped.

REFERENCES: 17th-cent. Index, p. 365

56

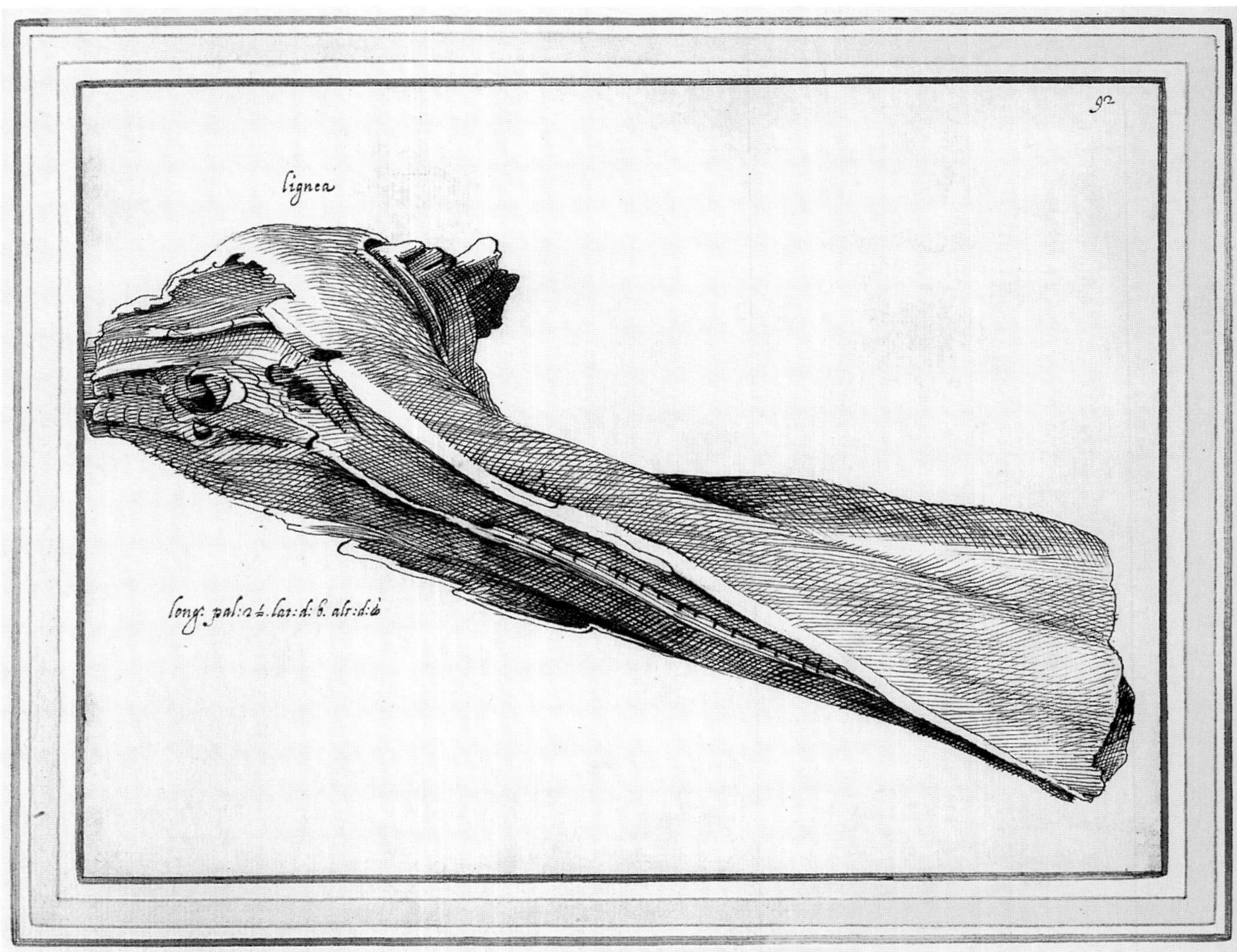

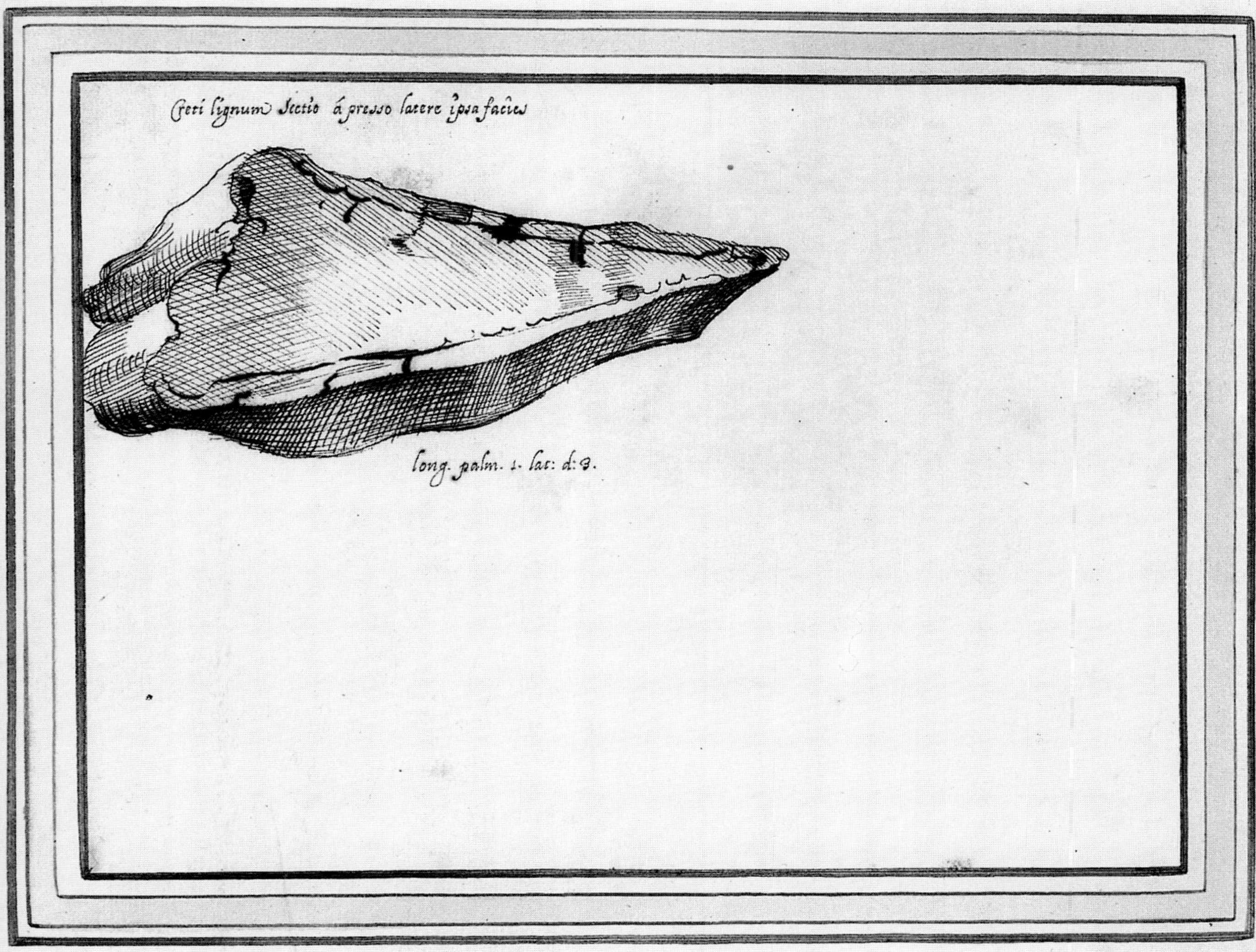

57

57. Fragment of fossil wood

Windsor, RL 25563, *Fossils*, vol. XIV, fol. 34

Pen and ink, over traces of black chalk

181 × 252 mm

NUMBERING: none; formerly *27* (see Appendix I)

ANNOTATION: [ink, upper left] *Creti lignum sectio a presso latere ipsa facies* [centre] *long. palm. 1. lat: d: 3.*

Described as a 'section through [a piece of] *cretilignum* viewed from the compacted side' this fragment appears to be an end-section of a piece of fossil wood. It may well be that another view of the same fragment was to have been included in the lower half of the sheet.

REFERENCES: 17th-cent. Index, p. 362

58. Fossil wood from Rosaro

Windsor, RL 25558, *Fossils*, vol. XIV, fol. 29

Pen and ink, over black chalk; retouched in ink on mount sheet

182 × 254 mm; small holes caused by ink

NUMBERING: [ink, upper right] *21*

ANNOTATION: [ink, top centre] *Creti lignum nigerrimum Acciavacci nitore in/ signum exiccatum ex matre magnorum Triūm / frustor*[um] *ad Rosarium*
[bottom] *Long: pal. 3. d: 3. lat. p. 1. alt: d: 3*

This piece of wood, described as 'a very black *cretilignum* with the shine of *acciavaccio*' (perhaps a kind of anthracite) is noted as 'coming from the very large pieces at Rosaro' (see above, pp. 88, 95). For further references to *acciavaccio* see under **60** and **95**, as well as Appendix III a.

REFERENCES: 17th-cent. Index, p. 362

OTHER DRAWINGS: **59**

59. Two views of fossil wood from Rosaro

Windsor, RL 25559, *Fossils*, vol. XIV, fol. 30

Pen and ink, over black chalk

185 × 261 mm

NUMBERING: [ink, upper right] *22*

ANNOTATION: [ink, centre] *Idem exiccatione rimosum*

The upper drawing shows a different angle of the same specimen depicted in **58** (where the measurements are given). The lower drawing is a view of the under surface, showing distinctive cracks. These, as the inscription indicates, could have been formed by desiccation. The 'black and shining' quality mentioned in **58** may imply that this surface has been charred by fire, but black coalified woods also occur in this area.

REFERENCES: 17th-cent. Index, p. 362

OTHER DRAWINGS: **58**

58

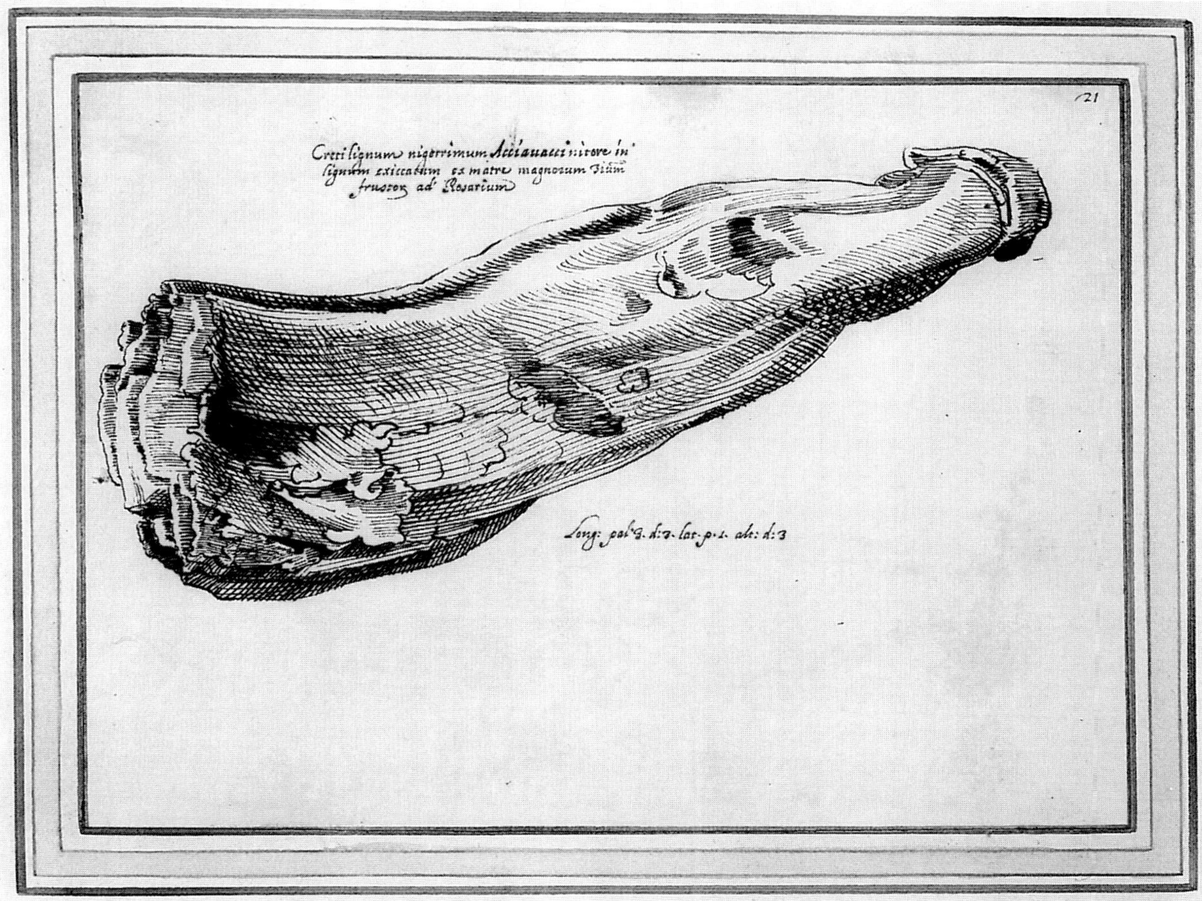
21
Creti lignum nigerrimum Acciauacci nitore in
lignum exiccatum ex matre magnorum

59

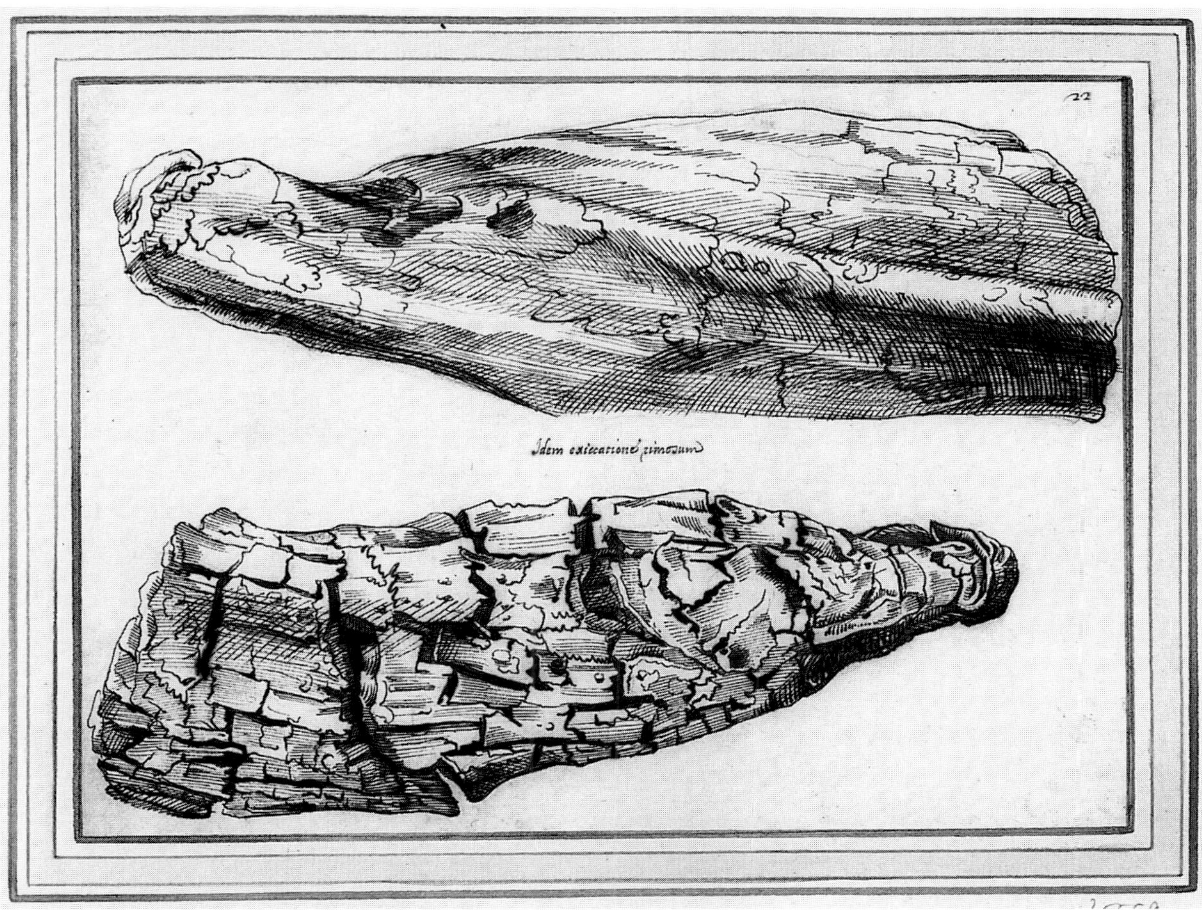
22
Idem exiccatione rimosum

60. Two pieces of fossil wood

Windsor, RL 25572, *Fossils*, vol. XIV, fol. 43

Pen and ink, over traces of black chalk

187 × 248 mm

NUMBERING: [ink, upper right] *34*

ANNOTATION: [ink, top left] *In ramum nitidum Acciav: fere*
[left] *long: pal: 1½ lat: d: 5. alt: d: 2*
[lower right] *Creti lignum Variæ porrectum / long: dit: 14. lat. d. 4. alt. d: 2½.*
[in Cesi's hand: vertical, lower left] *in r*[amum]/ *nit*[idum]/ *A*[cciav: fere]
[other fragments centre, top and bottom left]

The upper specimen is described as 'a branch with almost the shine of *acciavaccio*' (probably a kind of anthracite). This is almost certainly not anthracite as defined in modern usage. The lower specimen is yet another piece of *cretilignum* and appears to be a piece of wood coated with clay.

REFERENCES: 17th-cent. Index, pp. 362, 369

OTHER DRAWINGS: **58**

60

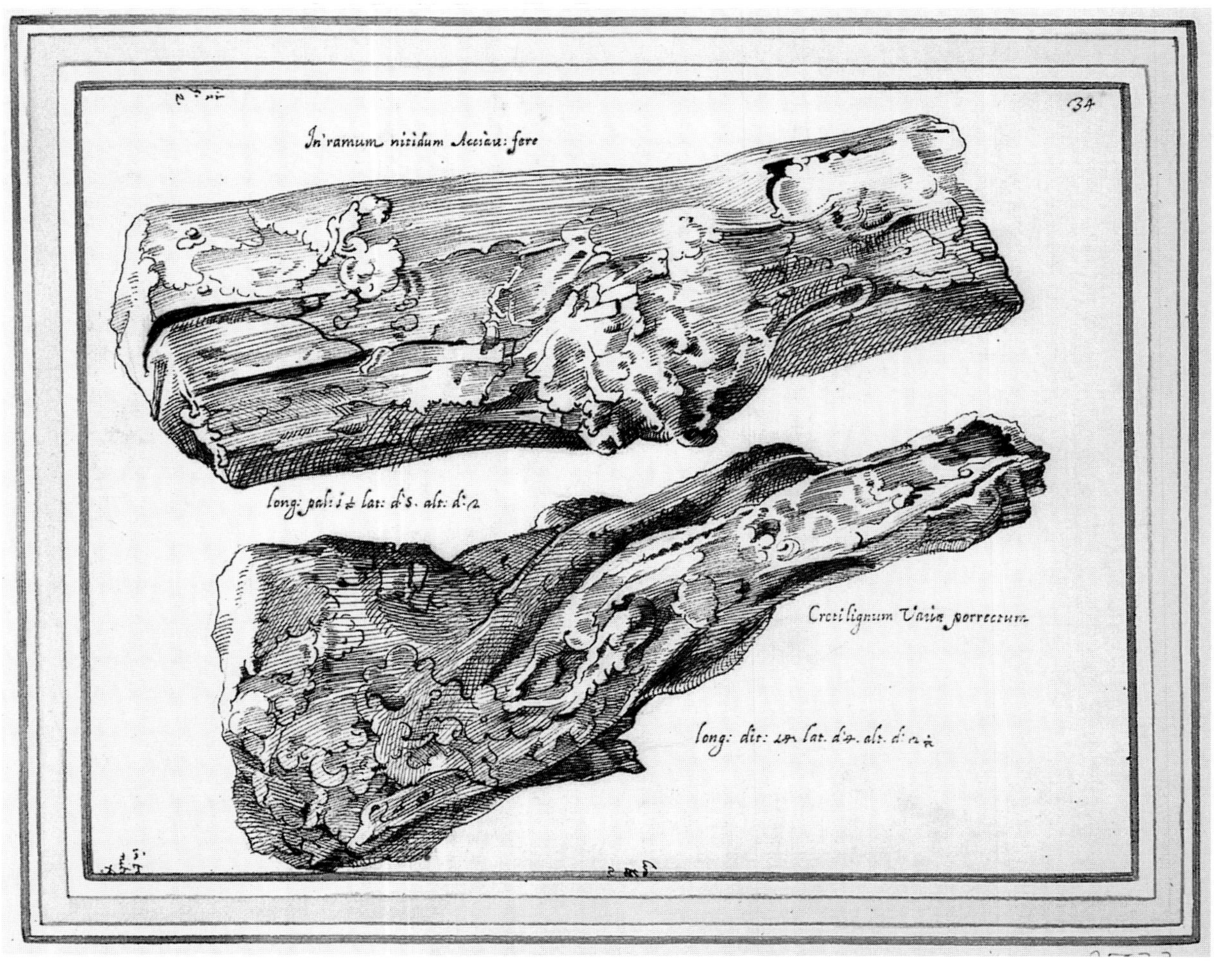

61

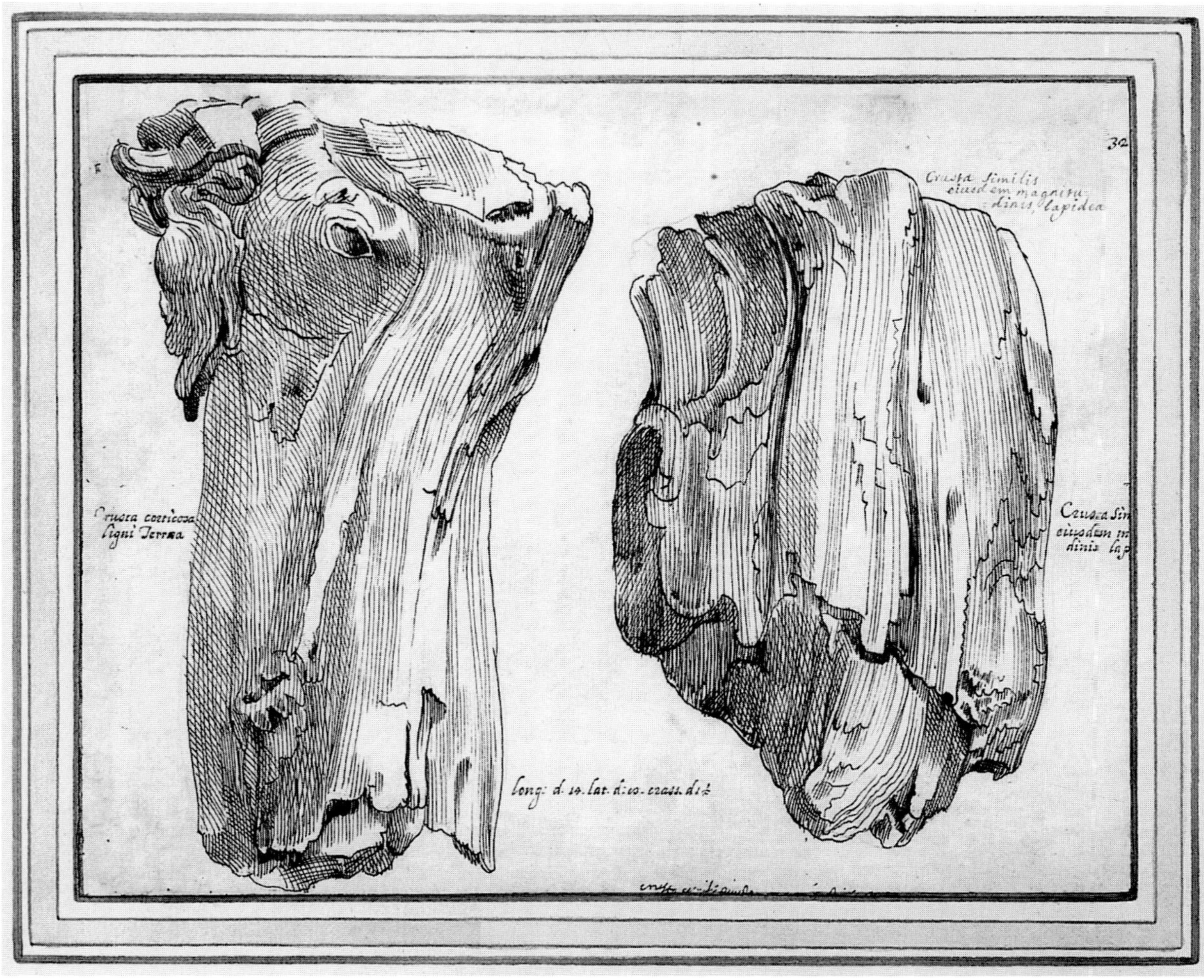

61. Two pieces of fossil wood

Windsor, RL 25570, *Fossils*, vol. XIV, fol. 41

Pen and ink, over black chalk (trimmed through inscription, right edge); measurements given for left-hand specimen only

206 × 265 mm. *Watermark*: Bird 41B

NUMBERING: [ink, upper right corner] *32*

ANNOTATION: [ink, left] *Crusta corticosa/ ligni Terrea*
[bottom] *long: d. 14. lat. d: 8. crass. d 1½*
[bottom edge] *crusta similis eiusdem*
[upper right in another hand] *Crusta similis/ eiusdem magnitu-/dinis, lapidea*
[right edge trimmed] *Crusta sim*[ilis]/ *eiusdem m*[agnitu-]/*dinis, lap*[idea]

Typical wood fragments, possibly of a branch. The right-hand specimen was described as 'stony' and was therefore probably petrified.

REFERENCES: 17th-cent. Index, p. 362

62. Three pieces of fossil wood, one shown in two views, from Ponte Caldaro

Windsor, RL 25577, *Fossils*, vol. XIV, fol. 48

Pen and ink, over black chalk; measurements are given for two of the specimens, while the third is described as of natural size

191 × 262 mm

NUMBERING: [ink, upper left] *39*

ANNOTATION: [ink, top left] *Ex P. cardar. ebenus fossil: frangibilis, gummosus/ ad Dini accedens / long: pal: 1. lat. d: 4 alt. d: 2*
[top right] *Dini ab alia Parte*
[centre right] *D. Dini nigrum nitidum Gagat./ Frangib. Gummos.*
[lower right] *del nāle.*
[lower left] *Rubens Squamosi ligni musculis com/pacta, lateritia potius quam/ lapidea*
[bottom left] *long: pal ½. lat. d. 3. alt. d. 1.*

The upper two drawings represent two views of the same specimen, which is noted as 'sticky (*gummosus*) and breakable'. The specimen lower right, drawn actual size, is described as 'black and glistening jet'. Gagate or jet fascinated both the ancient natural historians and Cesi's contemporaries. It is discussed at length both by Pliny (*Natural History*, XXXVI, 141–2) and by Aldrovandi 1648, pp. 418–24. Aldrovandi (p. 421) has a fine and large illustration of a specimen in fossil wood captioned *Lignum instar Gagatis* but it bears little resemblance to the present specimen.

Dana (1865, p. 760) notes that 'the Gagates (whence our word jet) occurred, according to Dioscorides and Pliny, at Gagas or Gages, a place in Lycia (Asia Minor). The former describes it as black, smooth and combustible to which Pliny adds that it was light, and looked much like wood, and that it emitted a disagreeable odor when rubbed, and burned with the smell of sulphur. It was in part at least, true lignite'. It is clear that the term Gagates does not equate to jet as currently used (a workable black coalified coniferous wood which is impregnated by bitumen usually found in marine sediments). The description by Dioscorides and Pliny could equally be identified with solid bitumens which had been carved into a range of objects in the Middle East for several thousand years (see Conan/ Deschesne 1996, p. 42). However, in the absence of the actual specimen, absolute identification is not possible.

The lower left-hand specimen on this sheet is described as baked rather than stony; this specimen, another *lateritium*, might have resulted from siderite impregnation.

Specimens collected by the author during 1999 fieldwork on gravel bars beneath the bridge of Ponte Caldaro included brown woods, black glistening woods and yellow-brown iron-impregnated woods. These were probably eroded out of the Fosso Bianco and Santa Maria di Ciciliano Formations upstream. Who *Dini* refers to in the inscription is not clear to us at the moment, although it could be that the reference is to Cesi's (and Galileo's) friend Pietro Dini, who was made Archbishop of Fermo in 1621, and died in 1625.

LITERATURE: Dana 1865; Conan/Deschesne 1996.

REFERENCES: 17th-cent. Index, p. 363

OTHER DRAWINGS: See **58** for other black shiny wood, and for comparative figure. See also **192**

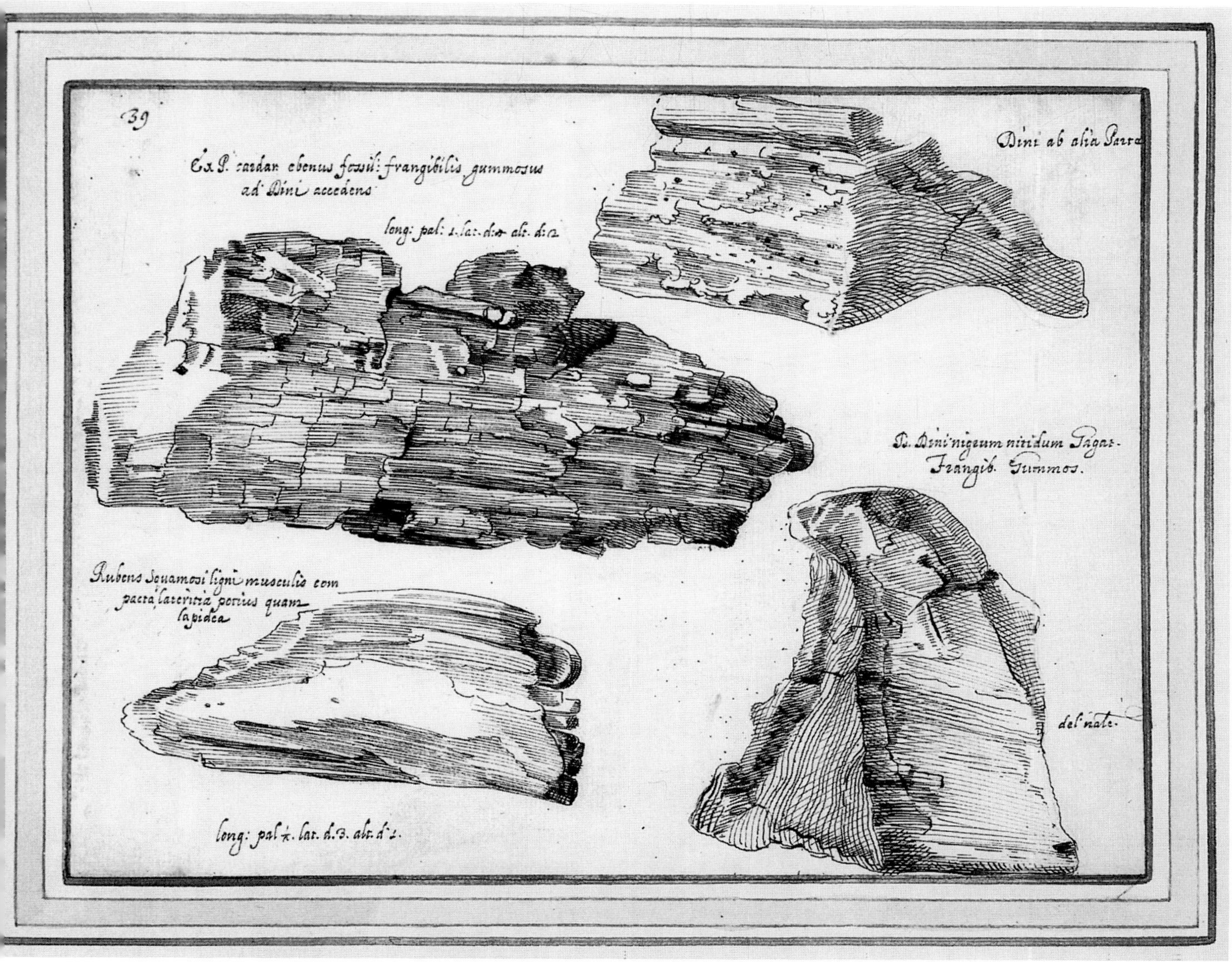

62

63. Fossil wood

Windsor, RL 25574, *Fossils*, vol. XIV, fol. 45

Pen and ink, over black chalk

181 × 278 mm

NUMBERING: [ink, upper right] *36*

ANNOTATION: [ink, top] *Ebenus fossilis* [bottom] *long. pal: 2½ lat. pal: 1. alt d. 2.*

A piece of fossil wood, presumably black, likened to ebony (see above, p. 131). The cracking on the lower right-hand part of the wood, together with the suggestion of black colour, may mean that the specimen was charred. Alternatively, the description may indicate a higher degree of coalification.

It is possible that the specimen came from Ponte Caldaro (cf. **62**).

REFERENCES: 17th-cent. Index, p. 362

63

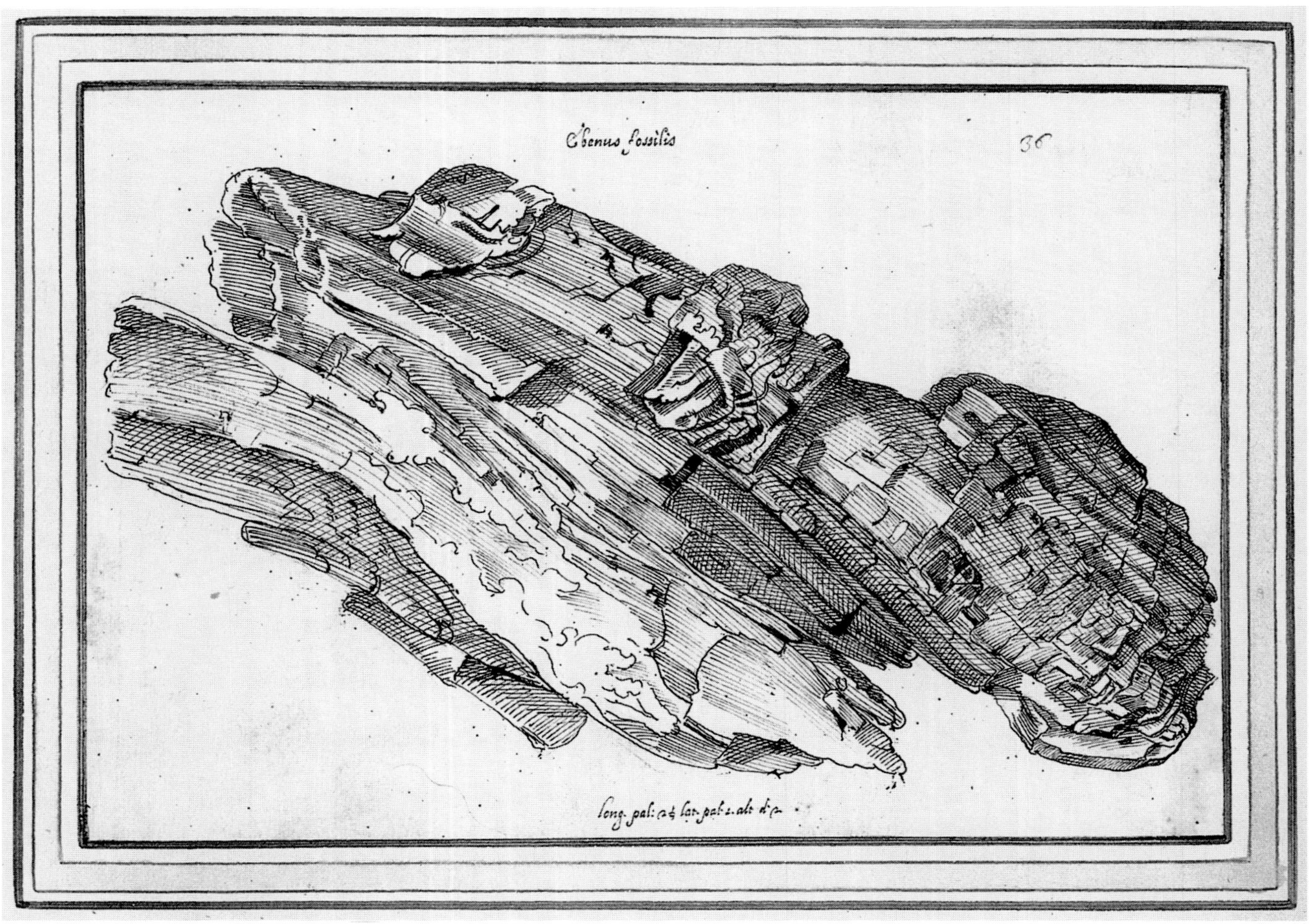

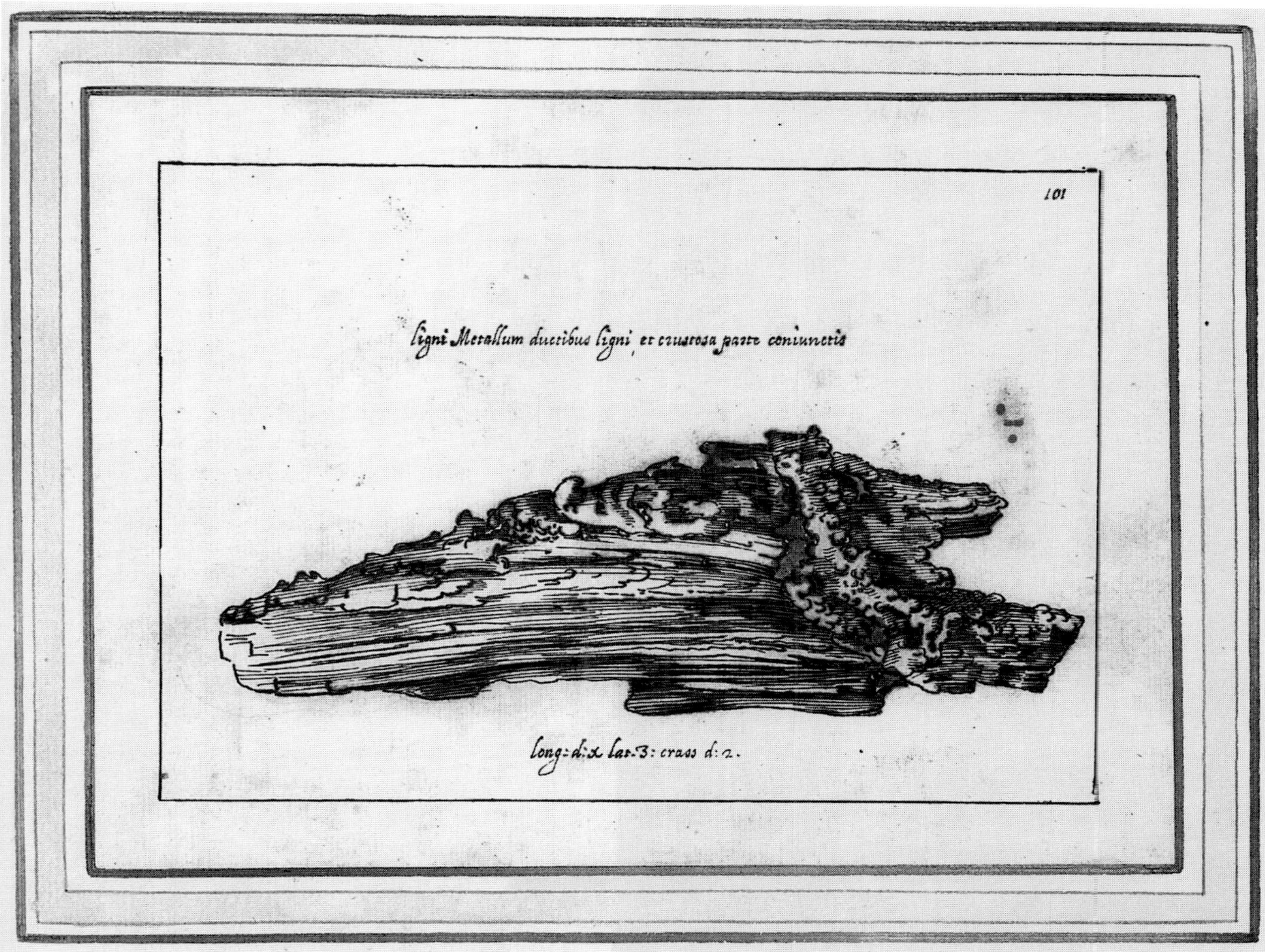

64

64. Fossil wood

Windsor, RL 25640, *Fossils*, vol. XVI, fol. 1

Pen and ink, over traces of black chalk; retouched in ink on mount sheet; ink frame line

164 × 230 mm; small holes caused by ink. *Watermark:* Bird 41D

NUMBERING: [ink, top right corner] *101*

ANNOTATION: [ink, top] *ligni Metallum ductibus ligni, et crustosa parte coniunctis* [bottom] *long: d: x lat. 3: crass d: 2.*

This specimen of fossil wood is described as 'metallized wood with ligneous veins joined to the crusty part'. It is probably pyritized.

REFERENCES: 17th-cent. Index, p. 366

65. Piece of fossil wood, possibly partially charred

Windsor, RL 25641, *Fossils*, vol. XVI, fol. 2

Pen and ink, over traces of black chalk; retouched in ink on mount sheet

187 × 266 mm; sheet trimmed through image on right and left sides; small holes caused by ink

NUMBERING: [ink, top left corner] *102*

ANNOTATION: [ink, top] *ligneum Nigrum cum ramo / ponderosum intus lapideum* [bottom] *long. pal. 2. lat. pal: 1½ alt: p ½*

As the inscription notes, this specimen of fossil wood is a 'black woody fragment with a branch, on the inside heavy and stony'. Once again (cf. **64**) the black colour together with the cracking seen on the lower half of the specimen may indicate charring of the wood.

REFERENCES: 17th-cent. Index, p. 366

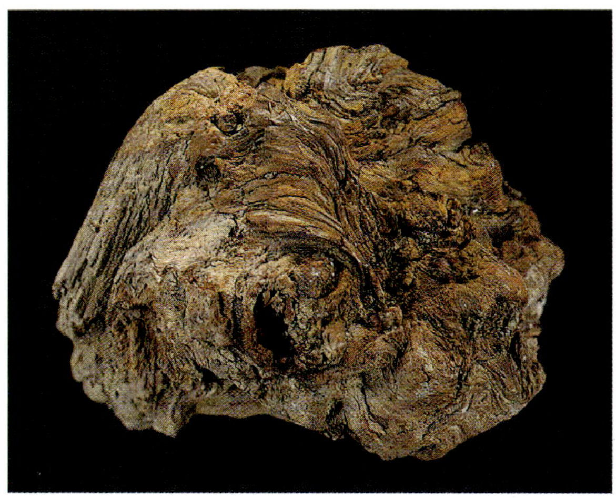

Comp. fig. 65. Brown wood (possibly charred) from Dunarobba

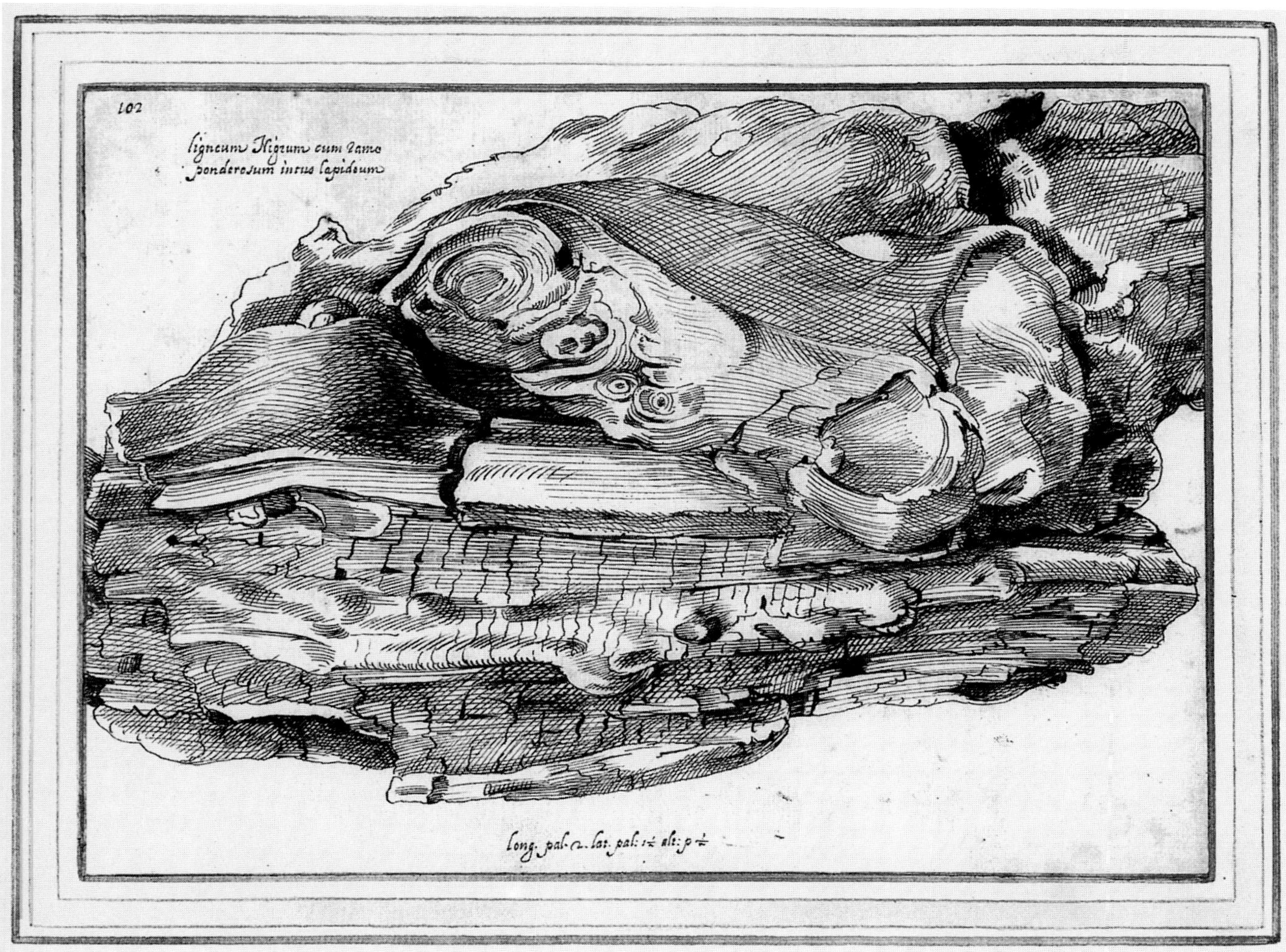
102
ligneum Nigrum cum Ramo
ponderosum intus lapideum
long. pal. 2. lat. pal: 1½ alt: p ½

65

66. Two views of a piece of fossil wood

Windsor, RL 25554, *Fossils*, vol. XIV, fol. 25

Pen and ink, over black chalk

185 × 234 mm

NUMBERING: none; formerly *18* (see Appendix I)

ANNOTATION: [ink, upper left] *Creti lignum exiccatione, in / lignum fere redactum*
[centre left] *long. palm: 1. lat. d: 6. alt. d: 3*
[centre right] *Sectio hic videtur*
[bottom] *long: d: 7. lat: d: 3. alt: d. 1½.*
[vertical inscription, top edge, in Cesi's hand, erased]

A piece of wood has been excavated from the clay; it has dried out and fragments of the wood are flaking off. The lower drawing represents a section of the same specimen. The inscription on the upper left records that by drying out, the [piece of] *cretilignum* has almost turned into wood (for an explanation of this process, see p. 133).

REFERENCES: 17th-cent. Index, p. 361

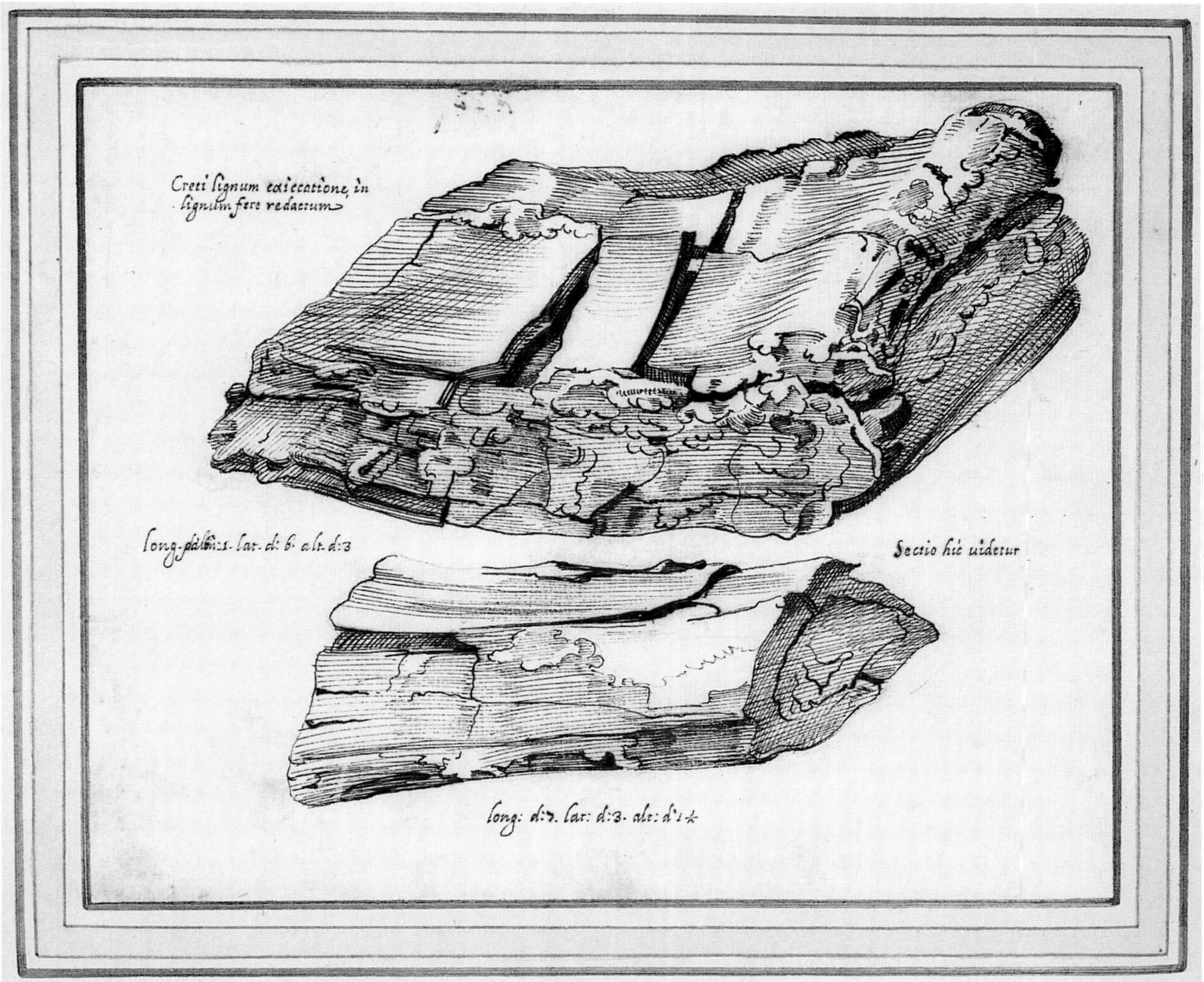
Creti lignum exiccatione in lignum fere redactum
Sectio hic uidetur

66

67. Five fragments of fossil wood

Windsor, RL 25592, *Fossils*, vol. XV, fol. 1

Pen and ink, over traces of black chalk; black chalk frame line

185 × 244 mm; small holes caused by ink

NUMBERING: [ink, upper left] *54*

ANNOTATION: [ink, upper left] *Tenue/ fasciæ/ Variæ/quasi / coriacæ/ sub nigræ*
[upper right] *Crassiuscula, crispata Tortuosa, quasi cocto corio alicubi/ subrubra;*
[centre left] *quasi cintu-/ræ Anulus*
[lower left] *Tenuissima quasi/ gladius plana prorsus*
[centre] *Corticosa.*
[in Cesi's hand, vertically: top] [Crassiusc]*ula crispata/* [Tortuo]*sa quasi/* [co]*rio/ subrubra;*
[bottom left] *tenuissi*[ma]/ *quasi/ gla*[dius]/ *plana p*[rorsus]
[bottom right] *Corticosa* [illegible at top right]

These five specimens of fossil wood are typical of lignitized woods which have dried out and peeled off in strips, and in some cases been flattened by compaction. They have extensive descriptions alluding to their 'bark-like' nature.

The upper right-hand specimen may have been pyritized as it is described as 'metallized' wood with 'brick-like' bark; alternatively the description may refer to the precipitation of other iron minerals.

REFERENCES: 17th-cent. Index, p. 364

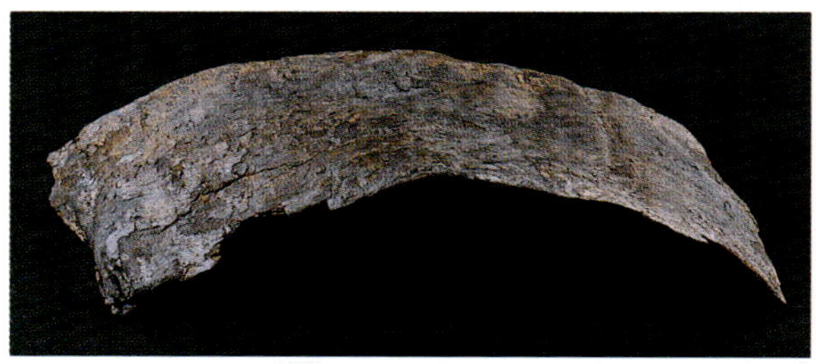

Comp. fig. 67. Flattened branch from Rosaro

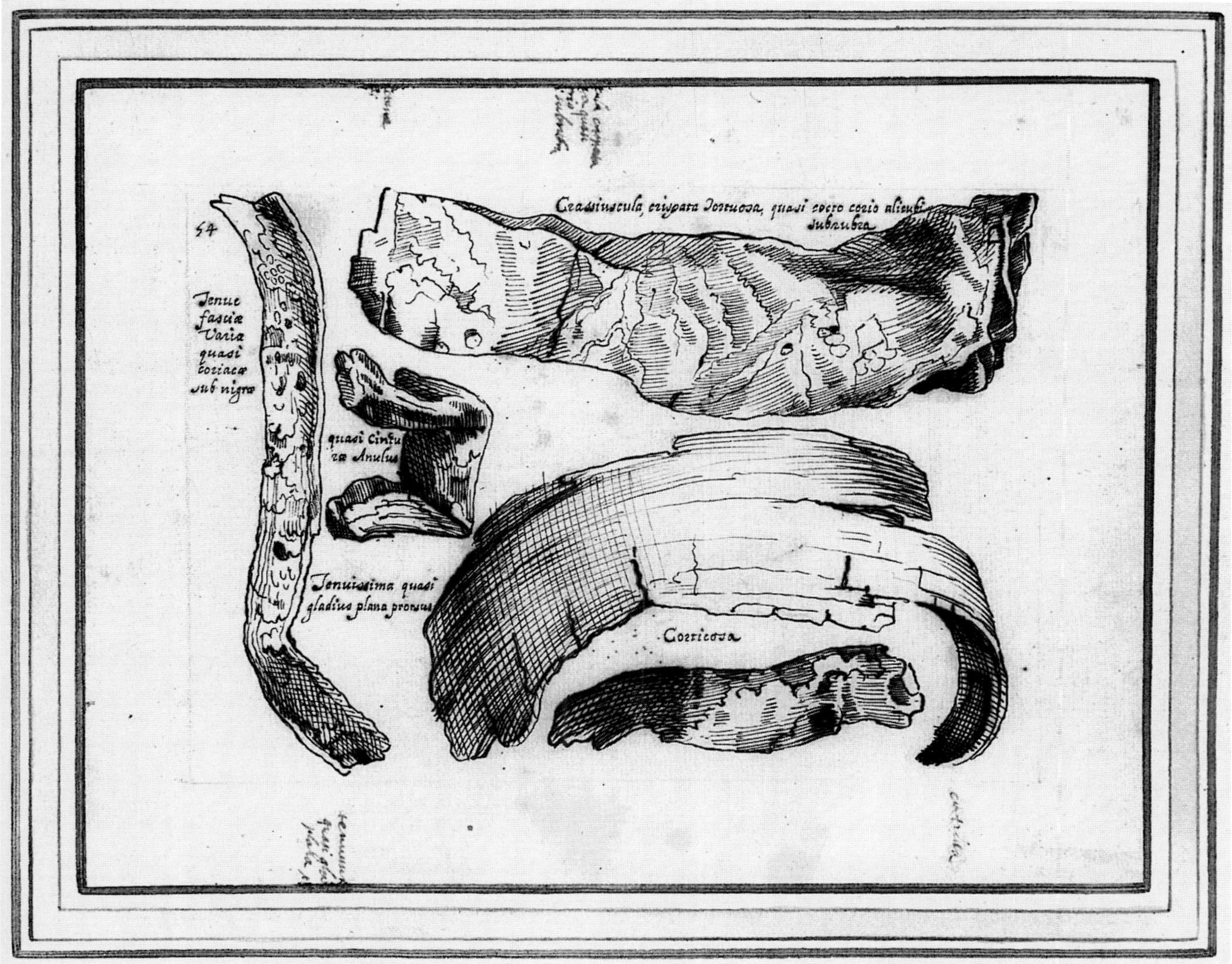
Crassiuscula, crispata tortuosa, quasi cocto corio alicui subrubra
54
Tenue fasciæ Variæ quasi coriaceæ sub nigræ
quasi cinctu
ræ Anulus
Tenuissima quasi gladius plana prorsus
Corticosa

67

68. Two pieces of fossil wood

Windsor, RL 25676, *Fossils*, vol. XVI, fol. 37

Pen and ink, over black chalk; measurements given for right-hand specimen only

190 × 248 mm; small holes caused by ink

NUMBERING: [ink, top right corner] *137*

ANNOTATION: [ink, left] *Ramus ligni cretosus ad pact, in ramum/ A. Creti lignum, et in Massa, B. Sub ala*
[right] *A: Sciunctum Ramum* [bottom] *long: pal: 1. d. 5. lat. d: 5. alt. d. 2.*
[Cesi's inscriptions at lower edge, cut]

Two pieces of fossil wood are shown with lumps of compacted clay and wood adhering to them. Part of the left-hand drawing seems to have been left incomplete.

REFERENCES: 17th-cent. Index, p. 369

68

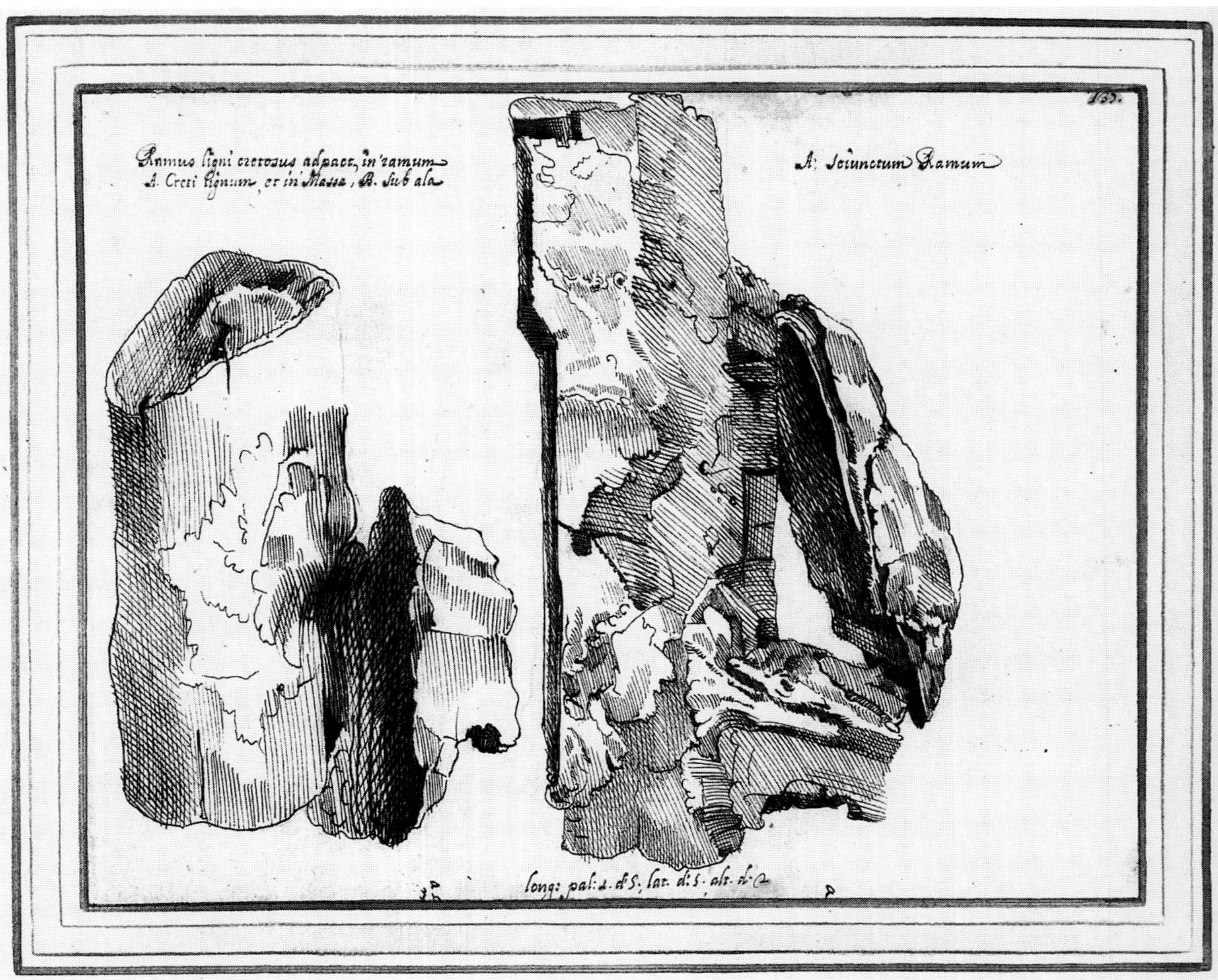

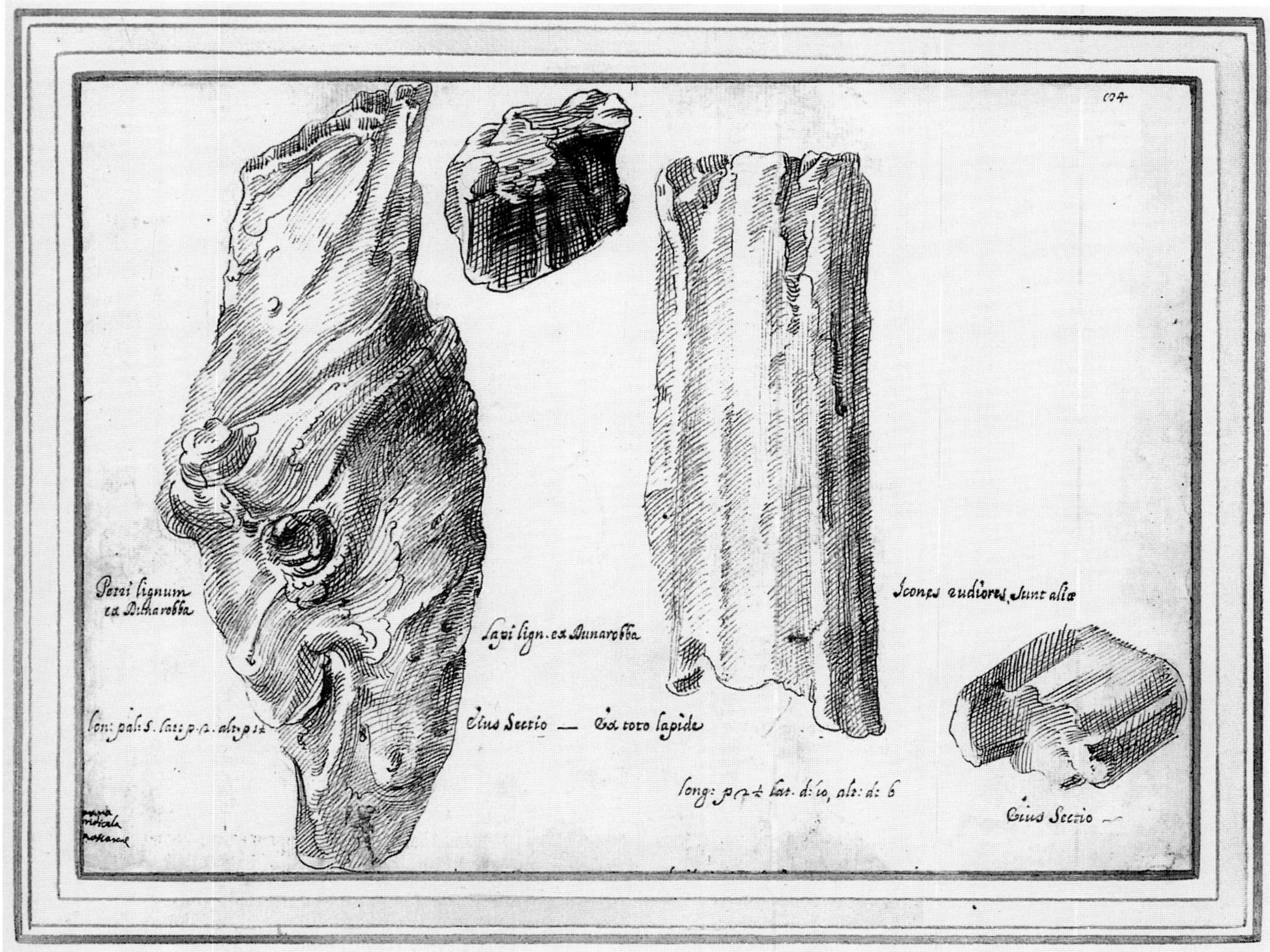

69

69. Two pieces of fossil wood from Dunarobba, with their sections

Windsor, RL 25623, *Fossils*, vol. XV, fol. 32

Pen and ink, over black chalk

196 × 273 mm. *Watermark:* Bird 41B

NUMBERING: [ink, upper right corner] *84*

ANNOTATION: [ink, left] *Petri lignum/ ex Dunarobba*
[bottom left] *long: pal: 5. lat: p. 2. alt. p 1½* [centre] *Lapi lign. ex Dunarobba*
[lower centre] *Eius Sectio—Ex toto lapide [right] Icones rudiores sunt aliæ*
[lower right] *long: p 2½ lat. d: 10, alt: d: 6* [lower right corner] *Eius Sectio.*
[Cesi's inscription, lower left corner, cut]

Different names are given to the two specimens (each shown with its cross-section), although both indicate the fact that they seem to be combinations of wood and stone. Unusually for this corpus, a clear reference is made to qualitative differences between different images (*Icones rudiores sunt aliæ*).

REFERENCES: 17th-cent. Index, pp. 365, 369

70. Petrified wood

Windsor, RL 25671, *Fossils*, vol. XVI, fol. 32

Pen and ink, over black chalk

184 × 281 mm; torn along left edge. *Watermark:* Bird 41B

NUMBERING: [ink, top right corner] *132*

ANNOTATION: [ink, top] *Petrefactum aliqua medio in nodo ligni particula album* [bottom] *long: pal: 2½ lat. p. 1. alt. pal. ½*

The meaning of the annotation is not totally clear but could be translated as 'a white petrified [piece] with some particles of wood in the middle of the knot'. The white colour may relate to the occurrence of silica in the cell spaces of the wood.

Small areas of the drawing seem to be unfinished.

REFERENCES: 17th-cent. Index, p. 368

70

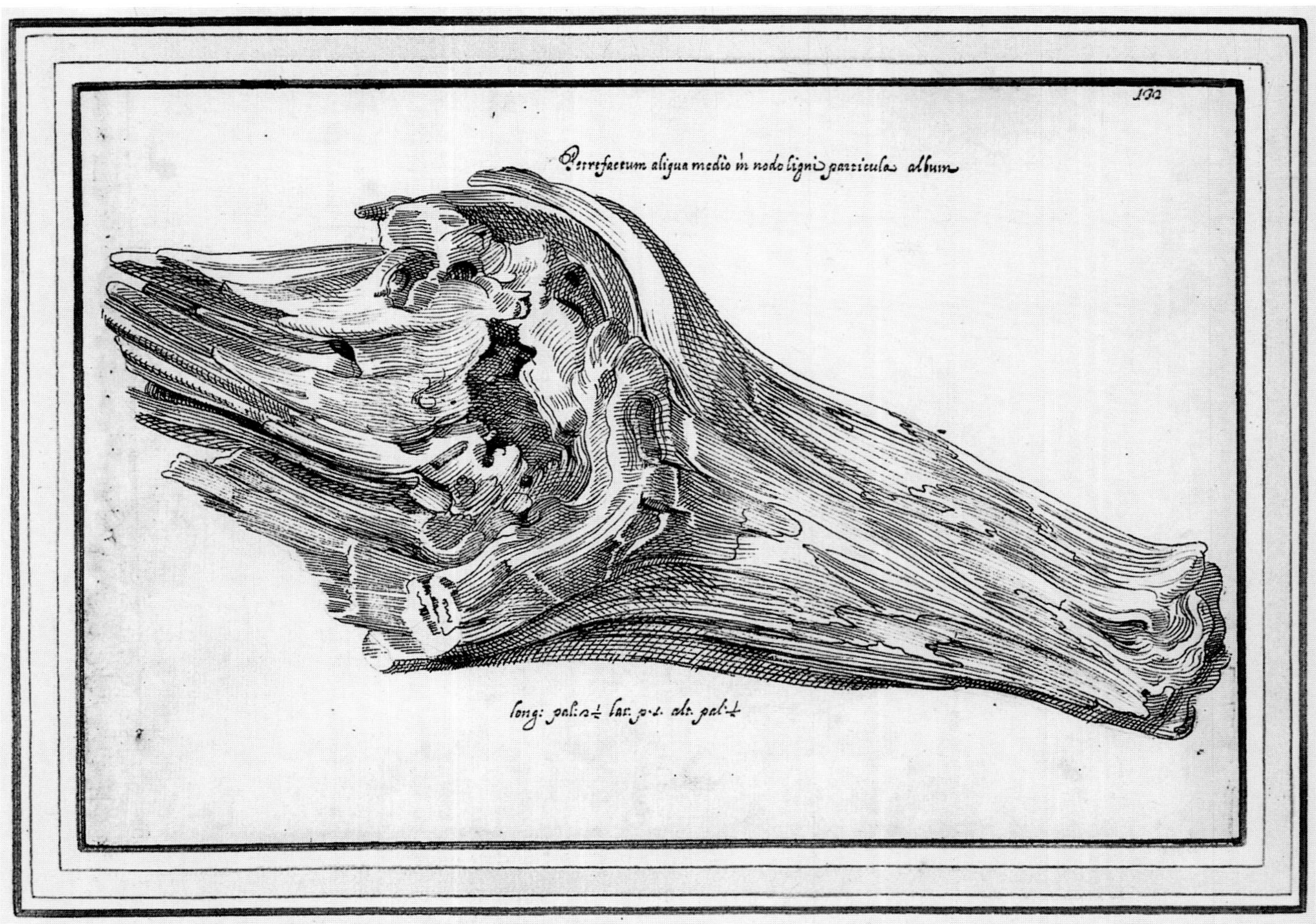

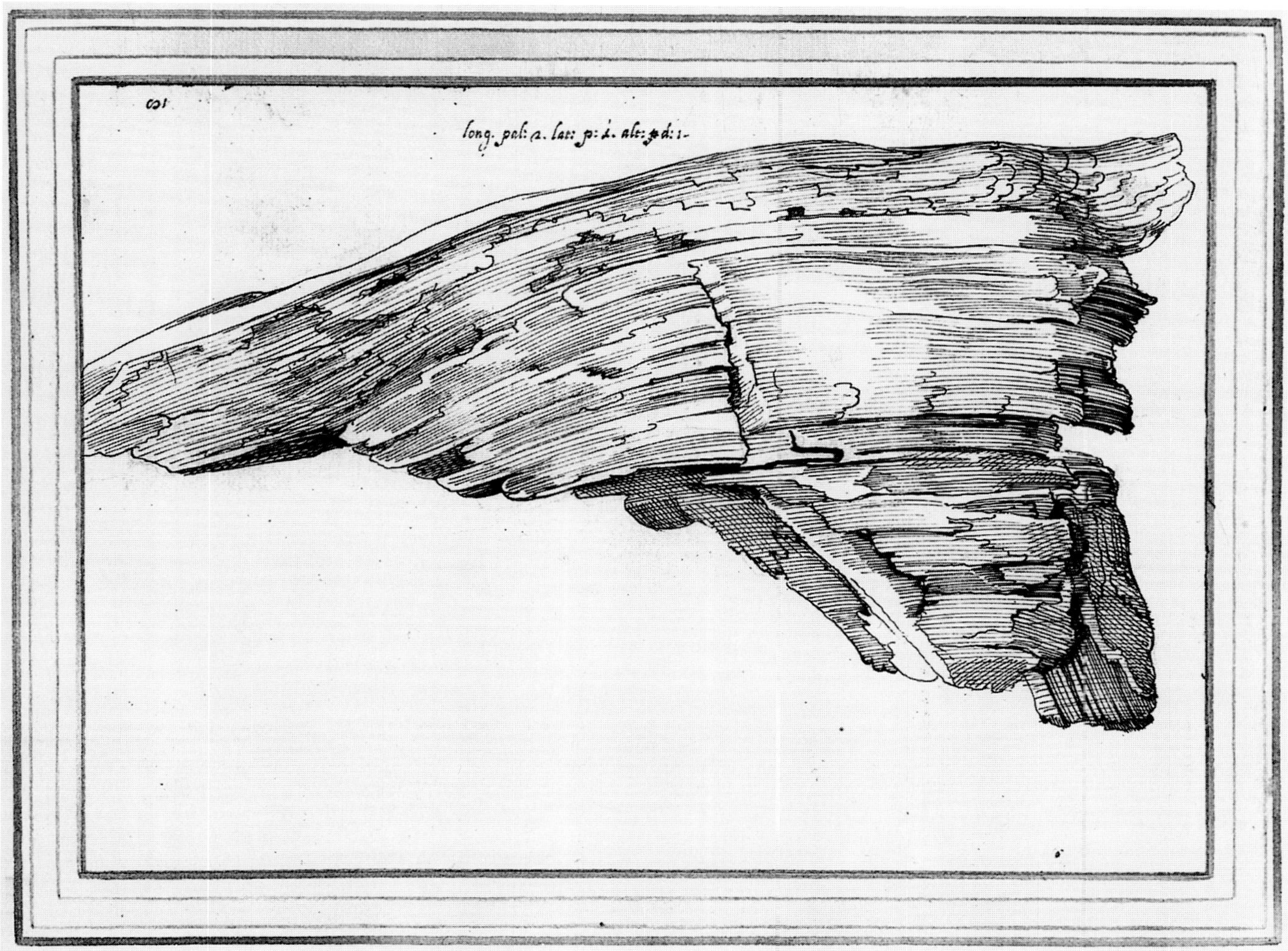

71

71. Piece of fossil wood

Windsor, RL 25620, *Fossils*, vol. XV, fol. 29

Pen and ink, over black chalk

186 × 261 mm

NUMBERING: [ink, top left corner] *81*

ANNOTATION: [ink] *long. pal: 2. lat: p: 1. alt: d. 1*

This piece of fossil wood shows part of a broken branch and knot. The drawing has been trimmed through the image on the left.

REFERENCES: 17th-cent. Index, p. 365

72. Two pieces of fossil wood

Windsor, RL 25668, *Fossils*, vol. XVI, fol. 29

Pen and ink, over black chalk

201 × 250 mm. *Watermark:* Bird 41B

NUMBERING: [ink, top right corner] *129*

ANNOTATION: [ink, upper left] *Nodo volutio Turbinata nigra*
[lower left] *long: pal: 3. lat. pal: 2. alt: d: 4.*
[lower right, cut] *long: pal: 3½ lat: d:* [in index *'8'* is given]/ *et crass:*
[Cesi's inscriptions trimmed along lower edge, left and centre]

The drawing shows two pieces of fossil wood. The left-hand specimen, described as a 'black whorled form', is probably a knot end of a branch. The right-hand specimen bears no description but is a branch with several branchlet scars. Perhaps the latter drawing was added a time after the first; in any case, it was done in a different coloured ink.

REFERENCES: 17th-cent. Index, p. 368

73. Two pieces of fossil wood

Windsor, RL 25621, *Fossils*, vol. XV, fol. 30

Pen and ink, over black chalk

186 × 262 mm; small holes caused by ink. *Watermark*: Bird 41C

NUMBERING: [ink, upper right corner] *82*

ANNOTATION: [ink, top] *lapifactum ex Albo, et Fusco varium*
[lower left] *long: pal: 2½ lat: p. 1. alt / circa pal: 1*
[lower right] *long: pal: 3. lat. p. 1. alt. pal: 1.*

The left-hand specimen is described as 'a [piece] turned into stone, variegated white and black' and is probably petrified. No description is given of the right-hand specimen, although it may well have been similar to its companion.

REFERENCES: 17th-cent. Index, p. 365

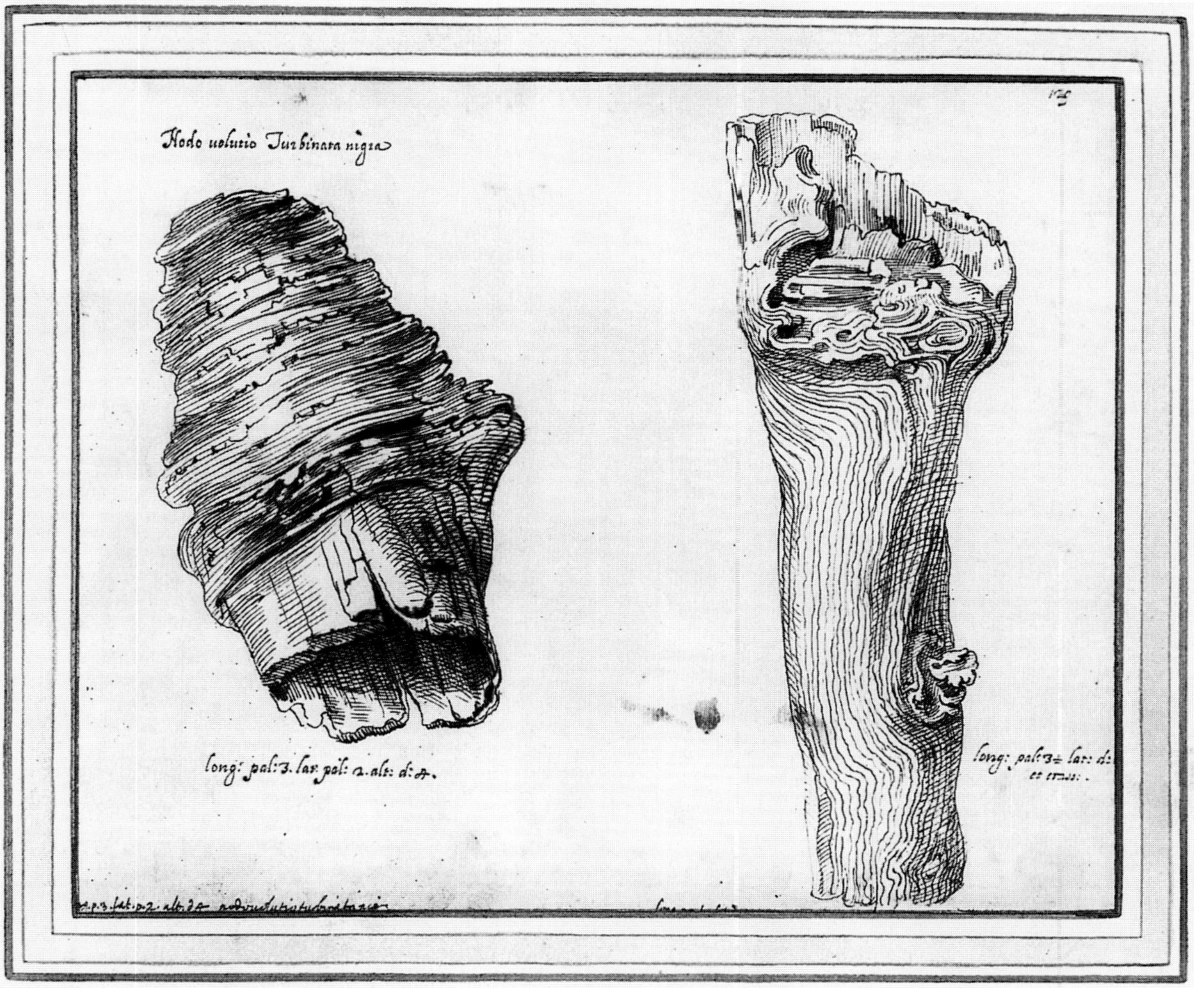

Nodo uolutio Turbinata nigra
long: pal: 3. lar: pal: 2. alt: di: 4.
long: pal: 3½ lar: di:
et trass:

72

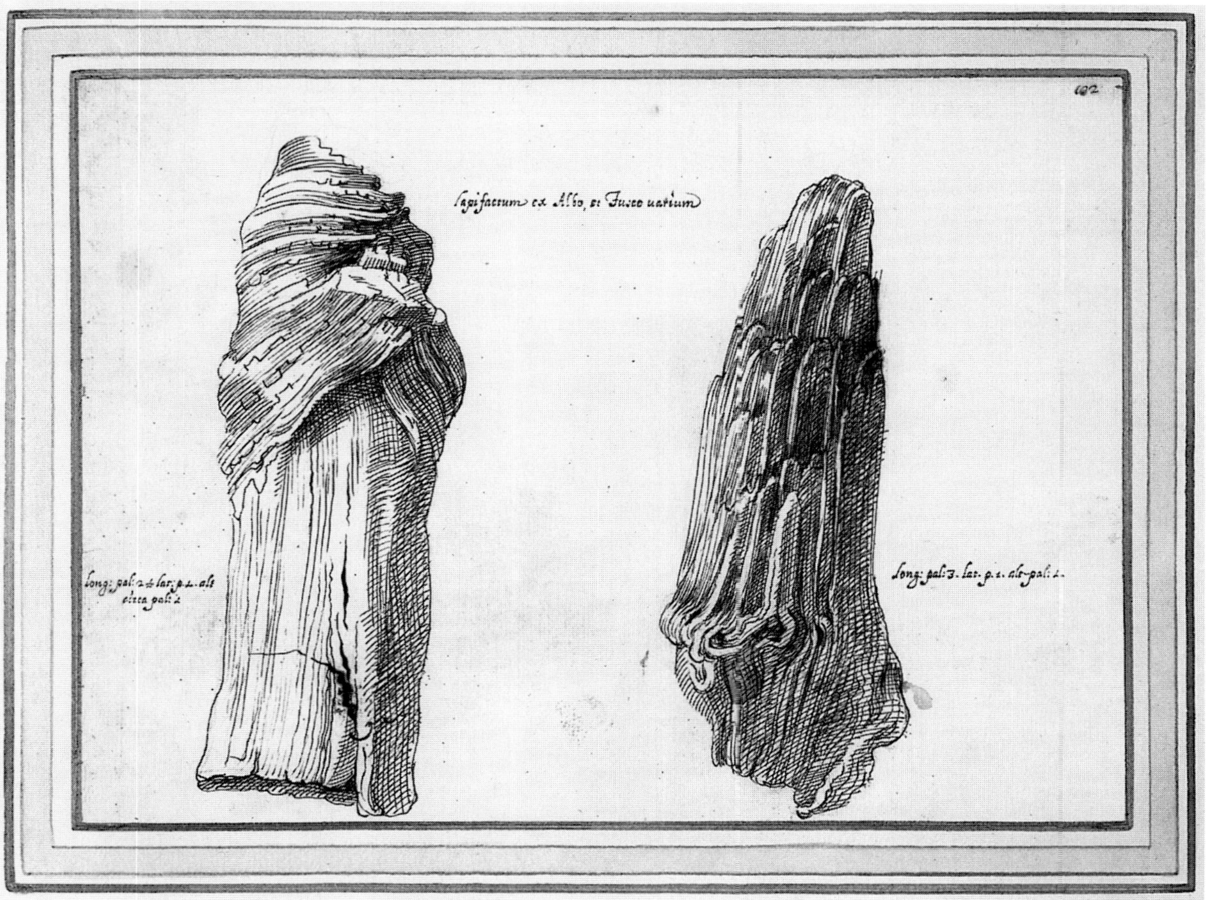

lapisfattum ex Albo, et Fusco uarium
long: pal: 2½ lar: p. 1. alt
circa pal: 1
long: pal: 3. lar: p. 1. alt pal: 1.

73

74. Fossil wood (knot)

Windsor, RL 25642, *Fossils*, vol. XVI, fol. 3

Pen and ink, over black chalk

208 × 267 mm; small holes caused by ink

NUMBERING: [ink, top right corner] *103*

ANNOTATION: [ink, top] *ligneum nodo Volutum* [bottom] *long: pal: 2½ lat: p: 1½ Crass. p. 1* [Cesi's inscriptions, cut lower right edge]

This convoluted piece of fossil wood, described as being 'wood wrapped around a knot', is the insertion point of a branch or knot. The drawing has been slightly trimmed through the image on either side.

REFERENCES: 17th-cent. Index, p. 366

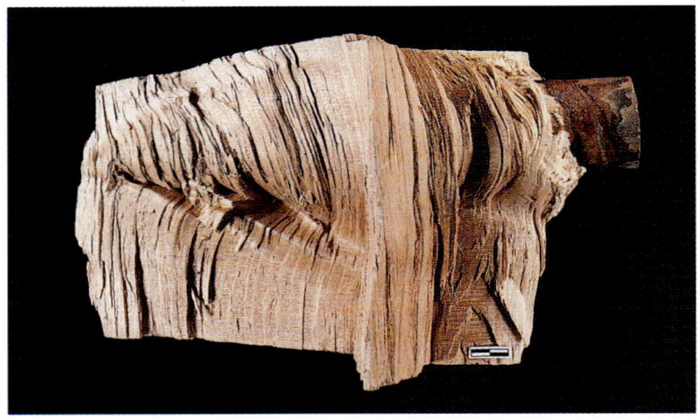

Comp. fig. 74. Knot end or branch insertion point into trunk of modern tree

74

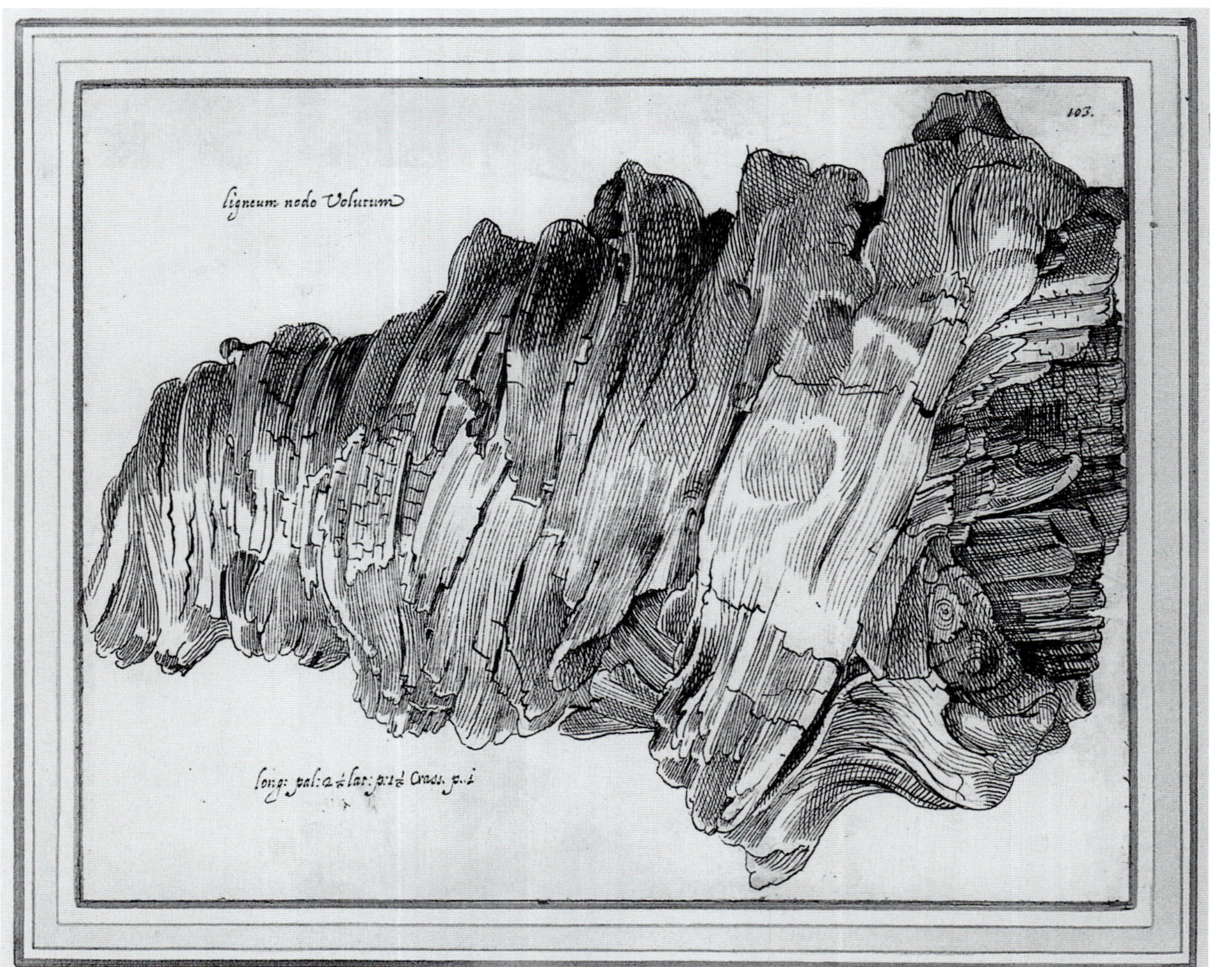

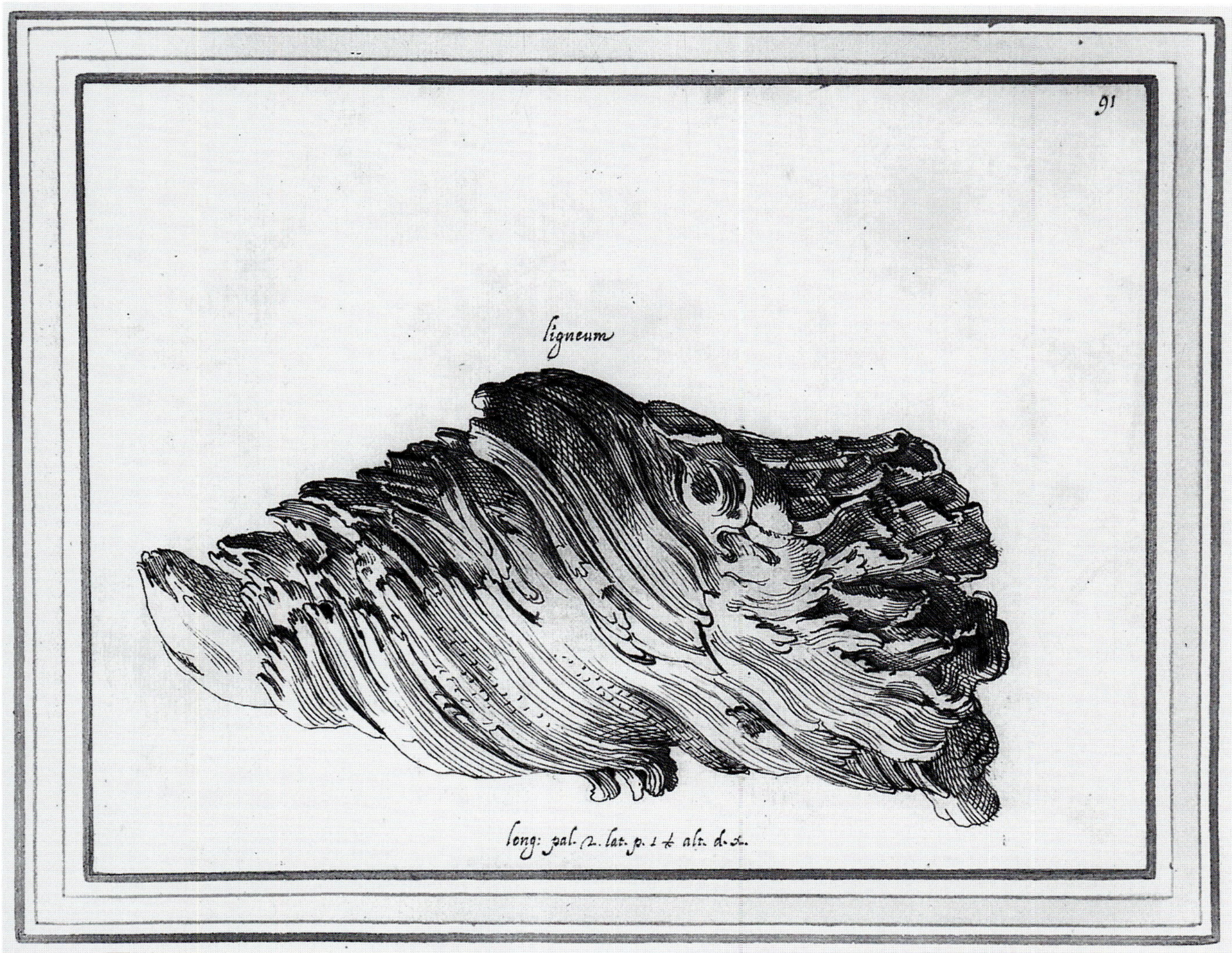

75

75. Piece of fossil wood

Windsor, RL 25630, *Fossils*, vol. XV, fol. 39

Pen and ink, over traces of black chalk

189 × 253 mm

NUMBERING: [ink, top right corner] *91*

ANNOTATION: [ink, top] *Ligneum*
[bottom] *long: pal. 2. lat. p. 1½ alt. d. x.*

This shows the typically twisted wood at the base of a branch. Space appears to have been left in the upper part for another drawing.

REFERENCES: 17th-cent. Index, p. 365

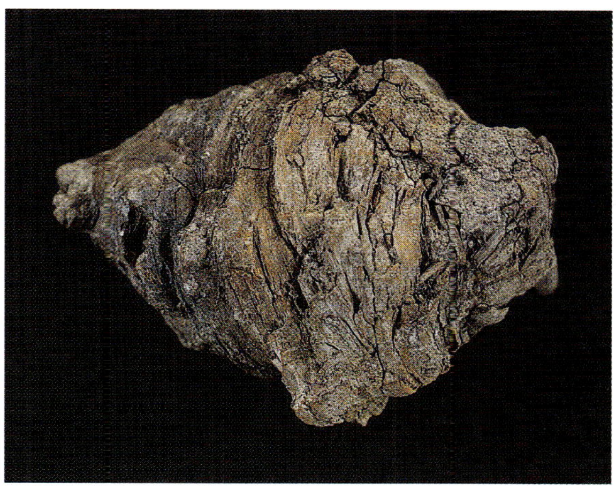

Comp. fig. 75. Branch insertion showing spiral character, from Dunarobba

76. Three specimens of fossil wood

Windsor, RL 25696, *Fossils*, vol. XVI, fol. 57

Pen and ink and blackened lead-white highlights, over traces of black chalk

196 × 255 mm; rubbed and scraped away, top and left; measurements of the two larger specimens are given. *Watermark*: Bird 41B

NUMBERING: [ink, upper right] *157*

ANNOTATION: [ink, upper left, another hand] *Voluta*
[scribal hand] [vo]*luta obrotunda lignea* [lower left] *long: d: 8: lat: d. 10, alt: d: 9.*
[upper right] *Voluta pressa pyramidalis/ lignea* [bottom right] *long: pal: 2½ lat. 6½ alt. d: 7.*
[bottom left] *Frustum ligneum Undatum* [Cesi's inscriptions, top and left, cut]

The specimen on the upper left is described as a 'round, woody volute', that on the right as a 'compacted pyramidal woody volute'. The latter shows the knot-end of a branch. The specimen on the lower left, described as a 'wavy, woody fragment', was engraved as fig. H of Stelluti's Plate 11. Stelluti noted that 'piece *H* was found in that form patterned with waves as its drawing illustrates'.

The drawing sheet was trimmed after the scribal inscriptions were added, and the affected words from each edge were supplied by a different hand.

REFERENCES Stelluti 1637, p. 10; 17th-cent. Index, pp. 364, 370

ENGRAVINGS: Stelluti 1637, Pl. 11 (Fig. H); for proof engraving in Montpellier, BEM, MS H.173, fol. 20, see p. 403

76

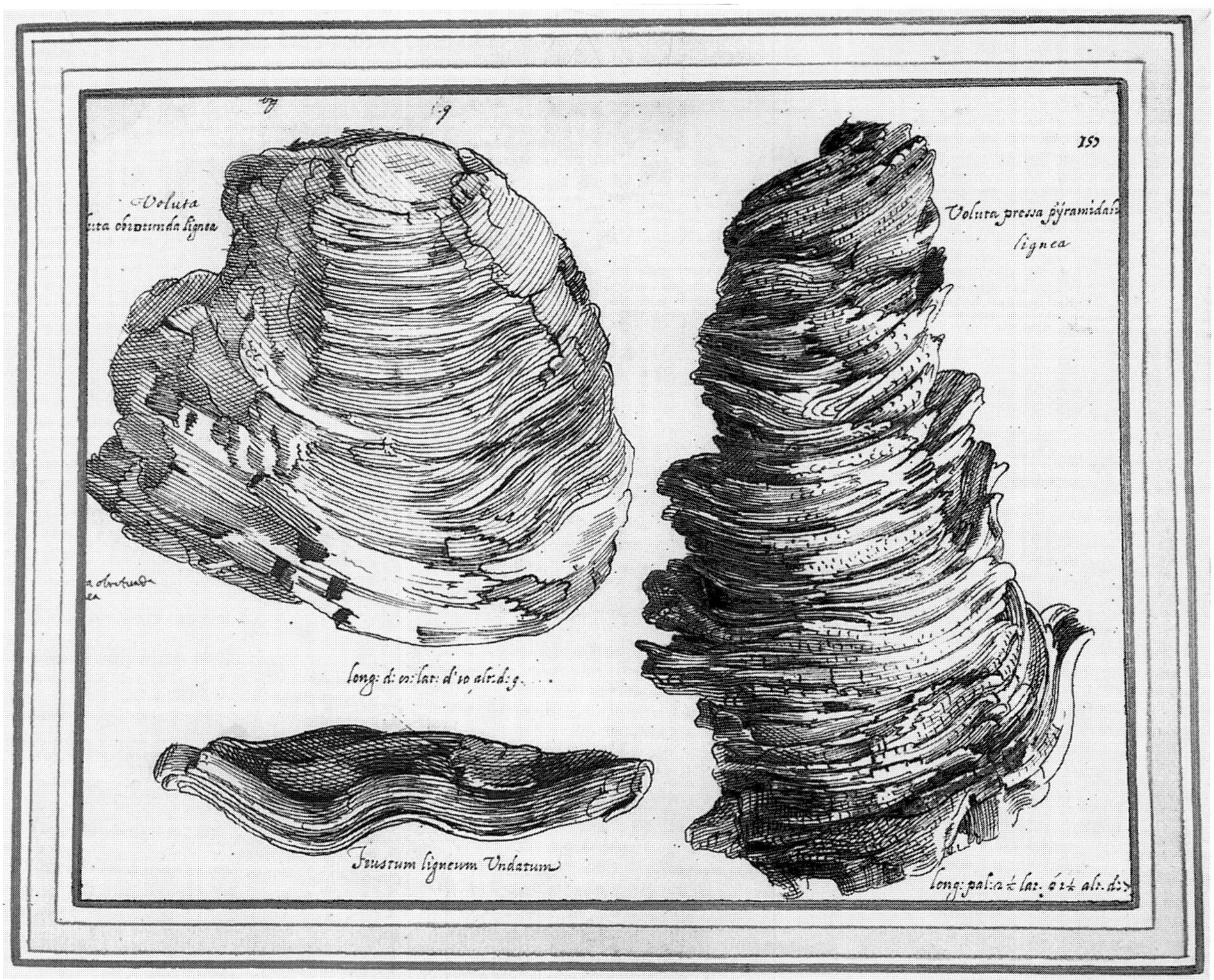

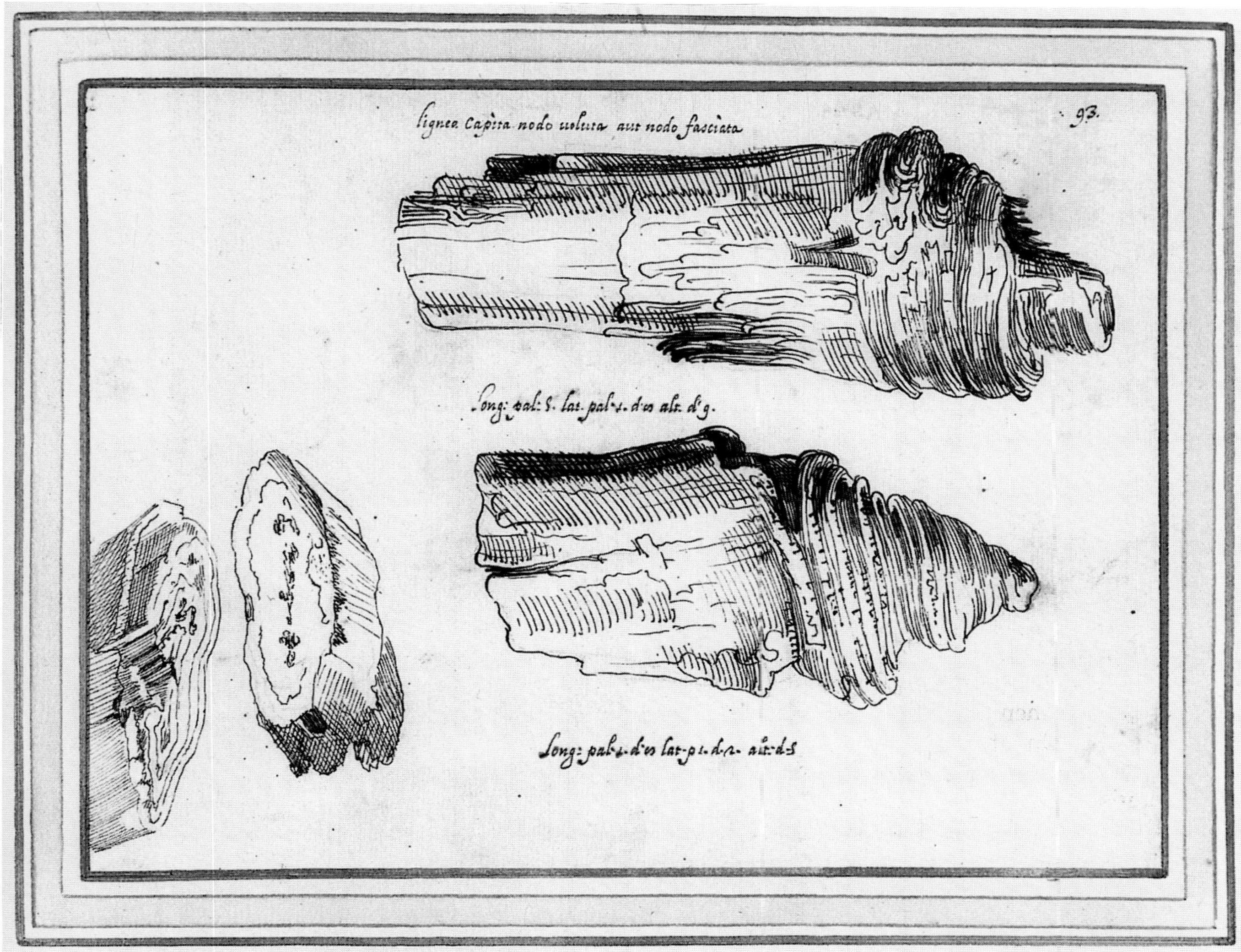

77. Two pieces of fossil wood, showing branches with knot-hole structures

Windsor, RL 25632, *Fossils*, vol. XV, fol. 41

Pen and ink, over traces of black chalk

186 × 254 mm

NUMBERING: [ink, top right corner] *93*

ANNOTATION: [ink, top] *lignea Capita nodo voluta aut nodo fasciata*
[centre] *long: pal: 5. lat. pal. 1. d. 8 alt. d. 9.*
[bottom] *long: pal. 1. d. 8 lat. p 1. d. 2. alt: d. 5*

Two specimens of fossil wood, both showing branches with knot-hole structures. These are noted in the inscription as 'woody heads wrapped around a knot or swathed in a knot'. The specimens are illustrated together with cross-sections, of which the left-hand one shows growth rings.

REFERENCES: 17th-cent. Index, p. 365

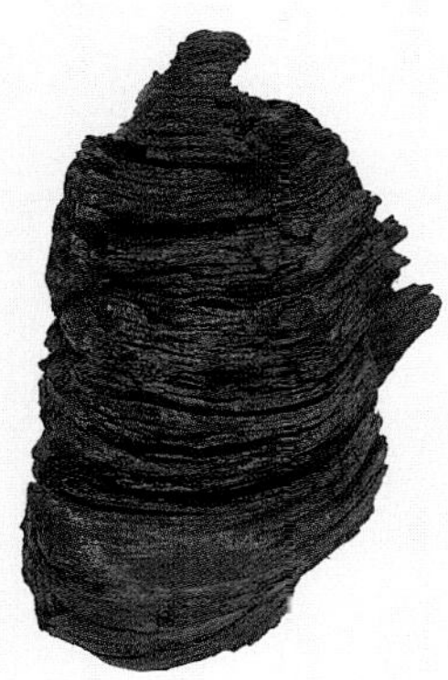

Comp. fig. 76. Branch insertion showing spiral character, from Dunarobba

78. Two pieces of fossil wood, one from Dunarobba

Windsor, RL 25662, *Fossils*, vol. XVI, fol. 23

Pen and ink, over black chalk

204 × 255 mm; small holes caused by ink; torn at lower edge

NUMBERING: [ink, top left corner] *123*

ANNOTATION: [ink, top left] *Nigrescens nodo volutum lapi lignum Dunarobbæ / forte aliquantulum Metallico pondere*
[lower left] *long: pal: 3. lat: p: 1. alt: pal: 1.*
[top right] *ligneum Nigreum*
[lower right] *long: pal: 3½ lat: pal: 2. alt: p.* [figure cut]
[inscriptions in Cesi's hand at lower edge, erased]

The left-hand specimen of fossil wood came from Dunarobba. It is described by the term *lapilignum* and is thus probably petrified. Its 'quite heavy, metallic weight' may well be caused by the occurrence of iron. The right-hand specimen is 'black and woody'. Both show prominent knot ends. They do not seem to be referred to in the Index.

REFERENCES: 17th-cent. Index, p. 368

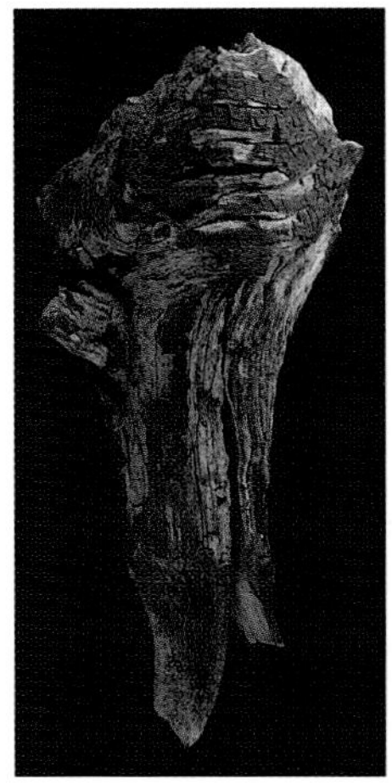

Comp. fig. 78. Branchlet from Dunarobba

78

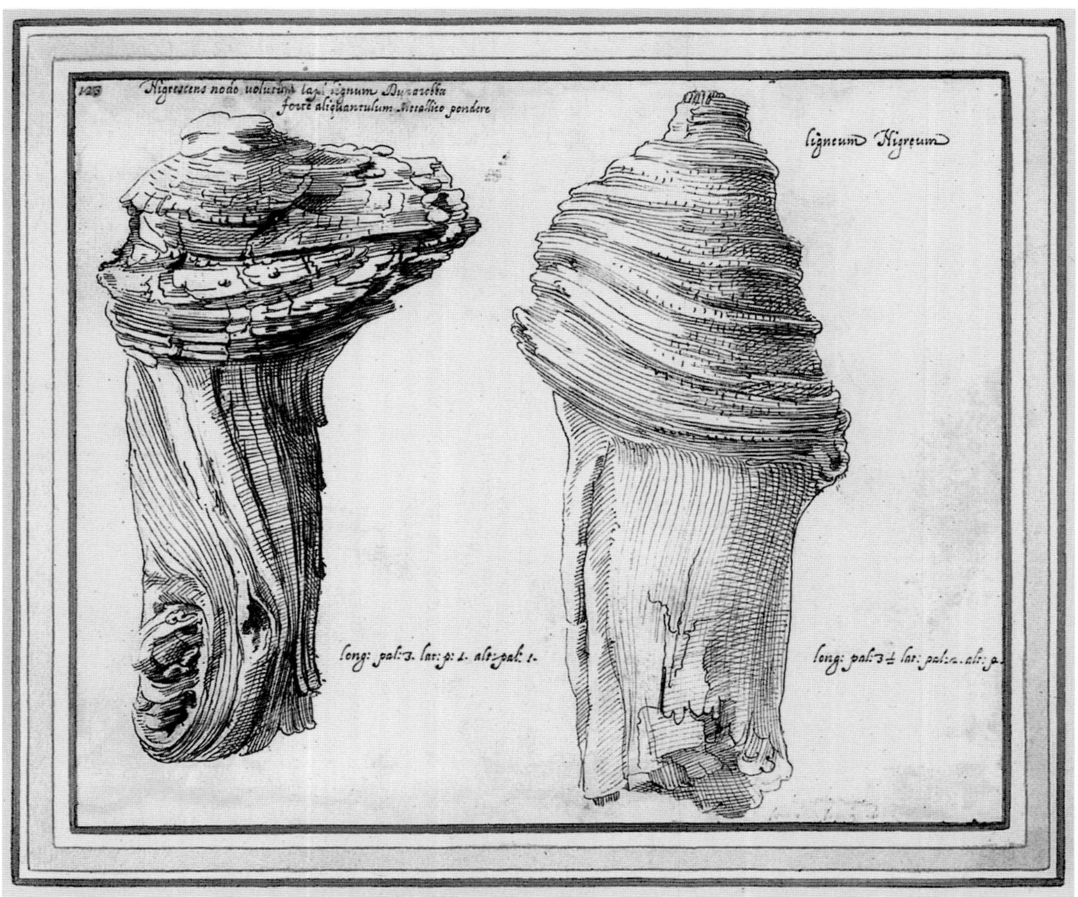

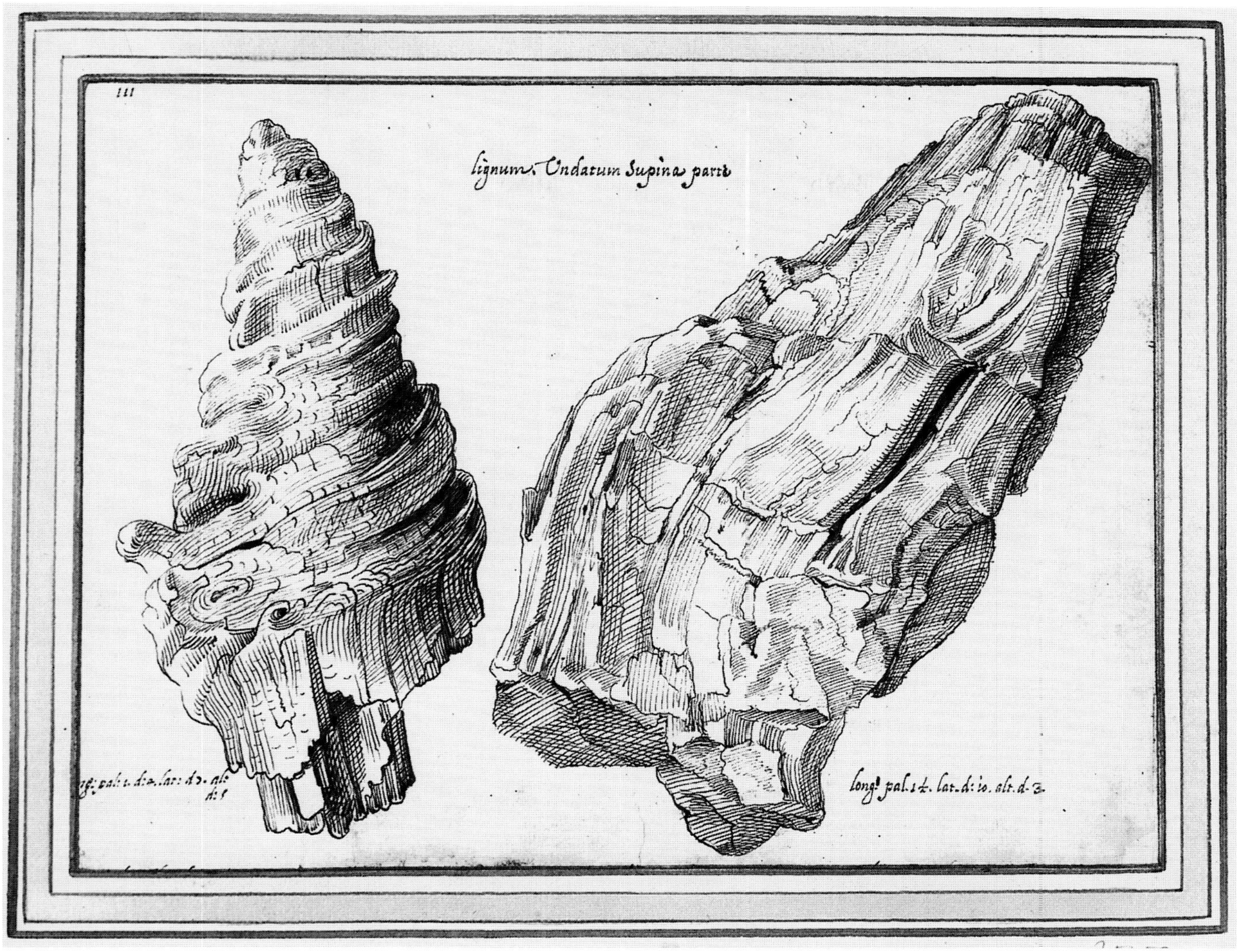

79

79. Two pieces of fossil wood, the left-hand one showing a knot end

Windsor, RL 25650, *Fossils*, vol. XVI, fol. 11

Pen and ink, over traces of black chalk

195 × 269 mm. *Watermark:* Bird 41B

NUMBERING: [ink, top left corner] *111*

ANNOTATION: [ink, top] *lignum Undatum Supina parte*
[bottom left] [lo]*ng: pal: 1. d: 4. lat: d 7. alt/d: 5*
[bottom right] *long. pal. 1½. lat. d: 10. alt. d. 3.*
[traces of Cesi's inscriptions at lower right and left edge]

The 'waviness' (*undatum*) of the wood referred to in the inscription is often commented upon in Cesi's letters about the fossil woods of Umbria, and in Stelluti's treatise (cf. pp. 373–87, 389–90 for several instances). It was clearly a subject of considerable interest—both aesthetic and scientific—for the members of the *Lincei*. The left-hand drawing clearly depicts a branch with a knot end and may be compared to several other specimens in the collection (see **76–78**).

REFERENCES: 17th-cent. Index, p. 367

80. Piece of fossil wood

Windsor, RL 25651, *Fossils*, vol. XVI, fol. 12
Pen and ink, over black chalk; marked with a cross
188 × 250 mm; small holes caused by ink
NUMBERING: [ink, top right corner] *112*
ANNOTATION: [ink, top] *lignum cum interiori exiguo lapideo nigrescens nodi Tuberati*
[bottom] *Alt: pal. 2½ lat: 2 Crass: 2½*

This contorted piece of fossil wood is described as 'blackish wood of a knotty excrescence with a slightly stony interior'.

REFERENCES: 17th-cent. Index, p. 368

80

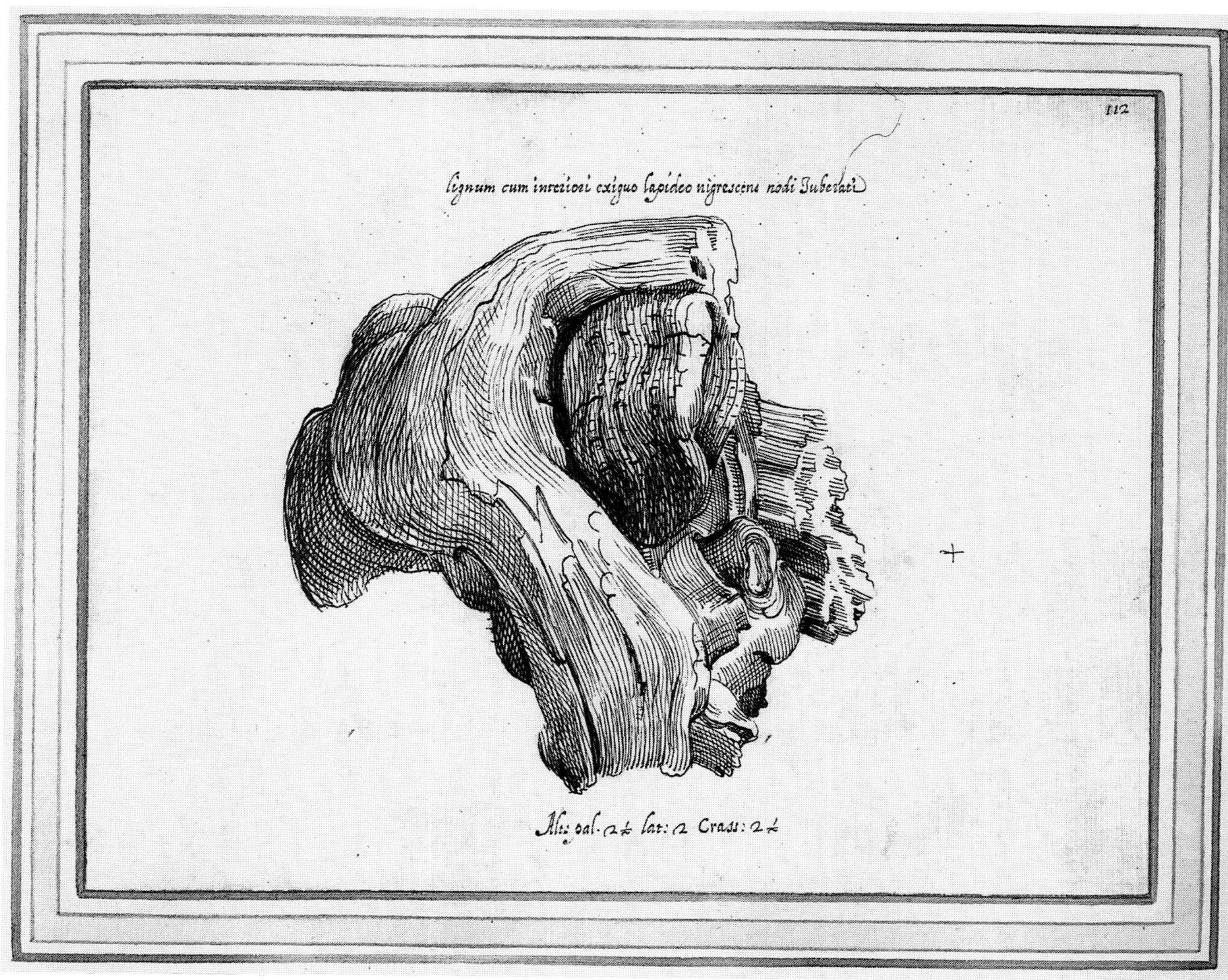

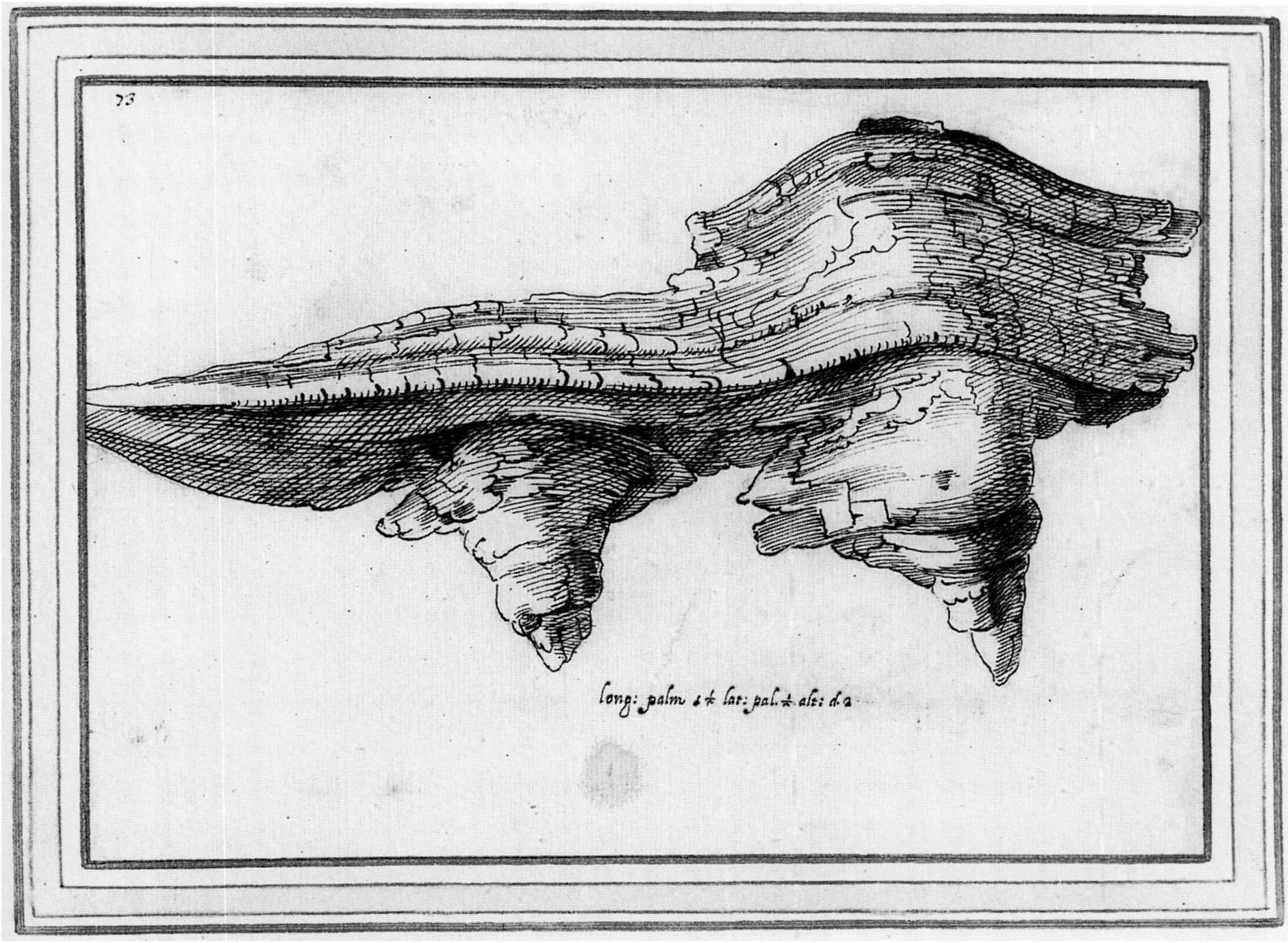

81

81. Conifer-like fossil wood with base of two branches

Windsor, RL 25612, *Fossils*, vol. XV, fol. 21

Pen and ink, over black chalk

181 × 261 mm; sheet trimmed through image, left edge

NUMBERING: [ink, top left corner] *73*

ANNOTATION: [ink] *long: palm 1½ lat: pal. ½. alt: d. 2*

A piece of fossil wood showing two prominent branches: it is possibly coniferous. This is one of a number of drawings in the corpus that contains no further description beyond the measurements of the specimen.

REFERENCES: missing from 17th-cent. Index

82. Fossil wood from Sismano

Windsor, RL 25663, *Fossils*, vol. XVI, fol. 24

Pen and ink, over black chalk

189 × 261 mm; small holes caused by ink; sheet trimmed through left and lower edges

NUMBERING: [ink, top right] *124*

ANNOTATION: [ink, top] *Nigrum lignum ex Sismano*
[bottom] *long. pal: 4. lat: pal: 1. alt. p ½*
[Cesi's vertical inscription at right edge, trimmed]

The drawing depicts a black piece of fossil wood from Sismano. The specimen shows a prominent branch and knot end.

REFERENCES: 17th-cent. Index, p. 368

82

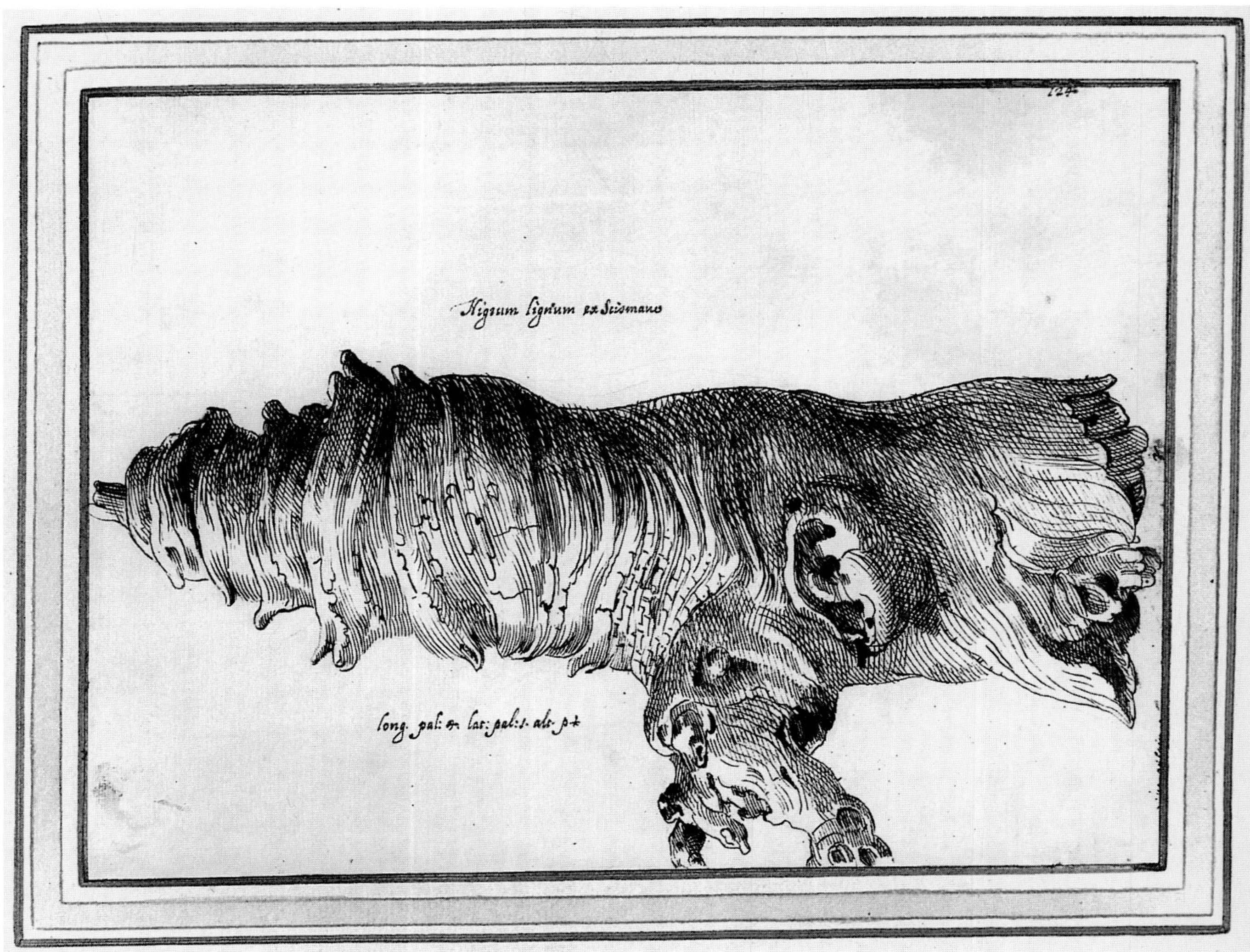

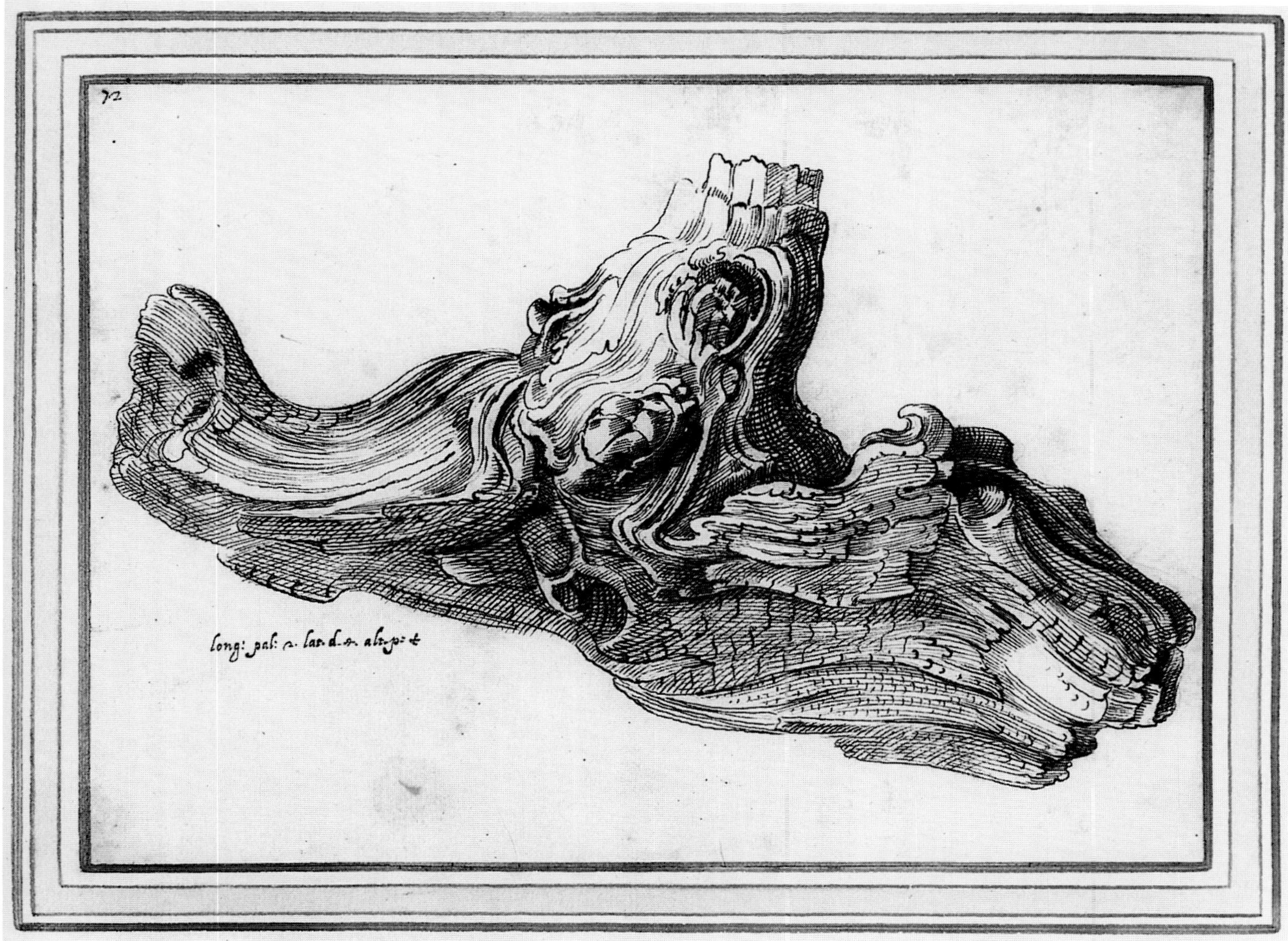

83

83. Fossil wood with branch and branch scars

Windsor, RL 25611, *Fossils*, vol. XV, fol. 20

Pen and ink, over traces of black chalk

184 × 262 mm; small holes caused by ink

NUMBERING: [ink, top left corner] *72*

ANNOTATION: [ink] *long: pal: 2. lat. d. 4. alt. p: ½*

This drawing shows a piece of wood with a branch and branch scars. There is no inscription describing it.

REFERENCES: missing from 17th-cent. Index

84. Two specimens of fossil wood

Windsor, RL 25541, *Fossils*, vol. XIV, fol. 12

Pen and ink, over black chalk

199 × 257 mm

NUMBERING: [ink, upper right] *5*

ANNOTATION: [ink, upper right] *Contuberatio an' fractuosa*
[lower left] [S]*quama ampla* [repeated in a later hand]
[lower centre] *long: pal: 5: lat: p: 1½. alt. d: 2*
[bottom centre] *long: pal: 3. lat. 1½ alt. d. 3*
[bottom right] *Nigricat planum ab alia parte*

The right-hand specimen is described as being a twisted knotty protuberance. The inscription notes that the other side is black.

The specimen on the left is noted as having 'broad scales or flakes', a description that may indicate either fragments of clay still attached to the wood or small pieces of wood flaking off the surface, caused by the specimen drying out.

REFERENCES: 17th-cent. Index, pp. 361, 369

84

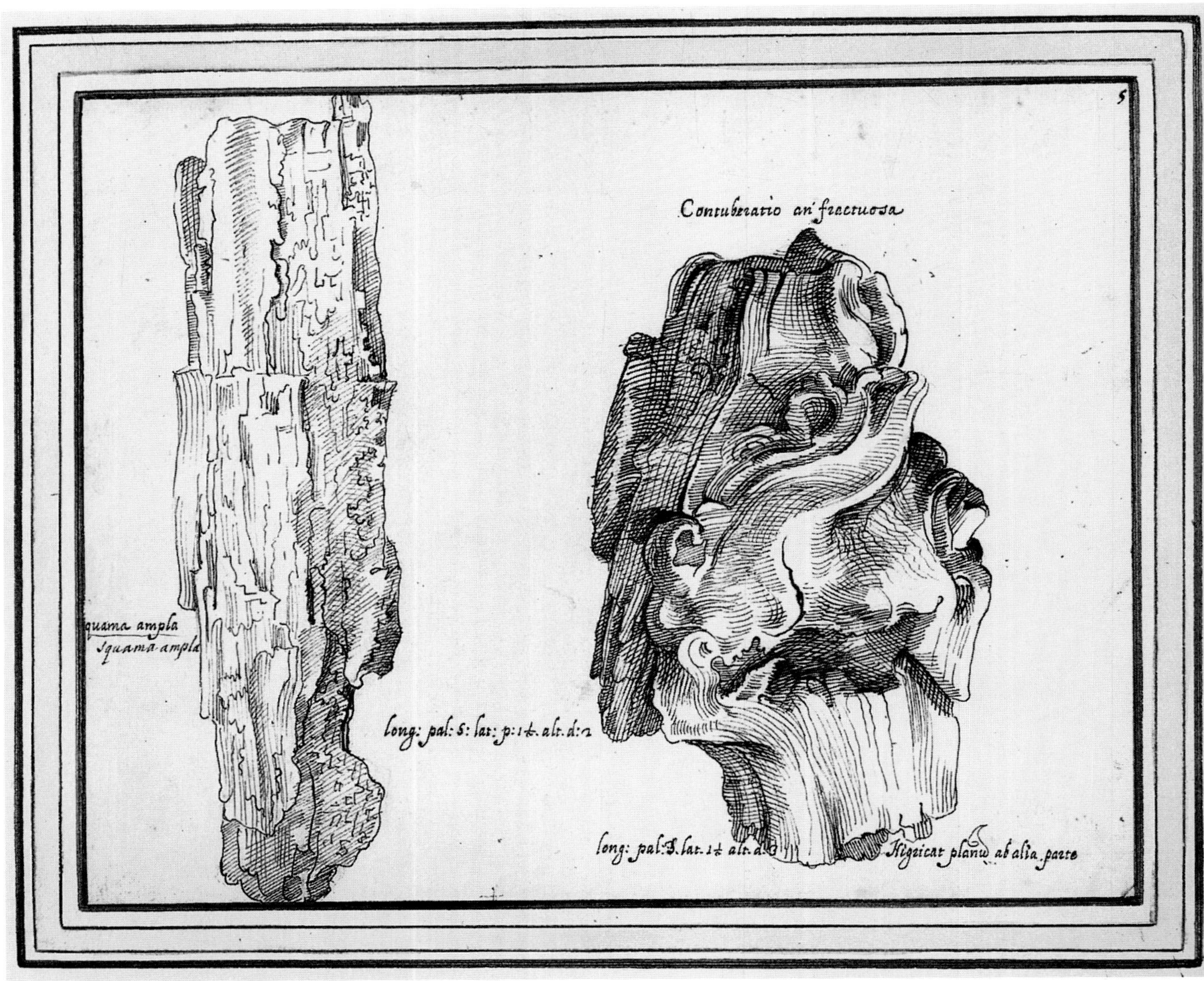

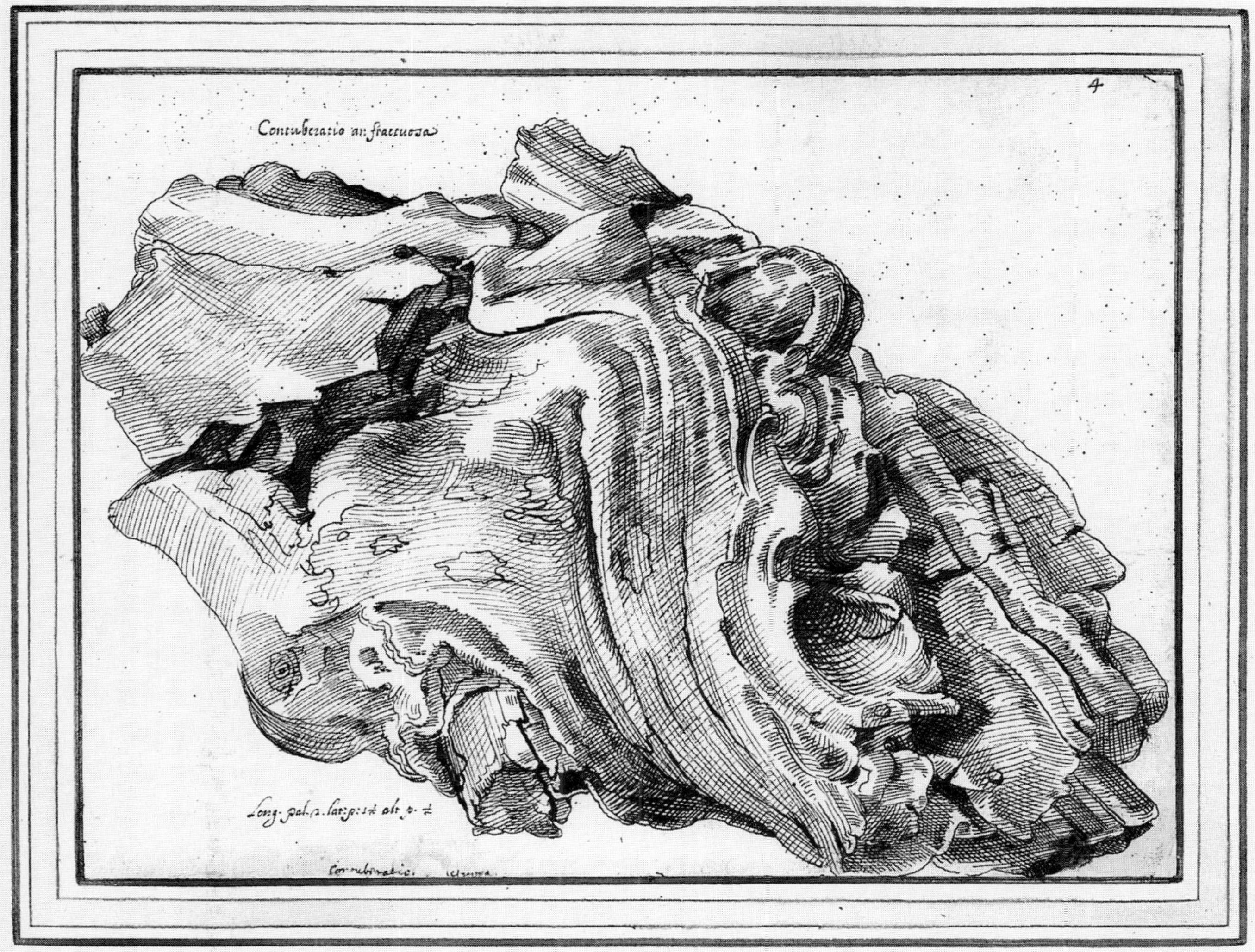

85

85. Fossil wood

Windsor, RL 25540, *Fossils*, vol. XIV, fol. 11

Pen and ink, over black chalk

200 × 271 mm; trimmed close to the lower edge of drawing

NUMBERING: [ink, upper right] *4*

ANNOTATION: [ink, upper left] *Contuberatio an fractuosa*
[bottom] *Long. pal. 2. lat: p: 1½ alt p. ½.*
[in Cesi's hand, erased and cut] *conturberatio* [an' fra]*ctuosa*

This specimen is also described, just as 84, as a '*contuberatio an' fractuosa*'. The wood may be part of a knotted branch; some of the branch is visible lower left.

REFERENCES: 17th-cent. Index, p. 361

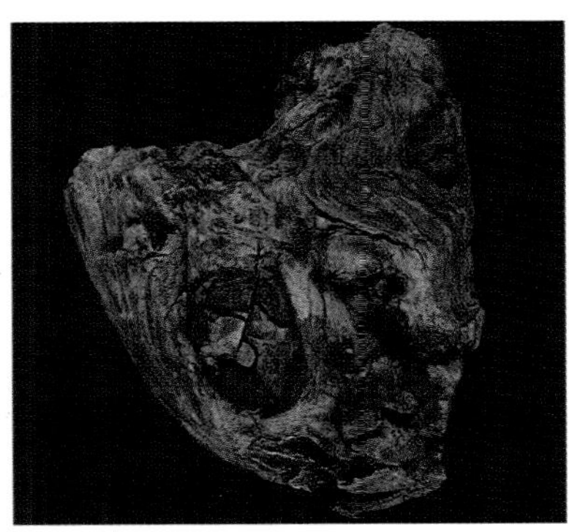

Comp. fig. 85. Knotty wood from Podere Casanova

86. Piece of fossil wood with branches

Windsor, RL 25665, *Fossils*, vol. XVI, fol. 26

Pen and ink, over traces of black chalk; ink frame line

179 × 241 mm

NUMBERING: [ink, top right corner] *126*

ANNOTATION: [ink, top] *Nodos. ligna Ramos promittens* [bottom] *long: pal: 1.½ lat: p. 1. crass. d. 4*

The specimen is described as 'knotty wood putting forth branches'. Another view is shown in **87**.

REFERENCES: 17th-cent. Index, p. 368

OTHER DRAWINGS: **87**

87. Another view of the preceding specimen

Windsor, RL 25666, *Fossils*, vol. XVI, fol. 27

Pen and ink, over traces of black chalk; retouched in ink on mount sheet; ink frame line; measurements given on previous drawing

171 × 241 mm; small holes caused by ink

NUMBERING: [ink, top right corner] *127*

ANNOTATION: [ink] *Idem*

A view of the other side of the specimen of fossil wood shown in **86**. Further areas of charring are shown.

REFERENCES: 17th-cent. Index, p. 368

OTHER DRAWINGS: **86**

86

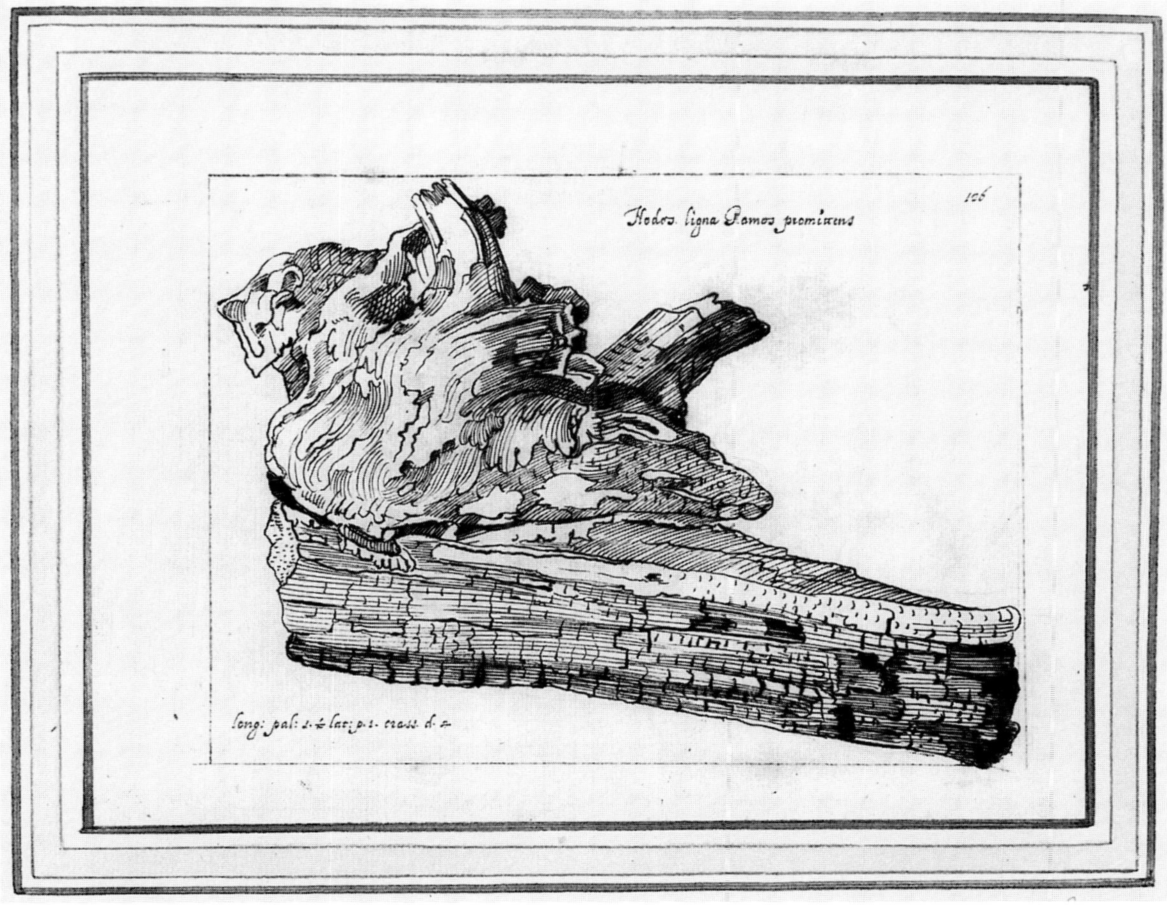
126
Nodos ligna Ramos promittens
long: pal: 1. ½ lat: p. 1. crass. d. 2

87

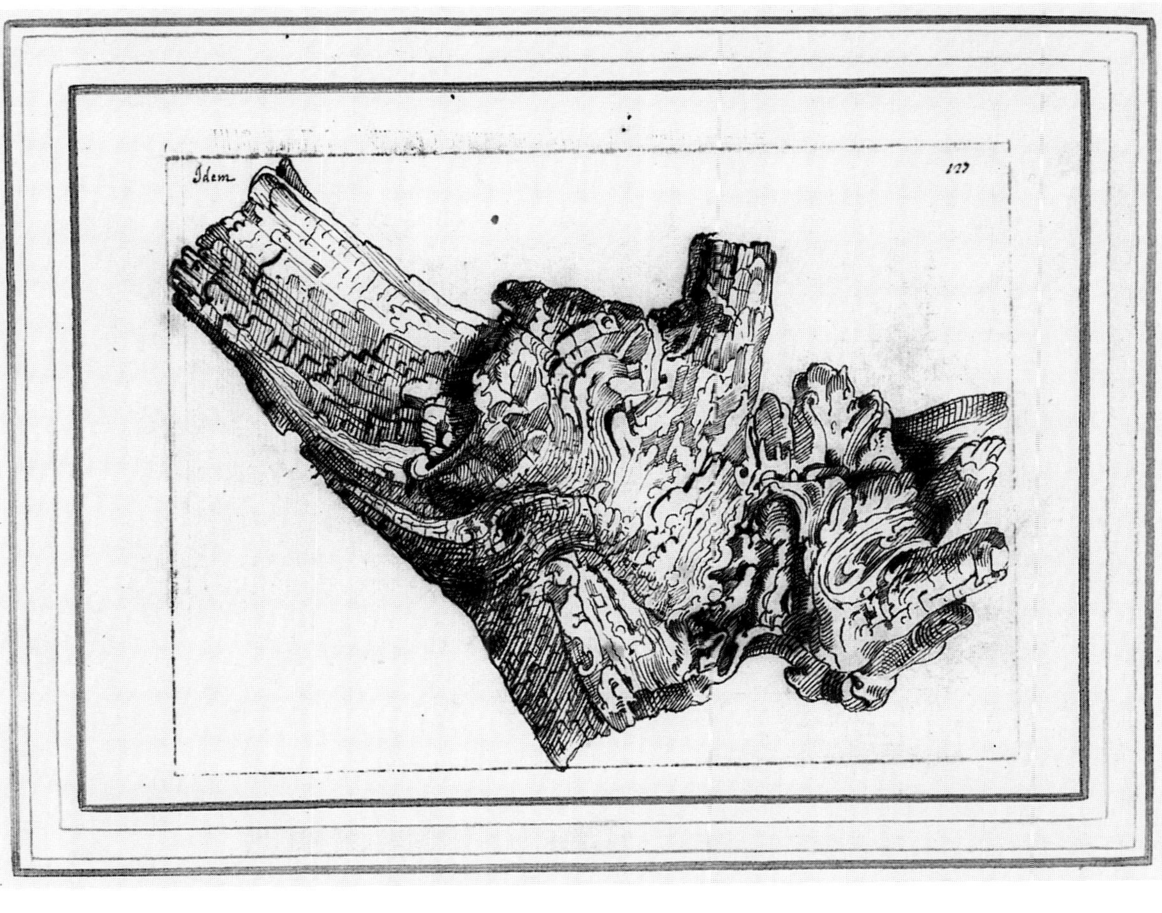
Idem
127

88. Fossil wood with insertion point of branch

Windsor, RL 25643, *Fossils*, vol. XVI, fol. 4

Pen and ink, over traces of black chalk; frame line in black chalk

181 × 249 mm

NUMBERING: [ink, top left corner] *104*

ANNOTATION: [ink] *ligneum nodo Volutum pressum-/ Navicula facie Turbine acuto emisso/ long: pal: 2½ et Turb./ crass. circ: pal. 1.*

This complex knotted wood specimen is imaginatively described as 'compacted wood turned in a knot, giving the appearance of a little boat with a sharp twirling shape projecting from it'. Compare **97** where the shape of the specimen is likened to that of a fish.

REFERENCES: 17th-cent. Index, p. 366

88

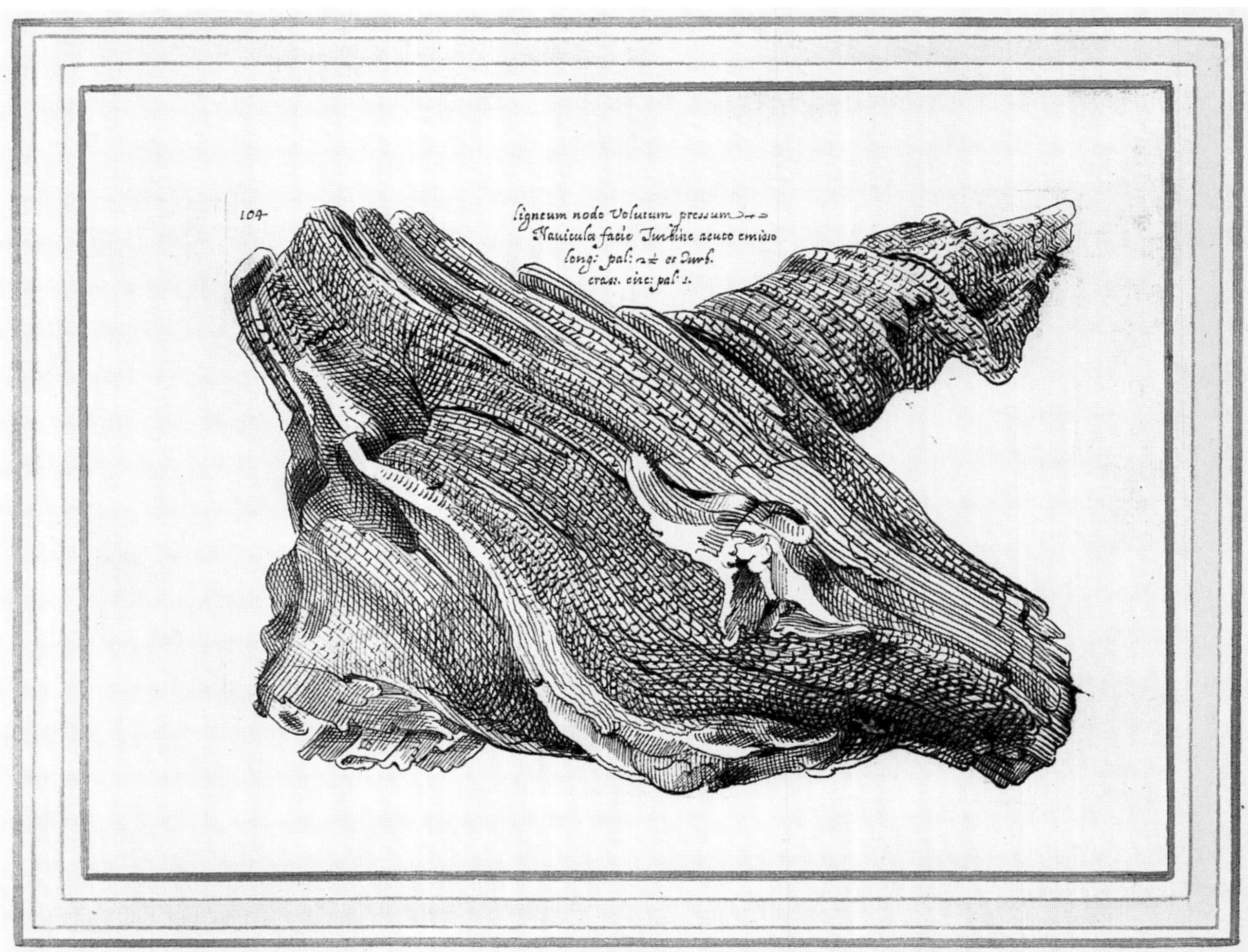

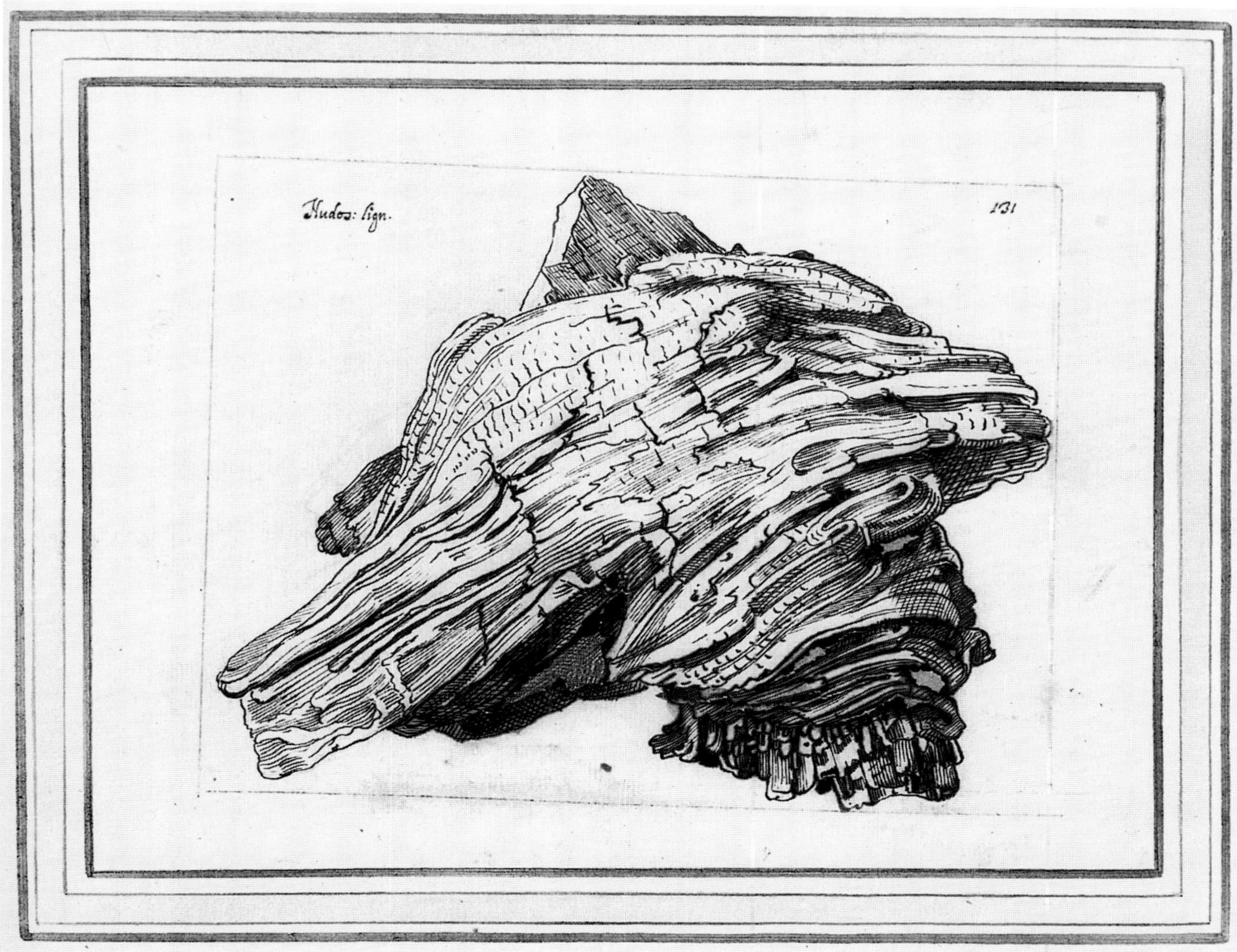

89

89. Knotted fossil wood

Windsor, RL 25670, *Fossils*, vol. XVI, fol. 31

Pen and ink, over traces of graphite; retouched in ink on mount sheet; no measurements given; ink frame line

180 × 245 mm; small holes caused by ink. *Watermark*: Bird 41B

NUMBERING: [ink, top right corner] *131*

ANNOTATION: [ink] *Nudos:* [sic] *lign.*

Another specimen of knotted wood. It is possible that this is another view of the specimen illustrated in **96–97**.

REFERENCES: 17th-cent. Index, p. 368

OTHER DRAWINGS: possibly **96–97**

90. Fossil wood branch with fragment of main trunk

Windsor, RL 25644, *Fossils*, vol. XVI, fol. 5

Pen and ink, over traces of black chalk; retouched in ink on mount sheet

164 × 229 mm; small holes caused by ink; frame line in ink over black chalk. *Watermark*: Bird 41B

NUMBERING: [ink, top right corner] *105*

ANNOTATION: [ink, top] *ligneum nodo volutum productum caudo culatum*
[bottom] *long: pal: 2½ lat: p: 2½ alt. p: 1*

The specimen, given the same seventeenth-century classification as **88** ('a compacted wood [specimen] in a knot') is also described as having 'a tail projecting from it'. It shows a fossil wood branch with a fragment of the main trunk attached. The carefully delineated hole in the upper part of the trunk fragment shows the point of insertion of a branch.

REFERENCES: 17th-cent. Index, p. 366

90

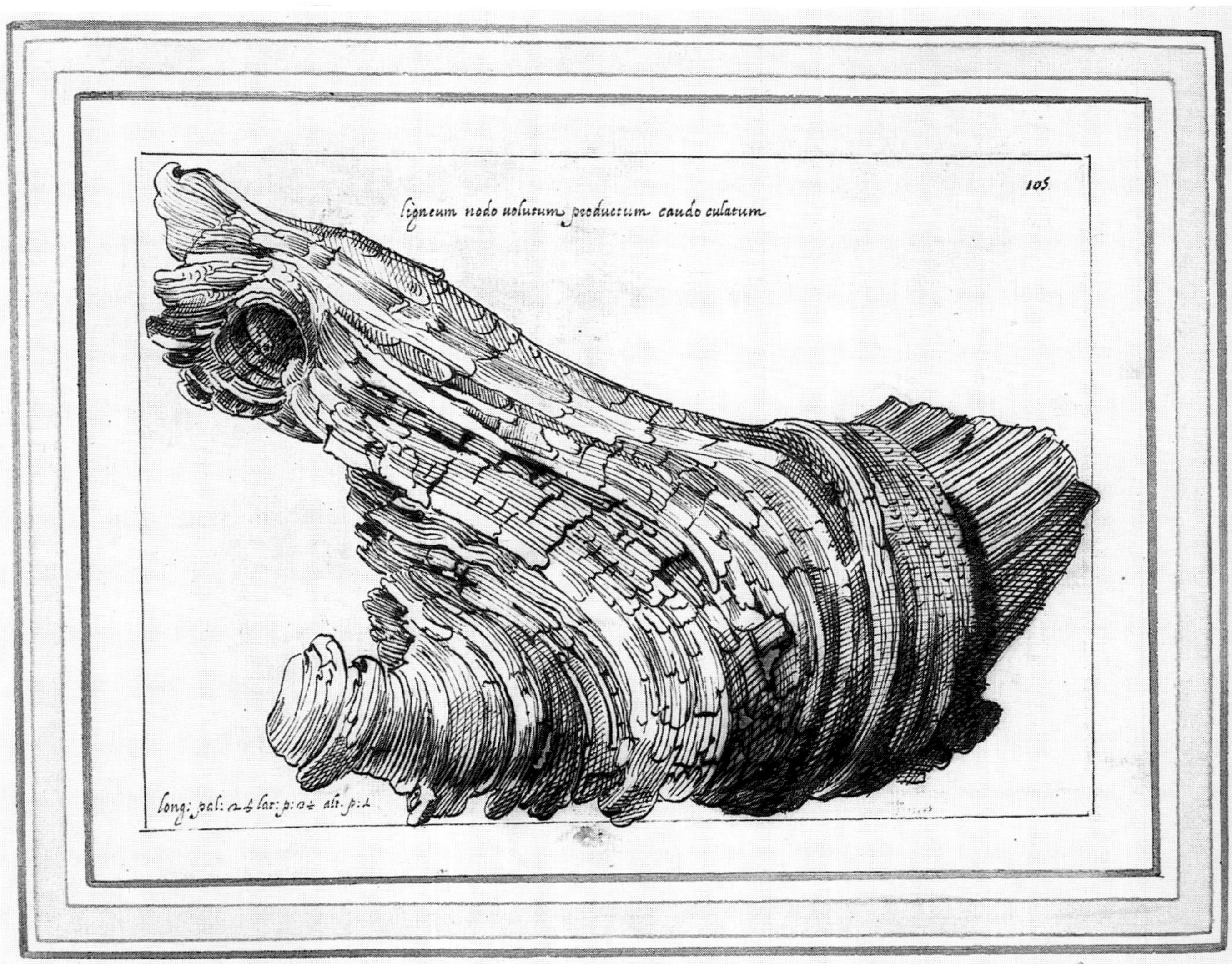

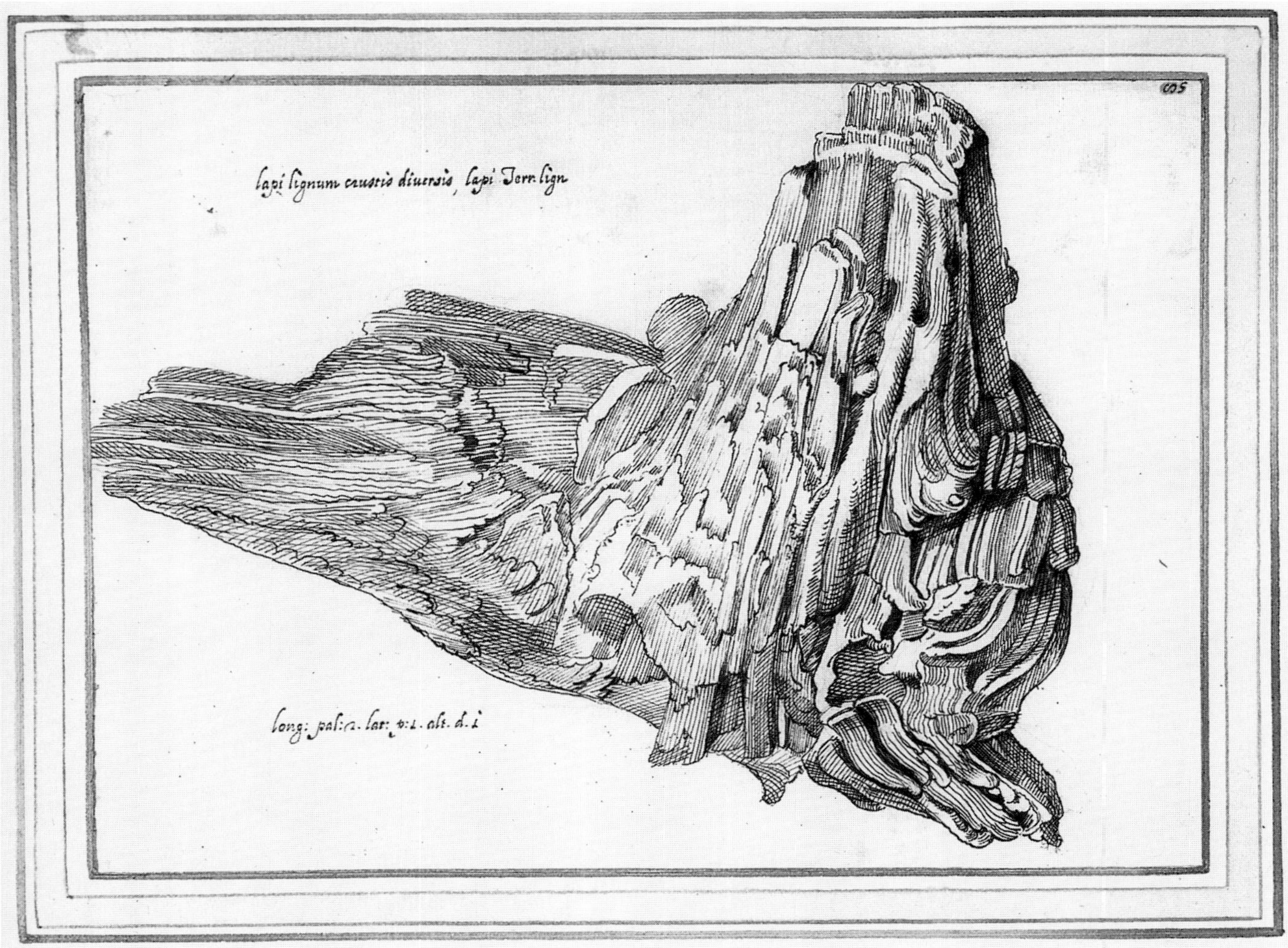

91

91. Piece of fossil wood

Windsor, RL 25624, *Fossils*, vol. XV, fol. 33

Pen and ink, over black chalk; retouched in ink on mount sheet

183 × 256 mm; small holes caused by ink

NUMBERING: [ink, top right corner] *85*

ANNOTATION: [ink, top left] *lapi lignum crustis diversis, lapi Terr. Lign* [bottom left] *long: pal: 2. lat: p: 1. alt. d. 1*

This specimen of fossil wood is described as a 'stone wood with diverse crusts, [or a] stone earth wood', terminology which may reflect the Lincean hypothesis regarding the generation of fossil wood from the earth itself (see above, pp. 58–60, 133). It is not possible to say what the reference to 'different crusts' might have meant. They could have been mineral deposits, clay, or a mixture of both. The description suggests the specimen may have been partially petrified.

REFERENCES: 17th-cent. Index, p. 365

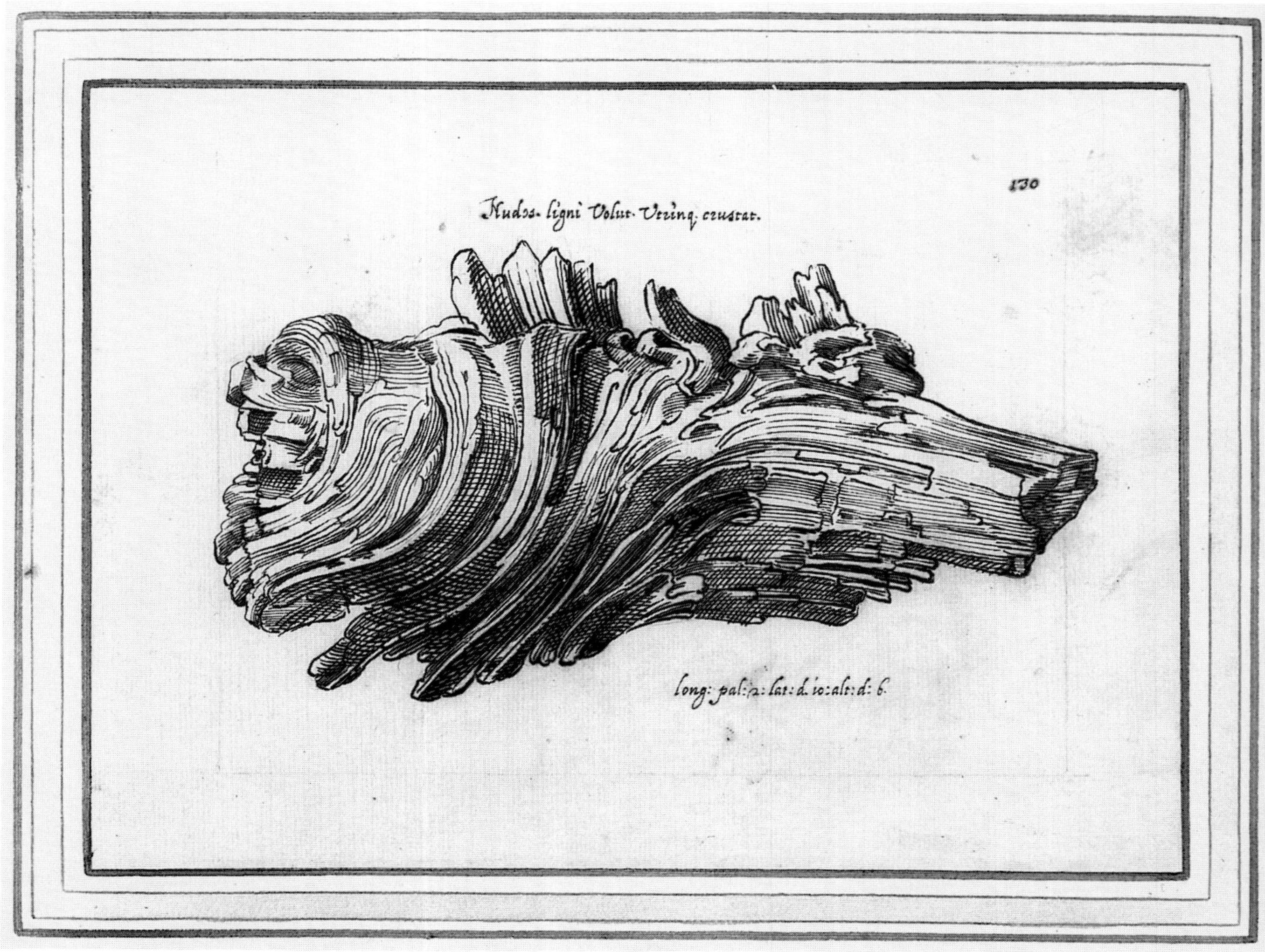

92

92. Knotted fossil wood

Windsor, RL 25669, *Fossils*, vol. XVI, fol. 30

Pen and ink, over traces of black chalk; black chalk frame line

179 × 248 mm

NUMBERING: [ink, top right corner] *130*

ANNOTATION: [ink, top] *Nudos. ligni Volut. Utrinq. Crustat.* [bottom] *long: pal: 2: lat: d. 10: alt: d: 6.*

Another specimen of knotted wood, described as a 'knotty volute of wood, encrusted on both sides'.

REFERENCES: 17th-cent. Index, p. 368

93. Fossil wood: view of 94 from above

Windsor, RL 25691, *Fossils*, vol. XVI, fol. 52

Pen and ink, over traces of black chalk

186 × 251 mm

NUMBERING: [ink, top right corner] *152*

ANNOTATION: [ink, top] *Turbinatum pressum ligneum* [bottom] *long: pal. 2. lat. palm. 1 alt: d. 3.*

A knot end of fossil wood, classified as a 'compacted wood whorl'. The drawing shows the exterior of the sectioned specimen depicted in **94**. Both views were engraved in Stelluti's Plate 11. The accompanying text to this plate notes that it 'is swirling in form [*F*], rather rough and bark-like, two feet long, and one foot wide [60 cm × 30 cm]. It was sawn in half to reveal the interior: hence piece *G* (**94**) shows the appearance within. Many pieces of wood of this swirling shape have been found in this area; but it will suffice to have shown this one, the others not differing greatly, except that some were hard as bone inside, which in time (I believe) would petrify.'

REFERENCES: Stelluti 1637, p. 11; 17th-cent. Index, p. 370

ENGRAVINGS: Stelluti 1637, Pl. 11, Fig. F; for proof engravings in Montpellier, BEM, MS H.173, fol. 20, see p. 403

OTHER DRAWINGS: **94**; for related drawings in BAV, MS Barb. Lat. 4355, fol. 37, and Montpellier, BEM, MS H.170, fol. 9, see p. 403 and Appendix II, Fig. 64

94. Longitudinal section of fossil wood specimen

Windsor, RL 25690, *Fossils*, vol. XVI, fol. 51

Pen and ink, over traces of black chalk

188 × 257 mm. *Watermark:* Bird 41B

NUMBERING: [ink, top right corner] *151*

ANNOTATION: [ink] *Turb: pres. Sectio media*

A longitudinal section of the 'compacted wood whorl' shown in **93**. Engraved in Stelluti's Plate 11 as figure G. See under **93**.

REFERENCES: Stelluti 1637, p. 11; 17th-cent. Index, p. 370

ENGRAVINGS: Stelluti 1637, Pl. 11 (Fig. G); for proof engraving in Montpellier, BEM, MS H.173, fol. 20, see p. 403

OTHER DRAWINGS: **93**

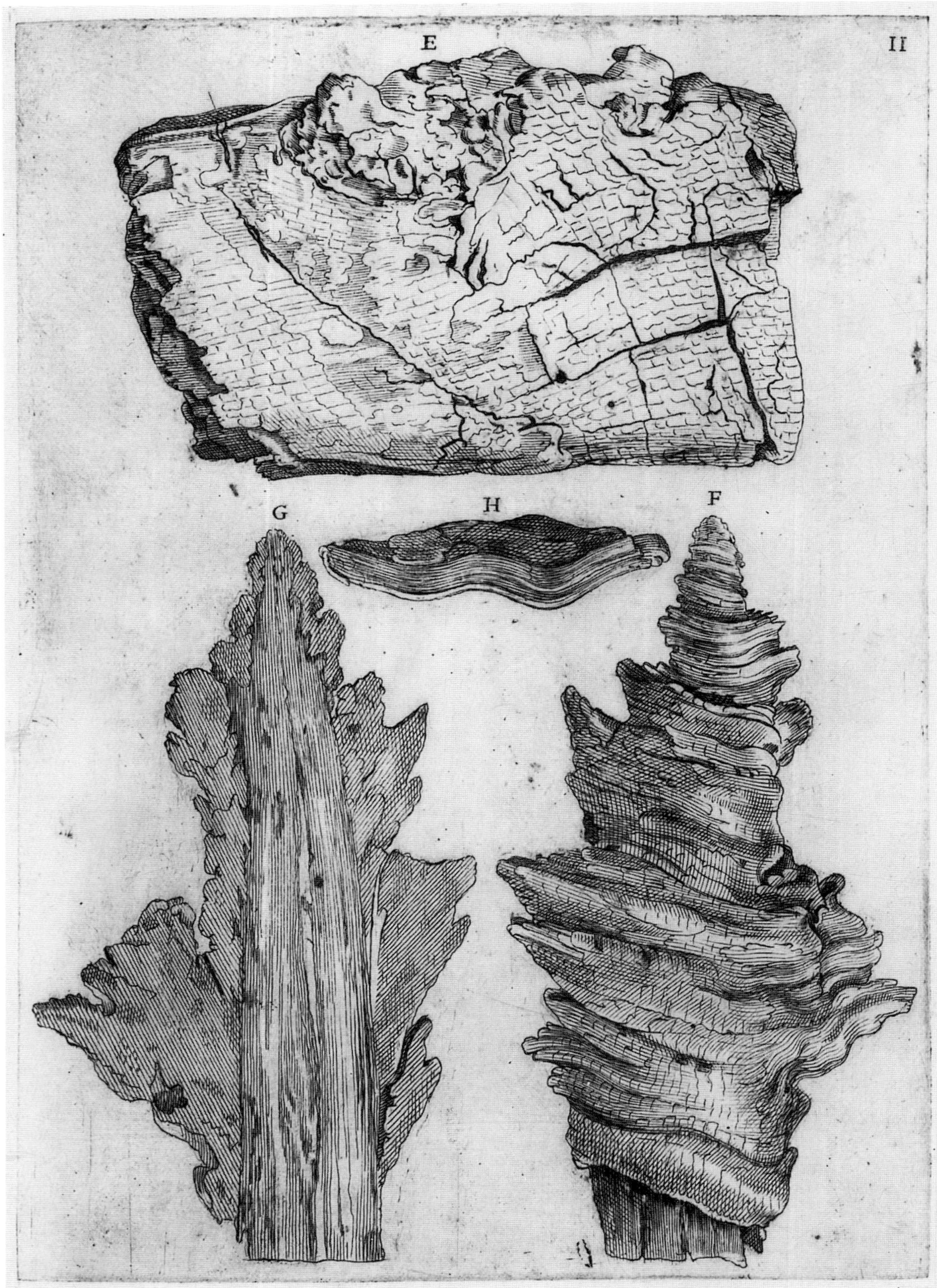

Comp. figs 93, 94 and 117. Stelluti 1637, Pl. 11, Figs F, G and E

93

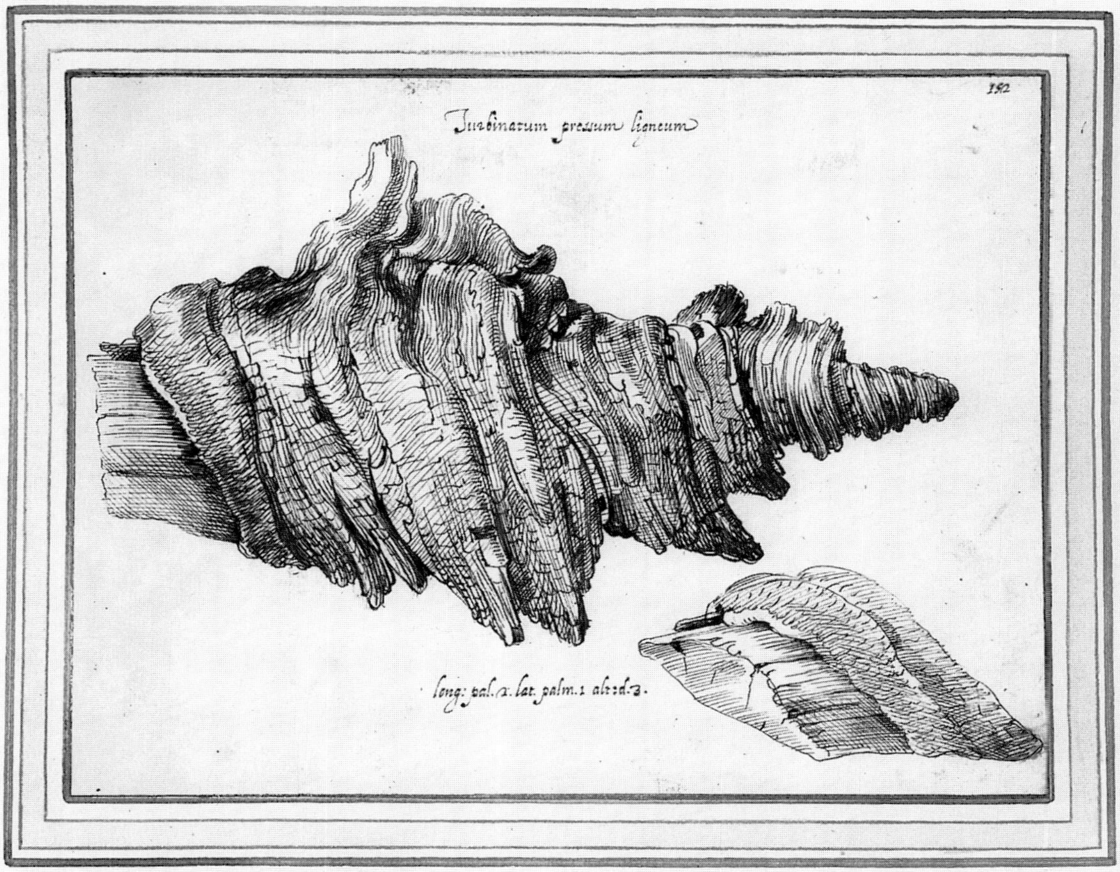
152
Turbinatum pressum ligneum
long: pal. 2. lat. palm. 1 alt: d. 3.

94

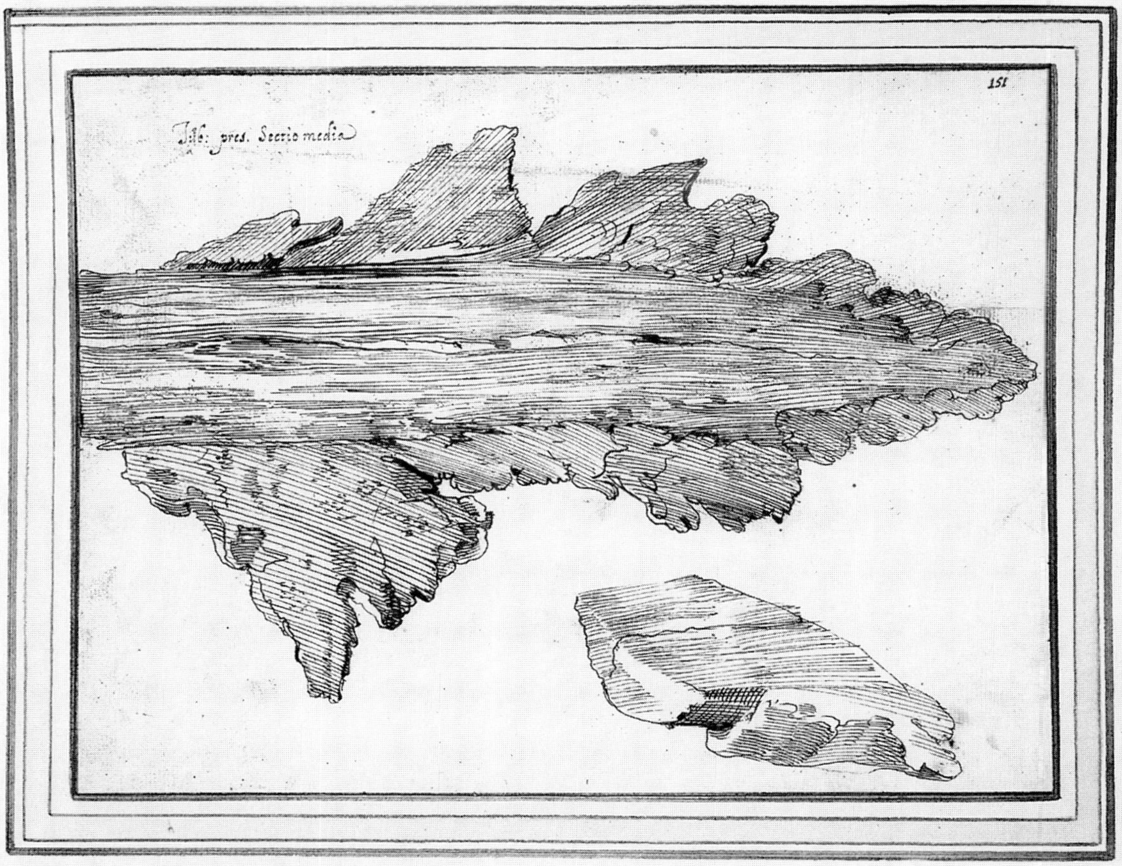
151
pres. Sectio media

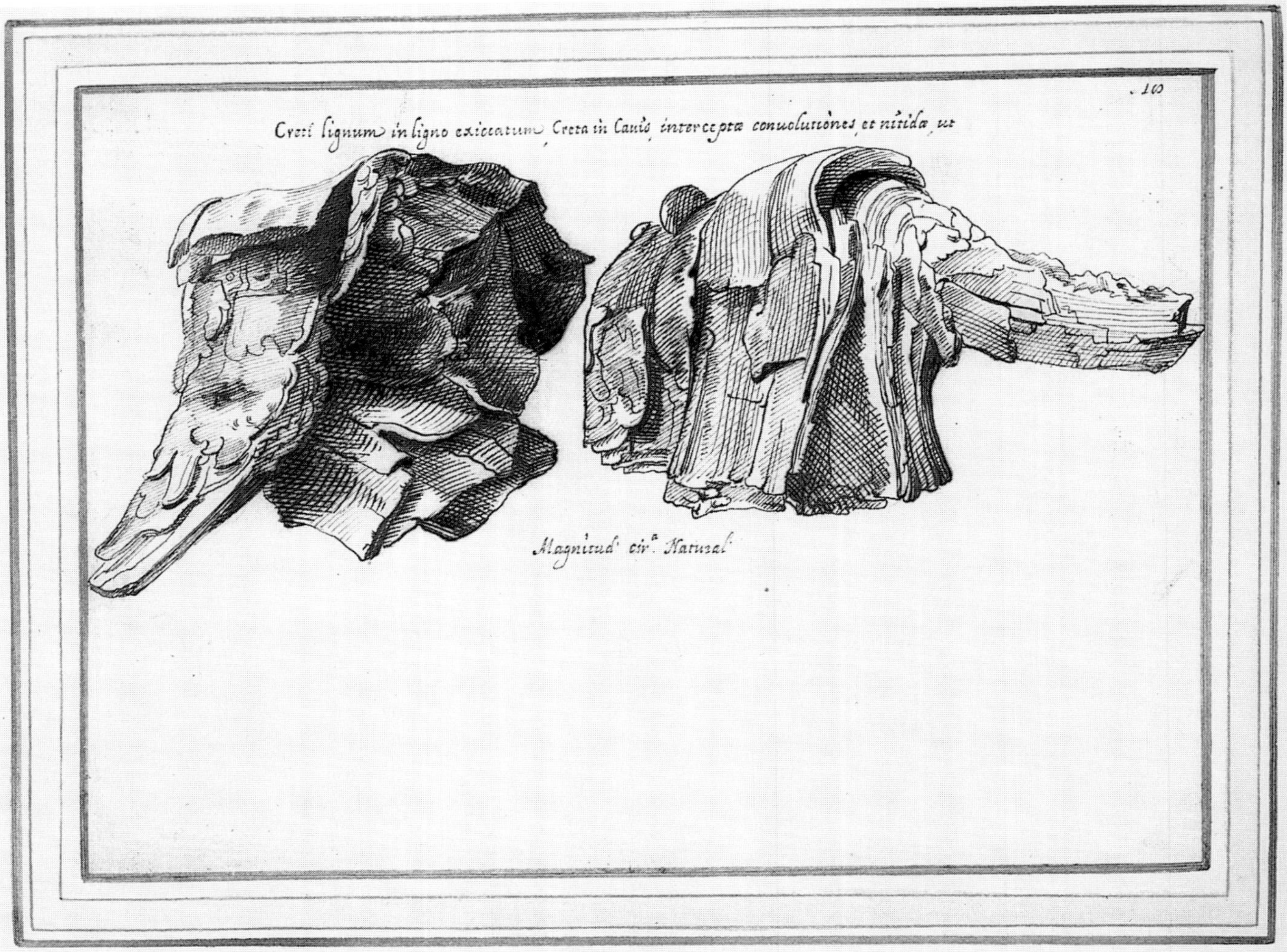

95. Fossil wood

Windsor, RL 25555, *Fossils*, vol. XIV, fol. 26

Pen and ink, over black chalk; retouched in ink on mount sheet; no measurements given

180 × 257 mm; small holes caused by ink

NUMBERING: [ink, upper right] *18*

ANNOTATION: [ink, top] *Creti lignum in ligno exiccatum, Creta in Cavis, interceptæ convolutiones, et nitidæ, ut* [centre] *Magnitud. cir.*[a] *Natural*[i]

This drawing shows two views of a fragment of fossil wood beginning to pull apart as a result of the drying that follows upon exposure to the air. Engraved as Plate 12, Figure M, in Stelluti, the accompanying text reads: 'Fragment *M* was cretaceous-wood in substance, that is, clay and wood, yet not exceedingly hard, and where it appears darker it was shiny and resplendent, like anthracite, but most brittle'. For anthracite, see **58**.

The arrangement on the page suggests that perhaps some space was left for further views of the same specimen.

REFERENCES: Stelluti 1637, p. 11; 17th–cent. Index, p. 361

ENGRAVINGS: Stelluti 1637, Pl. 12 (Fig. M); for proof engraving in Montpellier, BEM, MS H.173, fol. 21, see p. 403

96. Piece of fossil wood

Windsor, RL 25638, *Fossils*, vol. XV, fol. 47

Pen and ink, over traces of black chalk; retouched in ink on mount sheet; ink frame line

177 × 239 mm; small holes caused by ink

NUMBERING: [ink, top right corner] *99*

ANNOTATION: [ink, top left] *Ligni Metallum e' ferrugine Vergens/ ad colorem ferretti de Spagna/ dicti, et eo tingens.*
[bottom right] *long: d: 14. lat: p ½ crass. d: 5*

This specimen of fossil wood is described as [a piece of] metallized wood turning from iron to the colour of so-called Spanish iron [?], and tinged with it. It is not known what precisely is referred to by *ferretti de Spagna*. Another view of this specimen is shown in **97**.

REFERENCES: 17th-cent. Index, p. 366

OTHER DRAWINGS: **97**; possibly **89**

97. View from other side of 96

Windsor, RL 25629, *Fossils*, vol. XV, fol. 38

Pen and ink with corrections (to frame lines) in white heightening, touches of lead point; retouched in ink on mount sheet; frame lines in ink corrected in white bodycolour; measurements are given, but see **96**

176 × 235 mm; small holes caused by ink

NUMBERING: [ink, top right corner] *90*

ANNOTATION: [ink] *Ab alia parte*

This specimen annotated 'from the other side' clearly shows another view of **96**. The drawing is engraved in Stelluti's Plate 12, Fig. N. In the accompanying text it is described as having 'the shape of a fish, and was of a substance both ligneous and ferruginous, and in colour like *ferretto di Spagna*; and this wood was even heavier than the other. It was one palm and two inches in length, half a palm in width, and five fingers thick'.

It would appear from the description that the specimen was permineralized and impregnated with siderite.

REFERENCES: Stelluti 1637, p. 11; 17th-cent. Index, p. 365

ENGRAVINGS: Stelluti 1637, Pl. 12 (Fig. N); for proof engraving in Montpellier, BEM, MS H.173, fol. 21, see p. 403

OTHER DRAWINGS: **96**; possibly **89**

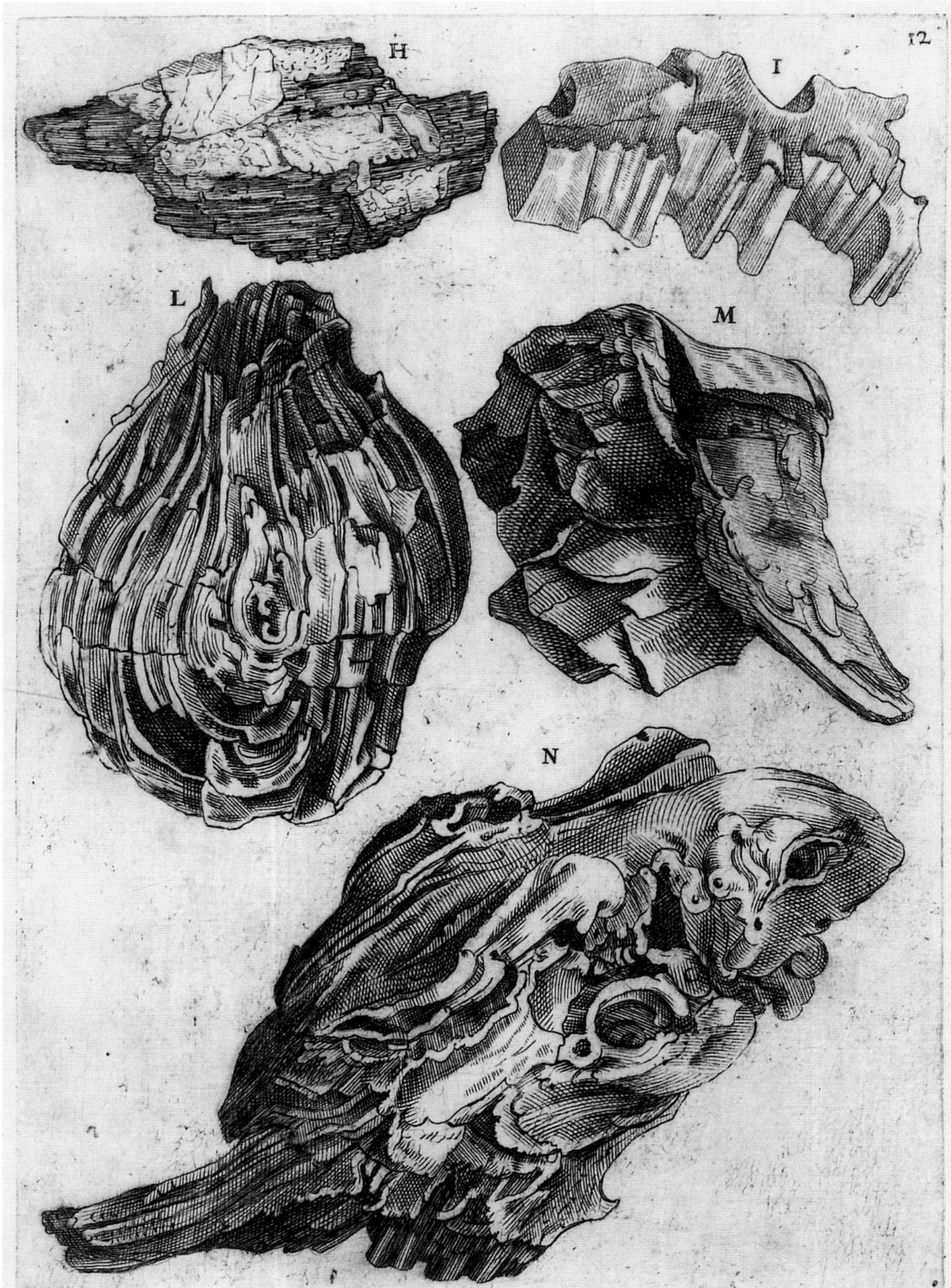

Comp. figs 95, 97, 104 and 112. Stelluti 1637, Pl. 12, Figs M, N, L and H

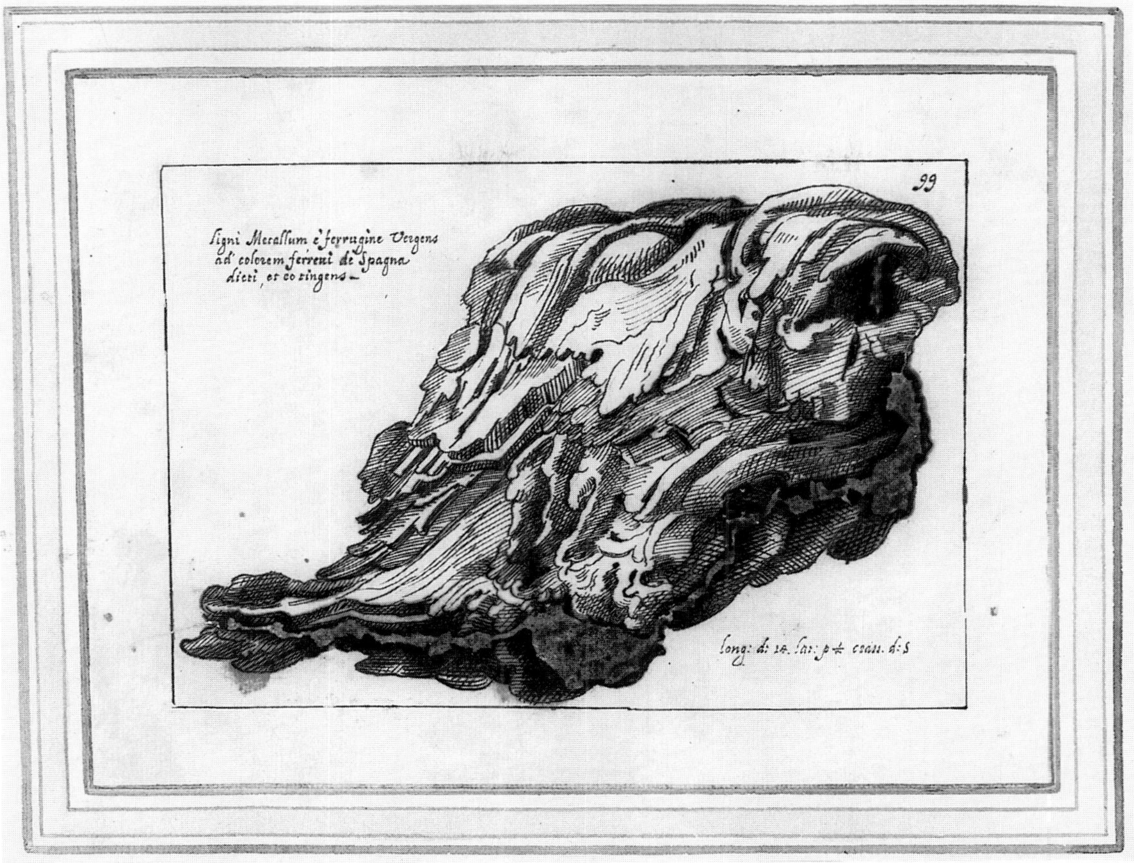
99
Ligni Metallum e ferrugine Vergens
ad colorem ferreni de Spagna
dicti, et co tingens
long: di 14. lat: p ÷ crass. di 5

96

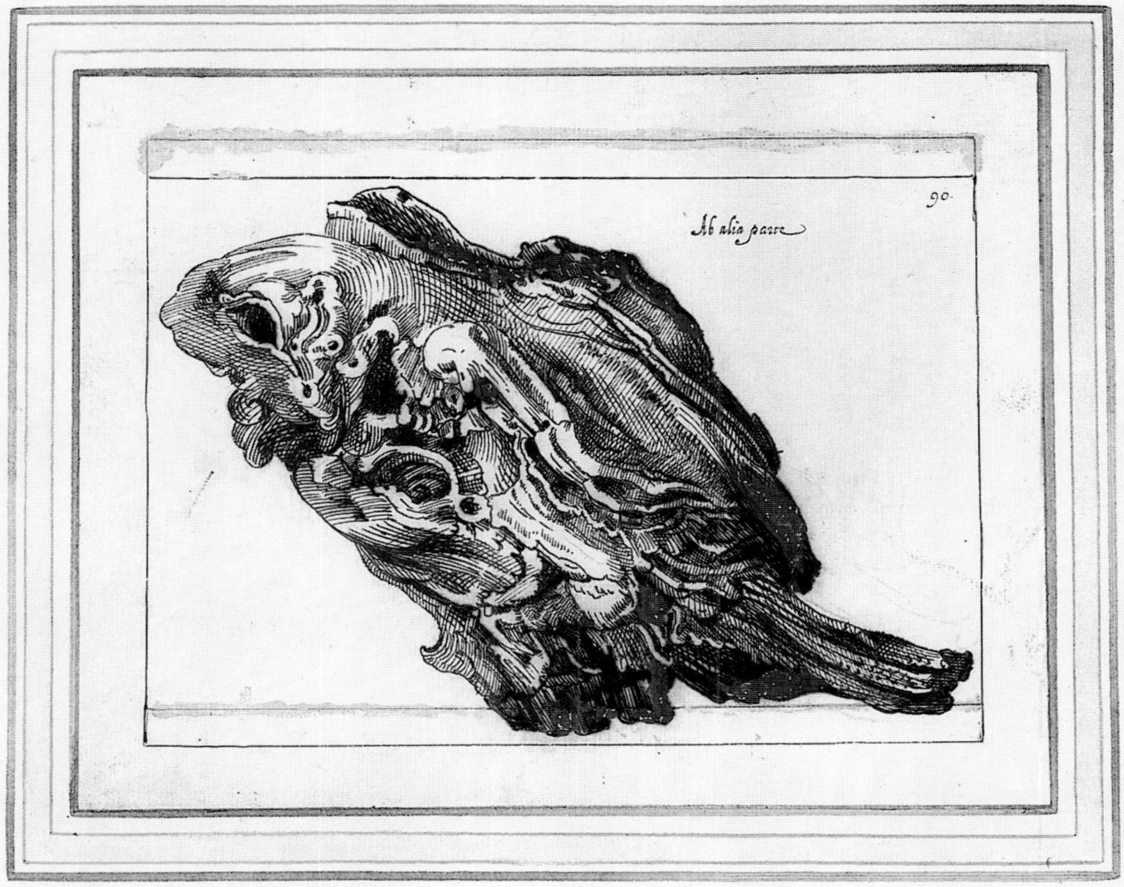
90.
Ab alia parte

97

98. Twisted piece of fossil wood

Windsor, RL 25626, *Fossils*, vol. XV, fol. 35

Pen and ink, over black chalk

189 × 256 mm

NUMBERING: [ink, top right corner] *87*

ANNOTATION: [ink, top] *Intus lapideum, extra ligneum Nigro fulvum* [bottom] *long: pal. 3: lat pal. 3. alt: pal: 1½*

This piece of fossil wood is described as being 'stony on the inside' with a 'woody blackish-yellow' exterior. In the index to the volume it is listed under the category '*Lapilignum* from Dunarobba' (see p. 365). The 'blackish- yellow' colour suggests some iron impregnation.

REFERENCES: 17th-cent. Index, p. 365

98

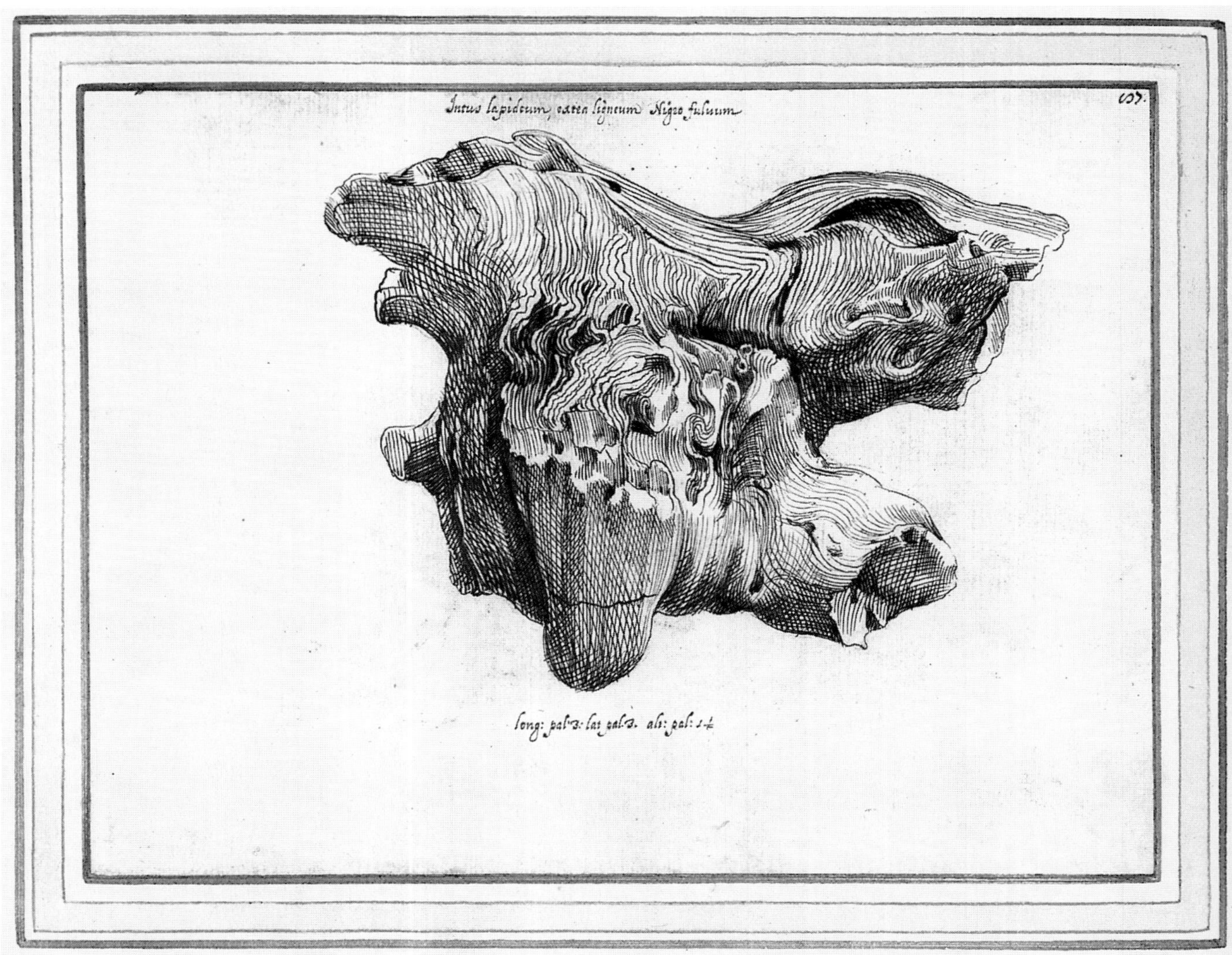

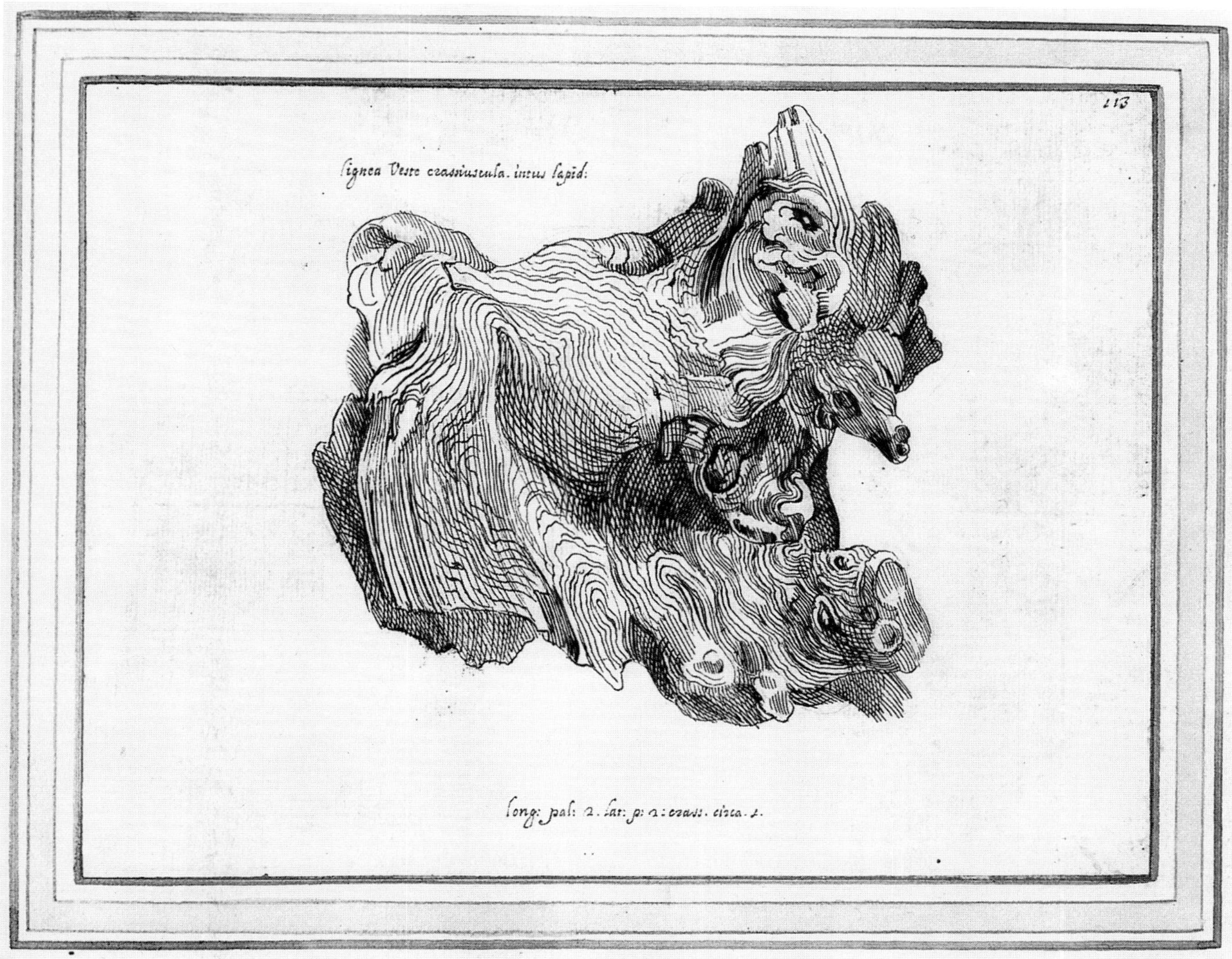

99

99. Piece of fossil wood

Windsor, RL 25652, *Fossils*, vol. XVI, fol. 13

Pen and ink, over black chalk

188 × 252 mm

NUMBERING: [ink, top right corner] *113*

ANNOTATION: [ink, top] *lignea Veste crassiuscula. intus lapid:* [bottom] *long: pal: 2. lat: p: 2: crass. circa 1.*

This densely knotted piece of wood is described as being '[a fragment] with a slightly thick woody covering, with a stony interior'. This suggests the specimen may have been partially petrified.

REFERENCES: missing from 17th-cent. Index

100. Specimen of knotted wood

Windsor, RL 25660, *Fossils*, vol. XVI, fol. 21

Pen and ink over black chalk

188 × 254 mm

NUMBERING: [ink, top right corner] *121*

ANNOTATION: [ink, top] *Nigra lignea crassa Veste intus aliquanto/ lapidea* [bottom] *long: pal. 3. lat. pal: 2: crass. pal: 1*

A specimen of black wood, possibly petrified in the centre (the inscription states 'on the inside somewhat stony') showing many knots and branch scars.

The wavy patterns are contorted growth lines which, when cut, give the appearance of 'waves'. Such patterns are described and illustrated by Stelluti in Plates 3–5 of his *Trattato* (see below, p. 380).

REFERENCES: 17th-cent. Index, p. 368

100

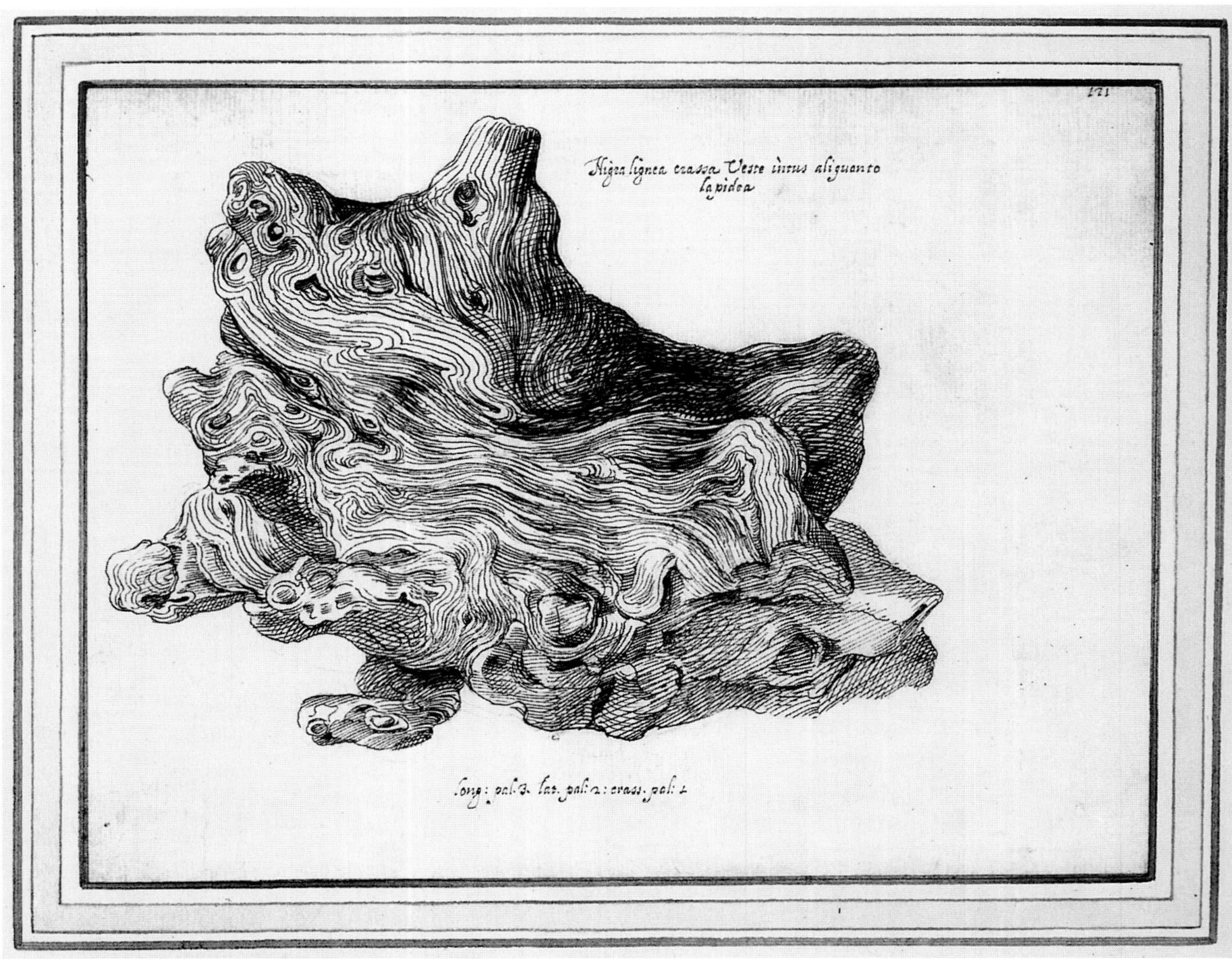

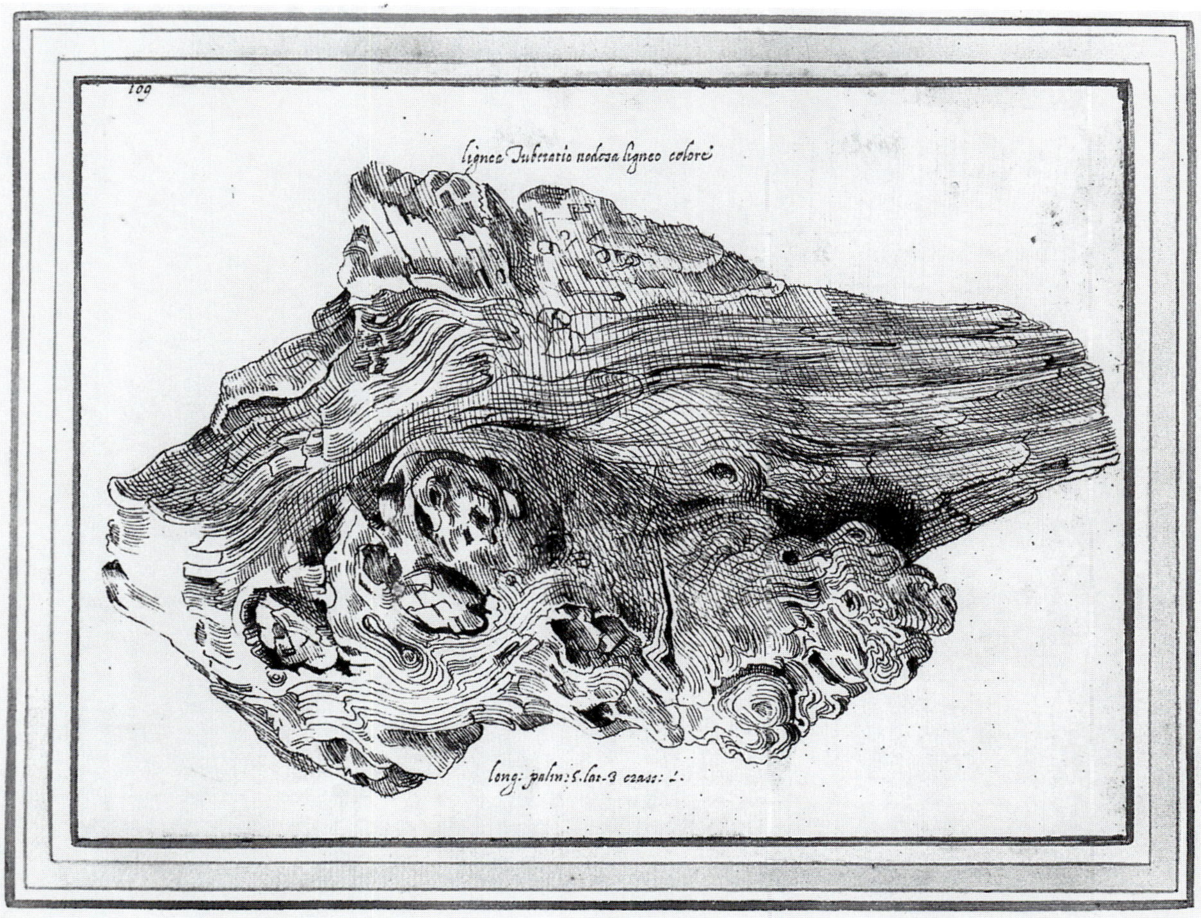

101

101. Piece of fossil wood

Windsor, RL 25648, *Fossils*, vol. XVI, fol. 9

Pen and ink, over traces of black chalk

190 × 261 mm; small hole caused by ink

NUMBERING: [ink, top left corner] *109*

ANNOTATION: [ink, top] *lignea Tuberatio nodosa ligneo colore* [bottom] *long: palm: 5. lat. 3 crass: 1.*

The description of this specimen as a 'knotty wood excrescence of a woody colour', is matched by the draughtsman's rendering of a complex mass of knots.

REFERENCES: 17th-cent. Index, p. 367

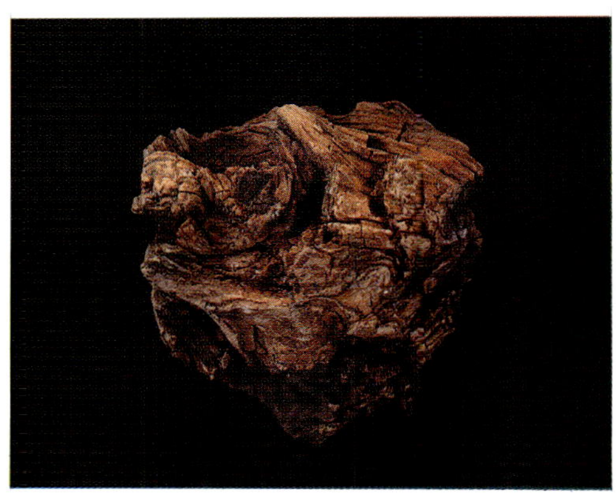

Comp. fig. 101. Knotty piece of wood from Dunarobba

102. Two views of a piece of fossil wood

Windsor, RL 25694, *Fossils*, vol. XVI, fol. 55

Pen and ink, over black chalk

188 × 258 mm; sheet trimmed through image, left and lower edges

NUMBERING: [ink, top left corner] *155*

ANNOTATION: [ink, centre] *Voluta lignea Sulphurea Scabie conspersa*
[bottom] *long: pal: 1. ½ lat. p. 1. alt. d: 4*

The specimen is classified as a 'woody volute', and is shown together with an end view. It was probably pyritized. The description 'sprinkled with sulphurous scales' suggests the presence of decay. A similar specimen (in the same category of 'woody volute') is shown at **103**.

REFERENCES: 17th-cent. Index, p. 370

OTHER DRAWINGS: **103**

102

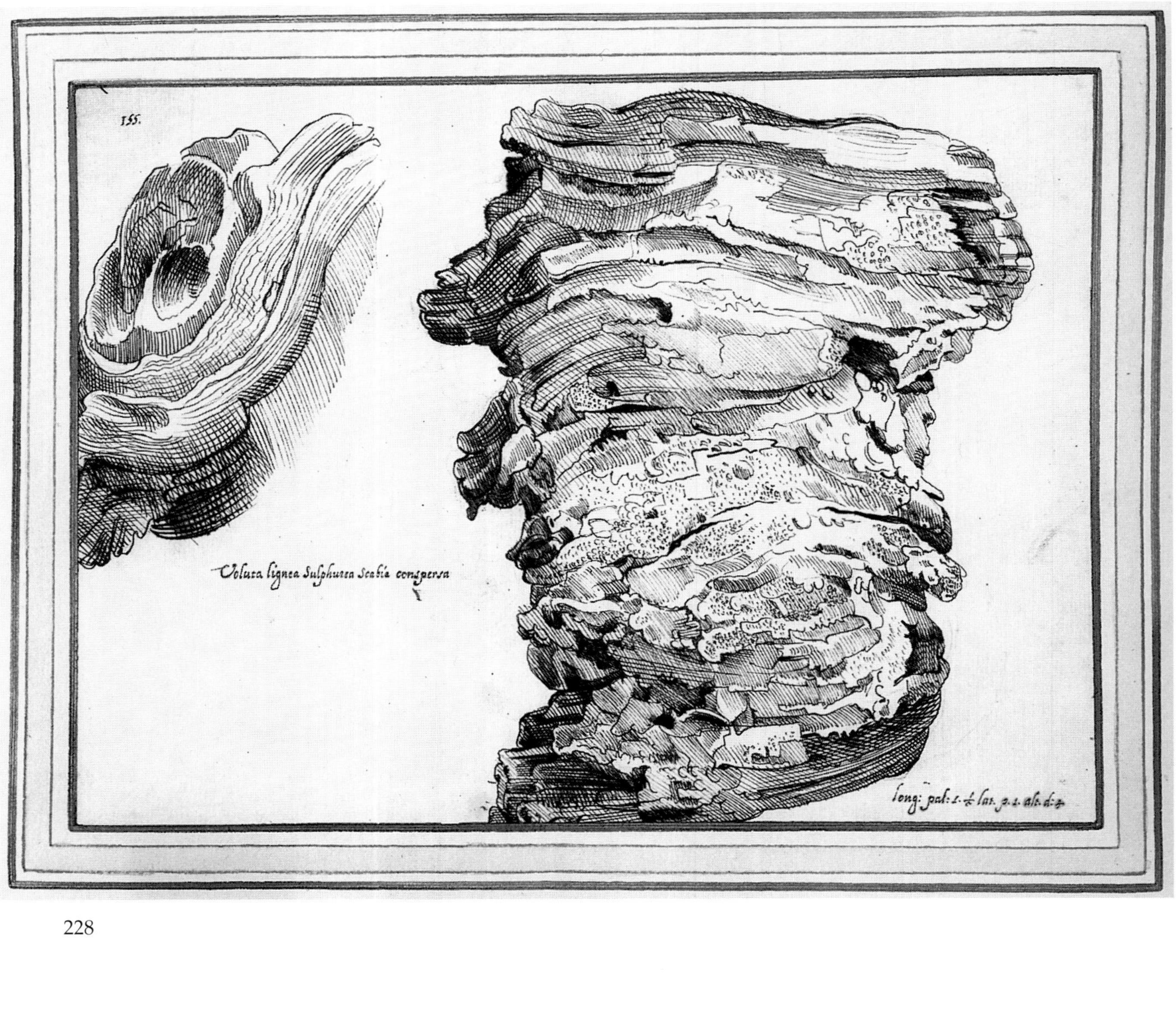

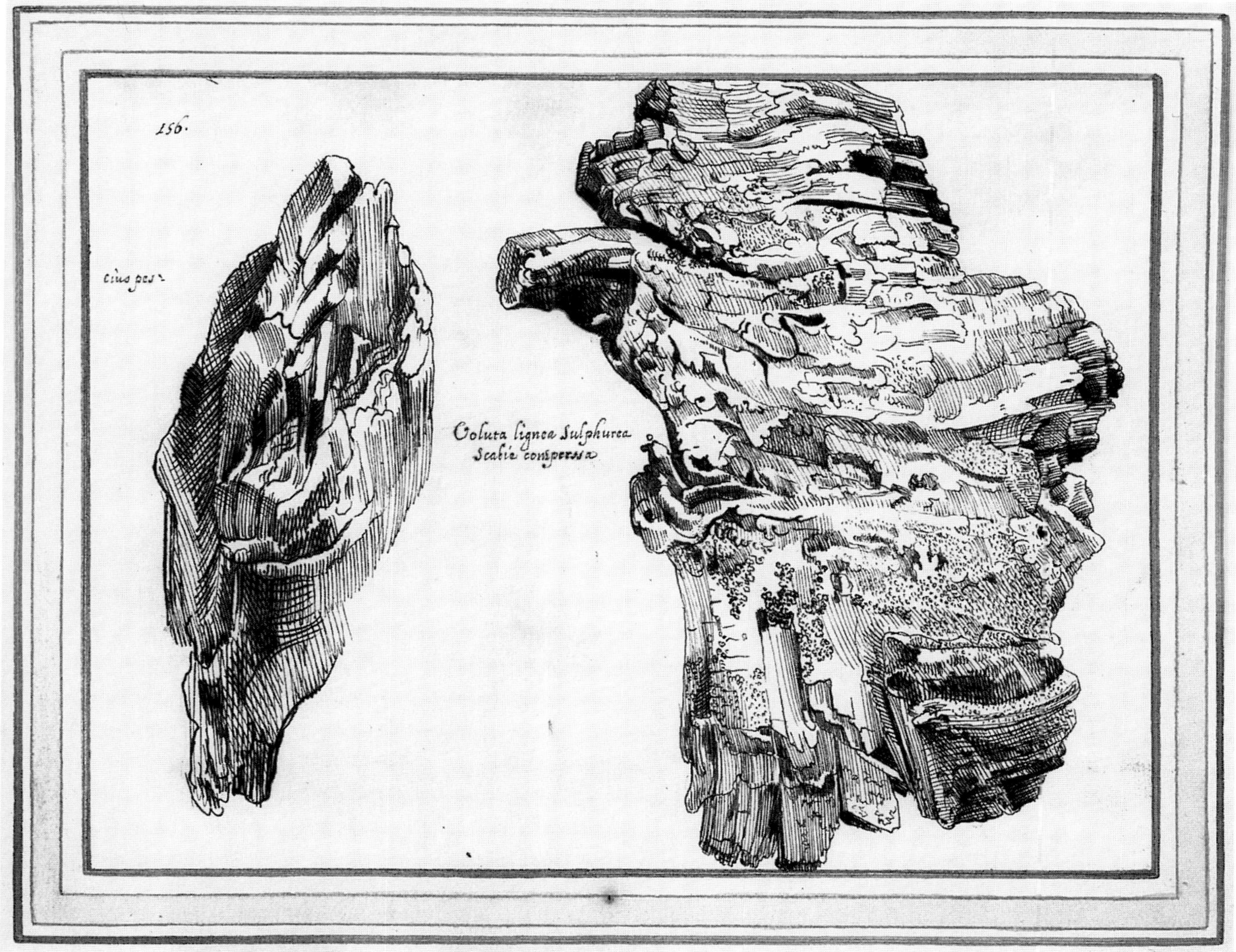

103

103. Two views of a piece of fossil wood

Windsor, RL 25695, *Fossils*, vol. XVI, fol. 56

Pen and ink, over traces of black chalk; measurements not given

186 × 249 mm; small hole caused by ink

NUMBERING: [ink, top left] *156*

ANNOTATION: [ink, centre] *Voluta lignea Sulphurea/ Scabiæ consperssa*
[upper left] *eius pes.*

The specimen bears an identical annotation to **102**, and has a very similar appearance.

REFERENCES: 17th-cent. Index, p. 370

OTHER DRAWINGS: **102**

104. Two pieces of fossil wood, possibly pyritized with knot holes

Windsor, RL 25645, *Fossils*, vol. XVI, fol. 6

Pen and ink, over traces of black chalk; retouched in ink on mount sheet

184 × 244 mm; small holes caused by ink; ink frame line

NUMBERING: [ink, top left corner] *106*

ANNOTATION: [ink, bottom-left corner] *ligni nudi / Metallificati / ligni Metallum/ Superius quadam / Tenus Spumosum*
[bottom centre] *Ponderosi ligni metall./ nod./ long: d: 8. Fere...*
[bottom right] *laxius aliud, solutius excoctius. long: d: 8: lat. d: 6: alt. d. 4*

These specimens of fossil wood are knot holes. The left-hand specimen bears two descriptions as a 'knot of metallized wood, metallic above, spongy below' and as a 'heavy knot of metallized wood'. Its measurements were not completed. The drawing was engraved as Stelluti's Plate 12, Fig. L (see Comp. fig. on p. 222). The accompanying text noted that it 'seemed to be entirely of bark, as it appears to the eye, but partook of the nature of metal, and some knots around it had already [become] completely metallized, its weight therefore being considerable'.

The specimen was probably pyritized or heavily impregnated with iron.

The specimen on the right is similar, and was described as 'a looser and more baked piece'.

REFERENCES: Stelluti 1637, p. 10; 17th-cent. Index, p. 366

ENGRAVINGS: Stelluti 1637, Pl. 12 (Fig. L); for proof engraving in Montpellier, BEM, MS H.173, fol. 21, see p. 403

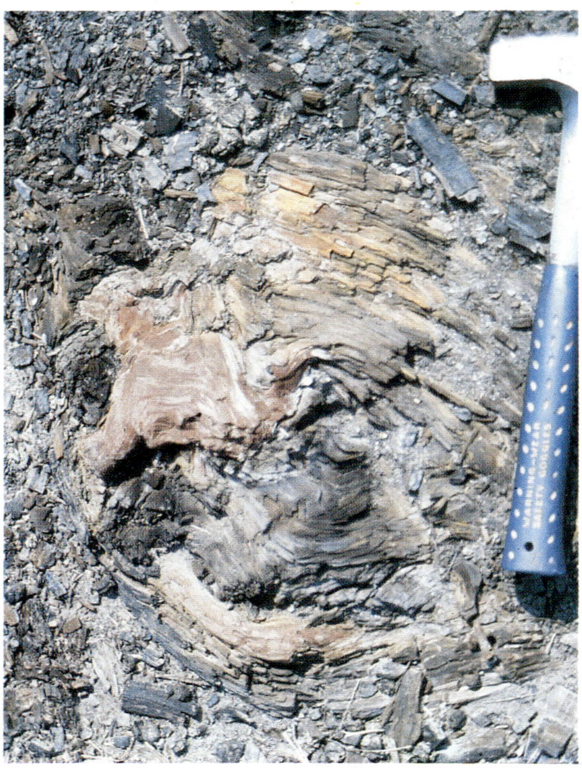

Comp. fig. 104a. Wood knot impregnated with iron, found near Dunarobba

Comp. fig. 104b. Fossil wood from Dunarobba showing iron (siderite and limonite) permineralization

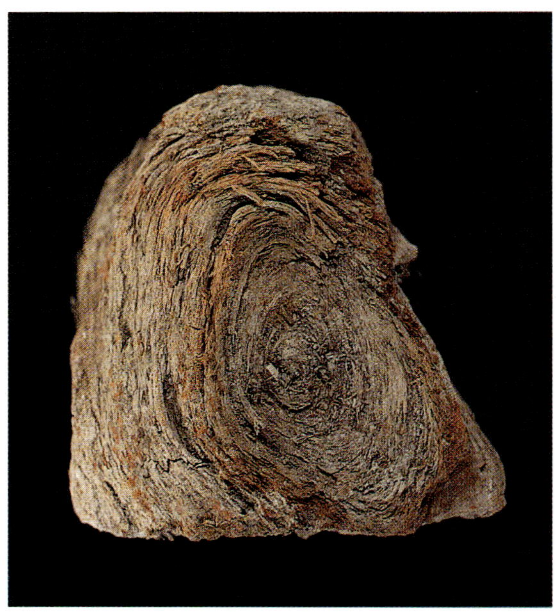

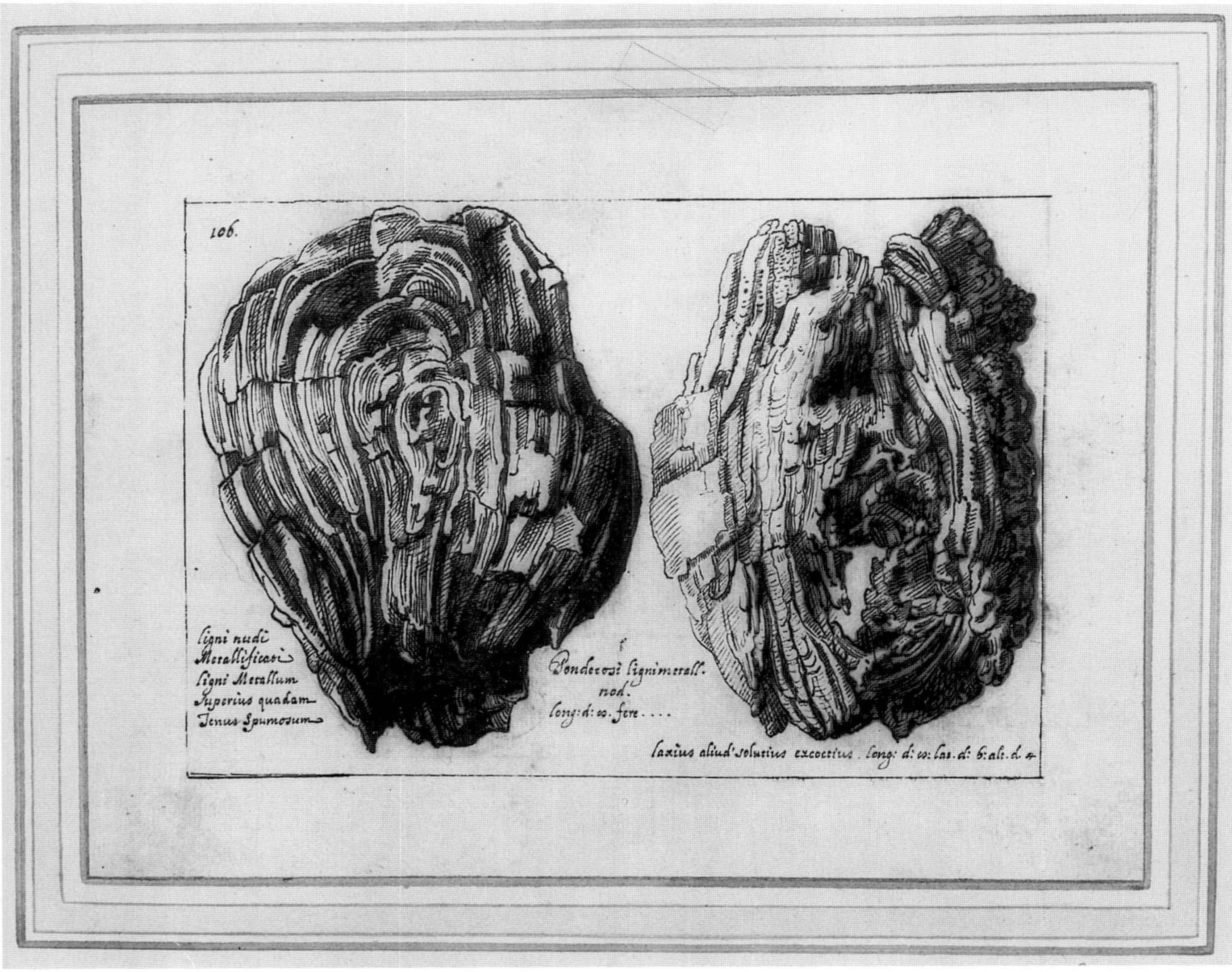
106.
ligni nudi Metallificati ligni Metallum Superius quadam Tenus Spumosum
Ponderosi ligni metall. nod. long: d: 10. fere....
laxius aliud solutius excoctius. long: d: 10: lat. d: 6: alt. d. 4

104

105. Possibly a piece of fossil wood

Windsor, RL 25625, *Fossils*, vol. XV, fol. 34

Pen and ink, over black chalk

186 × 257 mm; small holes caused by ink; marked with a cross

NUMBERING: [ink, top right corner] *86*

ANNOTATION: [ink, top] *lapi lignea fere tota petrefacta album ferens squammis crassiorib.* [bottom] *Long: d.x alt, et Crass. d.5*

This specimen, probably of petrified fossil wood, is described as being 'stone wood almost totally turned to stone, white, bearing thicker scales'. Space appears to have been left on the right of the sheet for further drawings.

REFERENCES: 17th-cent. Index, p. 365

105

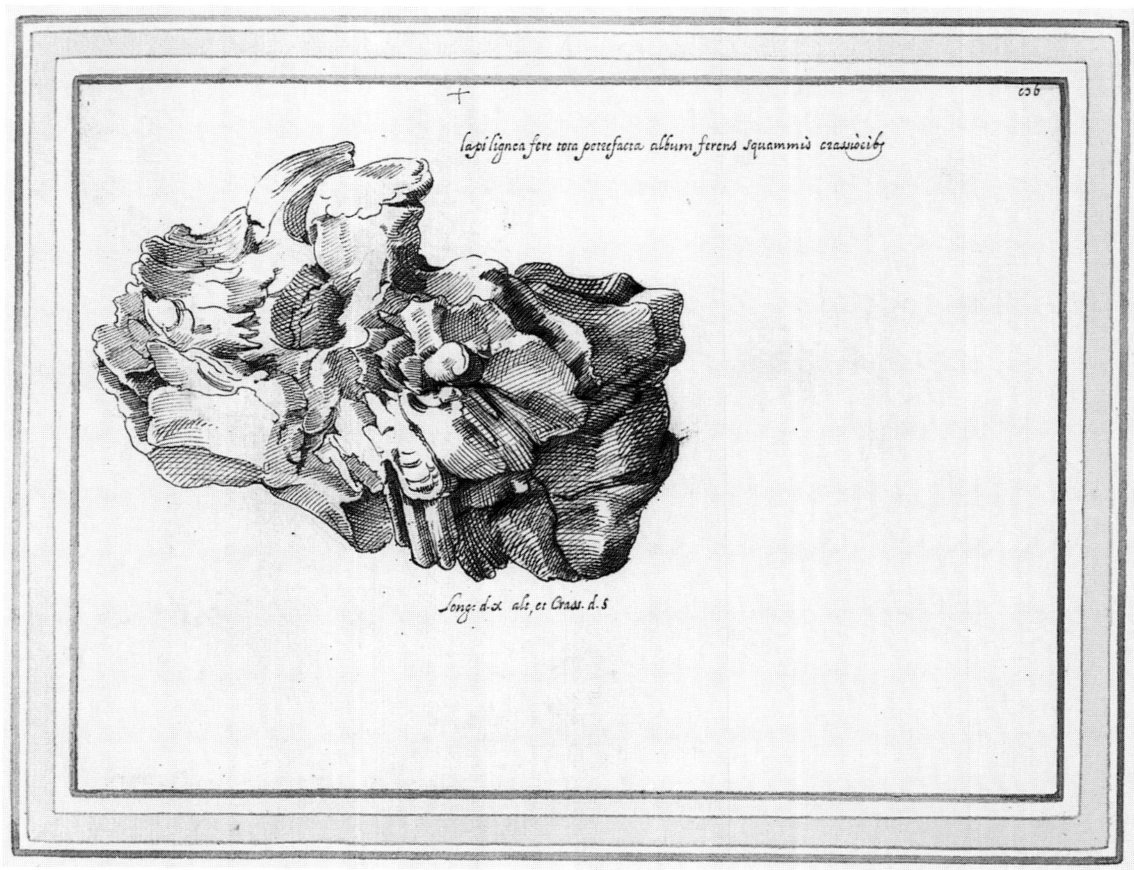

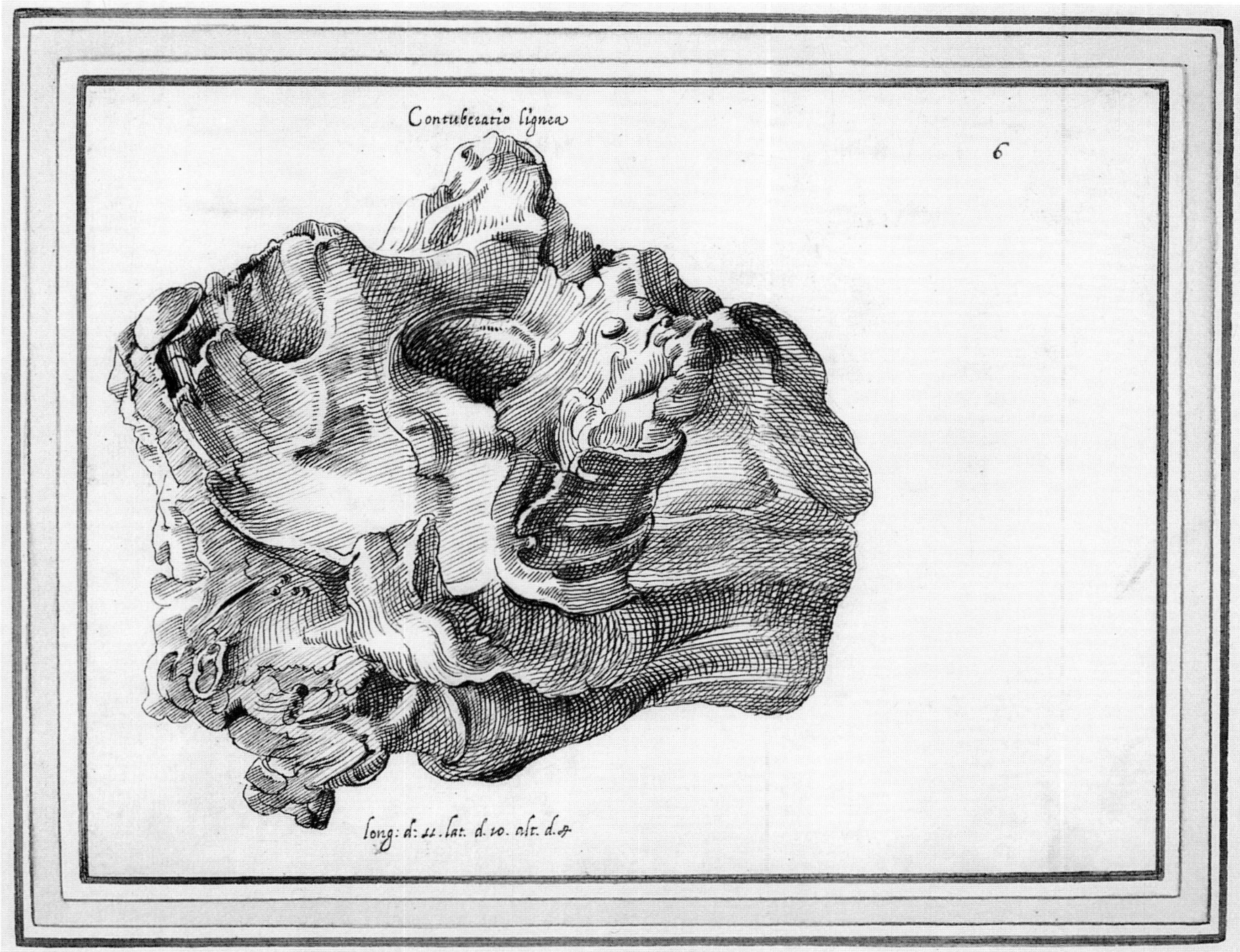

106

106. Fossil wood

Windsor, RL 25542, *Fossils*, vol. XIV, fol. 13

Pen and ink, over black chalk

185 × 250 mm

NUMBERING: [ink, upper right] *6*

ANNOTATION: [ink, top] *Contuberatio lignea* [bottom] *long: d: 11. lat. d. 10. alt.d. 4*

The specimen is described as a kind of convoluted woody mass (*contuberatio*).

It appears to be similar to those shown in **84** and **85**, which are also described as forms of *contuberatio*.

REFERENCES: 17th-cent. Index, p. 361

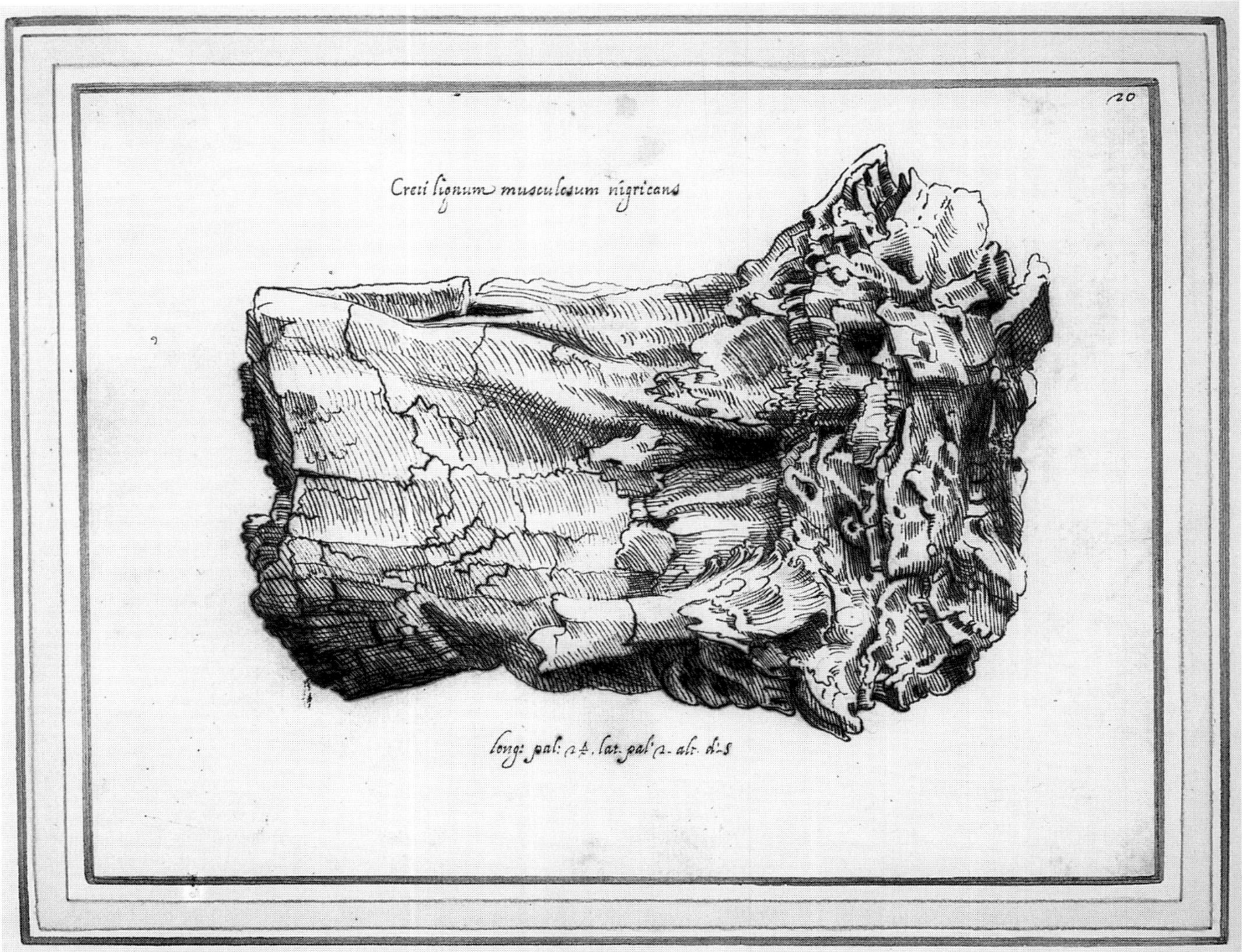

107

107. Fossil wood, possibly the junction between stem and root

Windsor, RL 25557, *Fossils*, vol. XIV, fol. 28

Pen and ink, over black chalk; retouched in ink on mount sheet

186 × 251 mm; small holes caused by ink

NUMBERING: [ink, upper right] *20*

ANNOTATION: [ink, top] *Creti lignum muscolosum nigricans* [bottom] *long: pal: 2½. lat. pal. 2. alt. d: 5*

The adjective 'muscle-like' reinforces the depiction of the form of the end of the branch, which has been ripped from the trunk leaving a mass of convoluted wood.

REFERENCES: 17th-cent. Index, p. 362

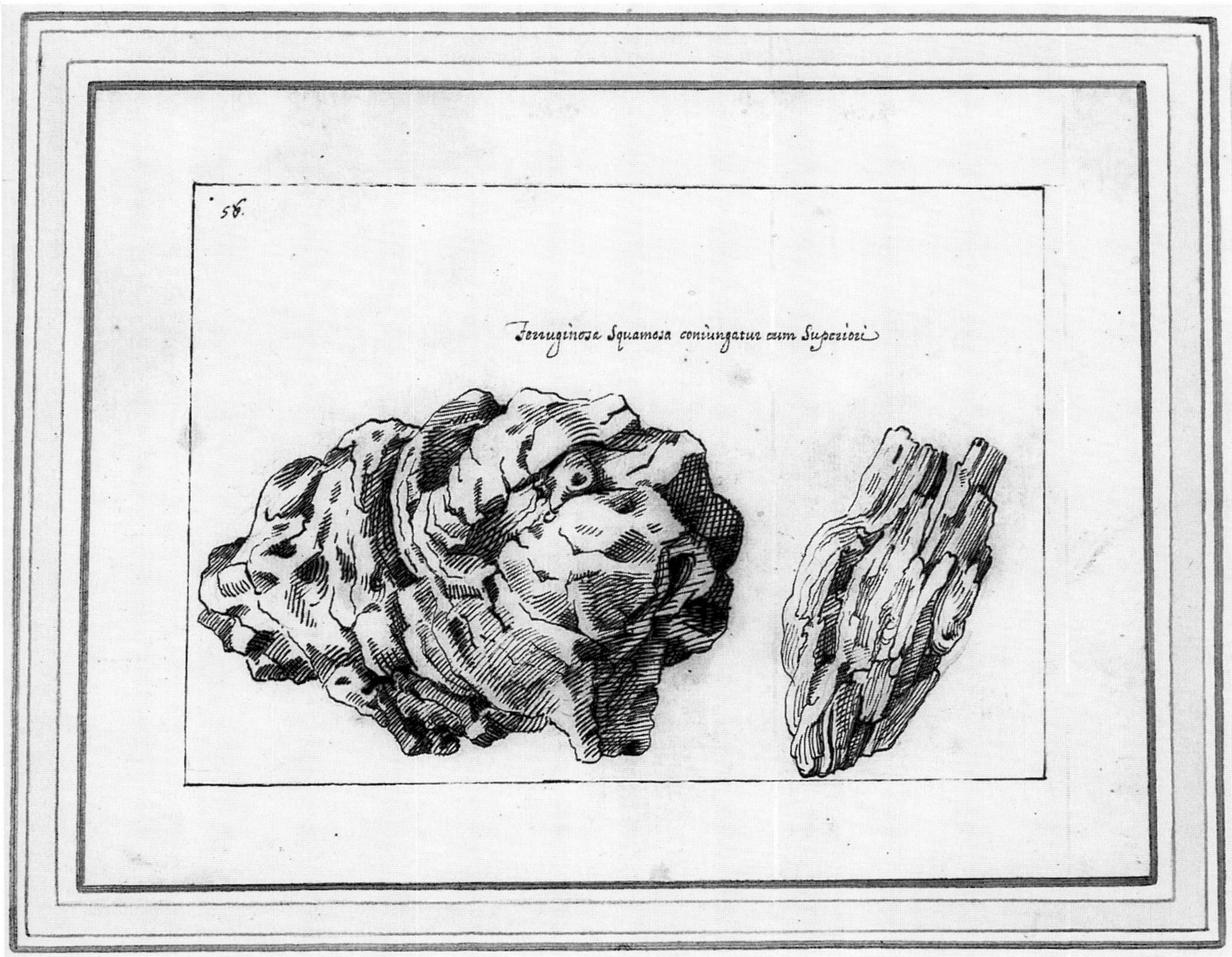

108

108. Two specimens of fossil wood

Windsor, RL 25595, *Fossils*, vol. XV, fol. 4

Pen and ink, over black chalk; retouched in ink on mount sheet

182 × 246 mm; small holes caused by ink; measurements not given; ink frame line

NUMBERING: [ink, top left] *57* over *56*

ANNOTATION: [ink] *Ferruginosa Squamosa coniungatur cum Superiori*

The drawing would appear to be incomplete as the inscription reads 'A squamous, ferruginous specimen, joined to the above'; it is possible that another drawing was to be added to the space at the top of the sheet. The two specimens depicted do not appear to be related. Alternatively the comment relates to another drawing on a different sheet.

REFERENCES: 17th-cent. Index, p. 364

109. Two specimens of fossil wood

Windsor, RL 25543, *Fossils*, vol. XIV, fol. 14

Pen and ink, over traces of black chalk; retouched in ink on mount sheet; marked with a cross

187 × 256 mm; small holes caused by ink

NUMBERING: [ink, upper right] *7*

ANNOTATION: [ink, top left] *Convolutio foliosa nodosa Umbi-/litis plena ligni terrosa subrufa*
[top right] *ligno lapidea nodosa Tuberatio*
[bottom left] *lon:palm. 1. lat. d. 5. crass/d: 3.*
[bottom right] *lon: 8.lat.d. 6. Crass.d. 5*

The specimen on the left is described as 'earthy-reddish, convoluted, leafy, knotty and full of nodules', the one on the right as a 'knotty and stony wood excrescence (*tuberatio*)'; it may be partially petrified.

REFERENCES: 17th-cent. Index, p. 361

109

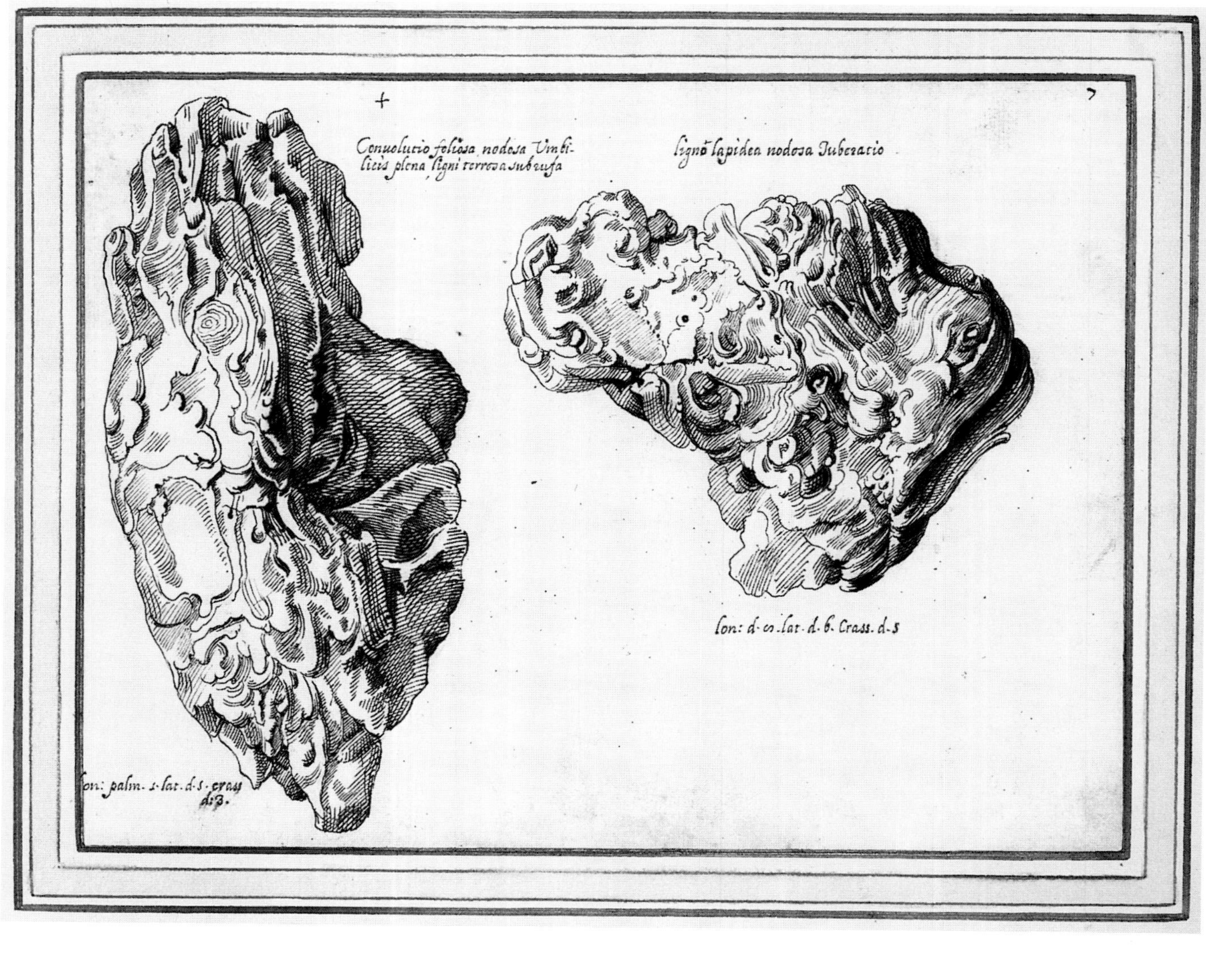

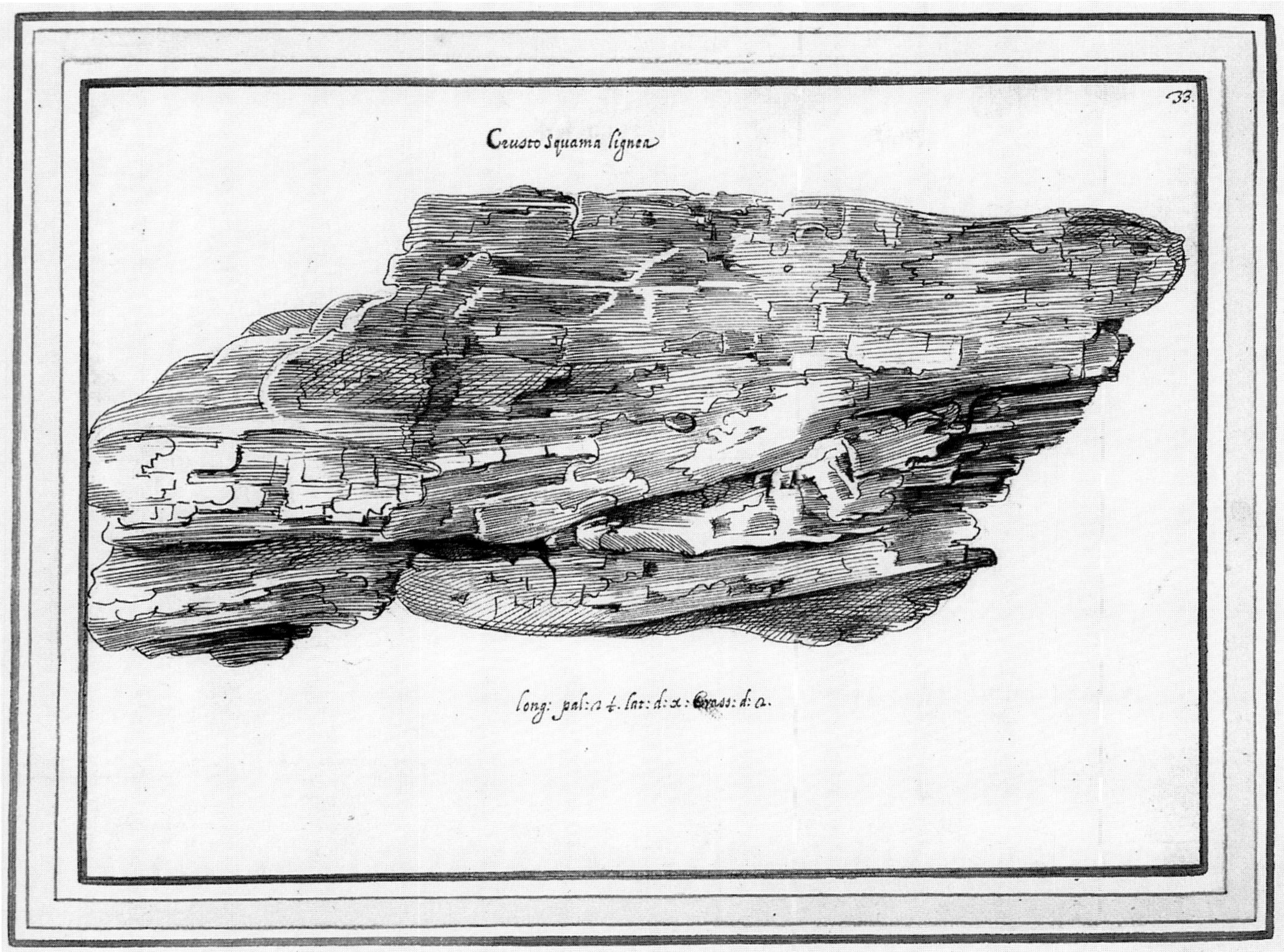

110

110. Fossil wood

Windsor, RL 25571, *Fossils*, vol. XIV, fol. 42

Pen and ink, over black chalk

184 × 262 mm; sheet trimmed through image, left

NUMBERING: [ink, upper right] *33*

ANNOTATION: [ink, top] *Crusta Squama lignea;* [bottom] *long: pal: 2½. lat: d: x : Crass: d: 2.*

A typically dried-out specimen of fossil with a cracked surface. The way its surface is encrusted with woody scales is noted by the inscription.

REFERENCES: 17th-cent. Index, p. 362

111. Fossil wood with pyrite

Windsor, RL 25657, *Fossils*, vol. XVI, fol. 18

Pen and ink, over traces of black chalk

180 × 256 mm

NUMBERING: [ink, top right corner] *118*

ANNOTATION: [ink, top] *Metalligo Convoluta* [bottom] *long. pal. med. lat. d. 4 alt. d. 4*

A specimen classified under the term *Metalligo*, presumably referring to its metal-like qualities (for another specimen in this category see **132**). It would appear to be fossil wood encrusted with iron pyrite.

REFERENCES: 17th-cent. Index, p. 367

111

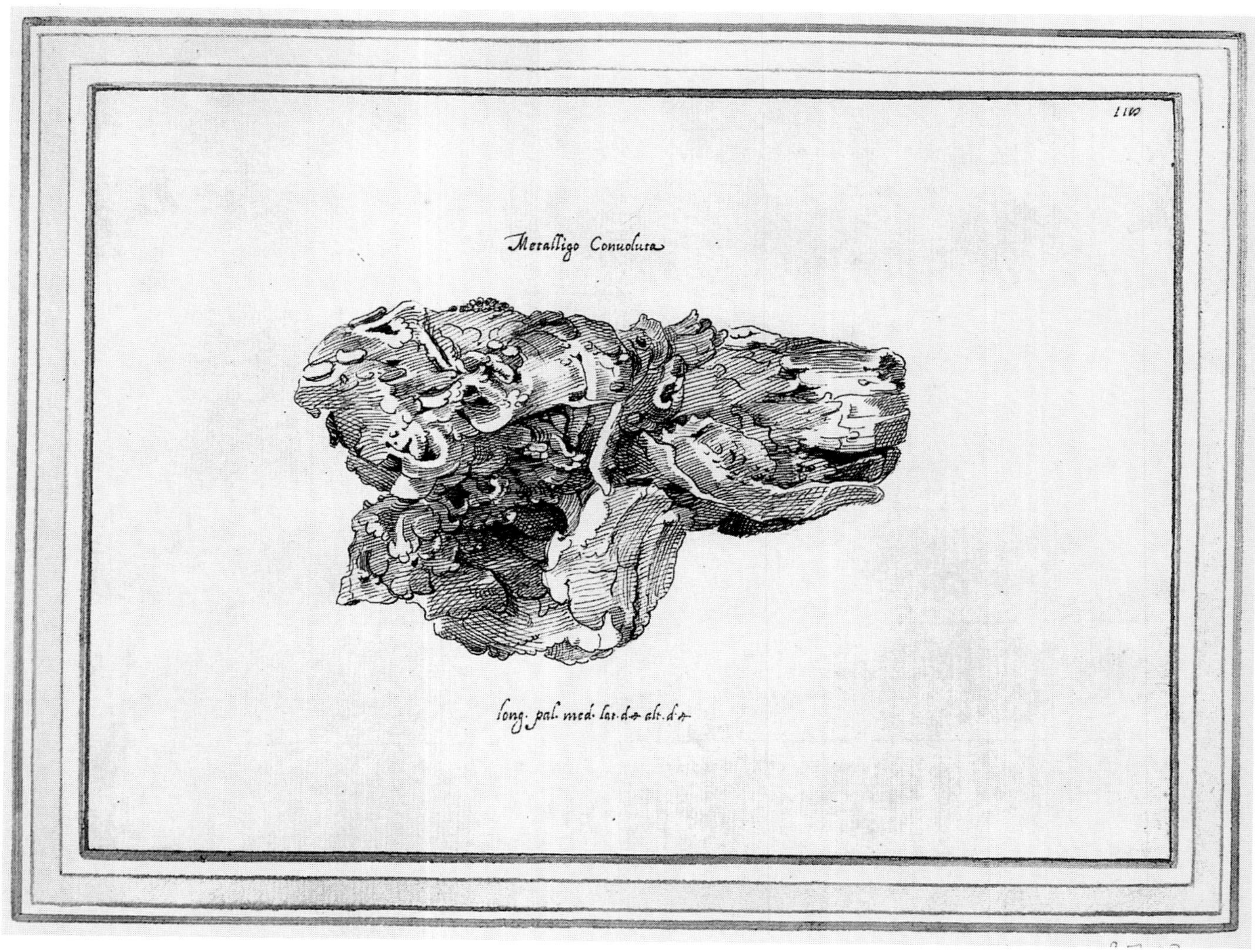

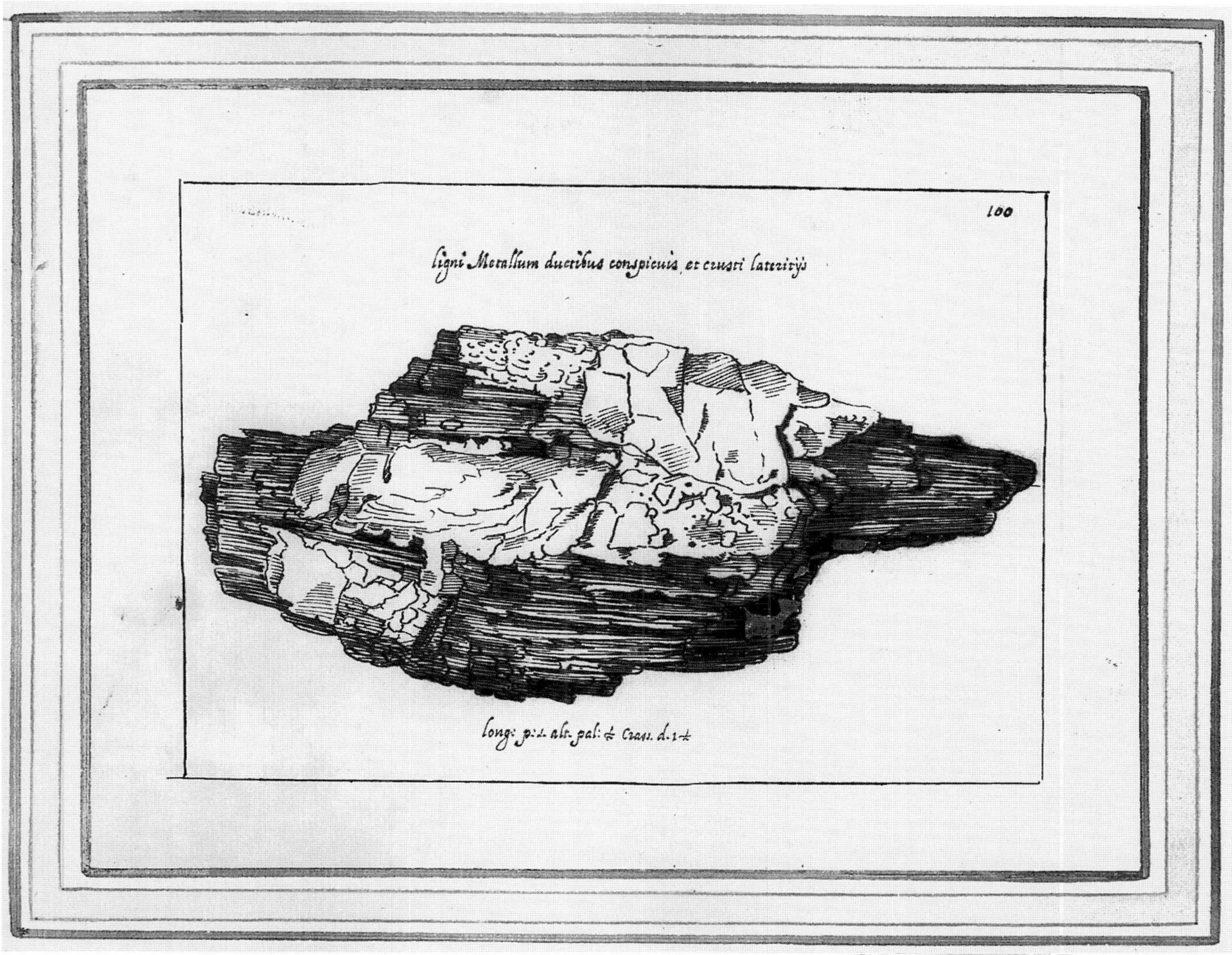

112

112. Fossil wood, possibly pyritized

Windsor, RL 25639, *Fossils*, vol. XV, fol. 48

Pen and ink, over traces of black chalk; retouched in ink on mount sheet

180 × 242 mm; small holes caused by ink; ink frame line

NUMBERING: [ink, top right corner] *100*

ANNOTATION: [ink, top] *ligni Metallum ductibus conspicuis, et crusti laterijs* [bottom] *long: p: 1. alt. pal: ½ Crass. d. 1½*

This specimen of fossil wood is described as 'metallized wood with conspicuous ducts and a brick-like crust'. It was engraved as Plate 12, Figure H in Stelluti. The accompanying text reads: 'Piece *H,* in that part which appears darker, was both wood and metal, and the outer, lighter crust was of laterite, that is, of baked earth, as bricks are'. This description suggests impregnation by iron carbonate (siderite).

REFERENCES: 17th-cent. Index, p. 366

ENGRAVINGS: Stelluti 1637, Pl. 12 (Fig. H)

113. Fossil wood fragments in matrix

Windsor, RL 25551, *Fossils*, vol. XIV, fol. 22

Pen and ink, over black chalk

185 × 250 mm; small holes caused by ink; no measurements given

NUMBERING: [ink, top left] *15*

ANNOTATION: [ink] *Creti lignum cinereo ceruleum Musculosum ligni squa-/ mis et corticosis Varie interceptis arenulis etiam Metallicis / nitens alicubi*

The specimen appears to be a block of plant-rich sediment packed with wood fragment and with thin layers of clay shown at the bottom of the specimen. Within these layers there may also be small pieces of pyrite. This intermix is reflected in the description where the specimen is classified as a 'muscle-like greyish-blue [piece of] *cretilignum* ('clayey wood'), with bark-like scales of wood variously interspersed with sandy granules that are even metallic and glistening here and there'.

The arrangement of the specimen on the page would seem to indicate space left for another drawing.

REFERENCES: 17th-cent. Index, p. 361

113

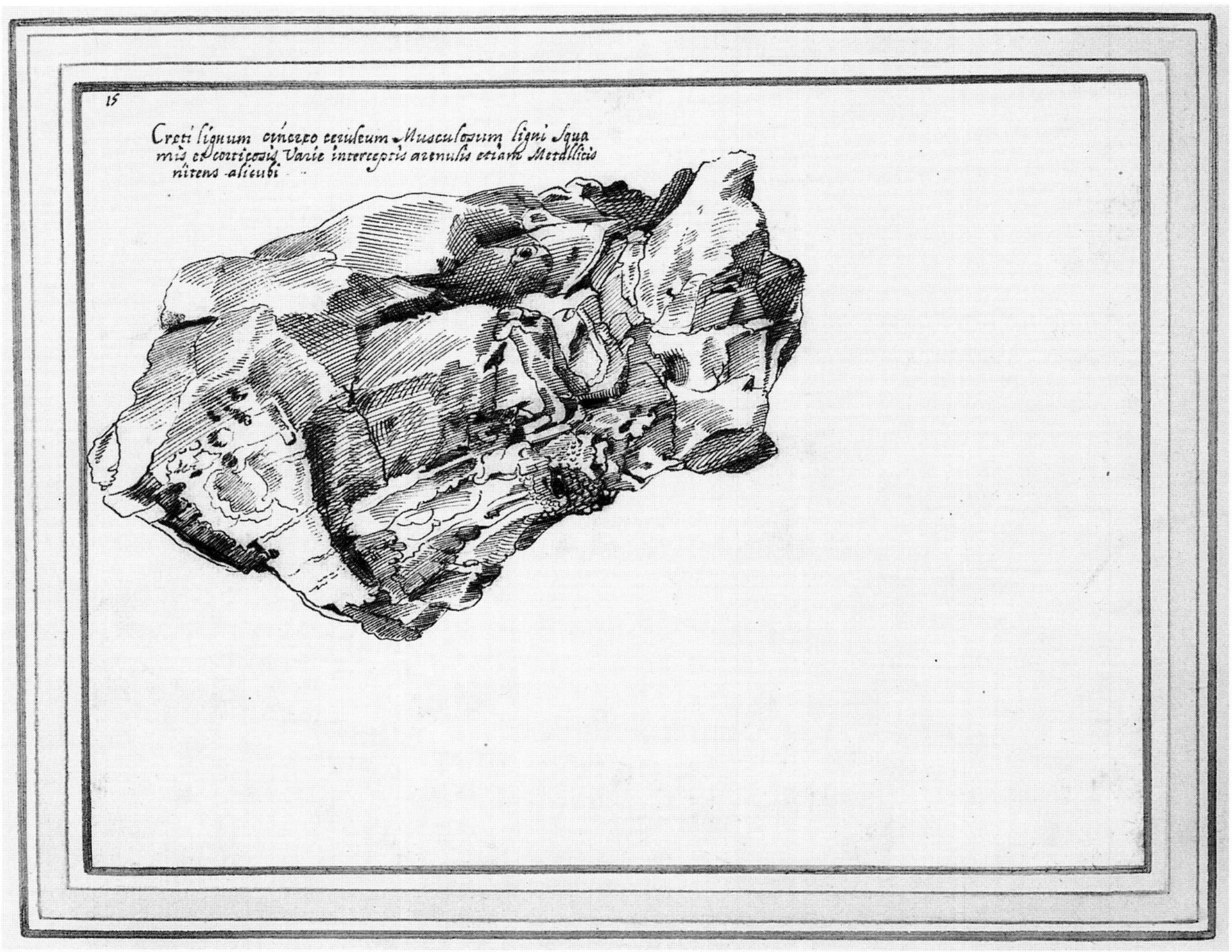

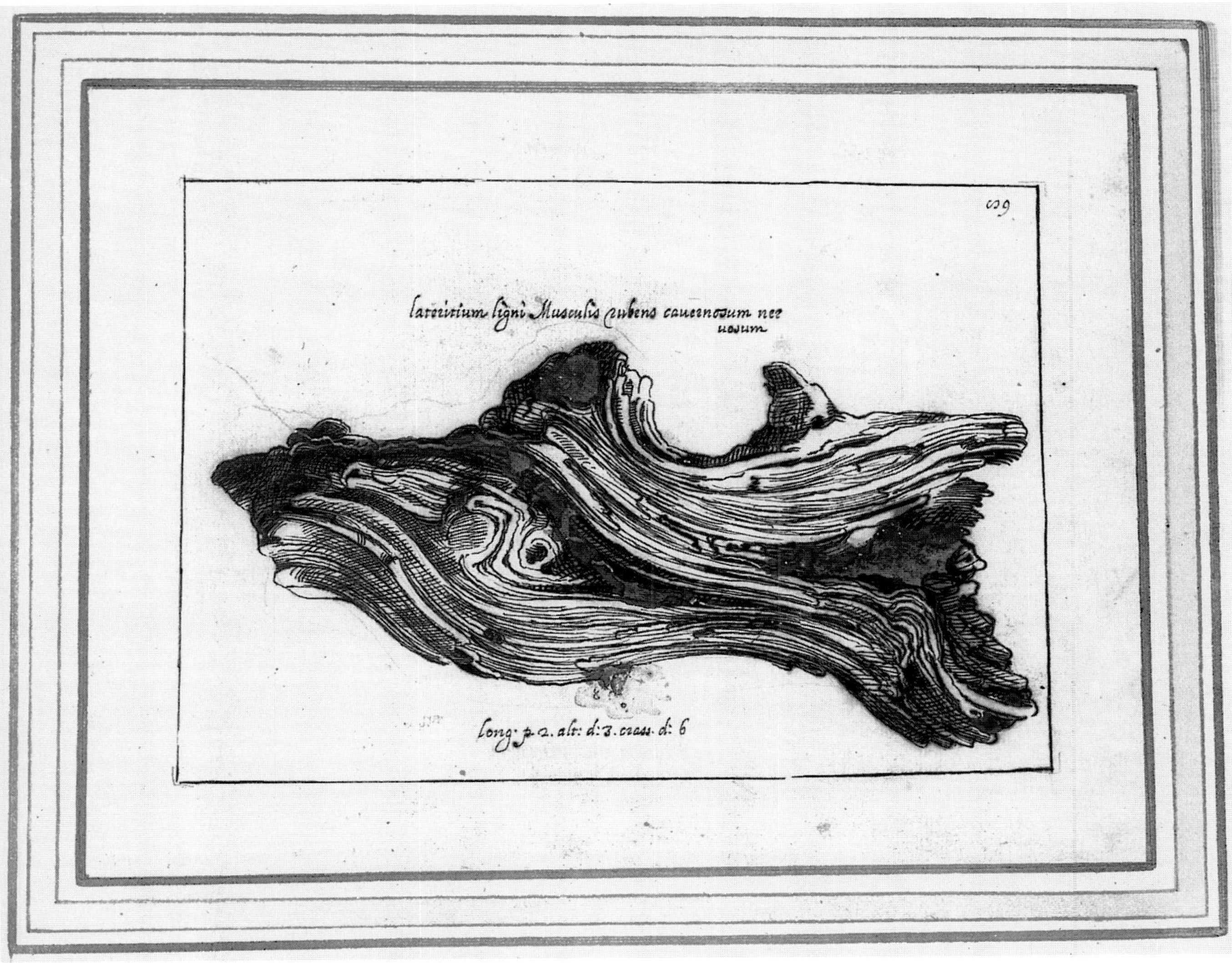

114

114. Piece of fossil wood from Dunarobba

Windsor, RL 25628, *Fossils*, vol. XV, fol. 37

Pen and ink with corrections in white heightening, over black chalk; retouched in ink on mount sheet; frame line in ink

175 × 242 mm; small holes caused by ink

NUMBERING: [ink, top right corner] *89*

ANNOTATION: [ink, top] *lateritium ligni Musculis rubens cavernosum ner/vosum* [bottom] *long. p. 2. alt: d: 3. crass. d: 6*

This specimen of fossil wood is described as a 'ruddy laterite with sinews of wood, with fibrous cavities'. The description suggests some impregnation with iron. According to the Index, the specimen was from Dunarobba.

REFERENCES: 17th-cent. Index, p. 365

115. Fossil wood

Windsor, RL 25560, *Fossils*, vol. XIV, fol. 31

Pen and ink, over traces of black chalk

186 × 229 mm

NUMBERING: [ink, upper right, in a different hand] *23*

ANNOTATION: [ink, top] *Creti lignum nigrum Squamis varijs crassioribus compactum ligneis quoq./ Squamis intermistis*

A compacted mass of wood and clay is described as a 'black *cretilignum* with rather thick scales compacted together and mixed with woody scales'. Space appears to have been left for another drawing below.

REFERENCES: 17th-cent. Index, p. 362

115

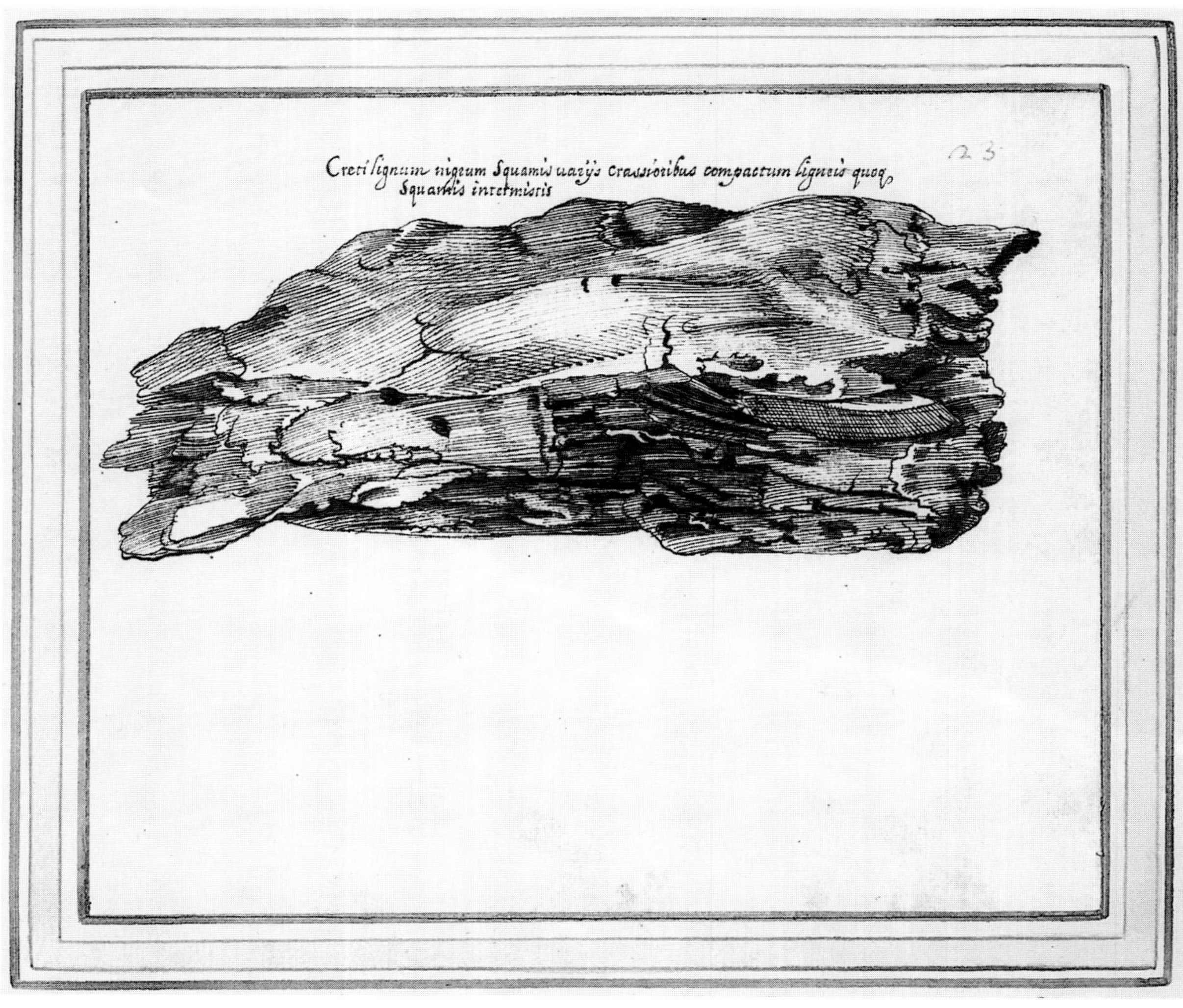

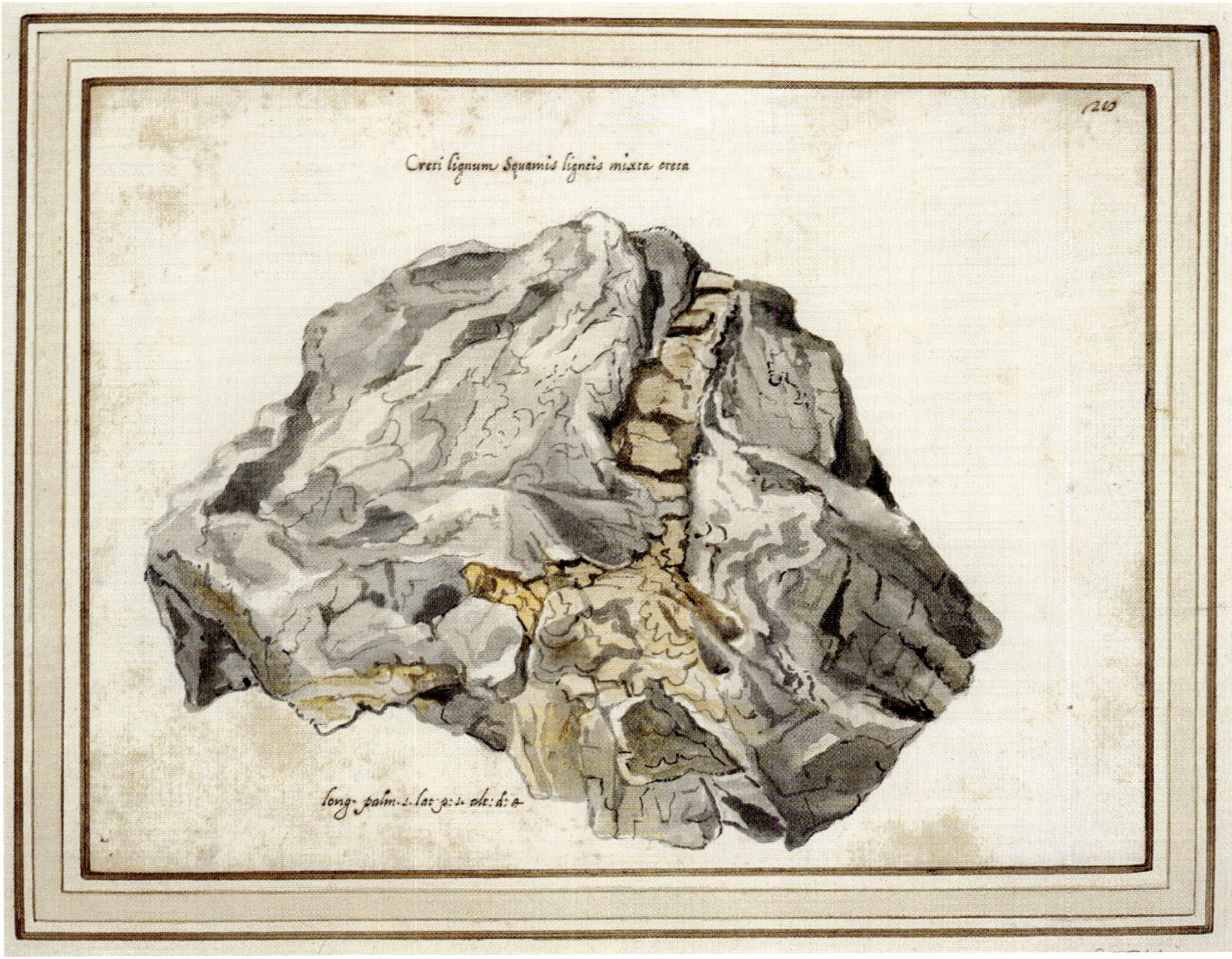

116

116. Possibly clay with a layer of plant fragments

Windsor, RL 25566, *Fossils*, vol. XIV, fol. 37

Watercolour, over traces of black chalk

186 × 254 mm

NUMBERING: [ink, upper right] *28*

ANNOTATION: [ink, top, centre] *Creti lignum Squamis ligneis mixta creta* [bottom] *long. palm. 1. lat. p: 1. alt: d: 4*

The specimen is described as a 'mixed clay [specimen of] *cretilignum* with woody scales'. The grey-coloured clay shows a layer with brownish plant fragments clearly depicted through the use of watercolour. It is one of only two coloured specimens in the album.

REFERENCES: 17th-cent. Index, p. 362

117. Piece of fossil wood

Windsor, RL 25598, *Fossils*, vol. XV, fol. 7

Pen and ink with corrections in white heightening, over traces of black chalk; retouched in ink on mount sheet; measurements not given; ink frame line

175 × 238 mm; small holes caused by ink

NUMBERING: [ink, bottom left corner] *60*

ANNOTATION: [ink, top left] *Frustum Anceps/ ligni Terreum/ ab una parte lignum/ ab alia creta*

Cesi's uncertainty about the true nature of this specimen is fully revealed by the inscription: 'An earthen piece, possibly of wood—in one part wood and in the other clay'. In keeping with the Linceans' consistent interest in everything that was ambiguous in nature, the specimen was selected for illustration by Stelluti (Plate 11), with the comment that '*E* is a fragment which cannot quite be defined as wood, but as ligneous earth, since one part, the foremost, can be seen once to have been all of wood, traces of bark being clearly distinguishable, and the other, the hindmost, all of hard clay not yet transformed into wood'.

The encrusting of wood by clay in a fossil specimen is by no means unusual, but was clearly important for both Cesi and Stelluti in their attempt to understand the origin of these fossils.

REFERENCES: Stelluti 1637, p. 10; 17th-cent. Index, p. 364

ENGRAVINGS: Stelluti 1637, Pl. 11 (Fig. E); for proof engraving in Montpellier, BEM, MS H.173, fol. 20, see p. 403

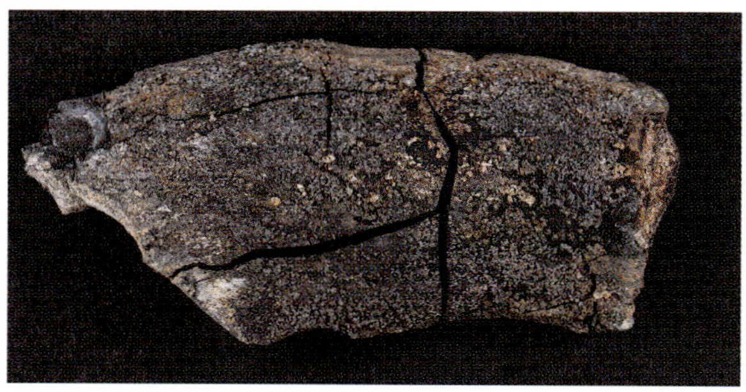

Comp. fig. 117. Black coalified wood from the Santa Maria di Ciciliano Formation

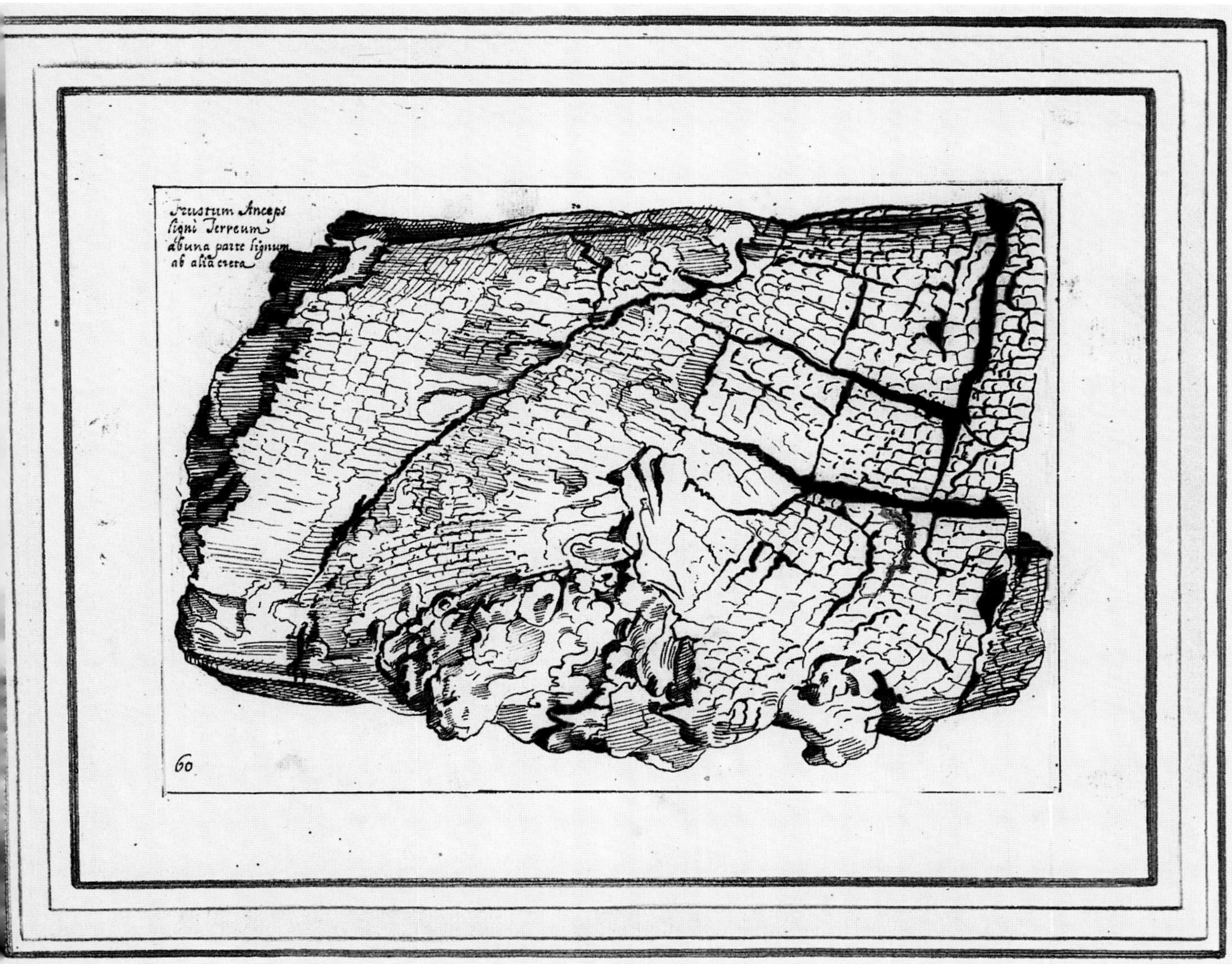
Frustum Anceps
ligni Terreum
ab una parte lignum
ab alia creta
60

117

118. Piece of fossil wood, partly pyritized

Windsor, RL 25599, *Fossils*, vol. XV, fol. 8

Pen and ink, over traces of black chalk; retouched in ink on mount sheet

165 × 226 mm; small holes caused by ink; measurements not given; ink frame line

NUMBERING: [ink, bottom left corner] *61*

ANNOTATION: [ink] *Frustum Metalliginis*

This 'metalliginous' piece of fossil wood shows partial encrustation by pyrite. Another view is shown in **119**.

REFERENCES: 17th-cent. Index, p. 364

OTHER DRAWINGS: **119**

118

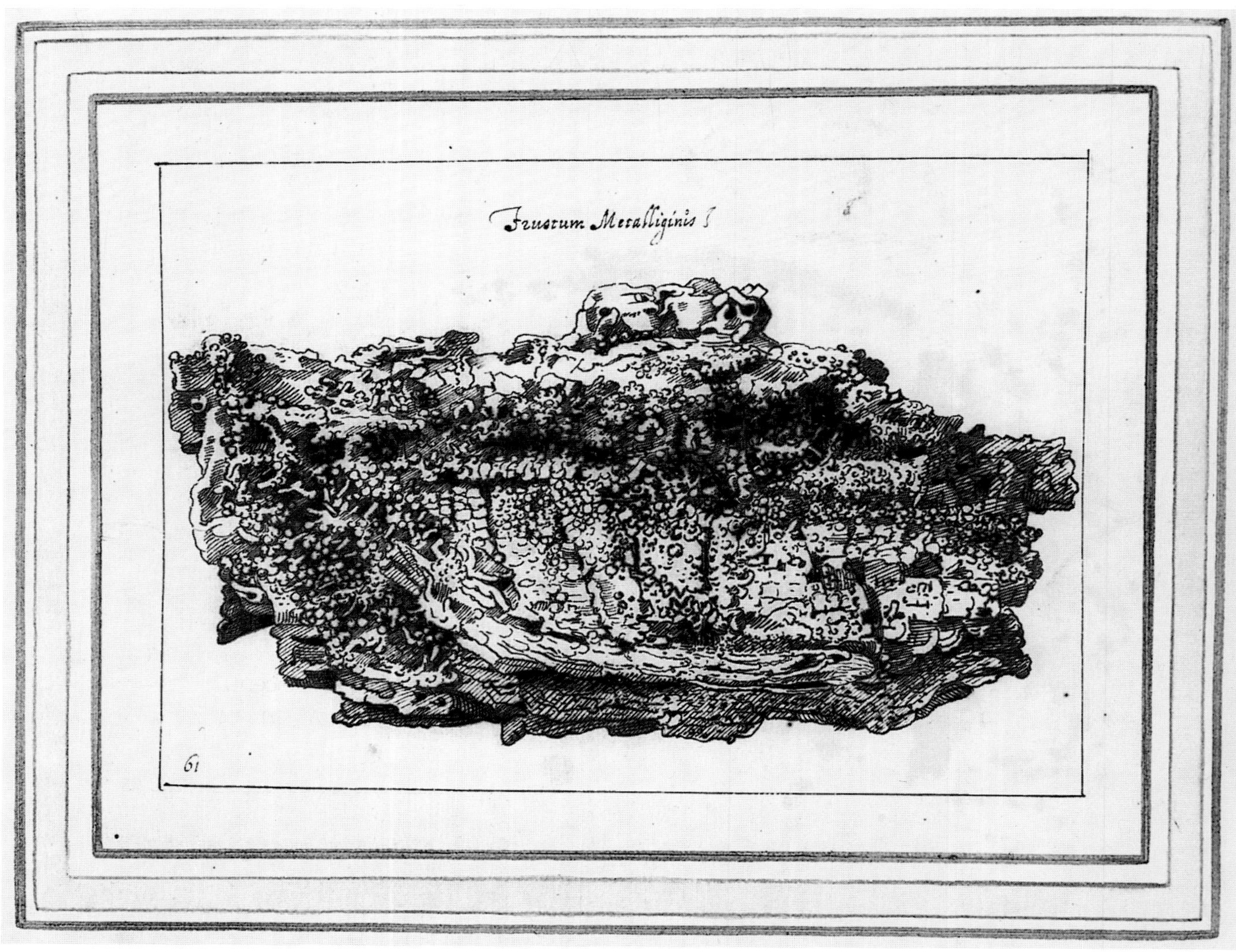

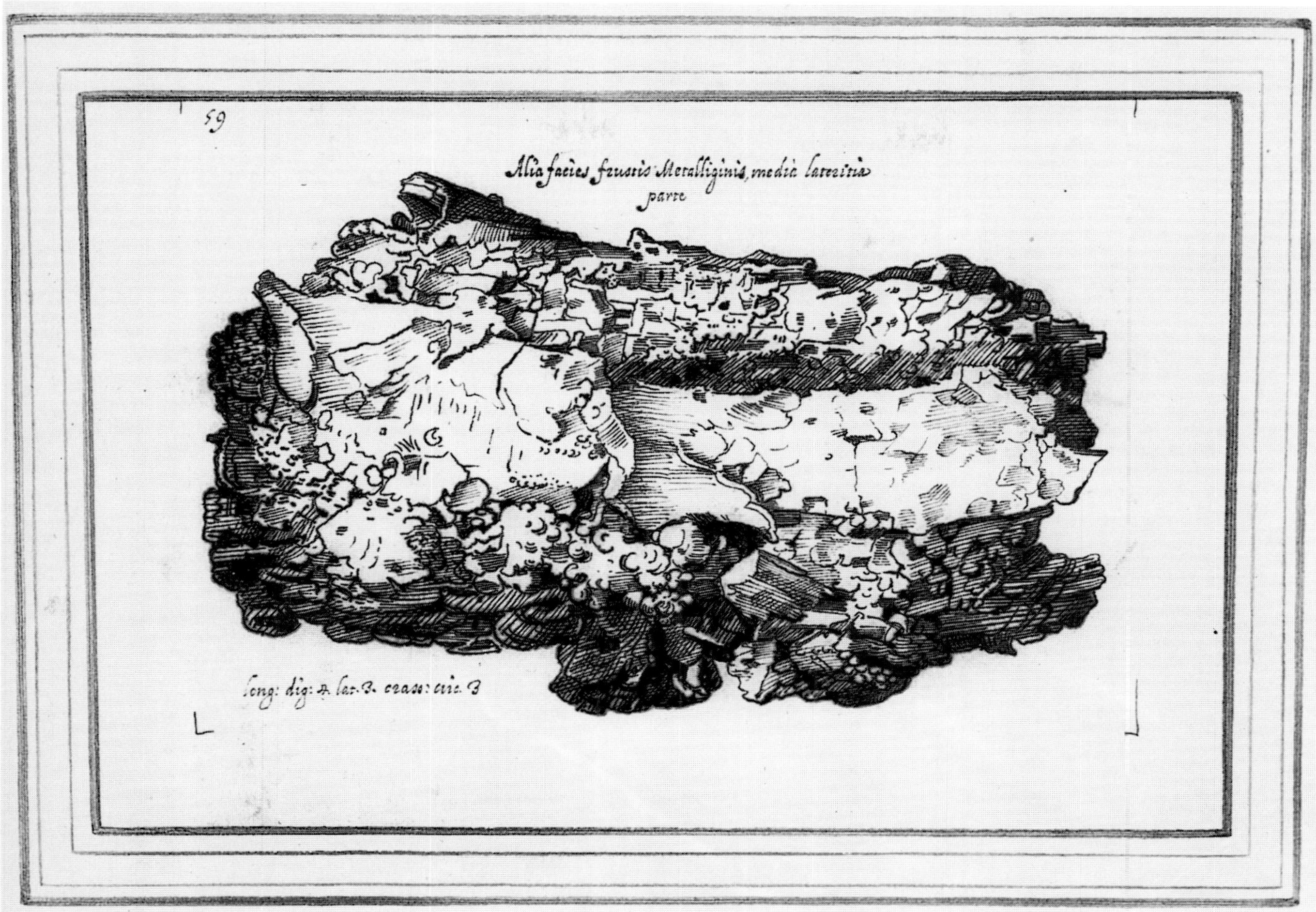

119

119. The other side of 118

Windsor, RL 25597, *Fossils*, vol. XV, fol. 6

Pen and ink, over traces of black chalk; retouched in ink on mount sheet; frame partly in ink, partly in black chalk

155 × 237 mm; small holes caused by ink

NUMBERING: [ink, upper left] *59*

ANNOTATION: [ink, top] *Alia facies frustis Metalliginis, media lateritia/ parte* [bottom left] *long: dig: 4. lat. 3. crass: circ. 3*

A view from the other side of **118**. In addition to its 'metalliginous bits', it also has a baked appearance in the middle (*lateritia*). The draughtsman clearly conveys a change of substance in the band-like area across the centre of the specimen, which may indicate iron impregnation.

REFERENCES: 17th-cent. Index, p. 364
OTHER DRAWINGS: **118**

120. Fossil wood, possibly pyritized

Windsor, RL 25544, *Fossils*, vol. XIV, fol. 15

Pen and ink, over traces of black chalk; retouched in ink on mount sheet

177 × 242 mm; small holes caused by ink

NUMBERING: [ink, upper right] *8*

ANNOTATION: [ink, top] *Cortic. ligni met. concavum ligni ductilo conspicuis* [bottom] *long: d: x. lat. 5. Crass. 1*

The inscription on this drawing does not yield its meaning easily; but it clearly refers to a flaky or 'barky' cavity of a piece of metallized wood with ducts running through it. The specimen may have been partially pyritized.

REFERENCES: 17th-cent. Index, p. 361

OTHER DRAWINGS: **121**

120

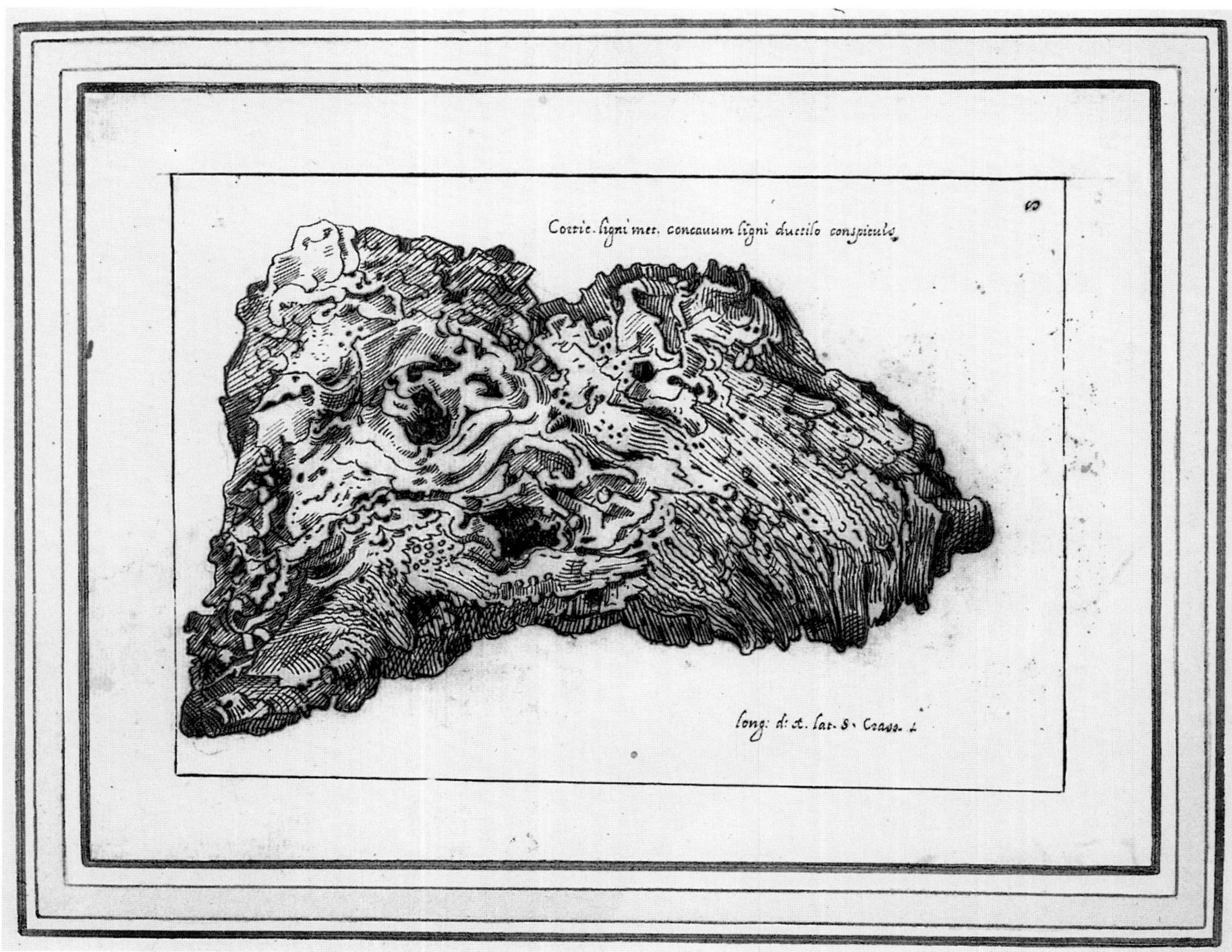

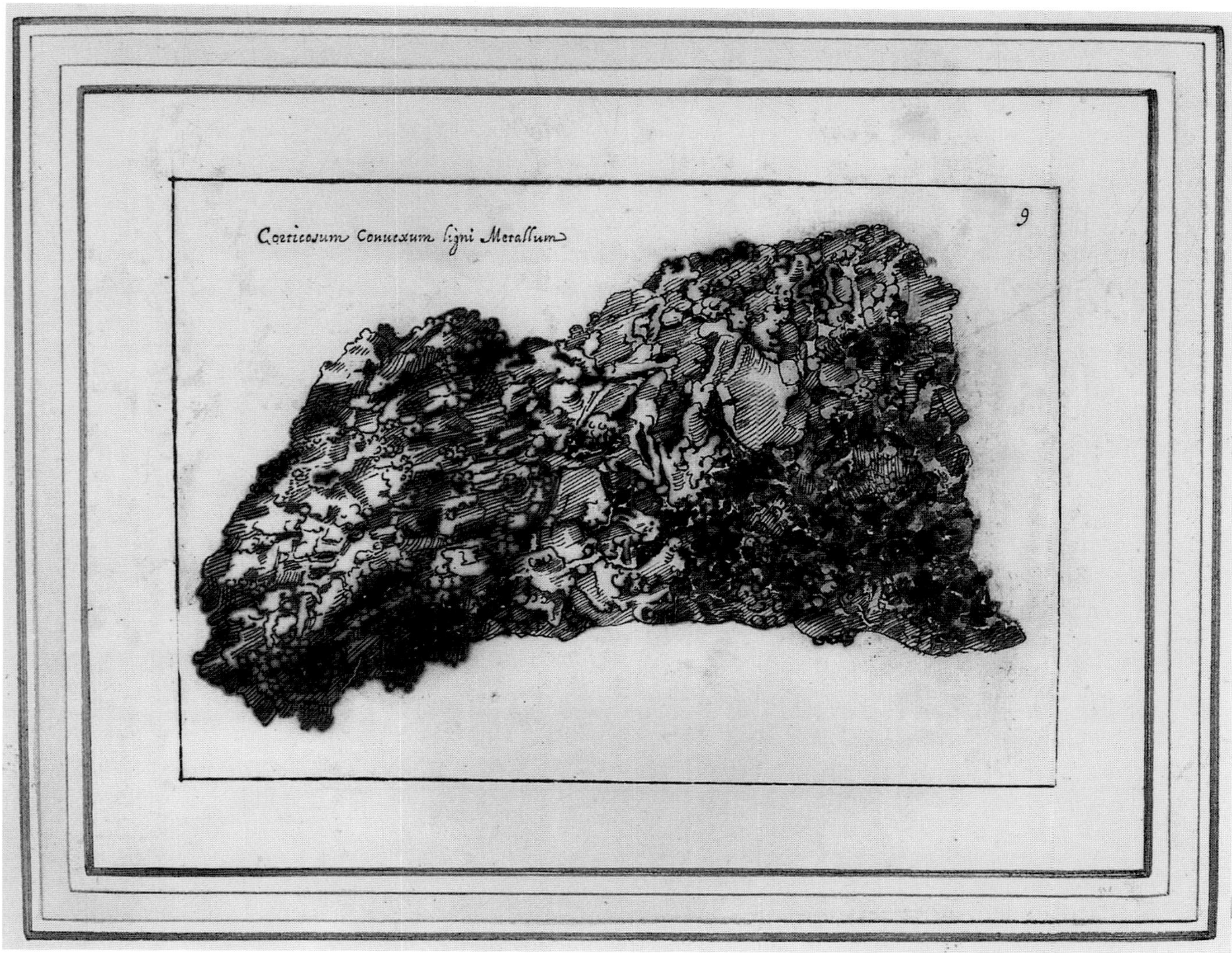

121

121. Fossil wood, possibly pyritized

Windsor, RL 25545, *Fossils*, vol. XIV, fol. 16

Pen and ink, over traces of black chalk; retouched in ink on mount sheet

177 × 239 mm; small holes caused by ink; measurements given in **120**; ink frame line

NUMBERING: [ink, upper right] *9*

ANNOTATION: [ink] *Corticosum Convexum ligni Metallum*

The description reads: 'The barky convex side of [the same piece of] metallized wood' (the same, that is, as **120**).

REFERENCES: 17th-cent. Index, p. 361

OTHER DRAWINGS: **120**

122. Fossil wood and lignite

Windsor, RL 25567, *Fossils*, vol. XIV, fol. 38

Pen and ink, over black chalk

190 × 258 mm; small holes caused by ink; measurements given of left-hand specimen only

NUMBERING: [ink, upper right, in another hand] *29*

ANNOTATION: [ink, top left] *ligni an' fractuosum*
[bottom left] *long. d: 11. lat. d. 10, alt. d. 4*
[top right] *Creti lignum Squamosum crassioribus Squamis*

The left-hand specimen, described as a 'convolution of wood', shows similar characteristics to the piece also called *an' fractuosum* in **85**.

The right-hand specimen is similar in appearance to **123** and, like it, is described as a *cretilignum*, with 'rather thicker scales'. It may be a piece of true lignite with thin clay layers.

REFERENCES: 17th-cent. Index, pp. 362, 366

122

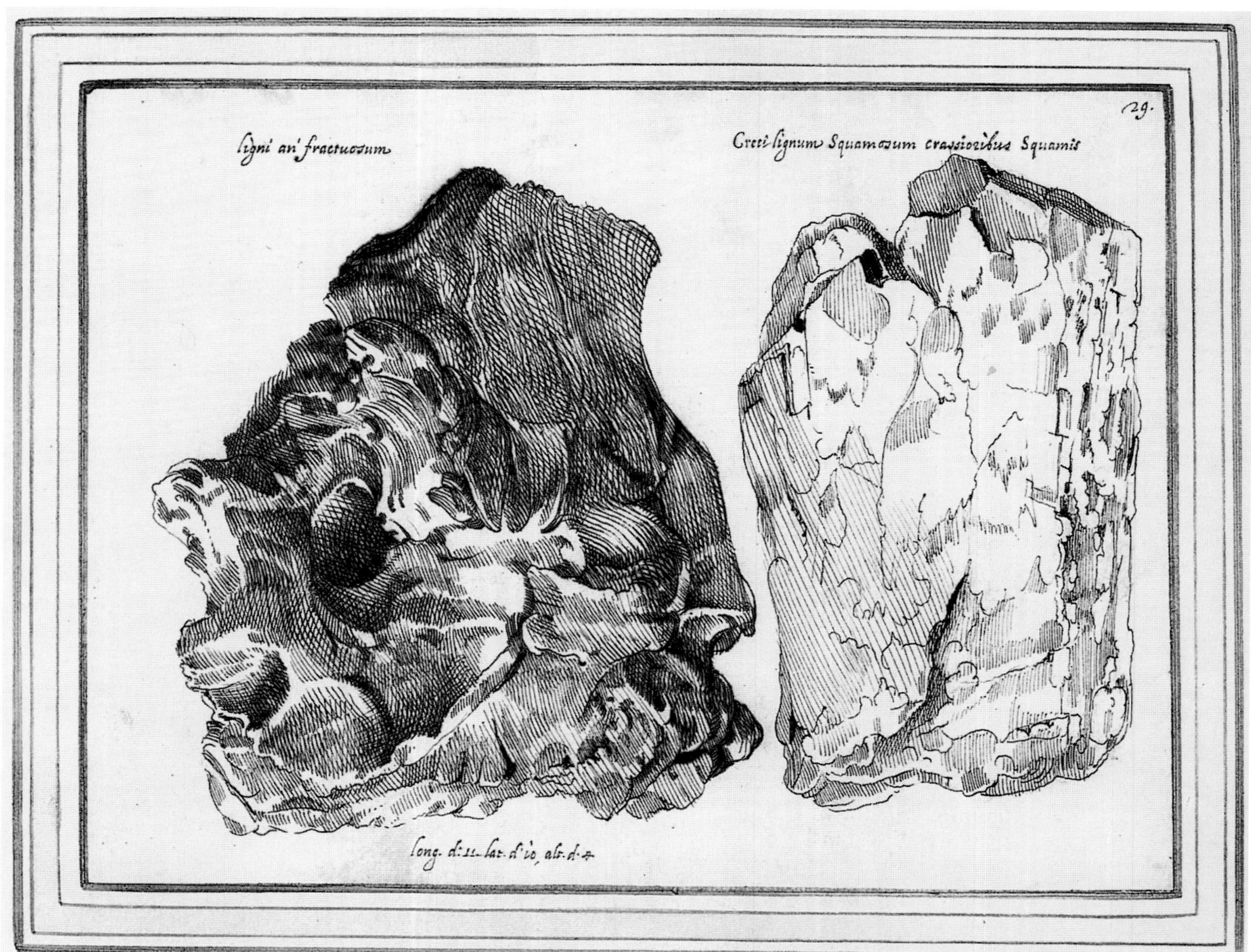

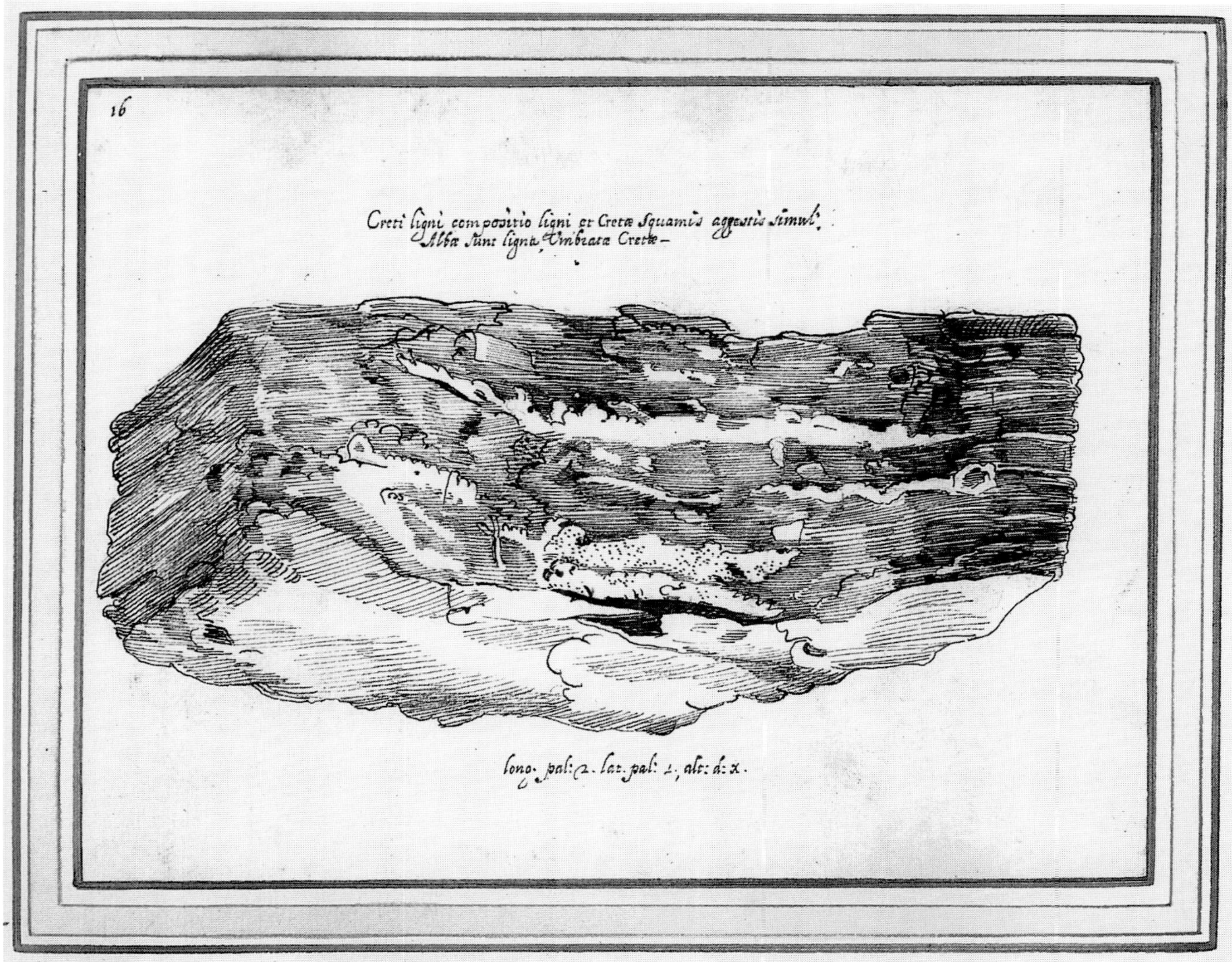

123

123. Possibly a block of true lignite

Windsor, RL 25552, *Fossils*, vol. XIV, fol. 23

Pen and ink, over black chalk

188 × 251 mm

NUMBERING: [ink, upper left] *16*

ANNOTATION: [ink, top] *Creti ligni compositio ligni et Cretæ Squamis aggrestis simul.'/ Albae sunt ligna Ombratæ Cretæ—*
[bottom] *long. pal: 2. lat. pal: 1, alt: d: x.*

This drawing shows 'a composition of wood and clay of *cretilignum*, with scales piled up on top of one another. The woods are white, the clays dark'. The specimen appears to be another block of organic-rich sediment with thin clay layers; perhaps it is a true lignite. Oddly enough, the wood is described as white and the clay as dark.

REFERENCES: 17th-cent. Index, p. 361

124. True lignite

Windsor, RL 25553, *Fossils*, vol. XIV, fol. 24

Pen and ink, over traces of black chalk; no measurements given

191 × 221 mm

NUMBERED: [ink, upper right] *17*

ANNOTATION: [ink] *Idem ab alia parte*

Despite the inscription indicating that this is a view from the other side of the specimen shown in **123**, it appears from its shape to be a different specimen altogether.

REFERENCES: 17th-cent. Index, p. 361

124

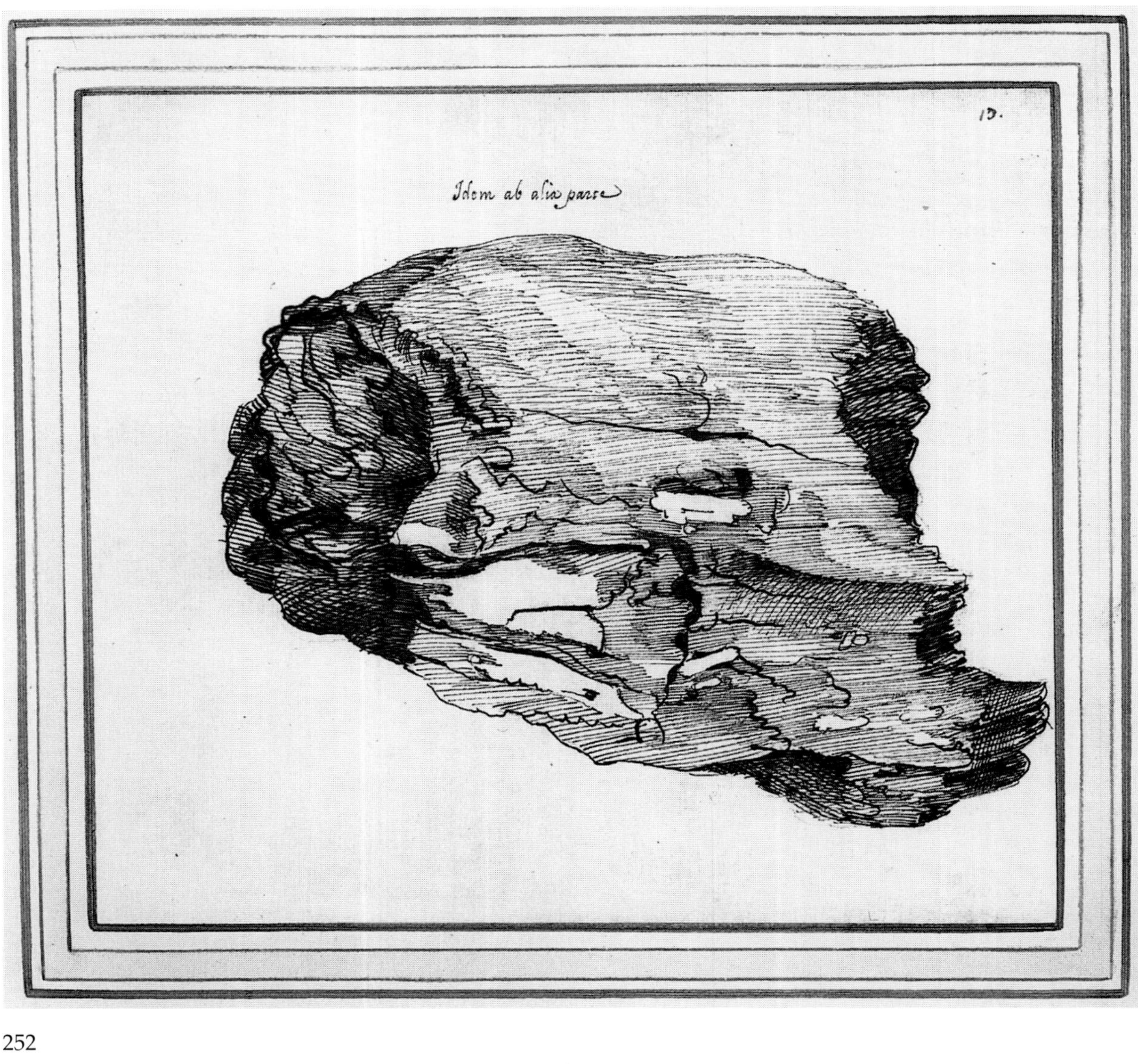

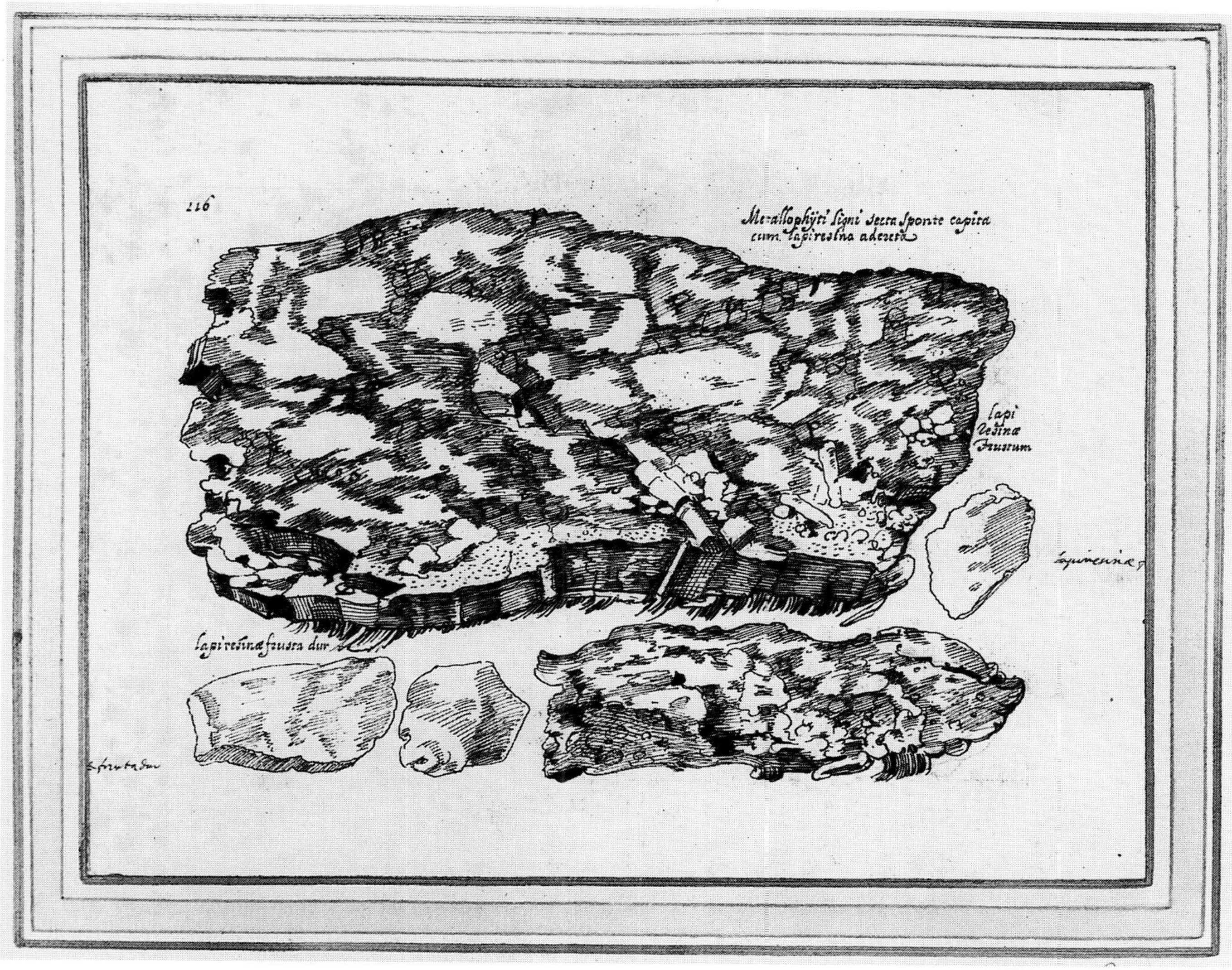

125. Broken piece of lignite, with three resin fragments 125

Windsor, RL 25655, *Fossils*, vol. XVI, fol. 16

Pen and ink, over traces of black chalk; no measurements given; black chalk frame line

182 × 242 mm; small holes caused by ink

NUMBERING: [ink, top left corner] *116*

ANNOTATION: [ink, upper right] *Metallophijti ligni secta Sponte capita/ cum lapiresina adcreta*
[right] *lapi/ resinae/ Frustum* [lower left] *lapi resinae frusta dur*
[in Cesi's hand: lower-right margin] *lapiresina* [lower-left margin] *frusta dur*

The main specimen is described as a metallophyte with a kind of stony resin accretion. The smaller pieces are described as 'pieces of stony resin'; the left-hand one is noted as being 'hard'. It is not clear what the *lapiresina* of the inscription refers to exactly. The filaments on the underside of the specimens could have been formed as a result of pyrite decay.

REFERENCES: 17th-cent. Index, p. 367

OTHER DRAWINGS: For another specimen with similar pyrite decay, see **131**

126. Fossil wood

Windsor, RL 25573, *Fossils*, vol. XIV, fol. 44

Pen and ink, over traces of black chalk; retouched in ink on mount sheet; sheet trimmed through image on left

184 × 255 mm; small holes caused by ink

NUMBERING: [ink, upper right corner] *35*

ANNOTATION: [ink, top] *Creta Mater Ebeni fossilis*
[bottom] *long: pal. 1½. lat. pal. 1. alt. d. 3.*

The specimen may represent a fragment of fossil wood, possibly coalified (explaining the epithet 'ebony'), with some clay still attached. Alternatively, this could be the black organic-rich matrix in which many of the fossil woods are found, as the inscription seems to suggest.

REFERENCES: 17th-cent. Index, p. 362

127. Possibly concretions around a fossil wood

Windsor, RL 25594, *Fossils*, vol. XV, fol. 3

Bodycolour and pen and ink, over traces of black chalk; retouched in ink on mount sheet

182 × 237 mm; small holes caused by ink; ink frame lines over black chalk

NUMBERING: [ink, top left corner] *56*

ANNOTATION: [top centre] *Ferrug.[a] Spumosa exprimens quasi ligni/ cortice colore suberaceo*
[ink, top left] *Corti/cosa* [bottom left] *long: d: 11./ lat: d: 7/ alt: d: 3½*
[top right] *Partes planæ/ lateritiæ* [upper right] *ligni Metallum/ cum corticeib./ lateritij s –*
[bottom left] *long: d. 14. lat. d: 8 crass fere 2.* [bottom right of both specimens] *6.*
[Cesi's inscriptions, cut and erased, top right corner and right margin]

The precise nature of the two specimens is unclear, but they are likely to represent some form of concretion, possibly around pieces of wood. The left-hand specimen is described as 'ferruginous and spongy, showing wood bark of a slightly ruddy colour': the upper-left corner bears the annotation *corticosa*, suggesting its bark-like appearance. The right-hand specimen, on the other hand, is described as 'metallized wood with brick-like bark', the upper area annotated as having 'flat brick-like areas'. The term *lateritia* suggests that it is a kind of baked clay.

REFERENCES: 17th-cent. Index, pp. 364, 366

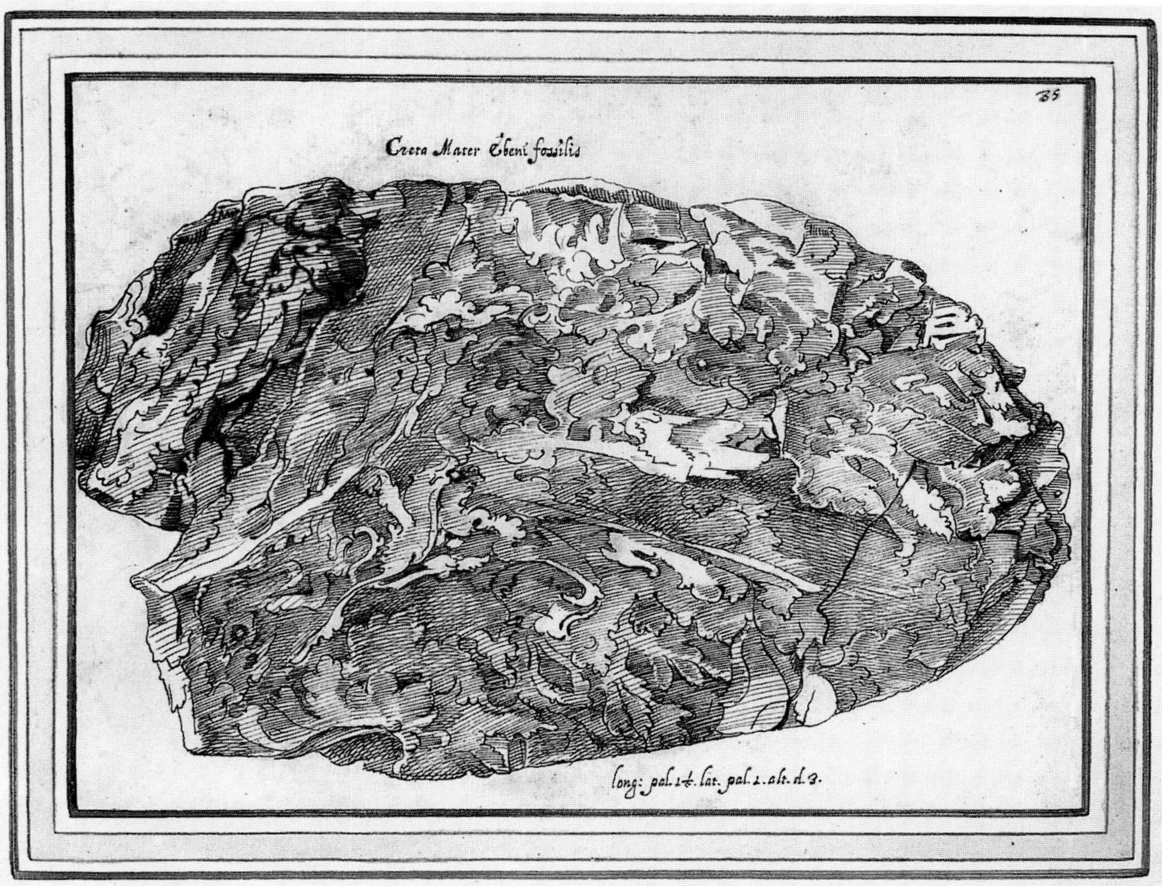
85
Creta Mater Ebeni fossilis
long: pal. 1½. lat. pal. 1. alt. d. 3.

126

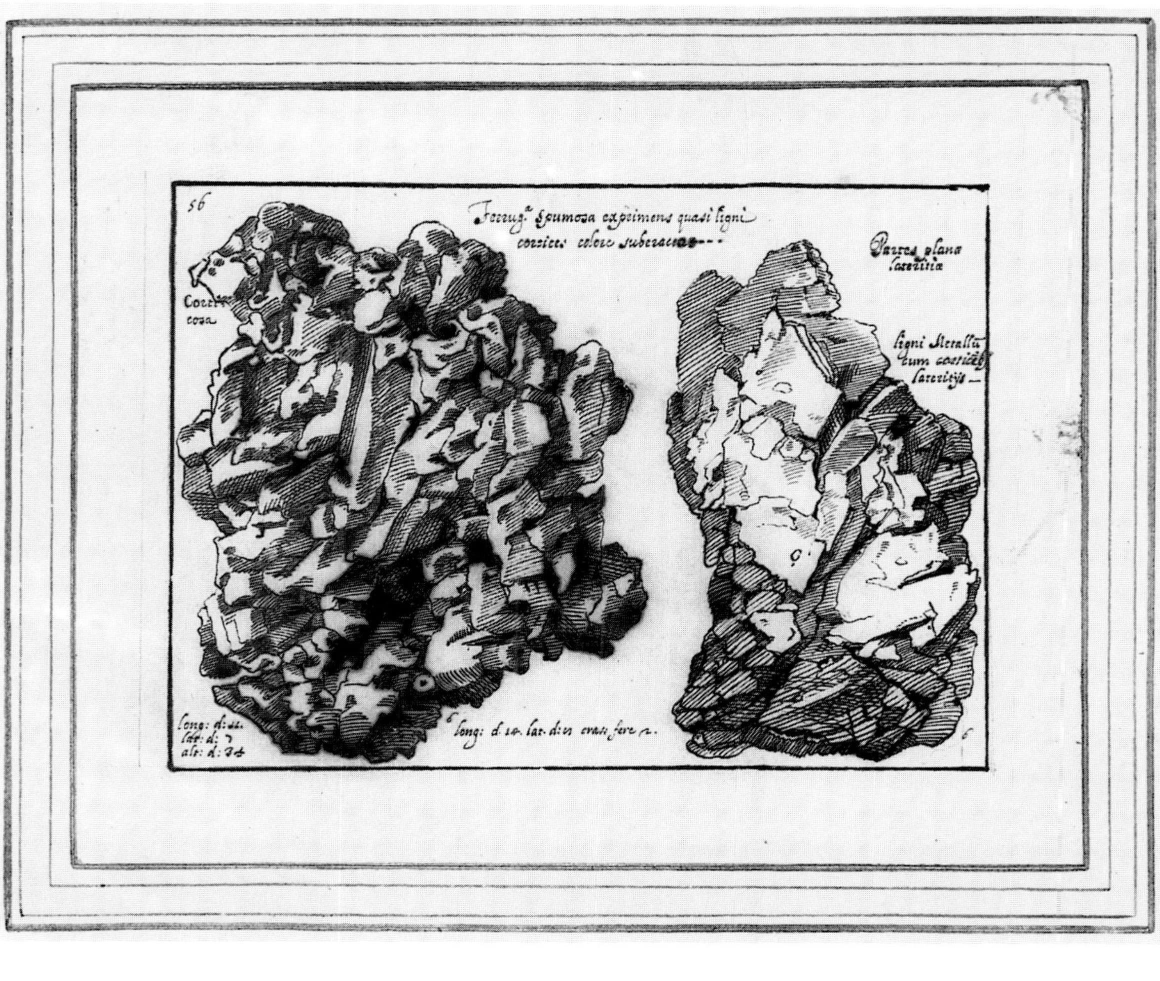
56
Ferrug. spumosa exprimens quasi ligni cortices colore suberaceo
Cortex cosa
Partes planæ lateritiæ
ligni Metallu cum cortice lateritijs
long: d. 10. lat. d. 15 crass. fere 2.
6

127

128. Fossil wood fragments in clay

Windsor, RL 25561, *Fossils*, vol. XIV, fol. 32

Pen and ink, over traces of black chalk

185 × 253 mm; small holes caused by ink

NUMBERING: [ink, upper right] *24*

ANNOTATION: [ink, top] *Creta nitenti Metallica Scabie aspersa æruginosa, Squamosa, nigra fulvosa*
[bottom] *long: d: 7. lat: d: 11. alt: d: 3.*

The specimen is described as 'clay sprinkled with a glistening metallic scaliness, coppery, scaly, yellow–black'. It would appear to be formed predominantly of clay with some wood fragments and large masses of fromboidal pyrite, giving a granular appearance in places.

REFERENCES: 17th-cent. Index, p. 362

128

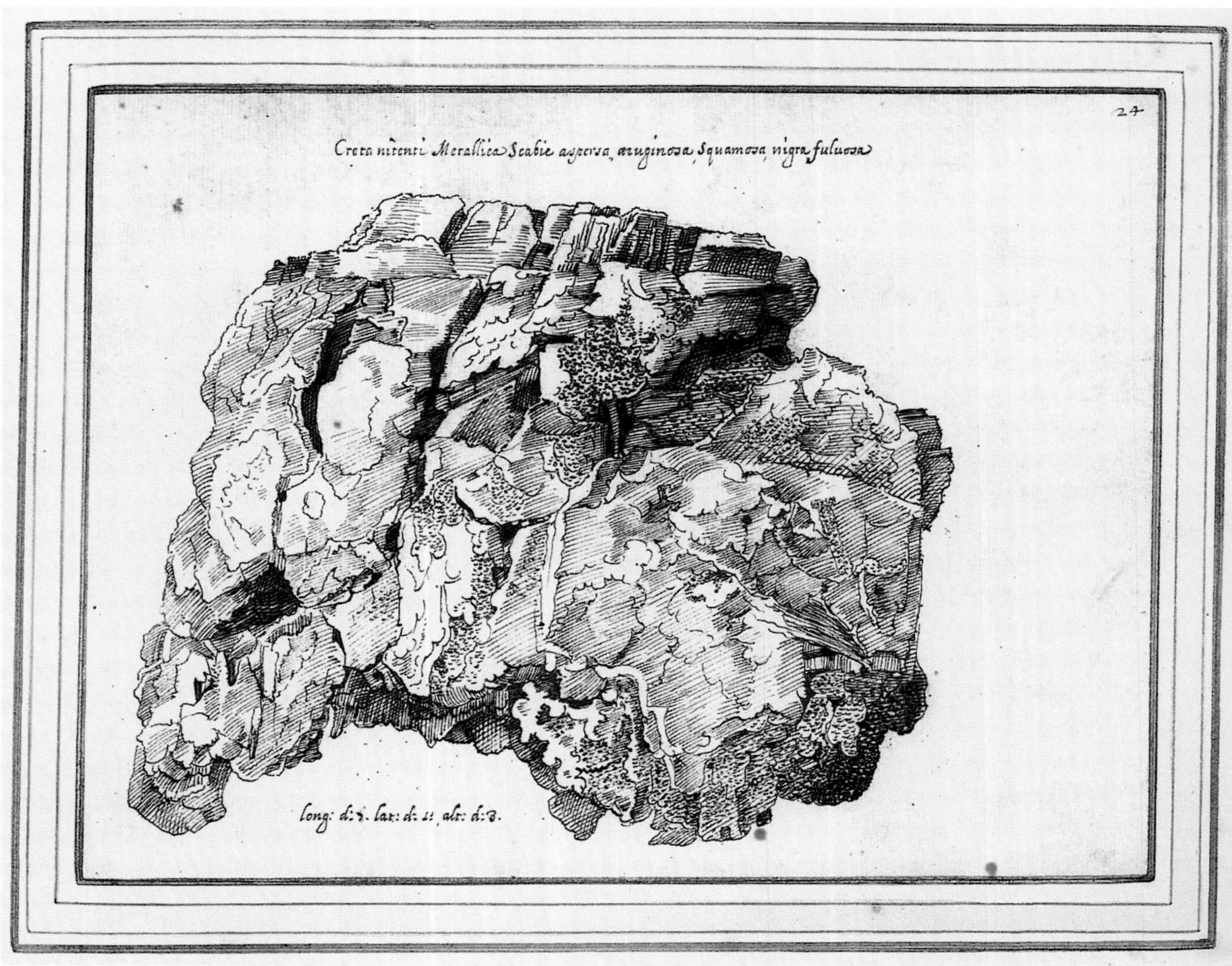

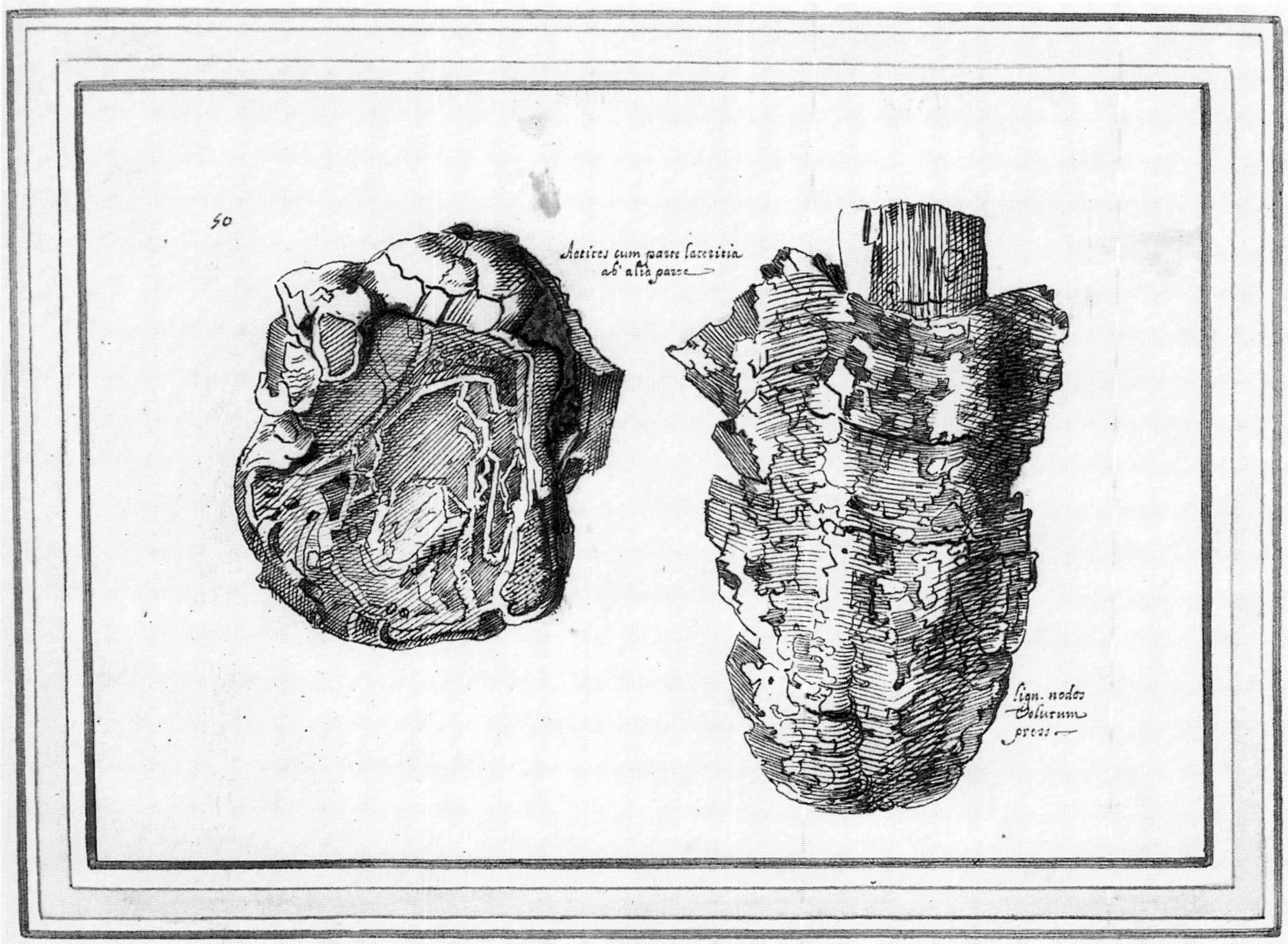

129

129. A concretion and a piece of fossil wood

Windsor, RL 25588, *Fossils*, vol. XIV, fol. 59

Pen and ink, and traces of black chalk; retouched in ink on mount sheet; measurements not given; black chalk frame line

170 × 243 mm; small holes caused by ink

NUMBERING: [ink, upper left corner] *50*

ANNOTATION: [ink, top] *Aetites cum parte lateritia/ ab alia parte—* [bottom right] *lign. nodos / Volutum / press—*

Cesi described the left-hand specimen as an 'aetite with a brick-like part (*cum parte lateritita*) seen from the other side'. On the right is a knotty piece of wood that seems, as the inscription suggests, to be pressed or turned around the core.

REFERENCES: 17th-cent. Index, p. 363

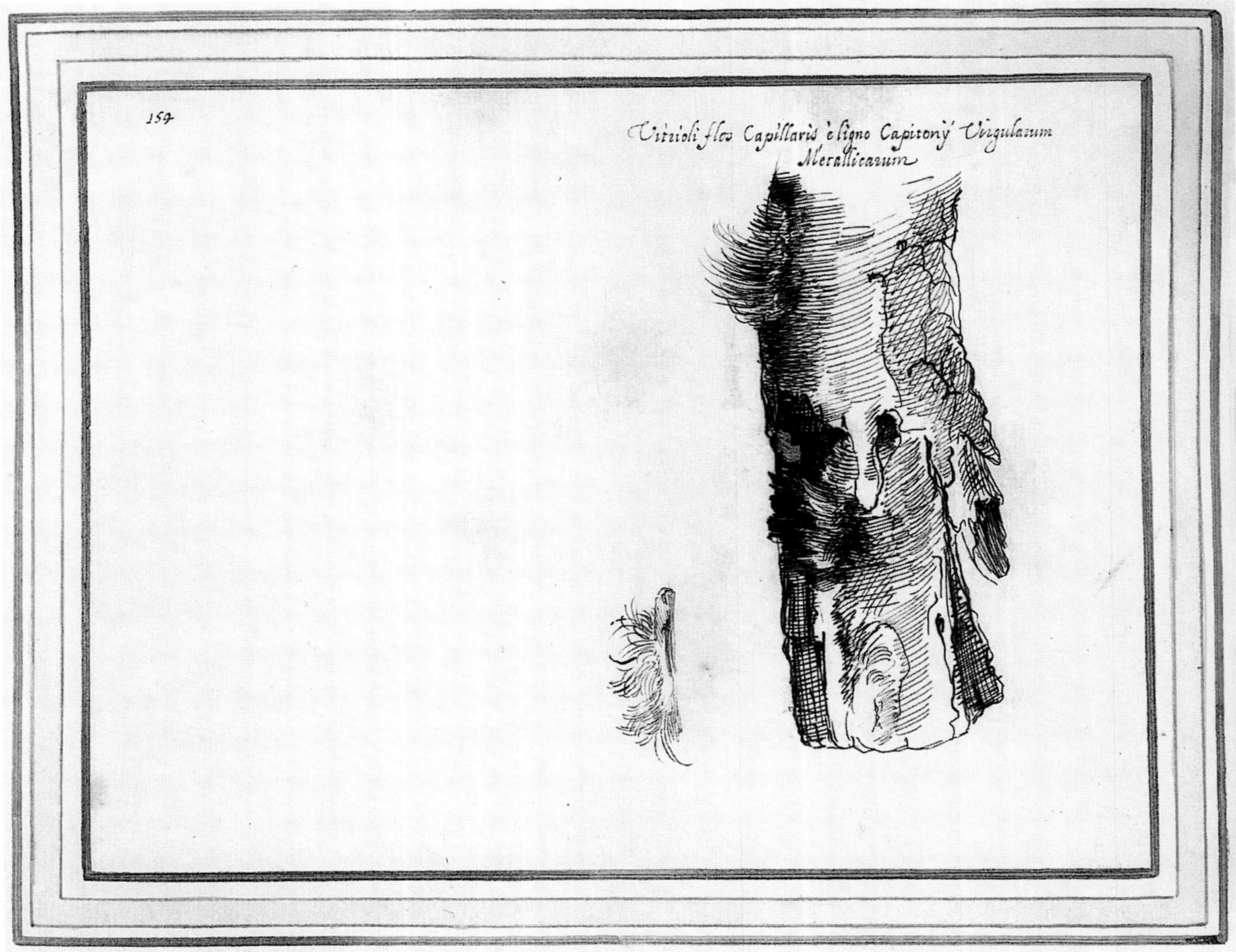

130

130. Specimen of fossil wood from Capitoni, showing pyrite decay

Windsor, RL 25693, *Fossils*, vol. XVI, fol. 54

Pen and ink; retouched in ink on mount sheet; measurements not given

183 × 247 mm; small holes caused by ink

NUMBERING: [ink, top left corner] *154*

ANNOTATION: [ink] *Vitrioli flos Capillaris è ligno Capitonij Virgularum/ Metallicarum*

The annotation on this drawing is somewhat obscure. It seems to read: 'Hair-like (capillaric) flowers of vitriol [sulphurous mineral] growing out of a base of wood with metallic twigs'.

This specimen of fossil wood—a typical copperas efflorescence—is probably pyritized and has undergone decay. Bromehead calls it a specimen of melanterite. Flowers of vitriol were well known to ancient writers such as Pliny and Theophrastus, as Caley and Richards have summarized. Its medicinal uses, typically enough, were also noted by writers such as Aldrovandi. The rest of the sheet was presumably left empty for another drawing.

LITERATURE: Caley/Richards 1956, p. 207; Bromehead 1947a, p. 74

REFERENCES: 17th-cent. Index, p. 370

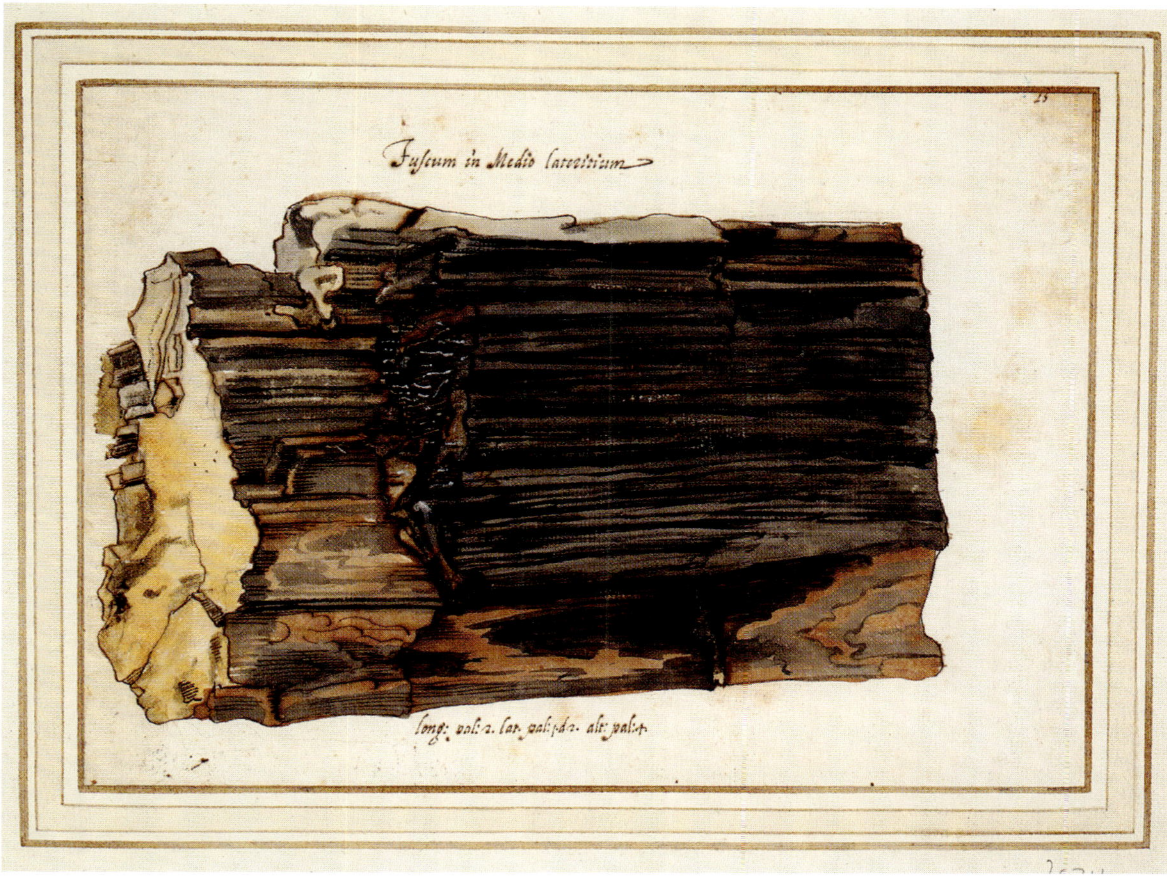

131

131. Fossil wood

Windsor, RL 25710, *Fossils*, vol. XVII, fol. 16

Watercolour and bodycolour, over traces of black chalk

175 × 250 mm

NUMBERING: [ink, top right corner] *14*

ANNOTATION: [ink, top] *Fuscum, intus Aureum, Punctis Argenteis foris*
[bottom] *long. pal: 1. d. 3. Alt: et lat : dig. 8*

A 'very dark [almost black]' specimen, 'gold inside, with silver specks on the outside'. It appears to have been pyritized. The gold suggests unaltered iron pyrite ('fool's gold'); the white specks described as 'silver' seem to be areas where there has been significant iron pyrite decay (see Fig. 35, p. 134).

REFERENCES: 17th-cent. Index, p. 371

OTHER DRAWINGS: For a specimen with similar pyrite decay, see **125**

132. Fossil wood with pyrite

Windsor, RL 25658, *Fossils*, vol. XVI, fol. 19

Pen and ink, over traces of black chalk

184 × 254 mm

NUMBERING: [ink, top right corner] *119* over *116*

ANNOTATION: [ink, top] *Metalligo floribus erectis quasi corallis cullis speciosa* [bottom] *nali*

A piece of fossil wood encrusted with iron pyrite, described by Cesi as a beautiful specimen of '*Metalligo* with flowers sticking up like coral' (see **111**). The specimen is drawn to actual size.

REFERENCES: 17th-cent. Index, p. 367

133. Possibly fossil wood in clay

Windsor, RL 25556, *Fossils*, vol. XIV, fol. 27

Pen and ink, over black chalk; no measurements given

157 × 232 mm

NUMBERING: [ink, upper right] *19*

ANNOTATION: [ink, top] *Creti lignum ferreis ærisquæ Mitis*

A specimen of '*cretilignum* with grains of iron and copper'. It appears to be a mixture of wood and clay with small pyrite fromboids (grains of iron and copper).

REFERENCES: 17th-cent. Index, p. 361

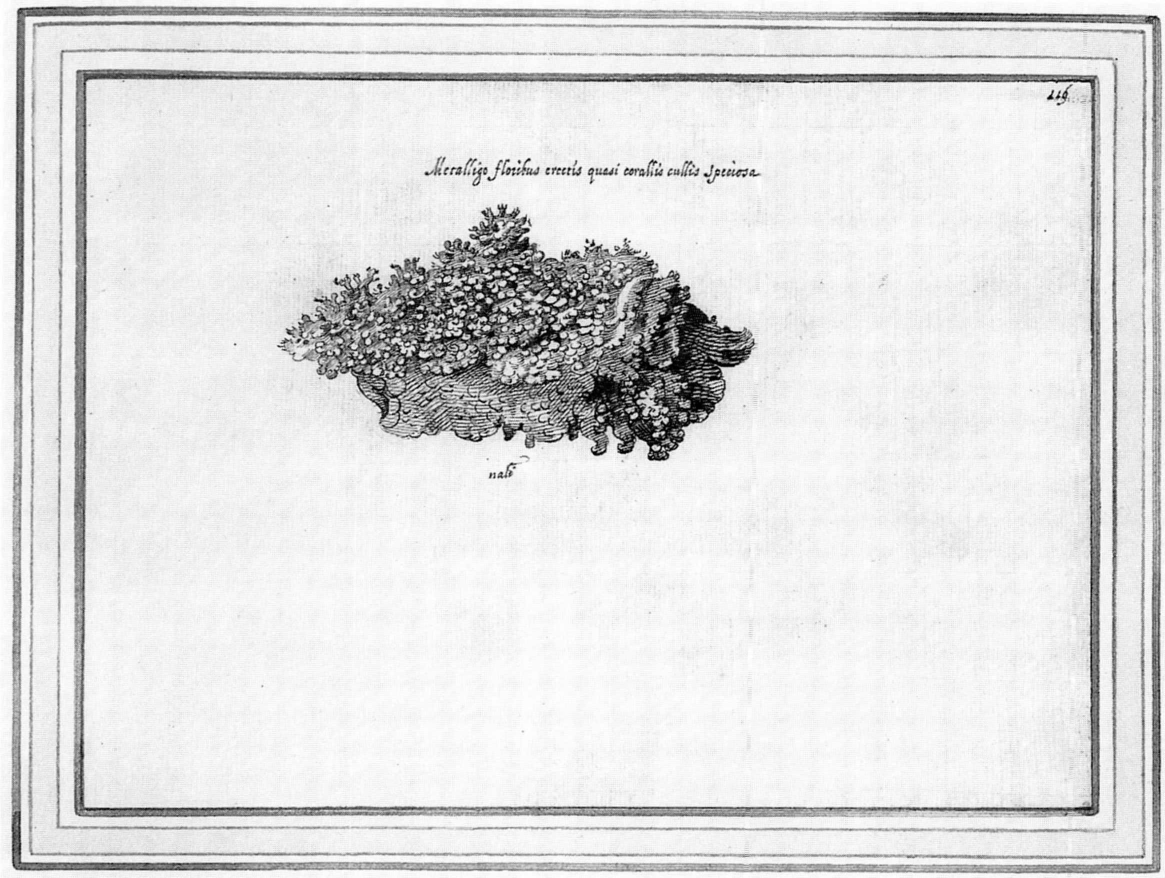
116
Metalligo floribus erectis quasi corallis cultis Speciosa
nali

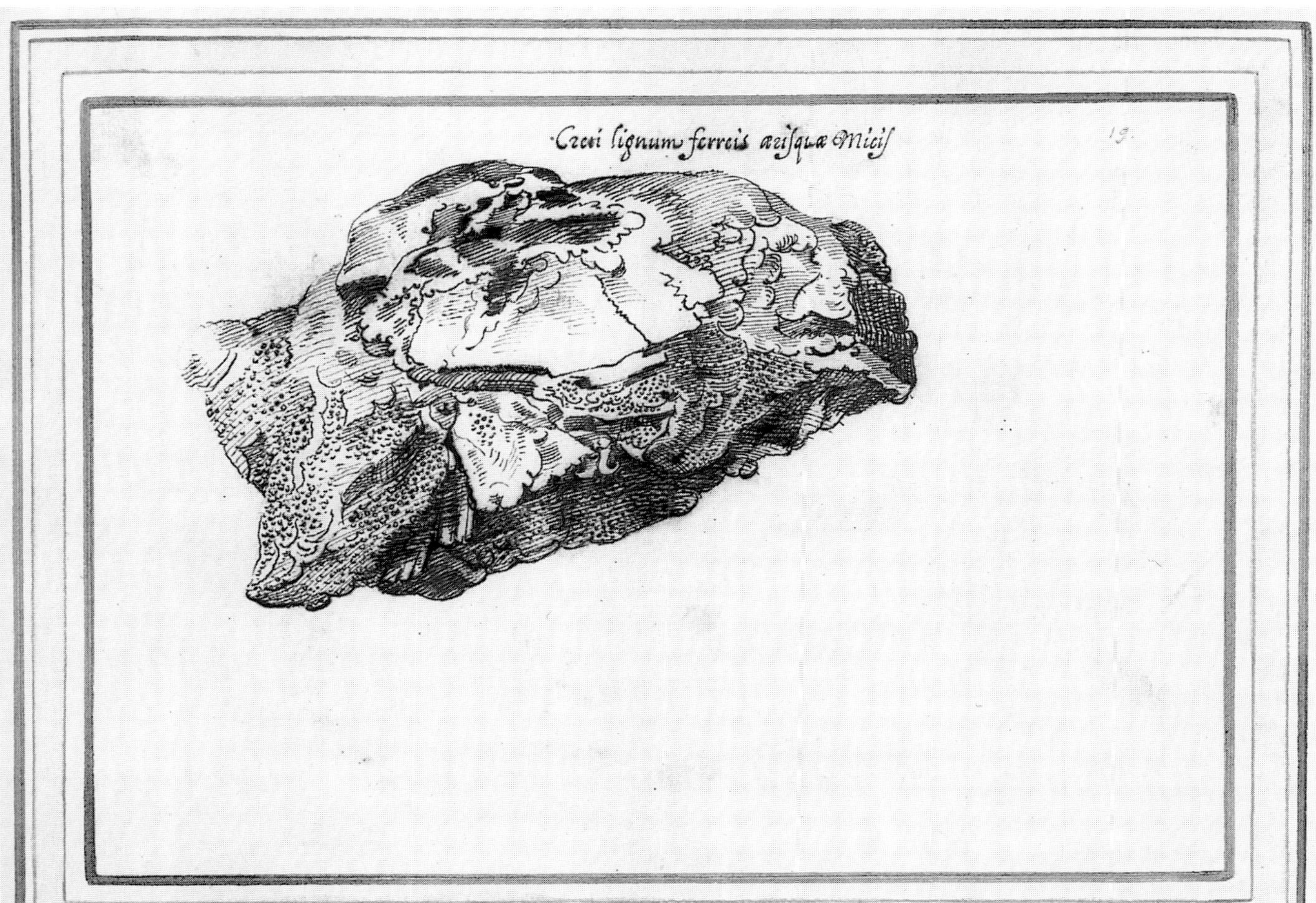
Creti lignum ferreis æisquæ Micis
19.

III. CONCRETIONS AND PYRITES [134–145]

Among the drawings are many 'aetites' or 'eagle-stones'; these are of clay–ironstone nodules. The history of such nodules is well documented by Bromehead[1] and by Van der Burgh,[2] and they had already attracted interest even before Pliny the Elder. Conrad Gesner (1516–1565),[3] whose work was well known to Cesi and Stelluti, recommended the use of aetites to promote pregnancy in animals; while Bernard Palissy (1510–1584)[4] believed that aetites were nothing more than petrified fruits, and that their contents were nuts (these objects sometimes rattle). Palissy also rejected the common belief that aetites had magical powers.[5] But what most interested Cesi, as always, was the problem of how these objects were produced. They belonged to the large class of phenomena that seemed to be 'pregnant', and thus offered some hope for the *Lincei* in their search for the basic principles of reproduction. It is in this context that we are to understand Stelluti's reference to Cesi: 'His Highness having through long and diligent observation discovered how the pregnant eagle-stone is generated'.[6]

Aetites are essentially siderite (iron carbonate) nodules or concretions. They form in clays through the concentration of iron associated with the decay of plant material. They are especially common in clays formed in the deposits and soils of the flood plains of rivers, as well as in those deposited in lakes. In some cases the concretions are moved from clays into sands and undergo a chemical alteration whereby the carbonate reacts with slightly acidic groundwater, and iron oxide is redeposited within the nodule. An insoluble residue often remains within the centre of the nodule and may be responsible for the rattle. Sideritic nodules are common in the sediments of the Acquasparta district and the specimens illustrated could have been found at several sites.[7]

No mention of a rattle is made in the inscriptions on the drawings of aetites, and they almost certainly represent ordinary siderite nodules. A collection of similar nodules made by the present author is here illustrated for comparison [Fig. 38]. As a whole, the concretions in these drawings represent a well-defined group, though in some specimens the interior of the aetites is simply described as having a 'clayey inside' (for example, **135**).

Iron pyrite, which has previously been mentioned, is clearly illustrated in two of the drawings (**144** and **145**).[8] In other cases, the material illustrated may be just cemented clay, perhaps the fossilized infill of animal burrows (**140**). Further specimens (**142, 143** and **154**) show growths of manganese oxide (dendrites).

The range of the concretions, as represented in the drawings, is very large. Many specimens are clearly identifiable, but others less so. The order of these drawings as they appear in the bound volumes has been retained.

NOTES

1. Bromehead 1947b, p. 16.
2. Van der Burgh 1975, p. 16.
3. Gesner 1565, as cited in Bromehead 1947b, p. 18.
4. Palissy 1580, in B. Palissy, *Œuvres complètes*, ed. A. France, Paris, 1880 edn, vol. II, p. 174; cf. also Bromehead 1947b, p. 18.
5. Bromehead 1947b.
6. See below, Appendix II b, p. 383
7. Basilici 1997, p. 79.
8. Bromehead 1947a, p. 74.

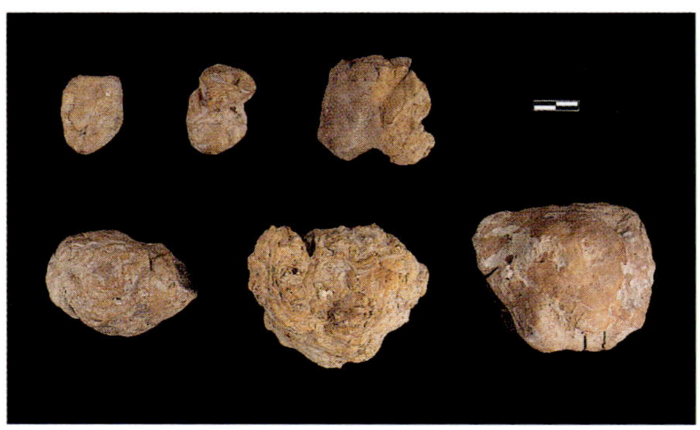

Fig. 38. Siderite concretions from Dunarobba

134. Various types of concretion

Windsor, RL 25579, *Fossils*, vol. XIV, fol. 50

Pen and ink, over traces of graphite; retouched in ink on mount sheet

192 × 255 mm; small holes caused by ink; measurements not given

NUMBERING: [ink, upper right] *40*

ANNOTATION: [ink, centre left] *Aetitis Multis/ Vacuis Septa*
[centre right] *Aetitis arte angularus*
[lower right] *Aetites scabro arenoso/ cortice*
[Cesi's inscriptions, lower and right edges erased]

A wide selection of ironstone concretions, three of which have accompanying descriptions. The largest specimen is shown in two views (but not described). The annotation to the centre specimen calls it a 'septa of aetite with many empty spaces', a modern interpretation of which would be 'aetite with numerous septa'.

The lower of the two small specimens on the right-hand side appears to have been chipped intentionally; it is annotated as 'angular by art'. The specimen beneath is described as an 'aetite with a scaly, sandy outer layer'. The remaining specimens are not described.

Although measurements are not given, the range of sizes seems to indicate that the specimens are reproduced actual size.

LITERATURE: Bromehead 1947a, pp. 77–8, pl. IV, fig. 1; Bromehead 1947b, pp. 16–22

REFERENCES: For a general discussion of aetites, see Stelluti 1637, p. 12; 17th-cent. Index, p. 363

135. Six specimens (some broken) of ironstone concretions

Windsor, RL 25580, *Fossils*, vol. XIV, fol. 51

Pen and ink, with corrections in bodycolour; measurements not given
but the specimens appear to be natural size; two specimens are marked with crosses

191 × 256 mm

NUMBERING: [ink, upper left corner] *42;* [ink, upper right corner] *41*

ANNOTATION: [ink, top centre] *Aetites Magnus*
[top right] *Aetit. Bombyc.* [upper right] *Aetites Minor.*
[lower right] *Intestina Aetit. rubentia/ ex Creta* [bottom right] *Christallus Intestina Aetitis*
[lower left] *Intestina Aetitis cretosa* [top left] *Aetit. Bombyc. eadem*
[Cesi's inscriptions, erased, at edges]

The central ironstone concretion is described as a 'large aetite'. On either side are views from above and below of a 'silk-worm aetite'. On the right is an example of a 'small aetite'. On the lower half of the sheet, right-hand corner, is a fragment described as the 'ruddy inside (*Intestina*) of an aetite of clay'; the lower fragments are the 'crystalline insides of an aetite' in contrast to the fragments on the lower left, which are 'clayey insides of an aetite'.

REFERENCES: 17th-cent. Index, p. 363

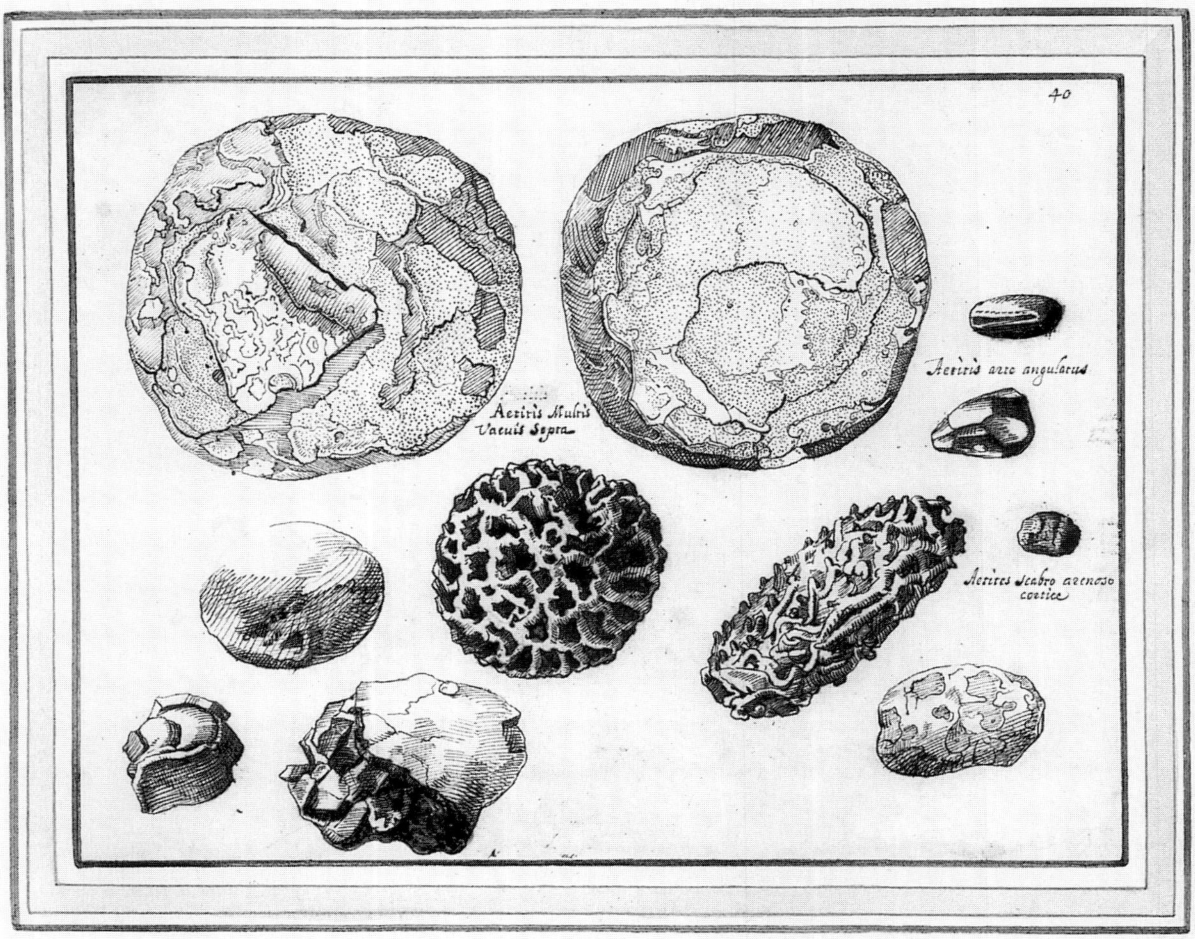
40
Aetitis Multis
Vacuis Septa
Aetitis arte angulatus
Aetites scabro arenoso
cortice

134

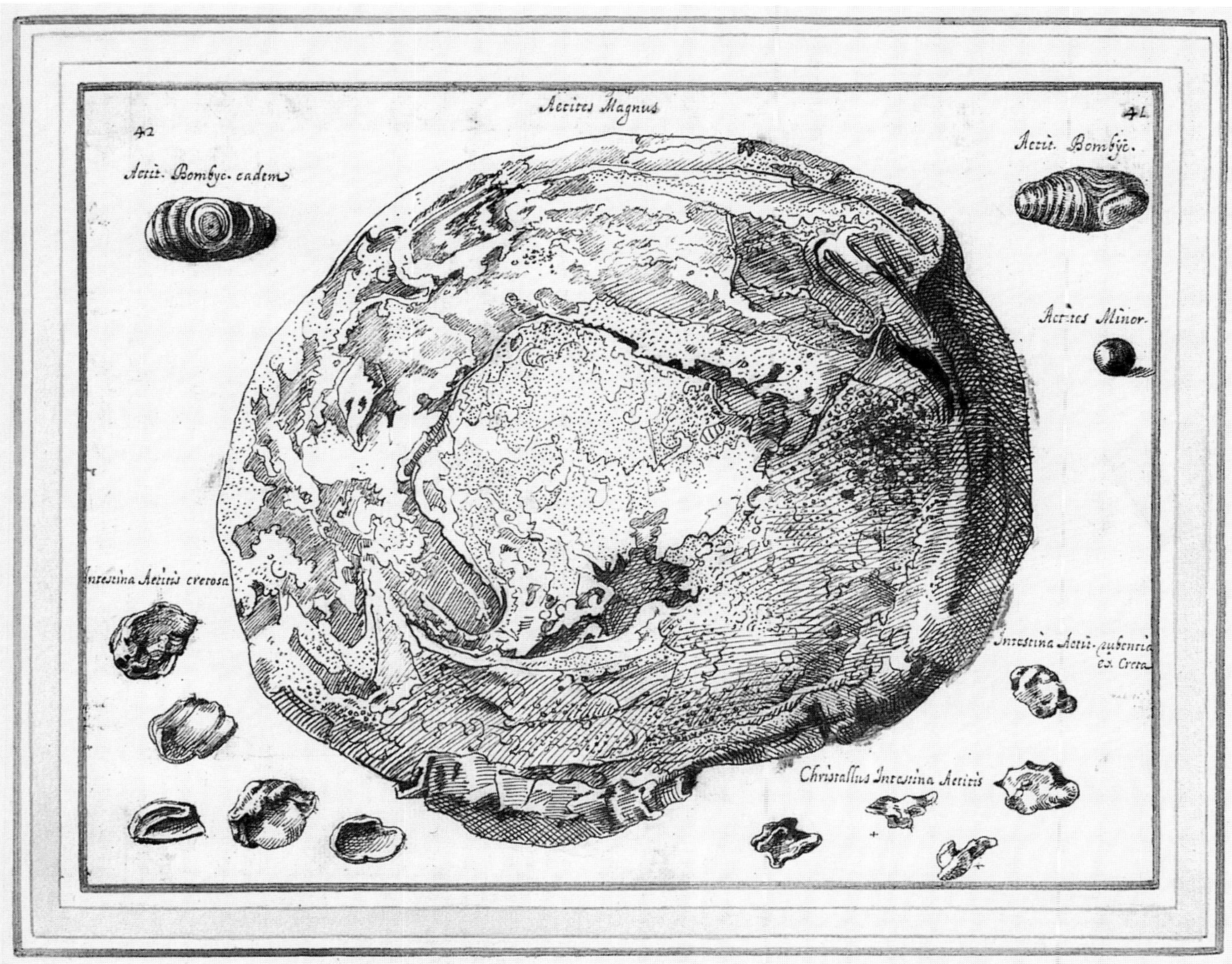
Aetites Magnus
42
41
Aetit. Bombyc. eadem
Aetit. Bombyc.
Aetites Minor
Intestina Aetitis cretosa
Intestina Aetit. rubentia ex Creta
Christallus Intestina Aetitis

135

136. Six ironstone concretions

Windsor, RL 25586, *Fossils*, vol. XIV, fol. 57

Pen and ink, over traces of black chalk; retouched in ink on mount sheet

193 × 252 mm; small holes caused by ink; measurements not given

NUMBERING: [ink, upper right corner] *48*

ANNOTATION: [ink, top left] *lateritia crusta rimosa inferior pars collecta Metallica*
[upper right] *Aetites Metallicus/ sub ceruleus*
[bottom right] *lapis solidus Macula Aetitis/ vacuus æmulatus—*
[left] *sub cerulea in Aetitem fere*
[Cesi's inscriptions at right and lower edge, cut]

The top left-hand specimen is annotated 'a laterite with a cracked crust, the lower part metallic'. The specimen clearly shows its concretionary nature with the brick-like crust probably indicating siderite. The indication of the presence of a metallic substance may imply the occurrence of a sulphide mineral.

The large concretion beneath is described as 'a bluish metallic aetite', a description which may indicate that the naturally blue clay has only a small impregnation of iron.

The round specimen at top right is shown in two views but without description. A small specimen below, labelled *A*, is described as being 'bluish and metallic' and may indicate sulphide mineralization. The specimen beneath is shown in two views without description. The bottom right-hand specimen is also shown in two views, with a hollow centre. The annotation notes that it is 'an empty (meaning, hollowed-out) solid stone emulating the *macula* of an aetite'. The object may well be a hollow concretion.

REFERENCES: 17th-cent. Index, p. 363

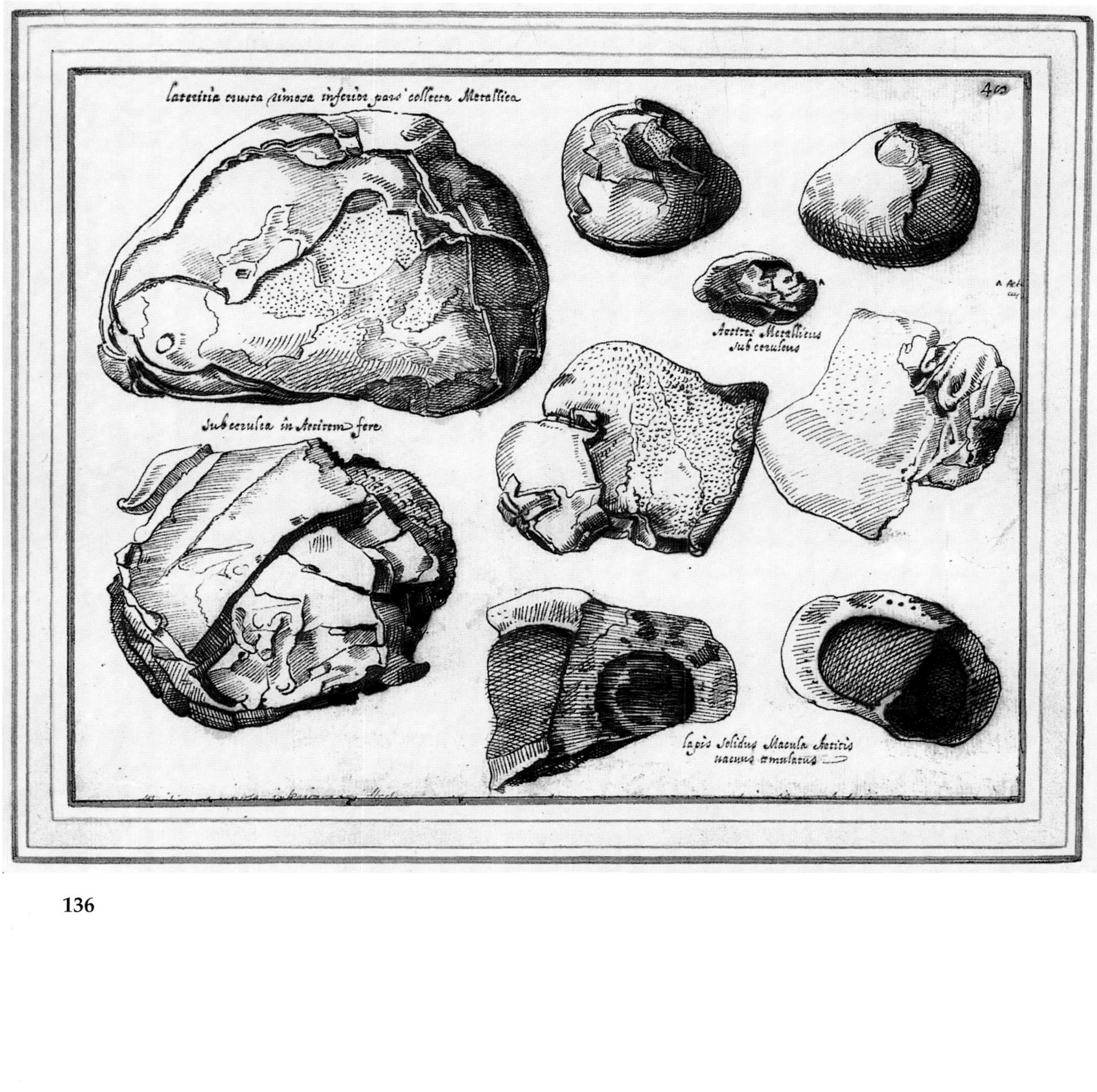
lateritia crusta rimosa inferior pars collecta Metallica
Subcerulea in Aetitem fere
Aetites Metallicus sub ceruleus
lapis solidus Macula Aetitis vacuus æmulatus

136

137. Three concretions

Windsor, RL 25589, *Fossils*, vol. XIV, fol. 60

Pen and ink, over black chalk; measurements not given; black chalk frame line; drawing marked with a cross

185 × 251 mm. *Watermark:* Bird 41B

NUMBERING: [ink, upper left corner] *51*

ANNOTATION: [ink, top centre] *Aetites Varie Crustosi*

These three aetites with different 'crusts' are typical siderite concretions. As with many of the other aetites, they are probably shown actual size.

REFERENCES: 17th-cent. Index, p. 363

OTHER DRAWINGS: **138, 139**

137

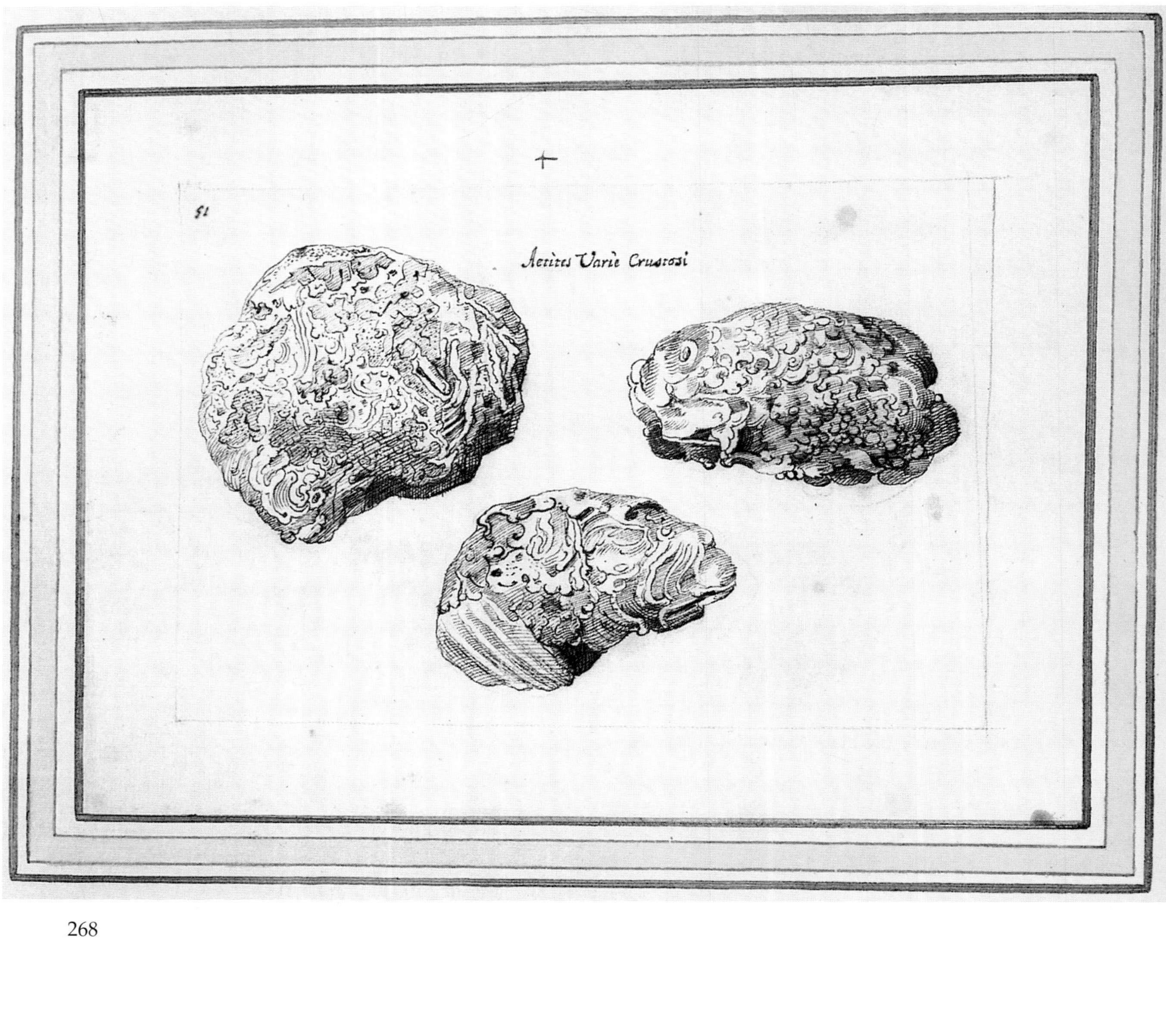

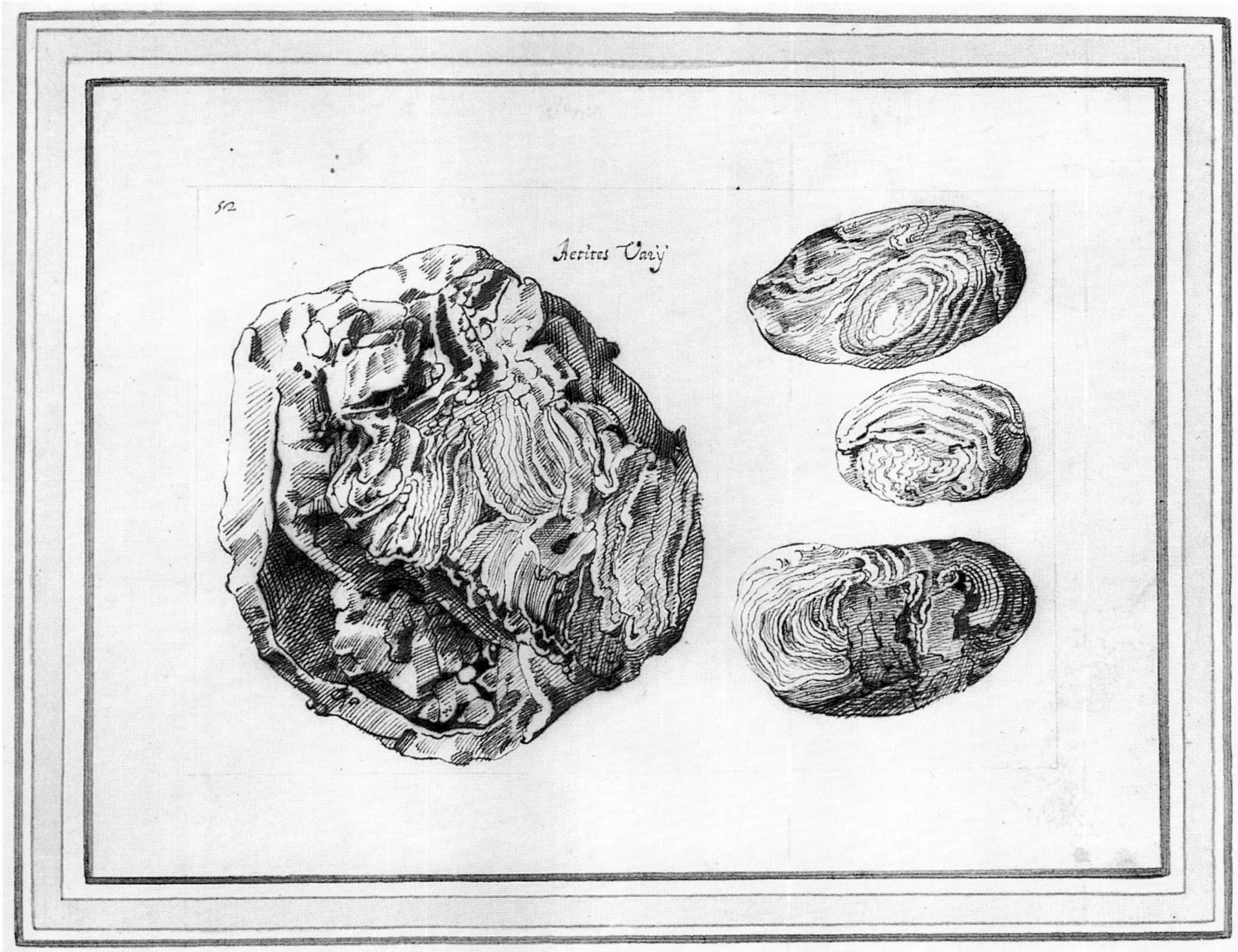

138

138. Four concretions

Windsor, RL 25590, *Fossils*, vol. XIV, fol. 61

Pen and ink, over traces of black chalk; measurements are not given; black chalk frame line

182 × 248 mm; small holes caused by ink

NUMBERING: [ink, upper left corner] *52*

ANNOTATION: [ink] *Aetites varij*

The sheet is headed 'Various Aetites' and shows four typical siderite concretions, again probably depicted life-size.

REFERENCES: 17th-cent. Index, p. 363

OTHER DRAWINGS: **137, 139**

139. Three concretions

Windsor, RL 25591, *Fossils*, vol. XIV, fol. 62

Pen and ink, over traces of black chalk; retouched in ink on mount sheet; measurements are not given; black chalk frame line

180 × 250 mm; small holes caused by ink

NUMBERING: [ink, upper left corner] *53*

ANNOTATION: [ink] *Aetites Varij*

Another sheet headed 'Various aetites', which shows three typical siderite concretions. They are probably depicted life-size.

REFERENCES: 17th-cent. Index, p. 363

OTHER DRAWINGS: **137, 138**

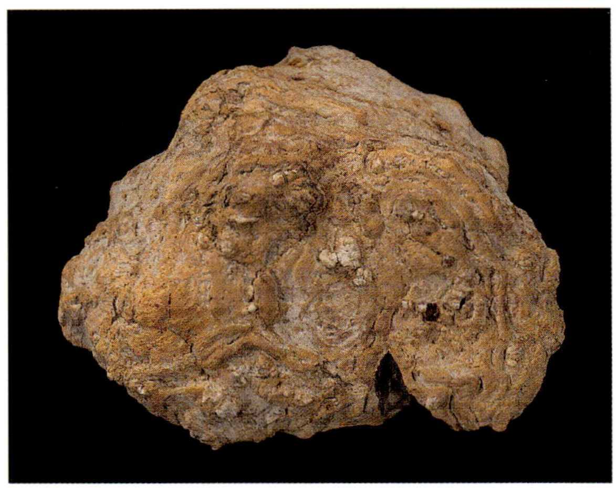

Comp. fig. 139. Siderite concretion from Dunarobba

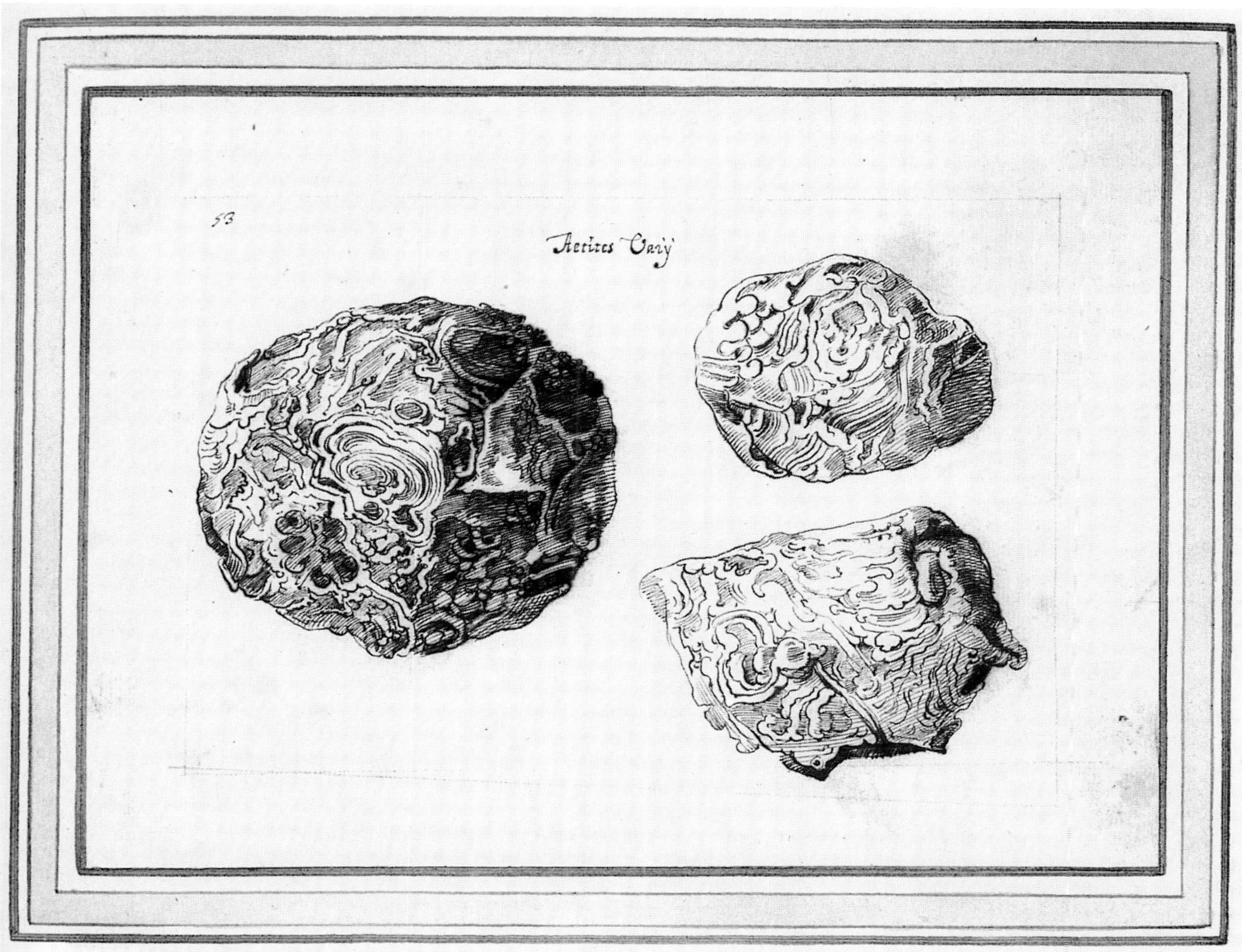

139

140. Two concretions

Windsor, RL 25581, *Fossils*, vol. XIV, fol. 52

Pen and ink, over traces of black chalk; retouched in ink on mount sheet; measurements not given; marked with a cross

186 × 251 mm; small holes caused by ink

NUMBERING: [ink, upper right corner] *43*

ANNOTATION: [ink, top left] *Collectitium, Rimosum Saxum Aetitis/ affine* [bottom right] *Globulosus contortus lapis/ ex Aetitis contubernio ipsi affinis*

The left-hand concretion is described as 'a rock full of cracks similar to an aetite' and probably does represent an ironstone nodule, perhaps the infilling of an animal burrow.

The unusually shaped concretion on the right is said to be a 'contorted and globular stone, closely associated with the aetite [that is, either found together with it or even within it] and very similar to it'

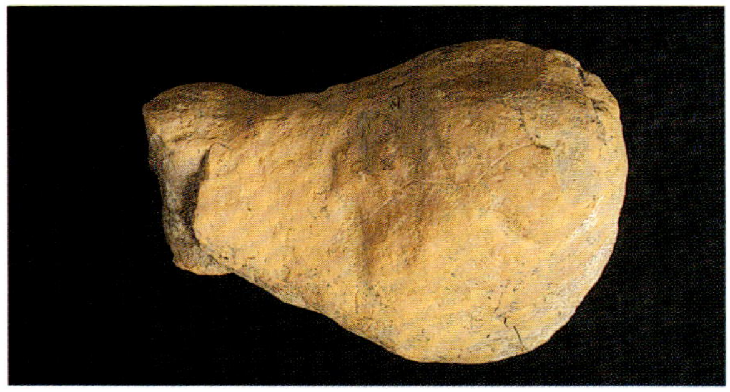

Comp. fig. 140. Ironstone concretion

REFERENCES: 17th-cent. Index, p. 363

140

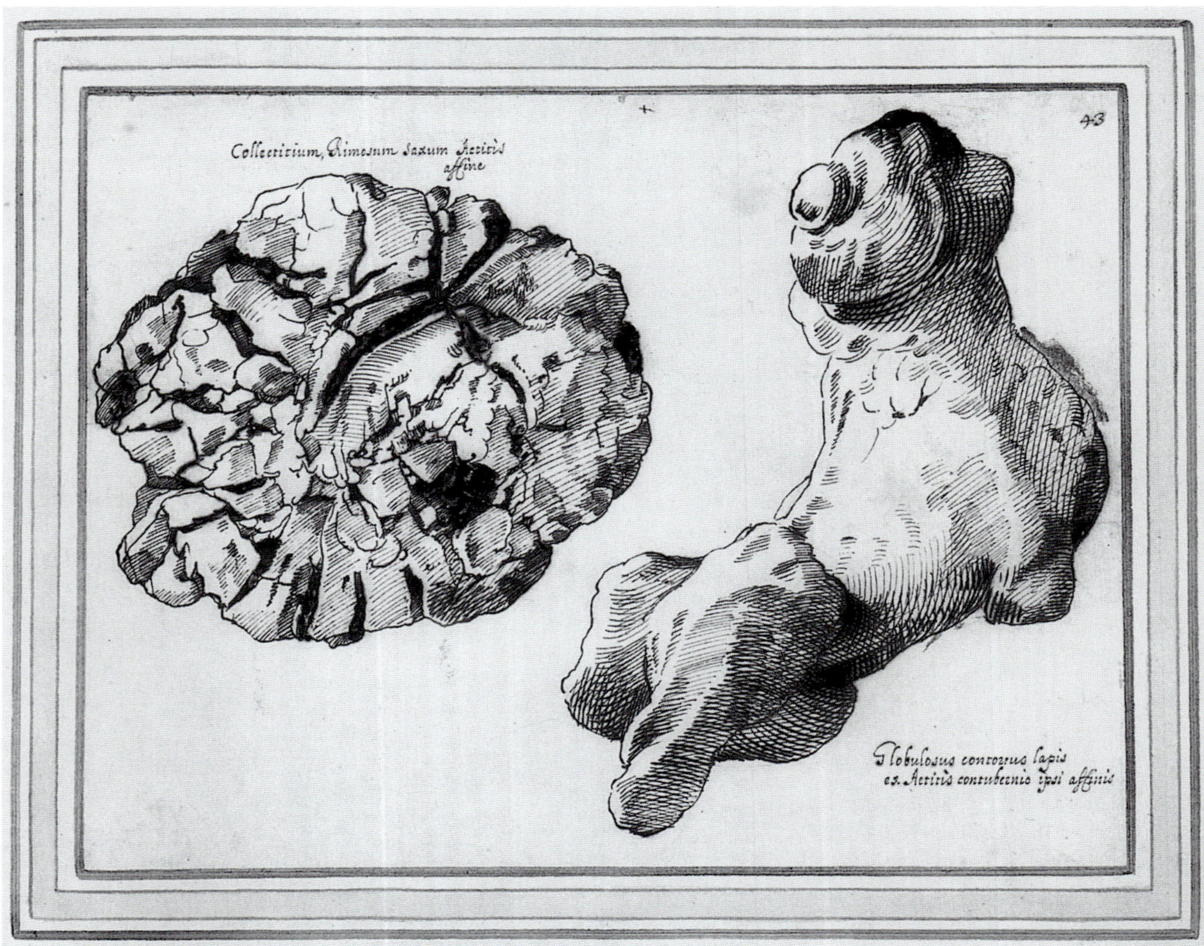

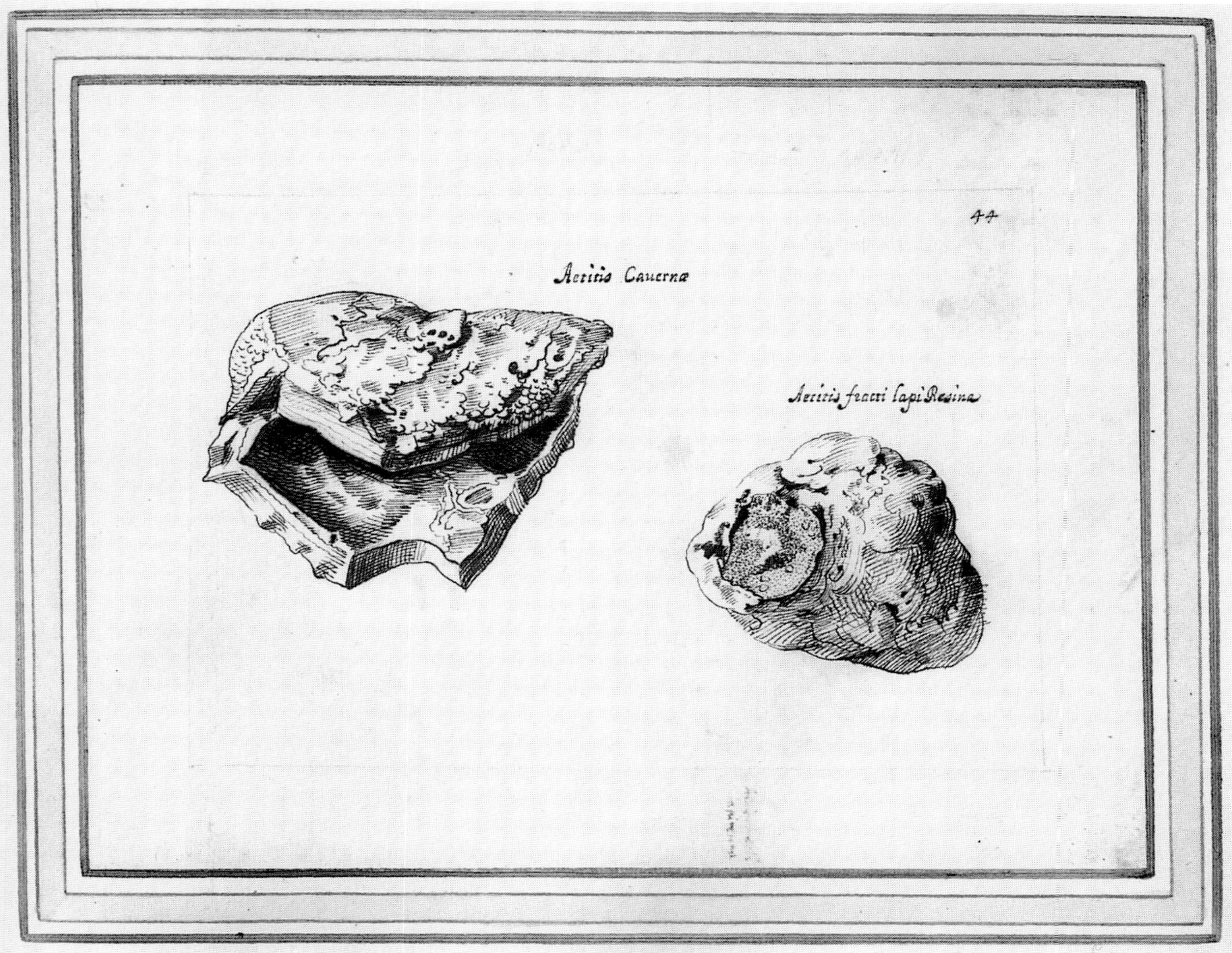

141

141. Two concretions

Windsor, RL 25582, *Fossils*, vol. XIV, fol. 53

Pen and ink, over black chalk; measurements are not given; black chalk frame line

186 × 249 mm

NUMBERING: [ink, upper right corner] *44*

ANNOTATION: [ink, centre] *Aetitis Cavernæ*
[right] *Aetitis fracti lapi Resina*
[vertical inscriptions in Cesi's hand, lower edge, erased]

The left-hand concretion shows the hollow interior of an aetite, as the inscription notes. On the right is the resin-like concretion from the interior of this or another aetite.

REFERENCES: 17th-cent. Index, p. 363

142. Concretions, two showing dendrites of manganese oxide

Windsor, RL 25587, *Fossils*, vol. XIV, fol. 58

Pen and ink, over black chalk; measurements given for the lower two specimens

183 × 258 mm

NUMBERING: [ink, upper left corner] *49*

ANNOTATION: [ink, top left] *Aetites Monstruosus Spongiosis fere/ Capitil-* [bottom left] *long. lat. alt. d. 3. in circa* [bottom right] *lon. pal. 1. lat. ½ alt. d. 3.*

The exact nature of the 'monstrous aetite with spongy elements in it' (top left-hand specimen) is not clarified by the apparently incomplete inscription.

The other two specimens clearly show the occurrence of dendrites of manganese oxide on their surfaces. The drawings are unfinished, have no inscriptions and are clearly unrelated to the 'monstrous aetite'. The only other specimens with dendrites in the collection are shown in **143** and **154**.

REFERENCES: 17th-cent. Index, p. 363

142

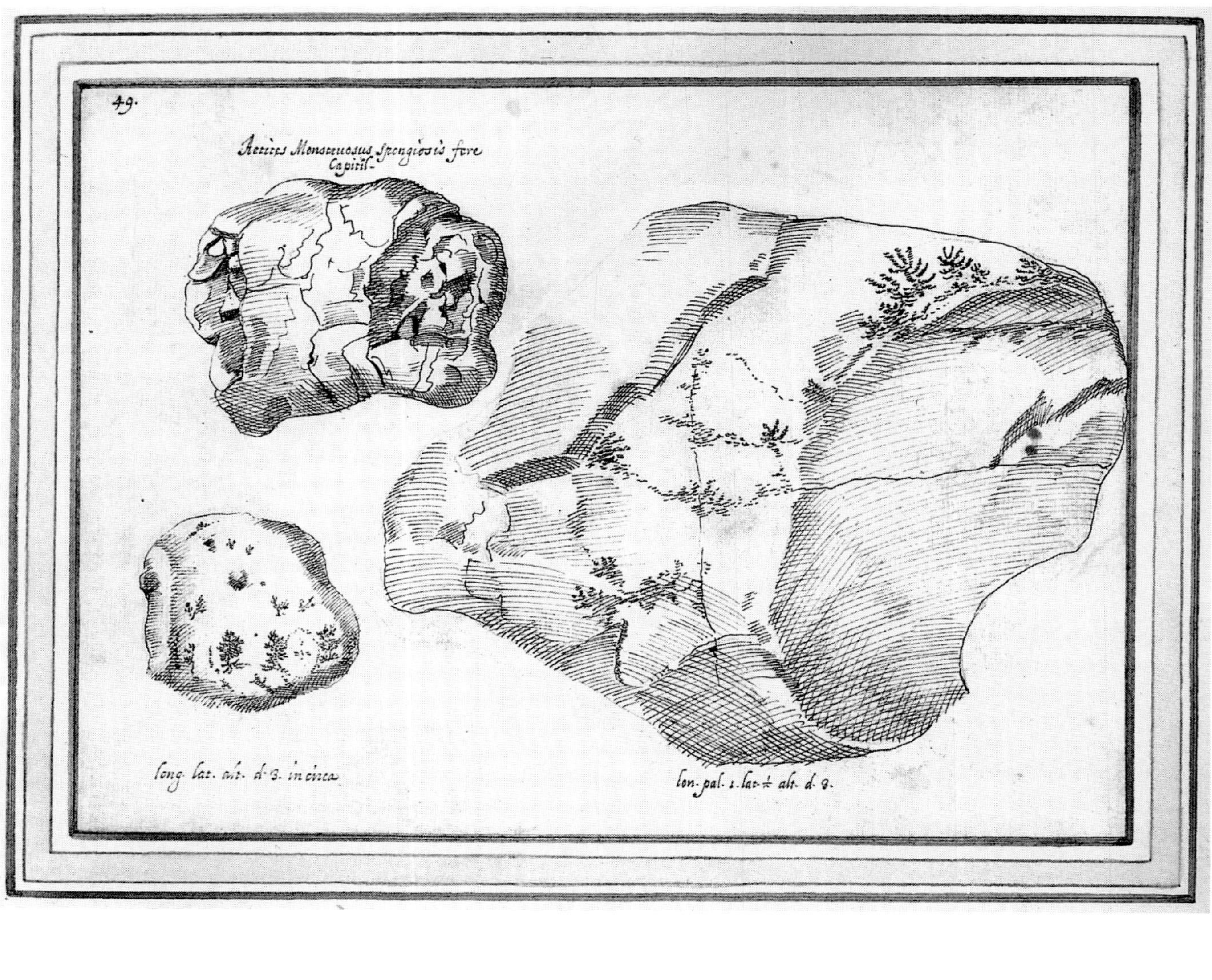

143. Dendrites

Windsor, RL 25675, *Fossils*, vol. XVI, fol. 36

Pen and ink, over black chalk; measurements not given; black chalk frame line

175 × 245 mm

NUMBERING: [ink, top right corner] *136*

ANNOTATION: [ink] *Pendroites: / Breviosa / Christallina / Tuberculosa*

The specimens represent dendrites of manganese oxide on a variety of materials, including possibly bone (top left-hand specimen). The annotation—which reads *pendroites* in error for dendroites—lists three different types of dendrites: 'short, crystalline, and tuberous'. The central specimen is probably formed of calcite crystals which can grow in cracks in limestones.

REFERENCES: 17th-cent. Index, p. 369

OTHER DRAWINGS: **142** and **154**

143

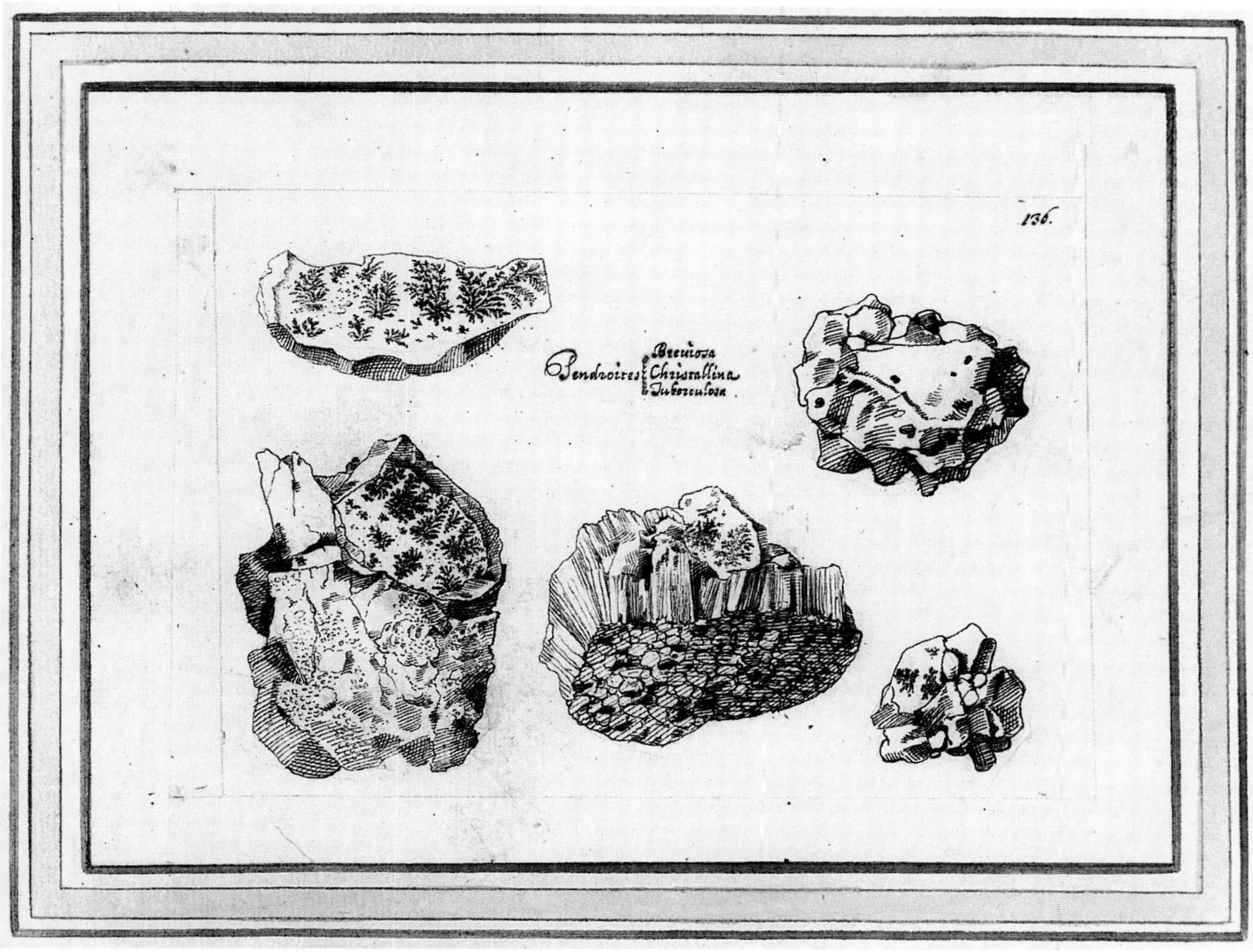

144. Iron pyrite, from Cesi's well

Windsor, RL 25562, *Fossils*, vol. XIV, fol. 33

Pen and ink, over traces of black chalk; no measurements given

182 × 255 mm

NUMBERING: [upper right] *25*

ANNOTATION: [ink, top] *Creta ex nr̄o. Puteo, Pyritis, Venis, Virgulis, Granis, Scabiequæ fræta*

The inscription records a 'piece of pyritic clay from our [Cesi's] well, with veins, nodules, granules and scaly'.

According to Bromehead, the geologist A. F. Hallimond considered the drawing to show a rock fragment with the columnar form of pyrites; he noted that such a formation was commonly found in wells. Bromehead compared this specimen with Stelluti's Plate 10, Figure A (see Fig. 63, p. 381). The present author however, does not believe the two illustrations to represent similar specimens, since the present drawing was described as pyritic clay, while the specimen illustrated by Stelluti was clearly said to be woody.

As noted by Bromehead, similar pieces of this type of material are illustrated in **145**.

LITERATURE: Bromehead 1947a, p. 74, Pl. 3, Fig. 2

REFERENCES: 17th-cent. Index, p. 362

OTHER DRAWINGS: **145**

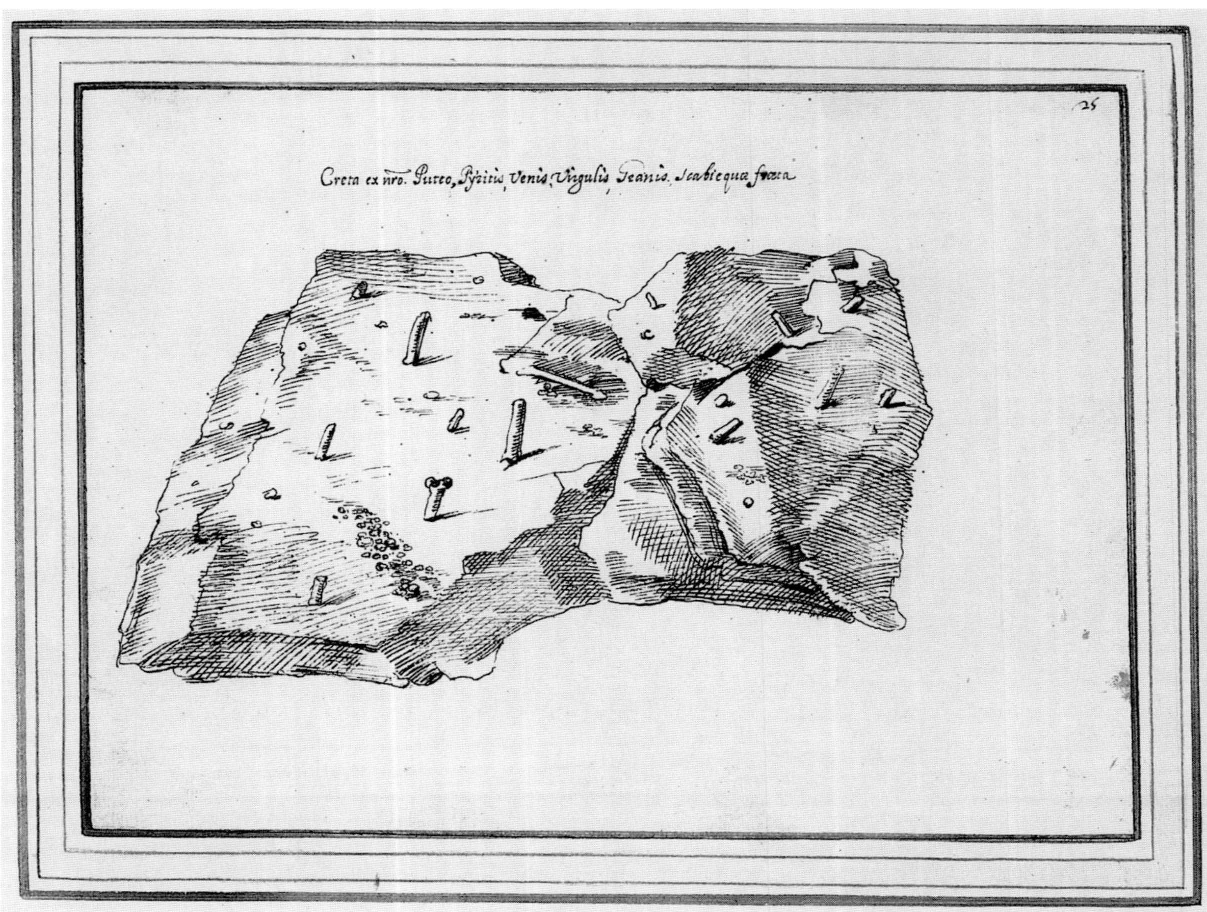

144

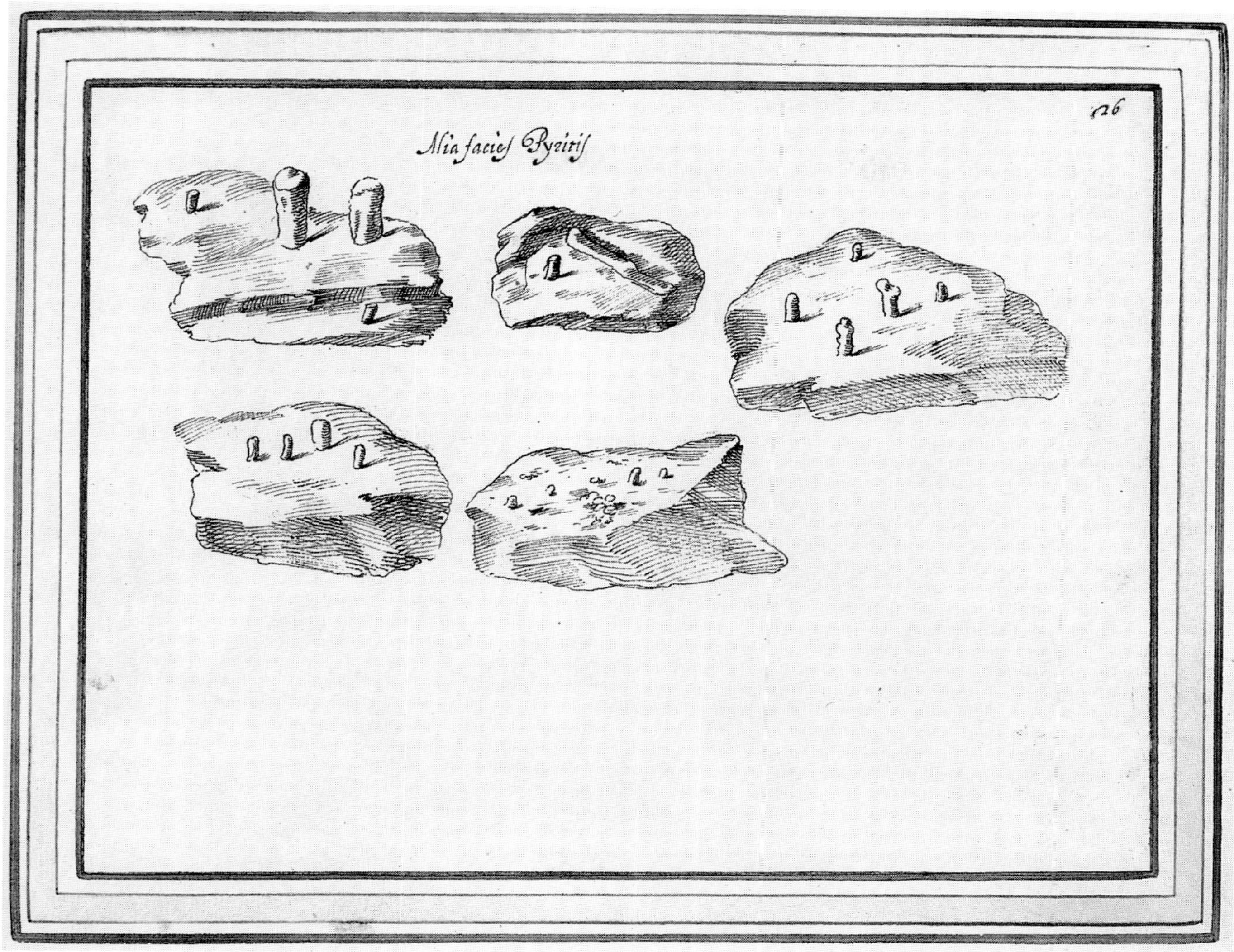

145

145. Iron pyrite, from Cesi's well

Windsor, RL 25564, *Fossils*, vol. XIV, fol. 35

Pen and ink, over traces of black chalk; black chalk outline of another specimen; no measurements given

185 × 252 mm

NUMBERING: [ink, upper right] *26*

ANNOTATION: [ink, top] *Alia facies Pyritis*

As noted in the caption, these are specimens of pyrite similar to those shown in **144**. A faint sketch in black chalk appearing in the original, lower right, would seem to represent the outline of the specimen shown in **144**, which was too large to fit on to the page. This would imply that these specimens were drawn to scale. (The fact that the outline of the large specimen is smaller than the completed drawing in **144** may imply that the latter was slightly enlarged.)

LITERATURE: Bromehead 1947a, p. 74

REFERENCES: 17th-cent. Index, p. 362

OTHER DRAWINGS: **144**

IV. FOSSIL ANIMALS [146–154]

While most of the shells illustrated in the drawings come from the area around Acquasparta, one (**153**) appears rather different and probably represents marine bivalves from elsewhere.

The Fosso Bianco Formation yields a variety of snails, both terrestrial and freshwater.[1] In particular, carbonate muds contain abundant snails. Such rocks are found throughout the vicinity of Dunarobba, Sismano and Rosaro [Fig. 39].[2] Some of these rocks with snails are included in the drawings, for example, **151**. In many places fossil fish also occur, but none is illustrated in the drawings. The Santa Maria di Ciciliano Formation above the Fosso Bianco Formation—both of late Pliocene/early Pleistocene age (1–2 mybp)—also yields freshwater shells. These rocks occur to the east of Rosaro. Cesi's well yielded numerous fossils which may be from this Formation (**151, 152**).

The overlying Pleistocene Acquasparta Formation also yields freshwater snails and clams. Small ivory fragments, identified as mammoth or elephant by Bromehead,[3] also occur in this Formation.[4] Some of the shells are identified on the drawings as coming from Acquasparta.

NOTES

1. Basilici 1997, p. 84; Ambrosetti 1995, p. 470. Aquatic snails and clams found at Dunarobba include the extinct forms: *Thoedoxus groyanus*, *Prososthednia ovata*, *Tournouerina belnensis*, *Micromelania zitteli*, *Emmericia umbra*, *Melanopsis affinis* and *Corbicula fluminalis* (information provided by Odoardo Girotti, 1998).
2. Basilici 1996, p. 76.
3. Bromehead 1947a, p. 73.
4. Venturi 1996, p. 35.

Fig. 39. Typical exposure of rocks yielding snails between Rosaro and Sismano

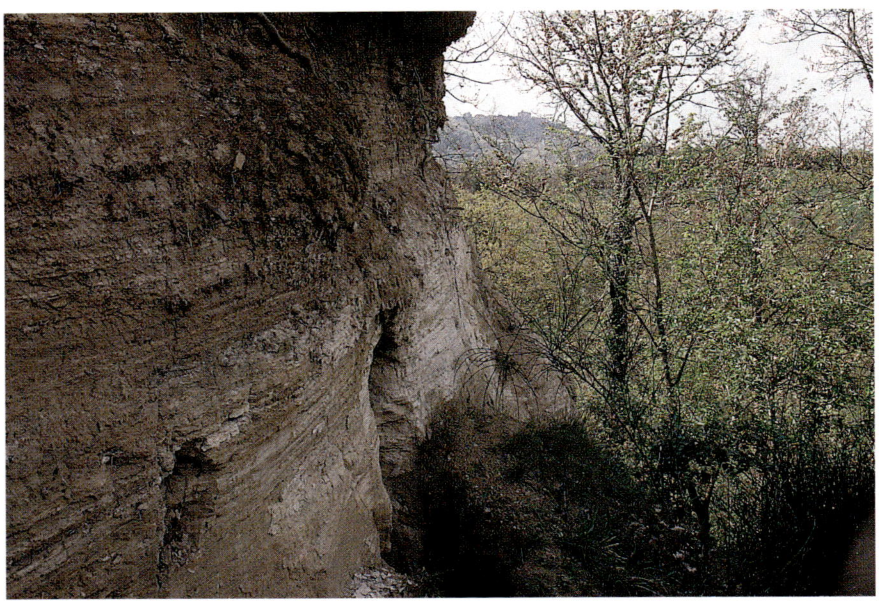

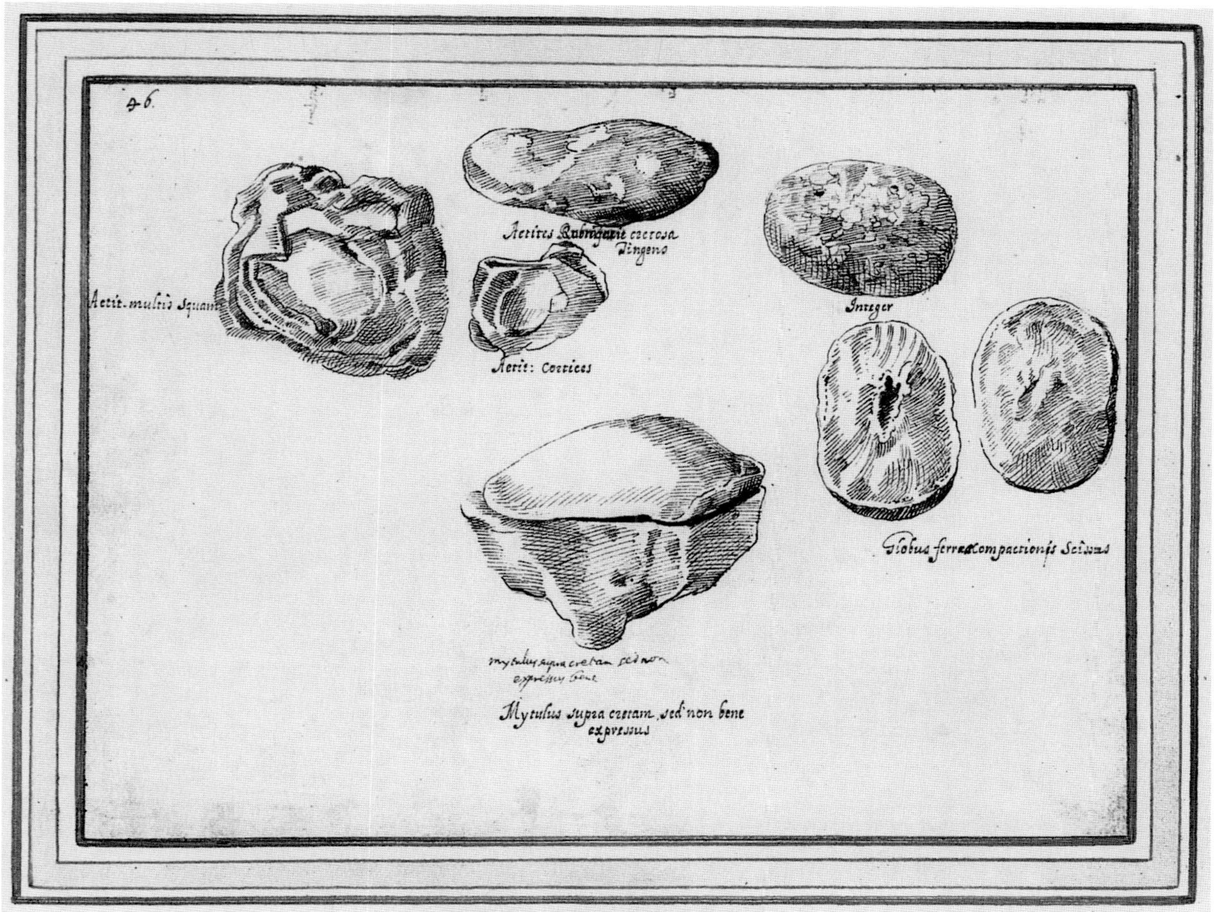

146. Five concretions, one with bivalve shell 146

Windsor, RL 25584, *Fossils*, vol. XIV, fol. 55

Pen and ink, over black chalk; measurements not given, although the specimens may well be actual size

182 × 254 mm

NUMBERING: [ink, upper left corner] *46*

ANNOTATION: [ink, upper left] *Aetit. multis squam:*
[top] *Aetites Rubra facie cretosa/ Tingens* [upper centre] *Aetit: Cortices*
[upper right] *Integer* [centre right] *Globus ferrae compactionis Scissus*
[bottom centre] *Mytulus supra cretam, sed non bene/ expressus*
[In Cesi's hand] *mytulus supra cretam sed non expressus bene*

The top left-hand specimen shows a typical aetite form with internal layering described as having 'many scales'. The specimen to the right—described as 'cortices of aetite'—may be a fragment of the same specimen. The top-centre specimen is described as 'clayey, of a ruddy appearance'.

The three drawings on the right appear to be an external view and two internal views of an oval concretion cut open to show a small hollow centre.

Cesi notes of the specimen at lower centre that it is a '*Mytulus* [bivalve shell] on top of clay, but not well expressed'. 'Expressed' here is more likely to refer to the relative lack of shell of the bivalve—or perhaps simply to its poor state of preservation—rather than the quality of the drawing.

REFERENCES: 17th-cent. Index, p. 363

147. Marble and shells from Acquasparta

Windsor, RL 25685, *Fossils*, vol. XVI, fol. 46

Pen and ink, over traces of black chalk; measurements not given

186 × 252 mm. *Watermark*: Bird 41B

NUMBERING: [ink, top left corner] *146*

ANNOTATION: [ink] *Tiburtinum cochleis Marmor, Aquae Spart:* [Cesi's inscriptions at lower left, cut]

The specimen on the left is clay with small gastropods and is described as 'Tiburtine marble with shells, at Acquasparta'. The right-hand specimens probably represent mould infills of a fossil bivalve shell. See also **146**.

REFERENCES: 17th-cent. Index, p. 369

Comp. fig. 147a. Clay with freshwater gastropods, found near Dunarobba

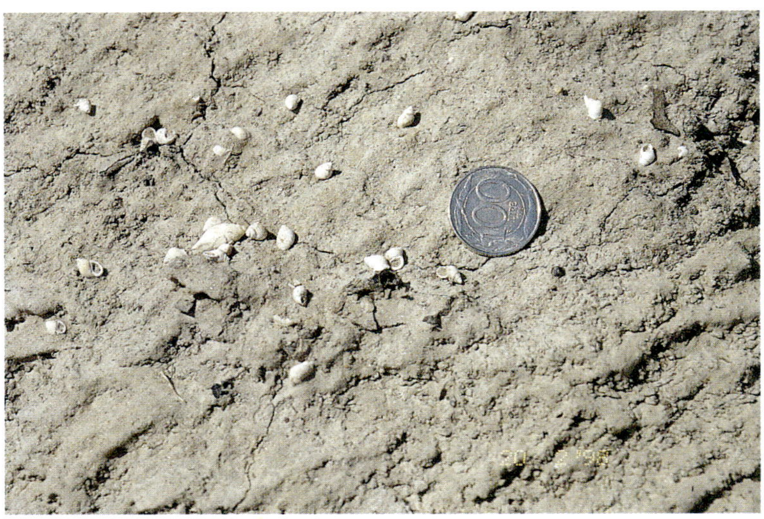

Comp. fig. 147b. Freshwater gastropod, found near Dunarobba

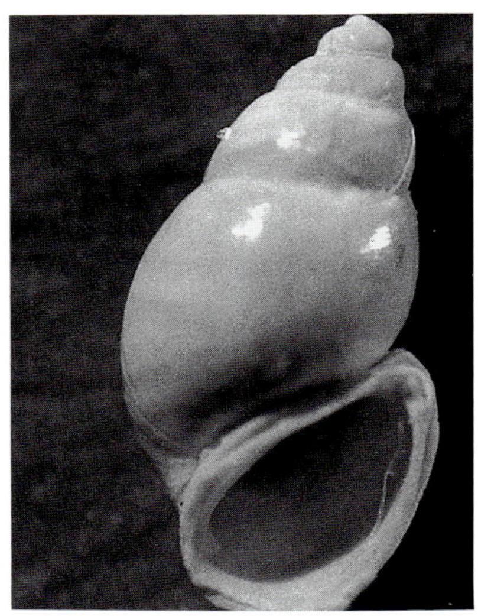

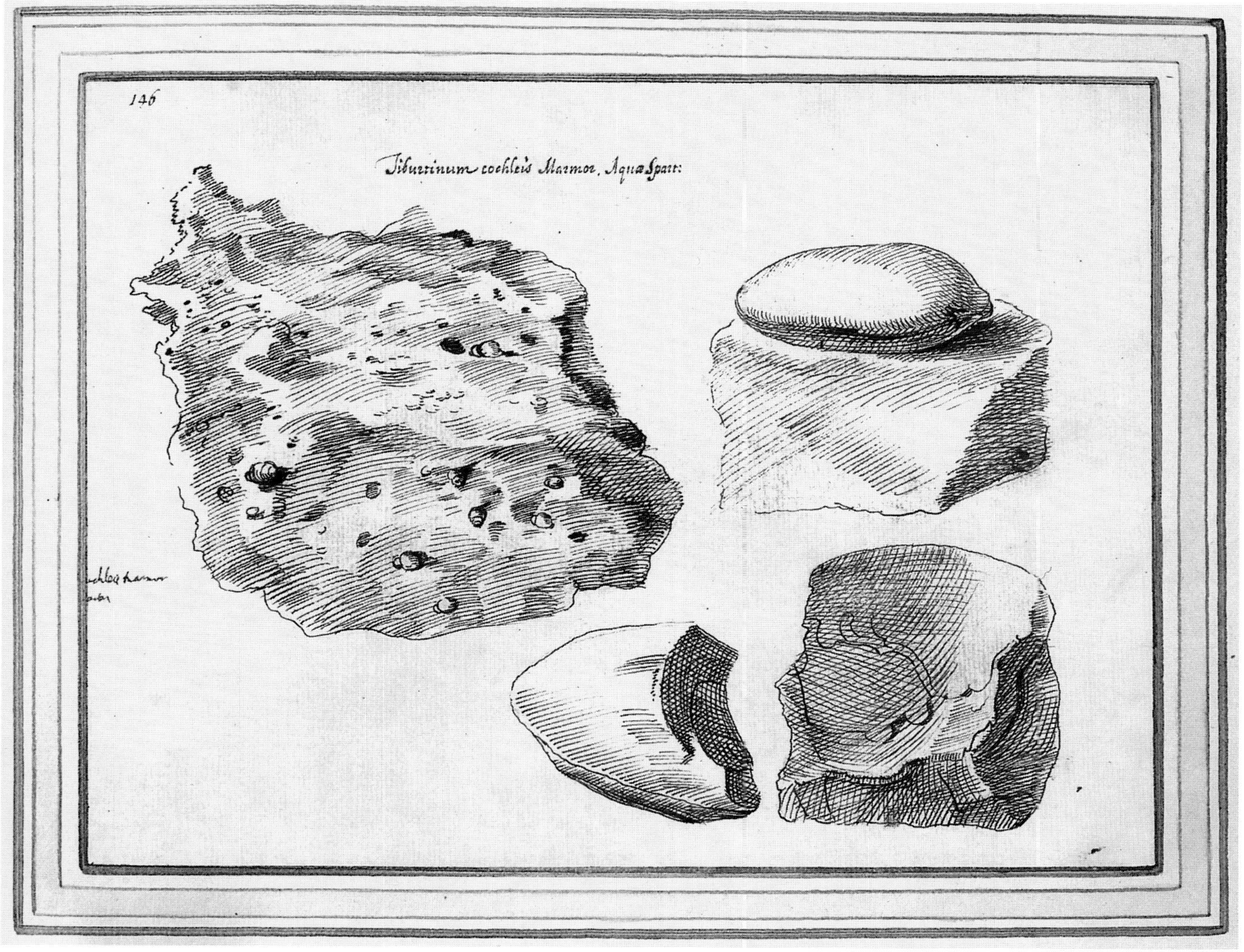

147

148. Piece of clay with small gastropods, from Dunarobba

Windsor, RL 25583, *Fossils*, vol. XIV, fol. 54

Pen and ink, over traces of graphite; measurements not given; ink frame line

188 × 240 mm

NUMBERING: [ink, upper left corner] *45*

ANNOTATION: [ink, top right] *Aetitis Crusta cochleis/ minimis inter-/ceptis, ex Duna/robba*

A specimen of organic, and possibly iron-rich, clay, with numerous small gastropod shells which appear to be aquatic forms, found at Dunarobba. The specimen is described as the 'crust of an aetite with many little shells interspersed'.

REFERENCES: 17th-cent. Index, p. 363

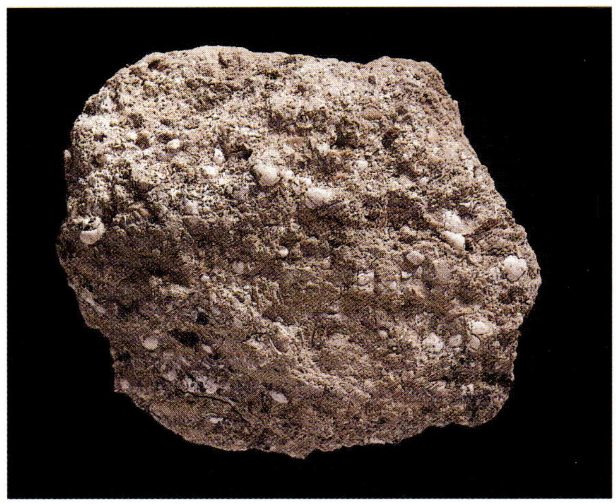

Comp. fig. 148. Limestone with aquatic gastropods, found near Dunarobba

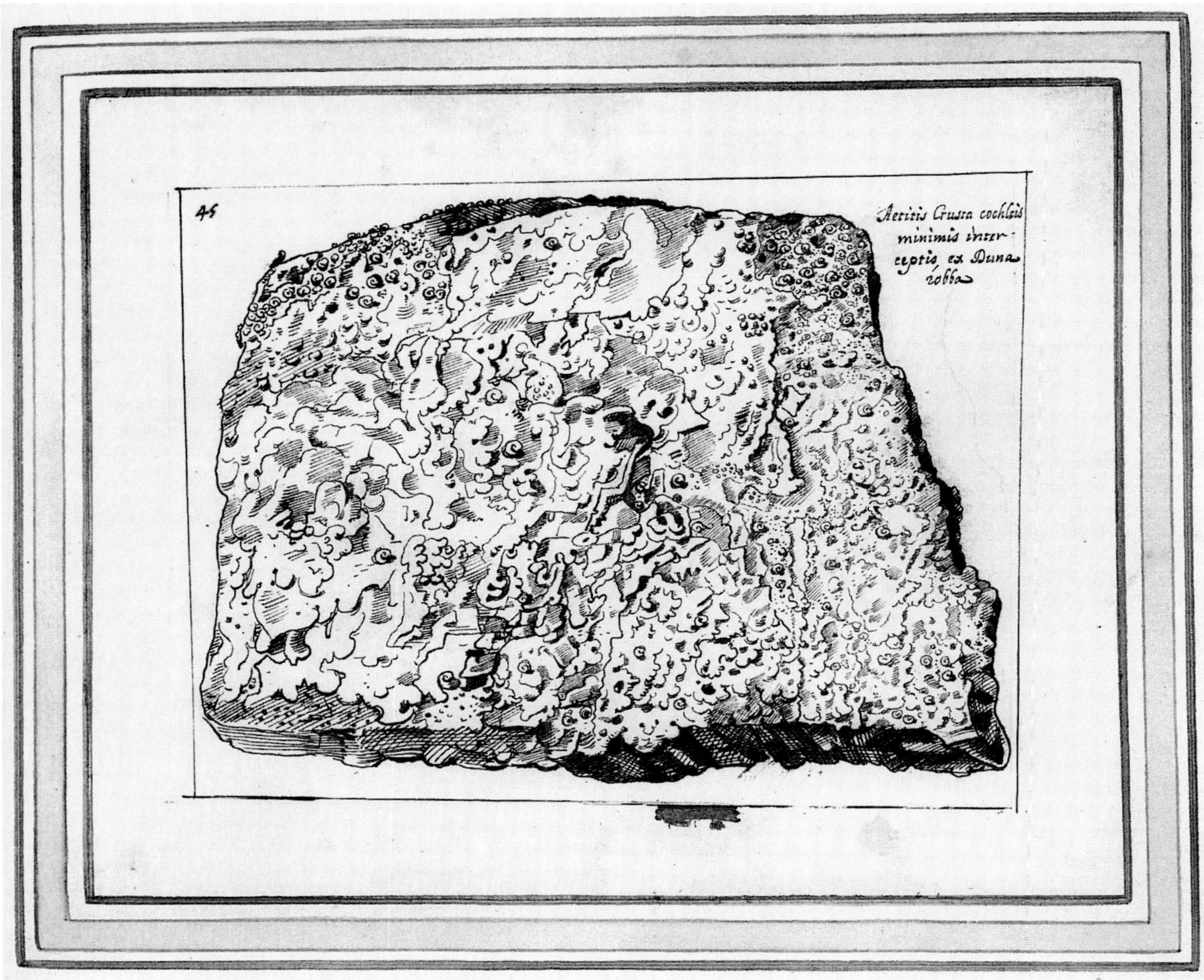
45
Aetitis Crusta cochleis
minimis inter-
ceptis ex Duna
robba

148

149. Piece of clay with small gastropods

Windsor, RL 25585, *Fossils*, vol. XIV, fol. 56

Pen and ink, over black chalk; measurements not given; marked with a cross

170 × 245 mm; old oil stains

NUMBERING: [ink, upper right corner] *47*

ANNOTATION: [ink, top] *Aetites crusta, et sua hic crassities expressum, et fractio cochleis/ minimis sempet intercept:*
[bottom] *lon: lat. pal. 1½ Crass. d. 2*

Another view of a 'crust of aetite' interspersed with small gastropods. It is listed in the Index as from Dunarobba and is quite possibly from the same locale and horizon as **148**.

REFERENCES: 17th-cent. Index, p. 363

149

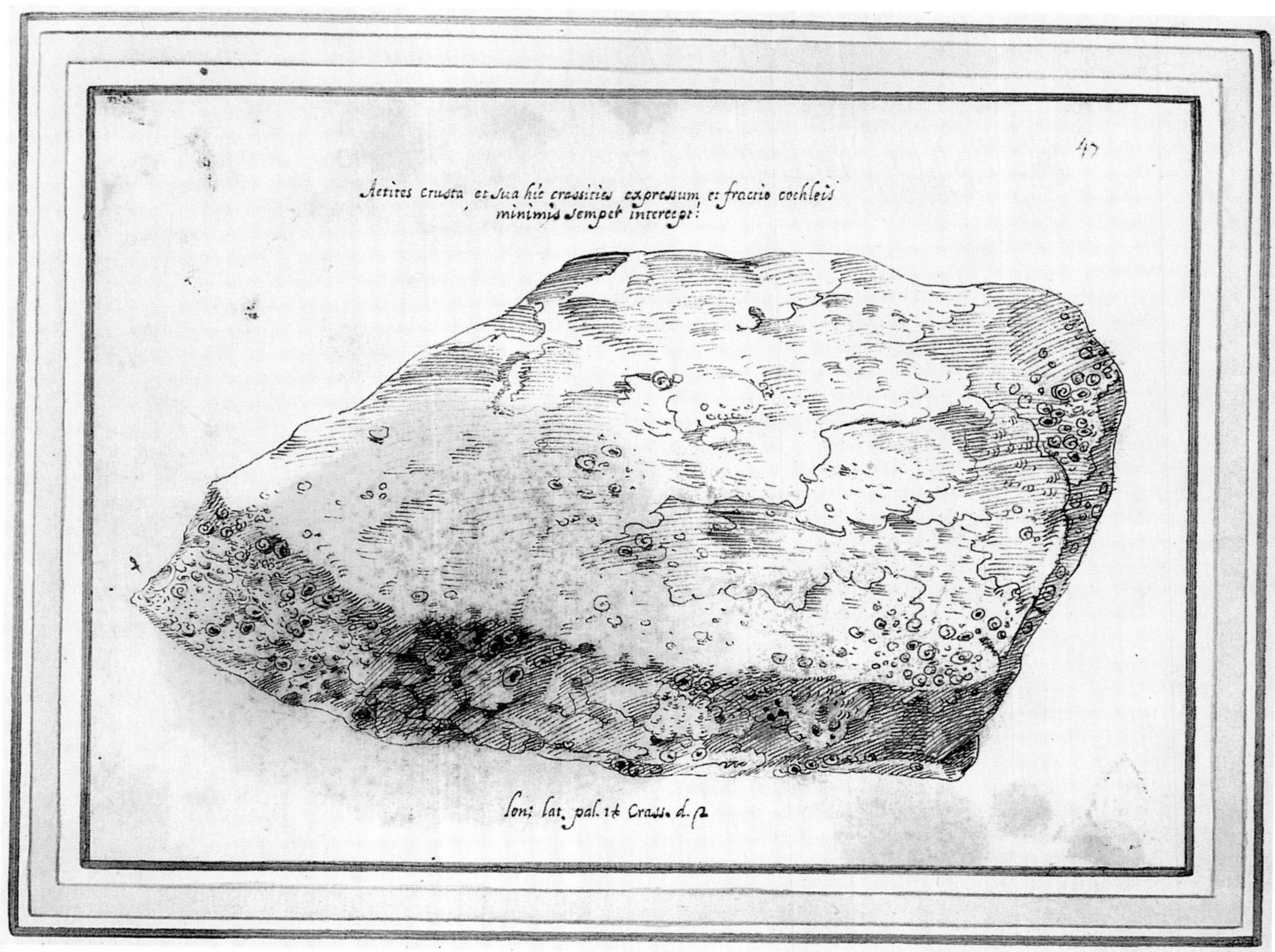

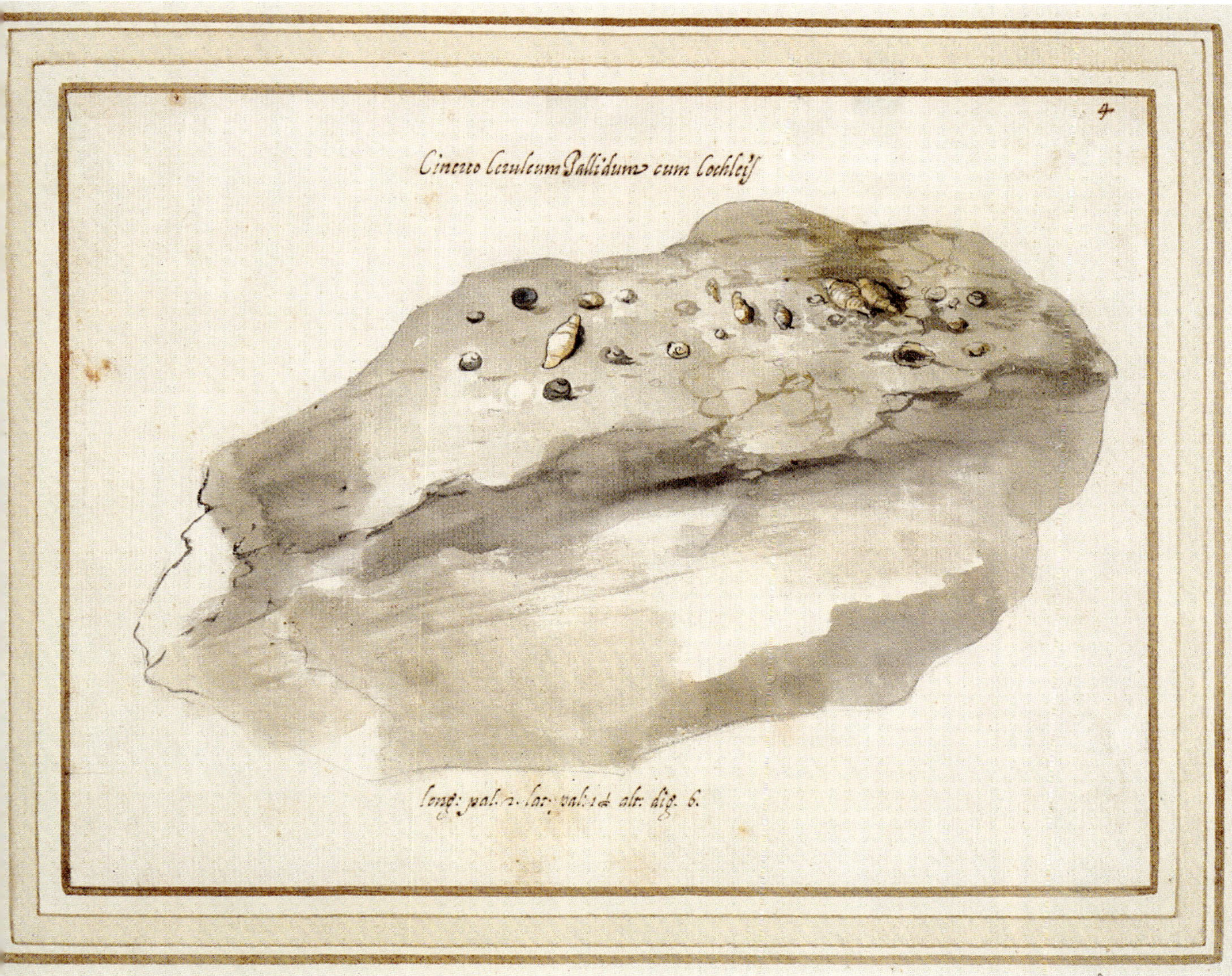

150

150. Clay with freshwater gastropods

Windsor, RL 25700, *Fossils*, vol. XVII, fol. 6

Watercolour, over black chalk

179 × 244 mm

NUMBERING: [ink, top right corner] *4*

ANNOTATION: [ink, top] *Cinereo Ceruleum Pallidum cum Cochleis*
[bottom] *long: pal. 2. lat. pal: 1½ alt: dig. 6.*

The specimen, described as 'pale ash-blue with shells', shows a lump of blue–grey clay with small freshwater gastropod shells. The shells appear similar to some shown in **151** and could be, like those in **152**, from Cesi's well (see p. 286). It is possible that these are gastropods from the Santa Maria di Ciciliano Formation (Pliocene–Pleistocene age).

REFERENCES: 17th-cent. Index, p. 370

OTHER DRAWINGS: See **151** and **152**

151. Five rock fragments containing fossil molluscs, from Cesi's well

Windsor, RL 25539, *Fossils*, vol. XIV, fol. 10

Pen and ink, over black chalk

181 × 260 mm

NUMBERING: [ink, upper right] *3*

ANNOTATION: [ink, top left] *Chama plana/ in Caerul. cret./ nostri Putei/.15 cann. Profund*
[top right] *Creta Putei nostri cum cochleis*
[centre right] *Creta Putei n.ri: cum cochleis*
[centre] *Coniuct, Valu 14 cann. dsup./ Terr.*
[bottom left] *Apertis Valvis fractus 12/ cann. d. Superficie Terræ*
[centre left] *Mytuli creta caerulea nr̄i. putei intercepti*
[Cesi's inscription upper left, at margin, trimmed]

The gastropods shown in the three right-hand specimens were identified as *Limnaea* by Bromehead. They are probably examples of *Viviparus* from the Acquasparta Formation (of Pleistocene age) which overlies the Santa Maria di Ciciliano Formation. The two specimens on the left are freshwater bivalves; Bromehead identified these as *Unio* and *Paludina*.

The flattened bivalve in the top-left drawing bears two inscriptions, the upper one reading 'flat bivalve found in the blue clay of our well 15 *canni* deep', and the lower one identifying the bivalve as a *Mytulus*, meaning horse mussel (cf. also 199 below). 'Our well' refers to Cesi's well, presumably at Acquasparta, in which he found all the specimens depicted here as well as those in **152**. The three specimens to the right are also clays with gastropods; the inscription states that the lower one was found 14 *canni* from the surface. The bivalve in the lower left of the drawing is depicted with its shells apart and still within the rock; it was collected at 12 *canni* from the surface.

LITERATURE: Bromehead 1947a, p. 69; Basilici/ Gentili 1992

REFERENCES: 17th-cent. Index, pp. 360, 362, 367

OTHER DRAWINGS: **152**

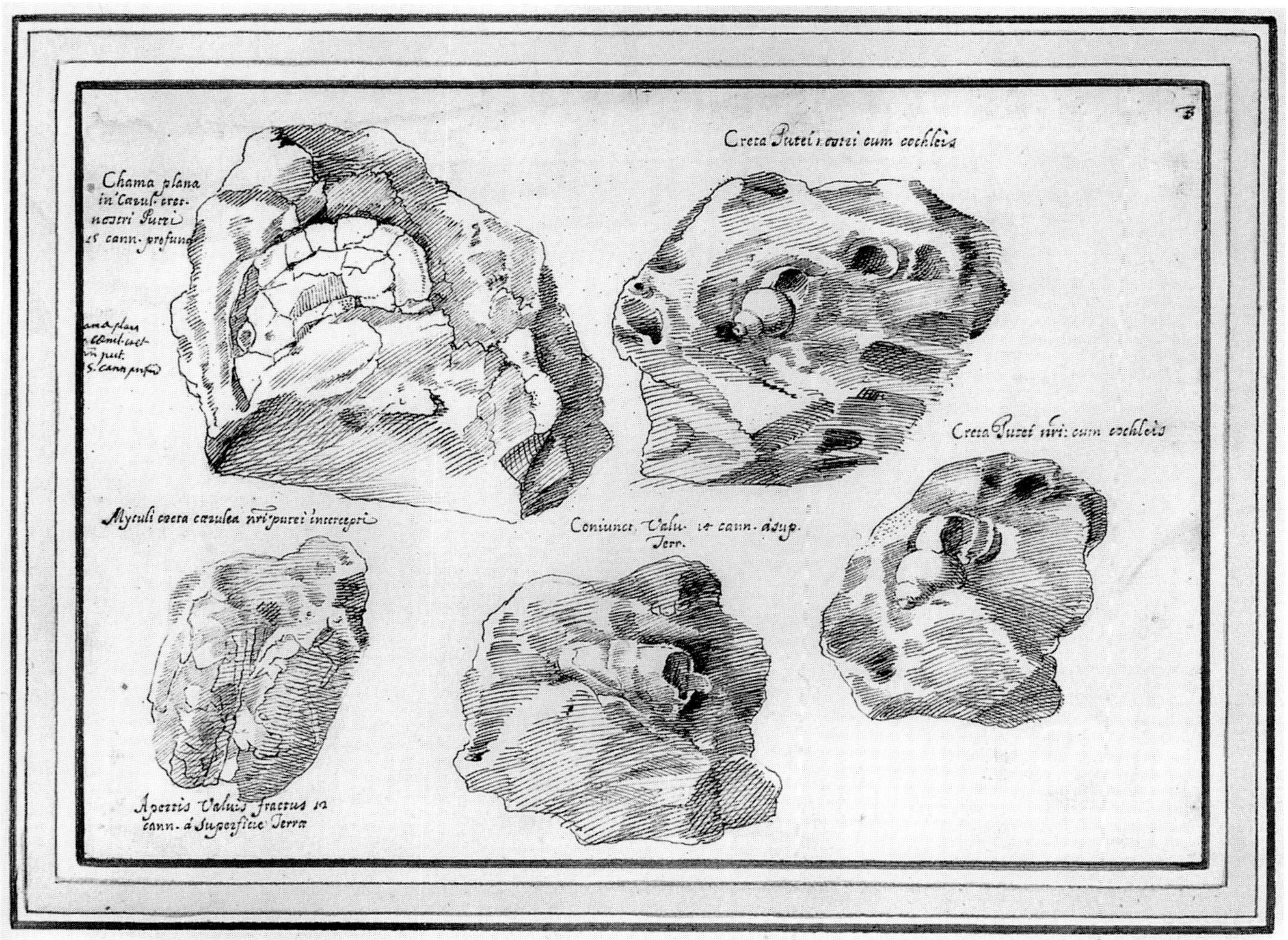
Chama plana
in Cavul. cret.
nostri Putei
15 cann. profund
Creta Putei nostri cum cochleis
Creta Putei nri: cum cochleis
Mytuli creta caerulea nri putei intercepti
Coniunct. Valu. 10 cann. d sup. Terr.
Apertis Valuis fractus 12
cann. d Superficie Terrae

151

152. Five concretions with freshwater shells, from Cesi's well

Windsor, RL 25692, *Fossils*, vol. XVI, fol. 53

Pen and ink, over traces of black chalk; left-hand specimen is drawn actual size, measurements of the others not given

182 × 249 mm; small hole caused by ink. *Watermark*: Bird 41B

NUMBERING: [ink, top left corner] *153*

ANNOTATION: [ink, top] *Tartar. ex Nr̄o: Puteo in profundo 18: Cann: cum Turbinibus et / cochleis, Rubens*
[upper right] *lapi surgens—*
[below larger drawing on left] *Nalī.*

Five concretions with freshwater gastropods which, according to Bromehead, belong to the genera *Clausilia* and *Helix*. The inscription appears to relate to all the specimens, and reads 'A spongy *Tartara* from our well at a depth of 18 canni [36 m] with whorls and shells, reddish'. The reddish colour probably means that the specimens are iron concretions.

The name *Tartara*, applied to the specimen, is not found elsewhere in the corpus and probably refers to the kind of spongy stone or *pietra spugnosa* that results from encrustation produced by sulphurous water (Gabrieli 1989, p. 45). The annotation *lapi surgens*, between the two upper-right specimens, appears to mean 'turning into stone' or 'with stones arising from it', a comment indicative of Cesi's attempt to analyse the 'middle nature' of these specimens.

LITERATURE: Bromehead 1947a, p. 69

REFERENCES: 17th-cent. Index, p. 369

OTHER DRAWINGS: **151**

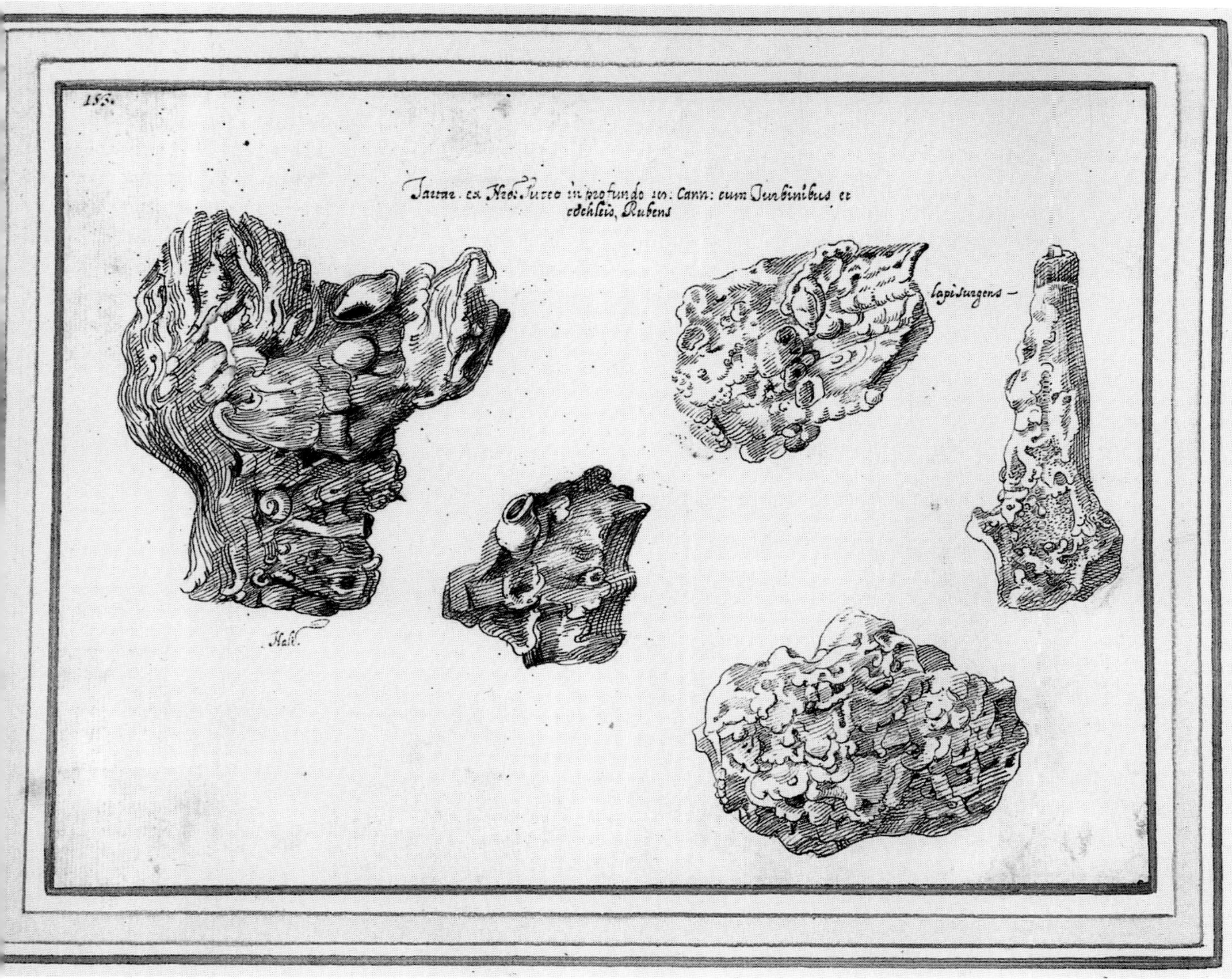

152

153. Fossil bivalve shells in rock

Windsor, RL 25618, *Fossils*, vol. XV, fol. 27

Pen and ink, over traces of black chalk

189 × 255 mm

NUMBERING: [ink, upper right corner] *79*

ANNOTATION: [ink, top] *Lapis conchites Firmanus* [bottom] *long: pal: 1. lat: p. ½. alt: pal. ½*

This specimen, of bivalves in rock, shows well-defined internal moulds; almost certainly these are of *Glycymeris* from marine Tertiary rocks. The inscription identifies it as a 'stone with shells' and is followed by the name *Firmanus*.

Bromehead believes that this is an early example of adding the author's name after the scientific identification. However, it is possible that *Firmanus* refers to the town of Fermo, not so far (50 km) from Stelluti's birthplace of Fabriano. Fermo was the diocese of which Cesi's friend Pietro Dini was made archbishop in 1621 (cf. **62**).

LITERATURE: Bromehead 1947a, p. 73

REFERENCES: 17th-cent. Index, p. 365

154. Pieces of bone with dendrites of manganese oxide

Windsor, RL 25596, *Fossils*, vol. XV, fol. 5

Pen and ink, over traces of black chalk; measurements not given

186 × 249 mm. *Watermark:* Anchor 16B

NUMBERING: [ink, top left] *58*

ANNOTATION: [ink] *In frustis Cornu Monocerotis, sive Terra Eretriæ Albæ Squamosæ dendroites*

Bromehead considered these four small specimens to be 'fragments of fossil bone or ivory with dendritic manganese markings'. As he noted, the inscription refers to them as 'dendroites on unicorn's horn', but he suggests they are probably 'pieces of a tusk of *Elephas* [mammoth] from the Upper Pliocene'.

Classical writers often referred to the white Eretrian earth, from that part of Tuscany around Arezzo (*Terra Eretriæ*). Theophrastus, Pliny and Celsus all discussed this white ash-coloured earth, the physical qualities of which are not dissimilar to the kind of bone or ivory referred to here. More contemporary accounts of its uses may be found in writers such as Palissy, Persius, Stelluti, Ceruti and Chiocco, who mention a variety of uses, including its use as a ground for painting.

The fact that much of the page is left blank may imply that other specimens were to be added. Alternatively, it is possible that these specimens might originally have been intended to be included with **142** and **143**.

LITERATURE: Ceruti and Chiocco 1622, pp. 124–5; Caley and Richards 1956, p. 208; Bromehead 1947a, p. 73

REFERENCES: 17th-cent. Index, p. 364

OTHER DRAWINGS: **142, 143**

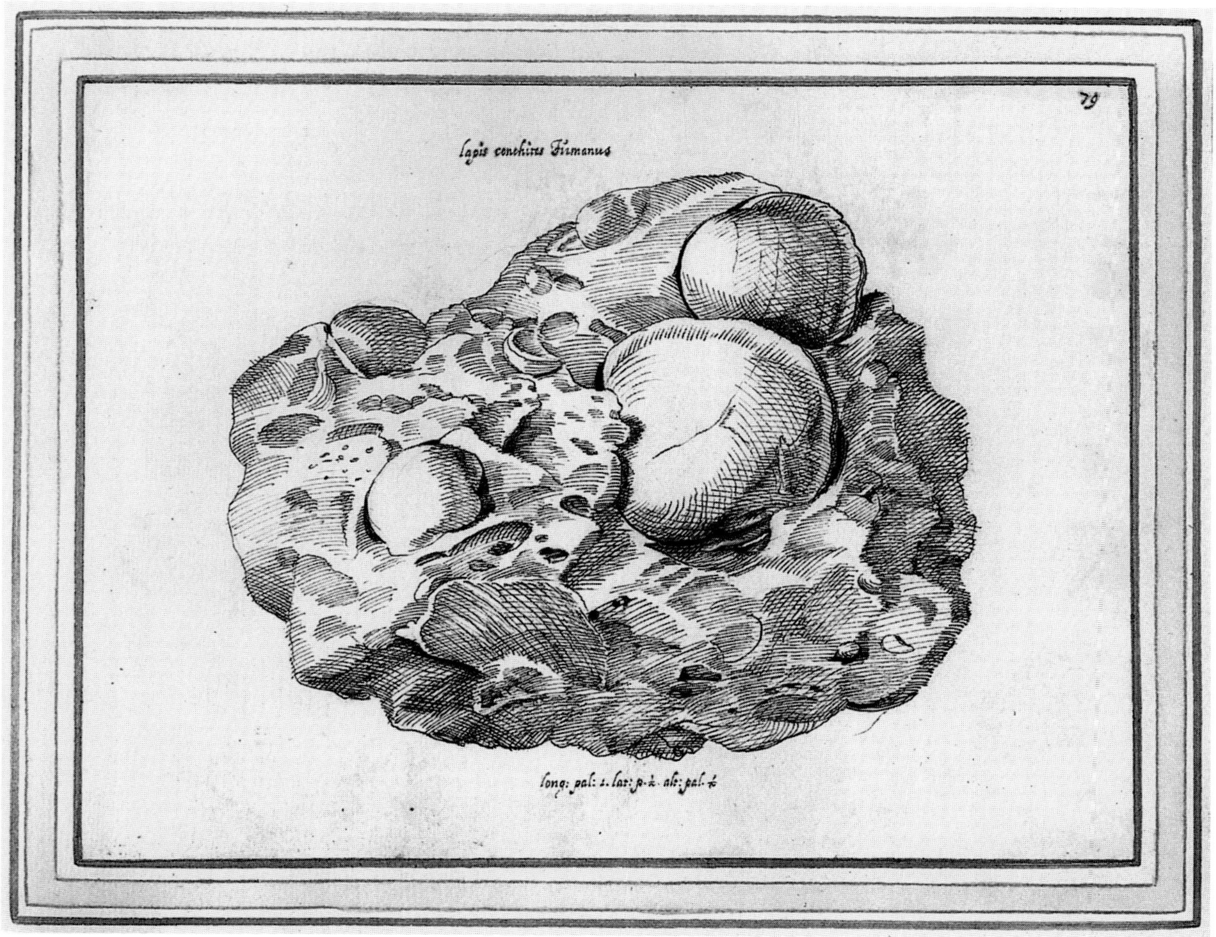
79
lapis conchites Firmanus
long: pal: 1. lat: p. 2. alt: pal. 4

153

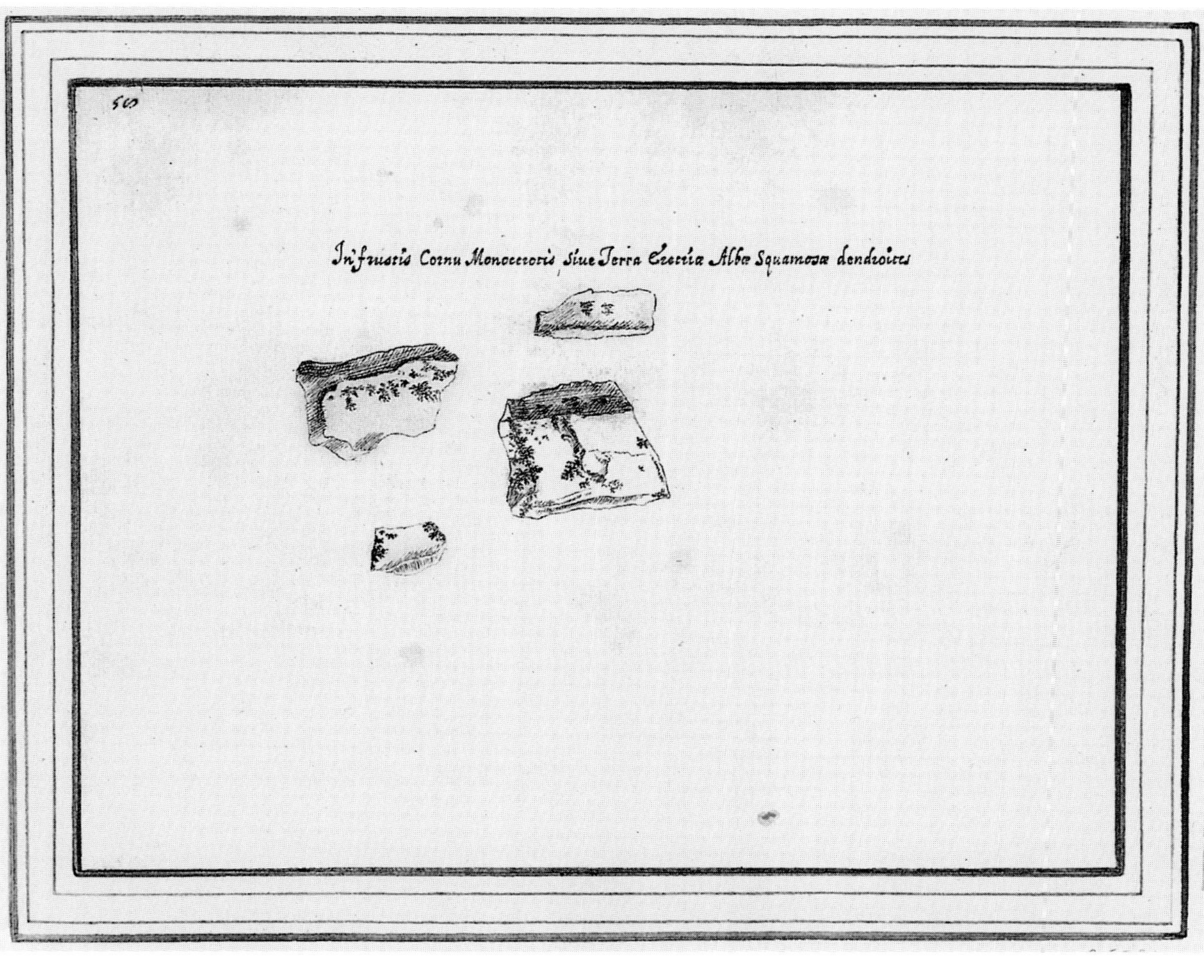
560
In frustis Cornu Monocerotis, siue Terra Cretica Albæ Squamosæ dendroites

154

V. BAKED CLAYS [155–183] AND UNALTERED SEDIMENTS [184–196]

Underground fires of the kind drawn in **13** and **14**, shown as Plate 2 of Stelluti's *Trattato*, and frequently referred to by Cesi and his friends, are often found burning in coal or lignite deposits. Once set alight, they may smoulder for many years (hence 'everlasting'). The effect on the surrounding rocks is interesting. If the sediments are clays with an organic content and some iron, the heat will have the effect of baking the clay hard and discolouring it to a variety of shades of yellow, brown, red, and even blue and grey. This type of effect is seen in most of the watercolours in Album XVII (see Concordance A, p. 394). Unfortunately the inscriptions do not specify the composition of the specimens depicted, nor where they are from, but it is almost certain that they represent material baked in this way. Stelluti's map of the region shows smoke rising from four localities [Fig. 17, p. 94].

Some of the fires may well have been started deliberately, rather than by spontaneous combustion caused by the decay of the iron pyrite which occurs in some of the rocks. (The decay of iron pyrite, described on p. 133, results in an increase in temperature. It is possible that this may lead to autocombustion.) While the weathering of clays under certain climatic conditions may produce similar colours, the apparent hardness of the specimens illustrated, combined with the range of colours depicted, seems to indicate baking (see Fig. 42, Comp. figs 169, 174). La Ferrière's letter to Peiresc (Appendix III b) would seem to support this. He wrote:

> My guide did show me a place where, he said, a fire had lasted for seven years, and the shepherds, warming themselves, had inadvertently set fire to the soil ... I suddenly remembered the treatise that the late Prince Cesi had written about this fossil wood, of which you have a copy, which states that there was a place where, the earth being naturally sulphurous, it had burnt for ten years and without doubt this was the place my guide had shown me in which I found that the soil broke off in what looked like terracotta scales resembling earthen pots or broken bricks.

If the watercolours of Album XVII do represent burnt clays from underground fires, then they are likely to come from the locality of Rosaro.

Several of the drawings indicate that they are of specimens that are hard, red and terracotta-like. (Cesi often used the term *lateritia*: see above, p. 60.) Stelluti noted that:

> And when the heat is sufficient, the wood is browned, or lightly scorched, and becomes similar to coal. If, then, the earthen material is not yet converted into wood, this fire bakes it, and it becomes similar to vases of terracotta baked in a kiln, or to bricks.

Fieldwork in 1998 identified several sites with burnt clays from underground fires, in the areas around Rosaro and Montecastrilli [Figs 40, 41]. Each of these was associated with 'lignite' horizons rich in fossil woods. The range of specimens matched those illustrated in the drawings[Fig. 42]. There is also some evidence that sulphur was given off during the burning, as small yellow crystals of sulphur occur on some of the specimens. The smoke from the fires would have smelt sulphurous. Similar fires have been seen over the second half of the twentieth century.[1]

While some of the drawings can be identified as baked clays, this identification is less certain for others; the material baked may have been the lignite itself. In these cases, gases would have been released and the remaining rock could have had a spongy texture [Comp. fig. 174]. This may have been the case where the description uses the term *spumosa* (**158**). A specimen of this kind was collected in 1998 [Comp. fig. 174]. In certain cases it is not clear from the drawings whether the red colours are due to baking or to iron impregnation.

In addition to the baked clays, the drawings include a selection of what appear to be unaltered sediments, which are placed at the end of this section. The term clay is used throughout to mean either clay (in the strict mineralogical sense) or mud (in the sedimentological sense). It is certain that the muds all contain clay minerals, and these when heated expand and harden to form the range of materials described by Cesi.

1. Personal communication from G. Basilici, 1998.

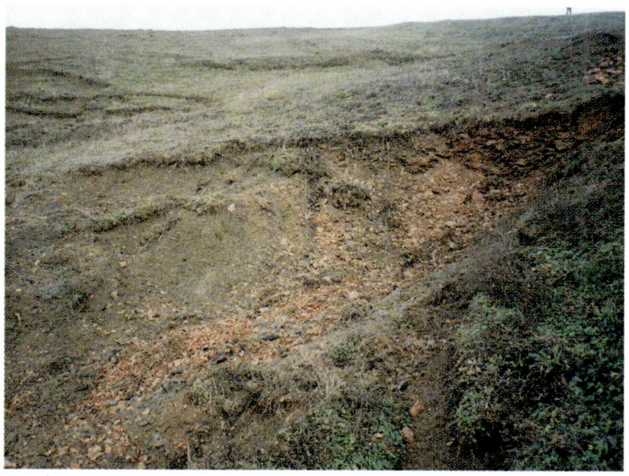

Fig. 40. Landslipped area between Rosaro and Sismano, showing red baked clays and fossil woods

Fig. 41. Ploughed field near Casanova, Collesecco, showing red baked clays

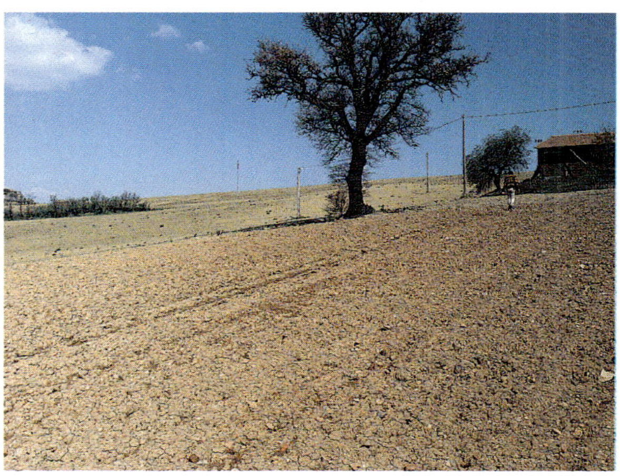

Fig. 42. Range of baked clays from Casanova

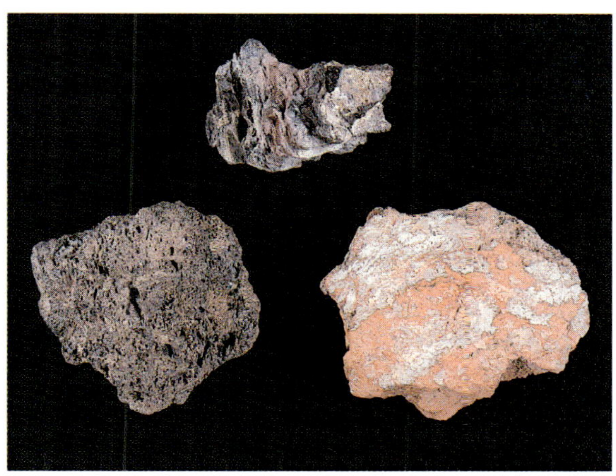

155. Two views of a piece of baked clay

Windsor, RL 25568, *Fossils*, vol. XIV, fol. 39

Pen and ink, over black chalk

188 × 258 mm

NUMBERING: [ink, upper right] *30*

ANNOTATION: [ink, top centre] *Creta in testum Cocta, rubens fissilis*
[centre] *long: d: 8: lat. d: 4. alt: d: 3*
[bottom] *long: pal: 1. lat. d. 4. alt. d. 3.*

The specimen is imaginatively described as 'clay baked to a pot, red and fissile'. An appearance of pottery is indeed characteristic of such baked clays (see above, p. 292).

REFERENCES: 17th-cent. Index, p. 362

155

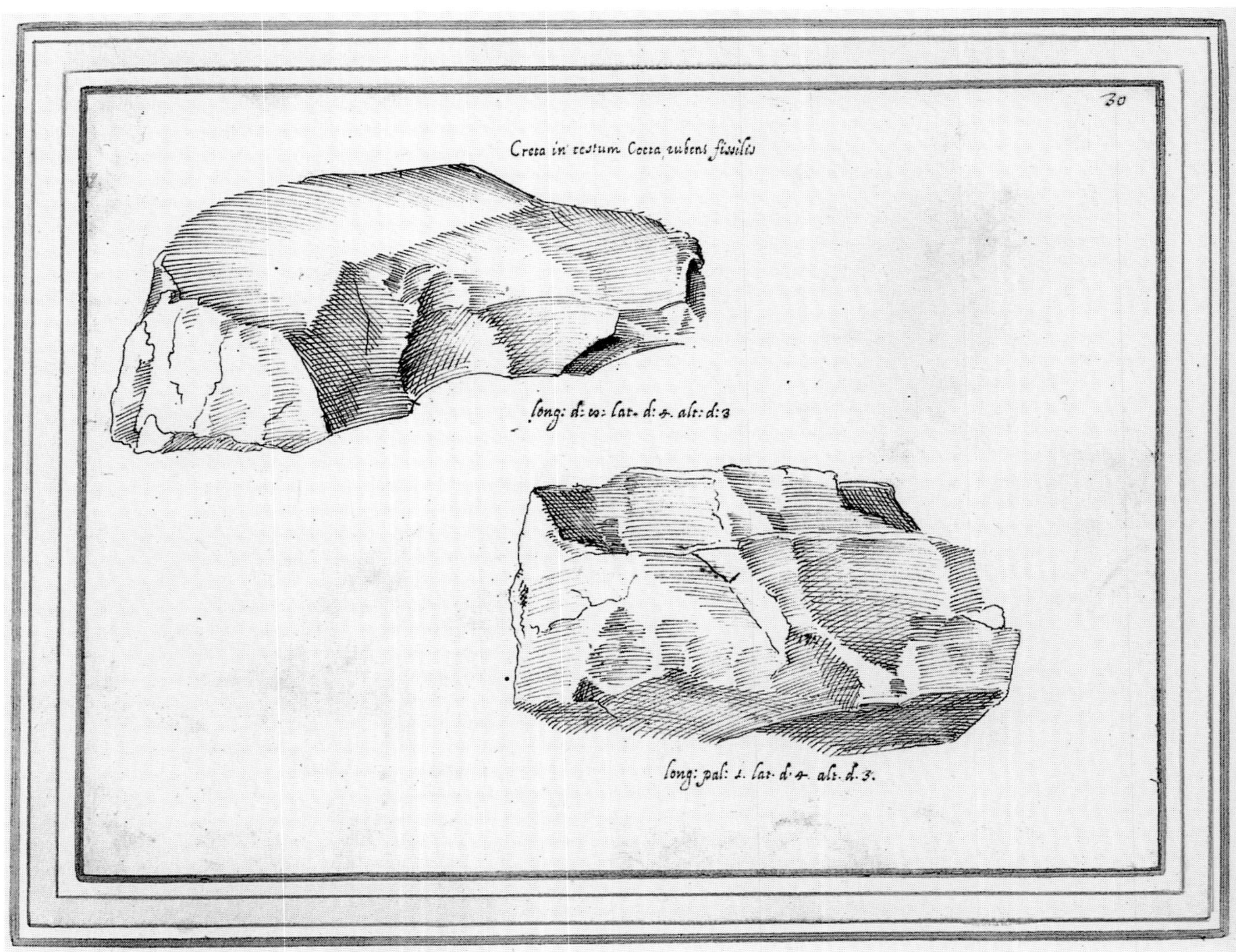

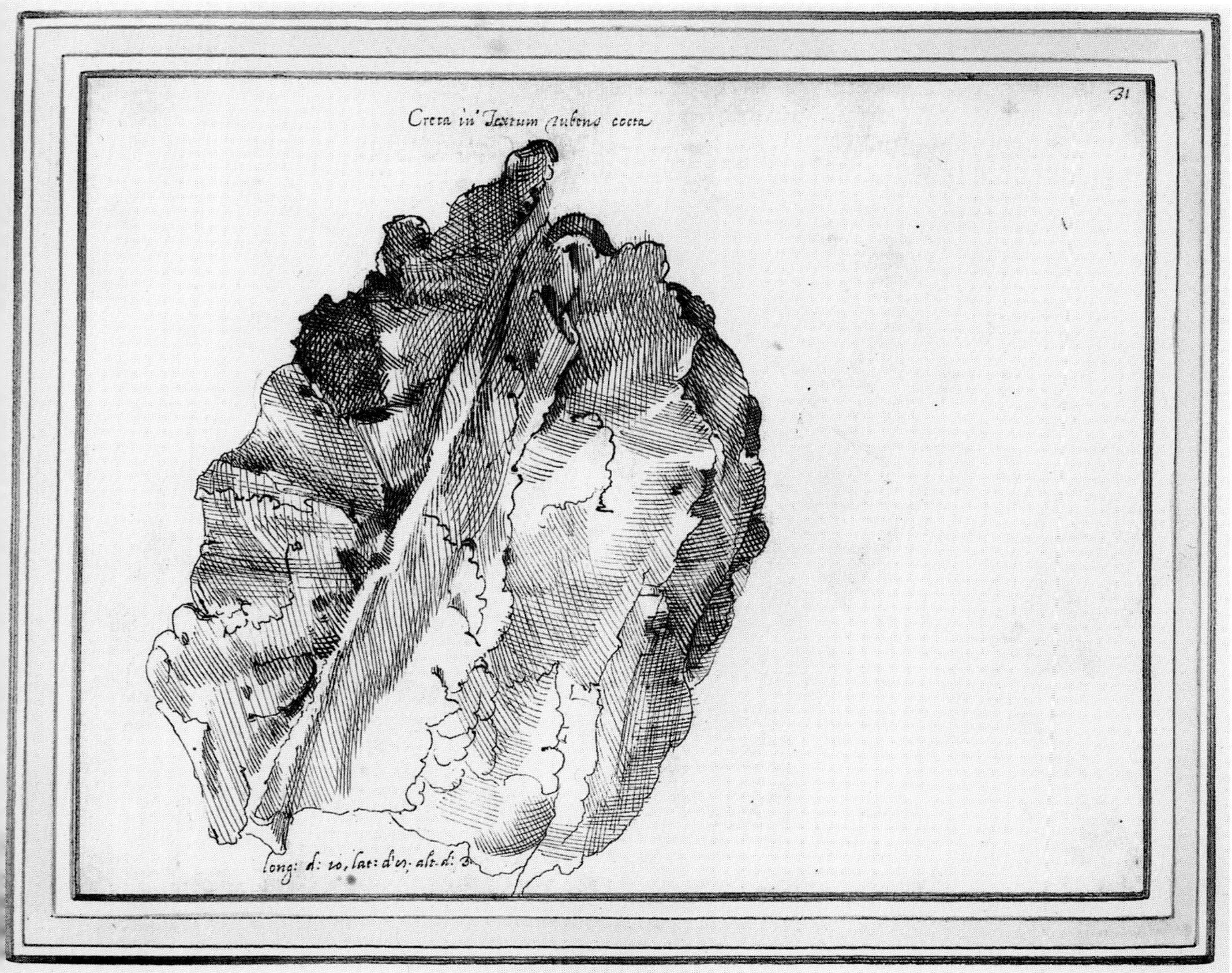

156

156. Baked clay

Windsor, RL 25569, *Fossils*, vol. XIV, fol. 40

Pen and ink, over black chalk

193 × 252 mm

NUMBERING: [ink, upper right] *31*

ANNOTATION: [ink, top] *Creta in Textum rubens cocta* [bottom] *long: d: 10, lat: d. 8: alt. d: 3.*

The drawing shows another piece of baked clay. The description notes that it is red and has been baked until it has the appearance of pottery.

REFERENCES: 17th-cent. Index, p. 362

157. Four specimens, possibly of baked clays

Windsor, RL 25548, *Fossils*, vol. XIV, fol. 19

Watercolour and bodycolour, over black chalk

179 × 246 mm

NUMBERING: [ink, upper left] *12*

ANNOTATION: [ink, top] *Cretæ coctae lateritiæ Coralliorum*
[right] *Stratis obscuris mixta;*
[left] *Aetitis affinia frusta crispa*

These coloured drawings are of rock fragments with colour variations from grey to red: this may indicate that they were originally organic-rich clays which were baked and thus reddened, or that they were impregnated with iron, as is the case with siderite nodules. The inscription at the top of the sheet identifies the specimens as 'Brick-like baked clays of coral [colour?]'. The right-hand specimen is described as 'mixed with dark strata' and the left-hand one as 'curly pieces, close to aetite'.

REFERENCES: 17th-cent. Index, pp. 361, 363

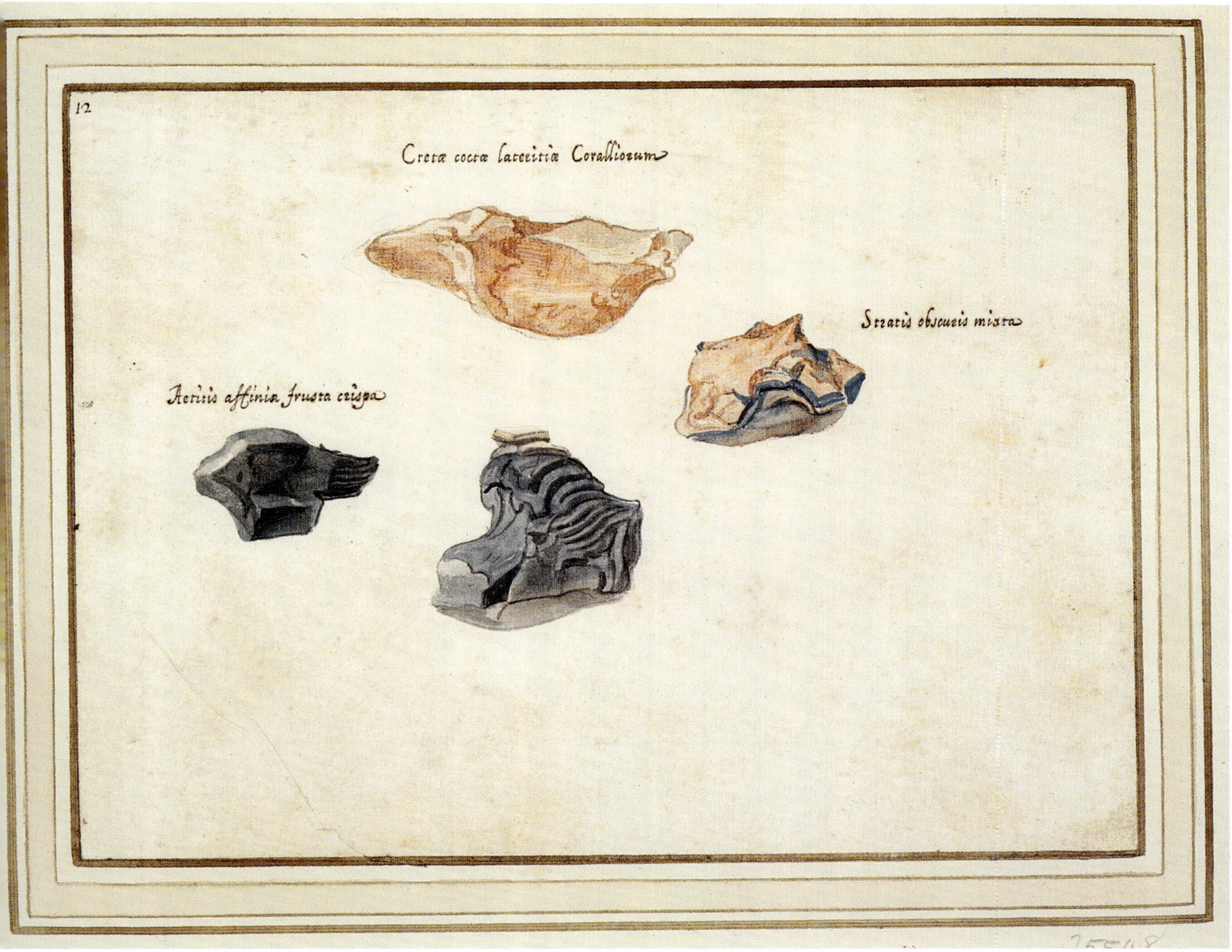
12
Cretæ coctæ lateritiæ Coralliorum
Stratis obscuris mixta
Aetitis affinia frusta crispa

157

158. Baked clay

Windsor, RL 25565, *Fossils*, vol. XIV, fol. 36

Pen and ink, and over traces of black chalk and graphite; retouched in ink on mount sheet; ink frame line

174 × 238 mm; small holes caused by ink

NUMBERING: [ink, top left, upper right] *27*

ANNOTATION: [ink, top left] *Spumosæ Ustæ Cretæ*
[lower right] *long: pal: 2. alt. et lat: p. ½*

The specimen is identified as 'spongy burnt clay'. It is possible that its vesicular nature was caused by the burning or baking of organic-rich clay or lignite.

REFERENCES: 17th-cent. Index, p. 362

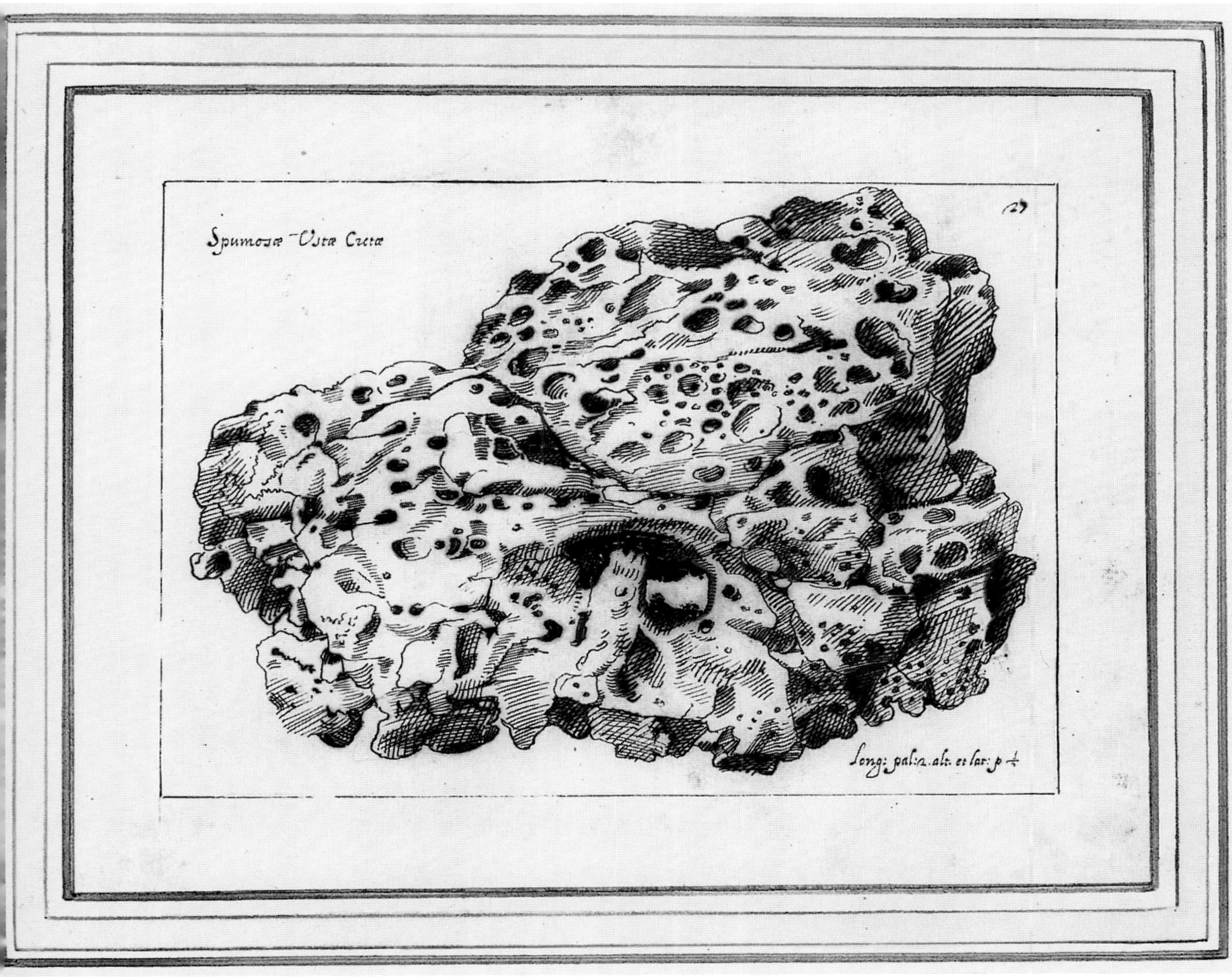

158

159. Two unrelated specimens: baked angiosperm leaf impression and unidentified rock fragment

Windsor, RL 25703, *Fossils*, vol. XVII, fol. 9

Watercolour and bodycolour and pen and ink, over traces of black chalk; measurements not given; the sheet is marked with a cross (lower left corner)

183 × 248 mm

NUMBERING: [ink, top right corner] *7*

ANNOTATION: [ink] *Coralloides*

The left-hand specimen, which is described as *coralloides*, represents an angiosperm leaf impression in a rock (see also **181** and **183**) which has probably been baked by an underground fire, and thus reddened.

The right-hand specimen, lacking an identification, is more enigmatic and may be a mixture of clay and ironstone. It is unclear why these two specimens have been placed together, but perhaps they were from the same locality.

It is likely that the word *coralloides* is an adjective referring to the colour of the left-hand specimen.

REFERENCES: 17th-cent. Index, p. 370

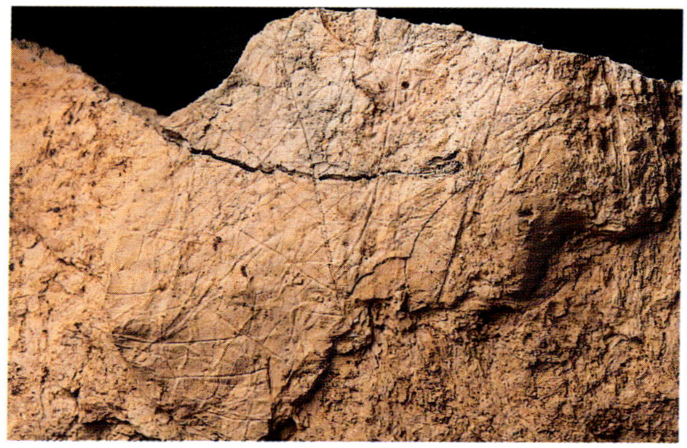

Comp. fig. 159. Baked clays with angiosperm leaf impression, from underground fire from Rosaro

160. Baked clay

Windsor, RL 25707, *Fossils*, vol. XVII, fol. 13

Watercolour and bodycolour, over traces of black chalk

188 × 252 mm

NUMBERING: [ink, top right corner] *11*

ANNOTATION: [ink, top] *Croceum Punctis Albis*
[bottom] *long: pal: 1. lat: pal: ½ alt: pal: 1.*

As noted in the inscription, this specimen shows an unusual mixture of colours from dark grey through to yellows and reds, with small white specks. Together with the pumice-like appearance of the lower part of the specimen, these colours may indicate that this was burnt clay from one of the underground fires (see p. 292). The small white specks may originally have been pyrite, altered by the burning process.

REFERENCES: 17th-cent. Index, p. 371

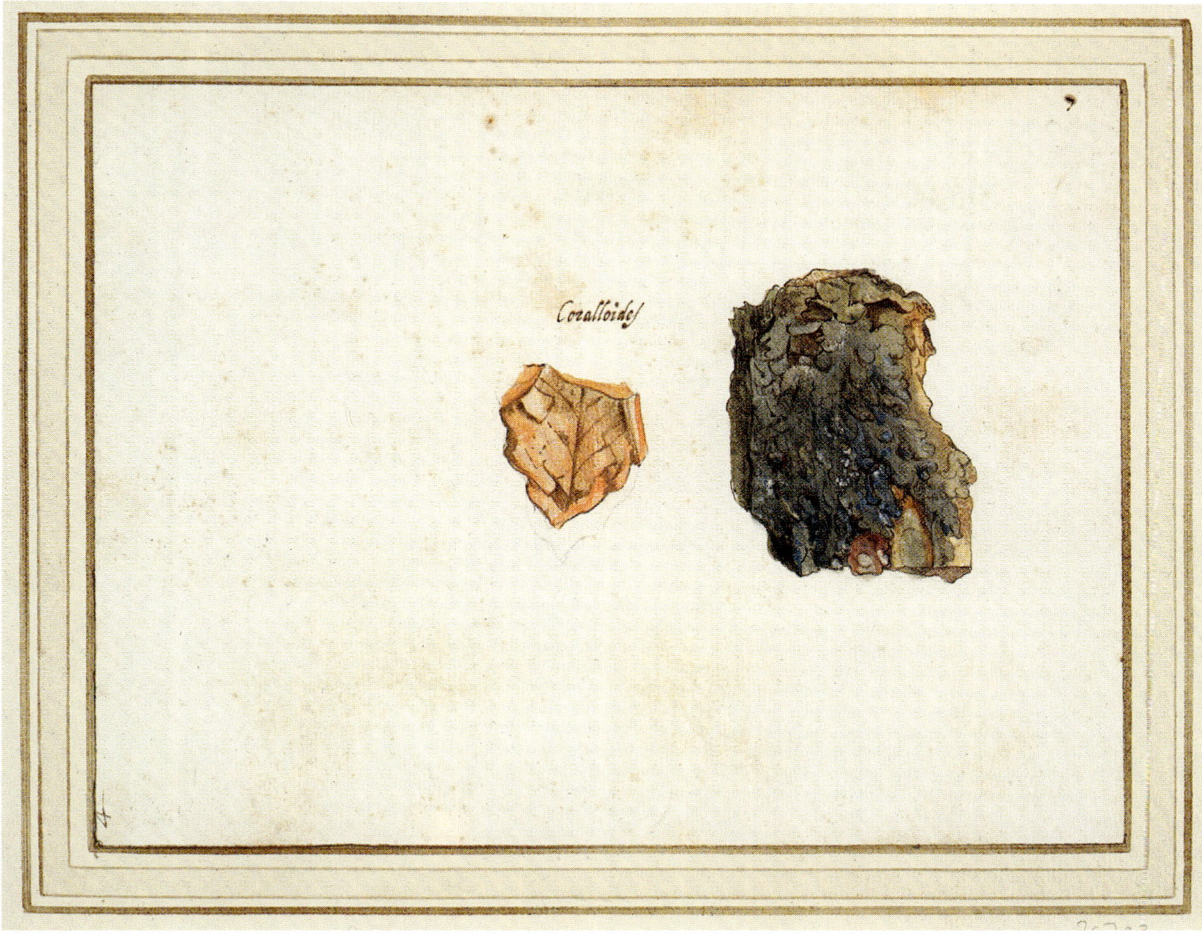

159

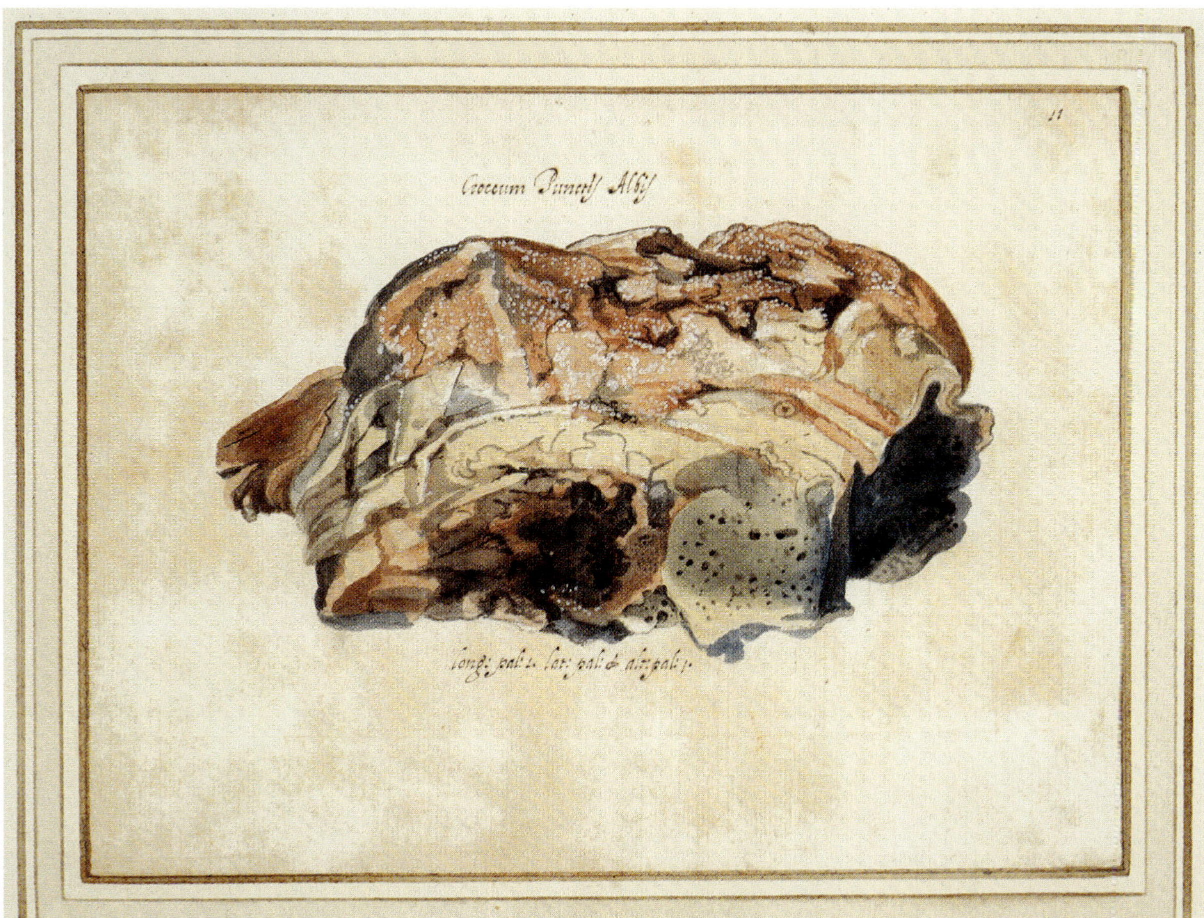

160

161. Baked clay

Windsor, RL 25714, *Fossils*, vol. XVII, fol. 20

Watercolour and touches of white heightening, over black chalk

184 × 247 mm

NUMBERING: [ink, top right corner] *18*

ANNOTATION: [ink, top] *lateritium Foraminosum*
[bottom] *long. lat: pal: 1. Crass. dig. 4*

A much pitted laterite. All these instances of *lateritia* are probably baked clays from underground fires. This example may originally have been clay with organic fragments and some iron, which has hardened, cracked and reddened as a result of baking. It is also described as having holes.

REFERENCES: 17th-cent. Index, p. 371

161

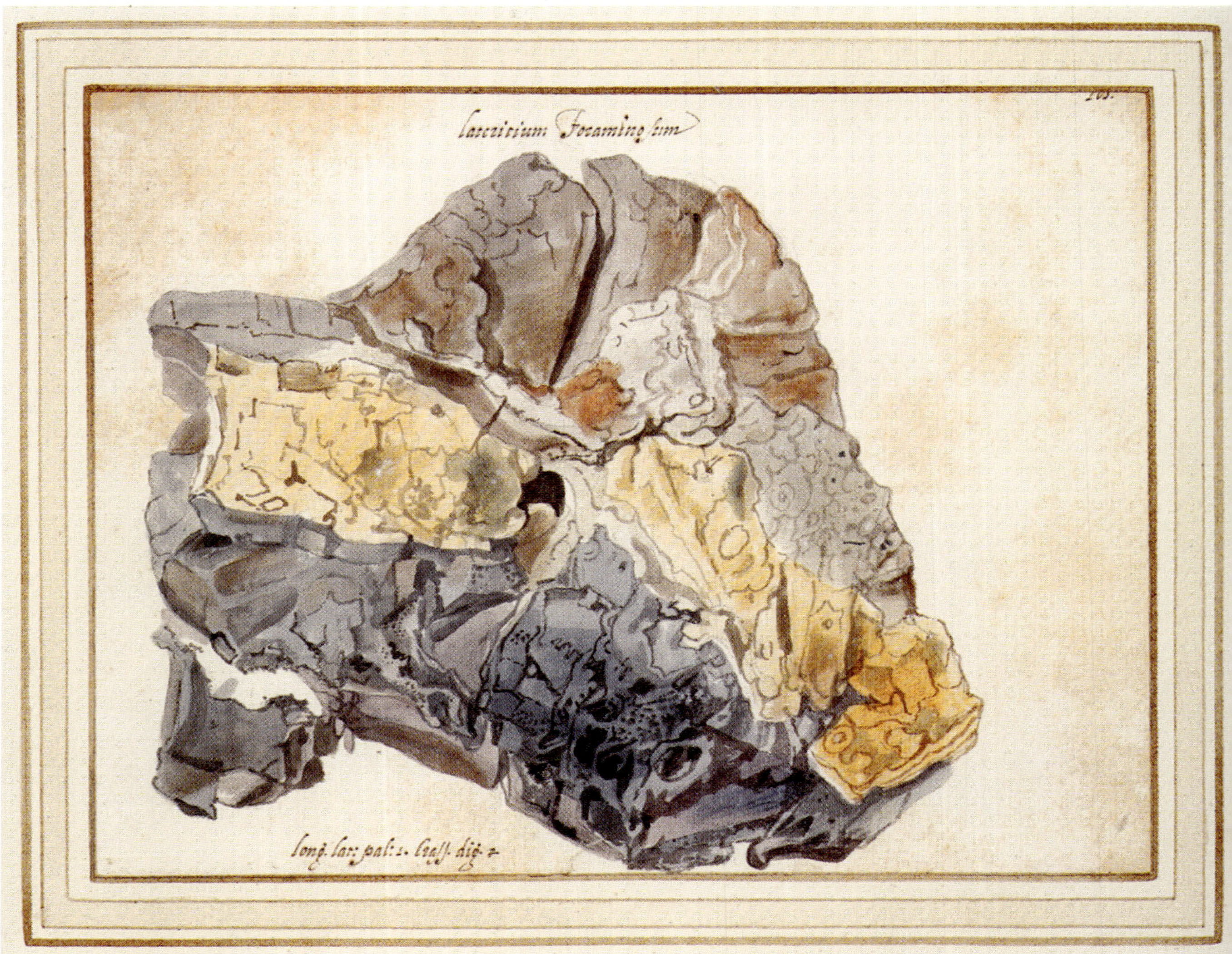

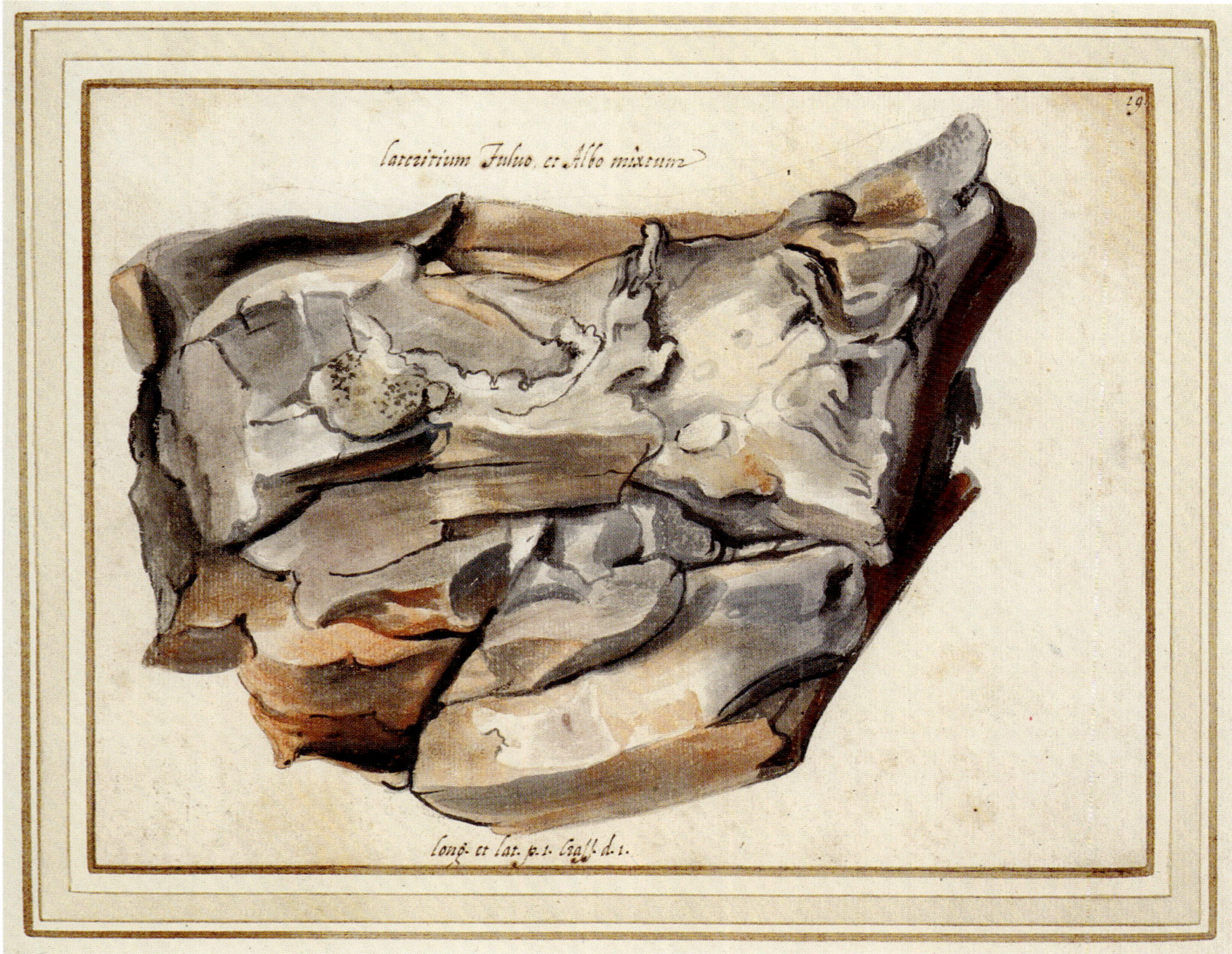

162

162. Baked clay

Windsor, RL 25715, *Fossils*, vol. XVII, fol. 21

Watercolour and bodycolour, over black chalk

180 × 243 mm. *Watermark:* Bird 41B

NUMBERING: [ink, top right corner] *19*

ANNOTATION: [ink, top] *lateritium Fulvo et Albo mixtum* [bottom] *long. et lat. p. 1. Crass. d. 1.*

This specimen of *lateritium* is described as being 'mixed with yellow and white'. It is probably a clay which has been baked red by the underground fire.

REFERENCES: 17th-cent. Index, p. 371

163. Baked clay

Windsor, RL 25716, *Fossils*, vol. XVII, fol. 21

Watercolour and bodycolour, over black chalk

184 × 248 mm

NUMBERING: [ink, top right corner] *20*

ANNOTATION: [ink, top] *lateritium ad' Gagat: Tendens* [bottom] *long: latitudo p. 1. Crass. dig. 5*

A specimen of *lateritium* described as 'tending towards jet' (compare **62** and **192**). It appears originally to have been a plant-rich clay which, as a result of baking by the underground fire, has not only reddened, but also vesiculated, that is, developed holes (or become pumice-like).

REFERENCES: 17th-cent. Index, p. 371

163

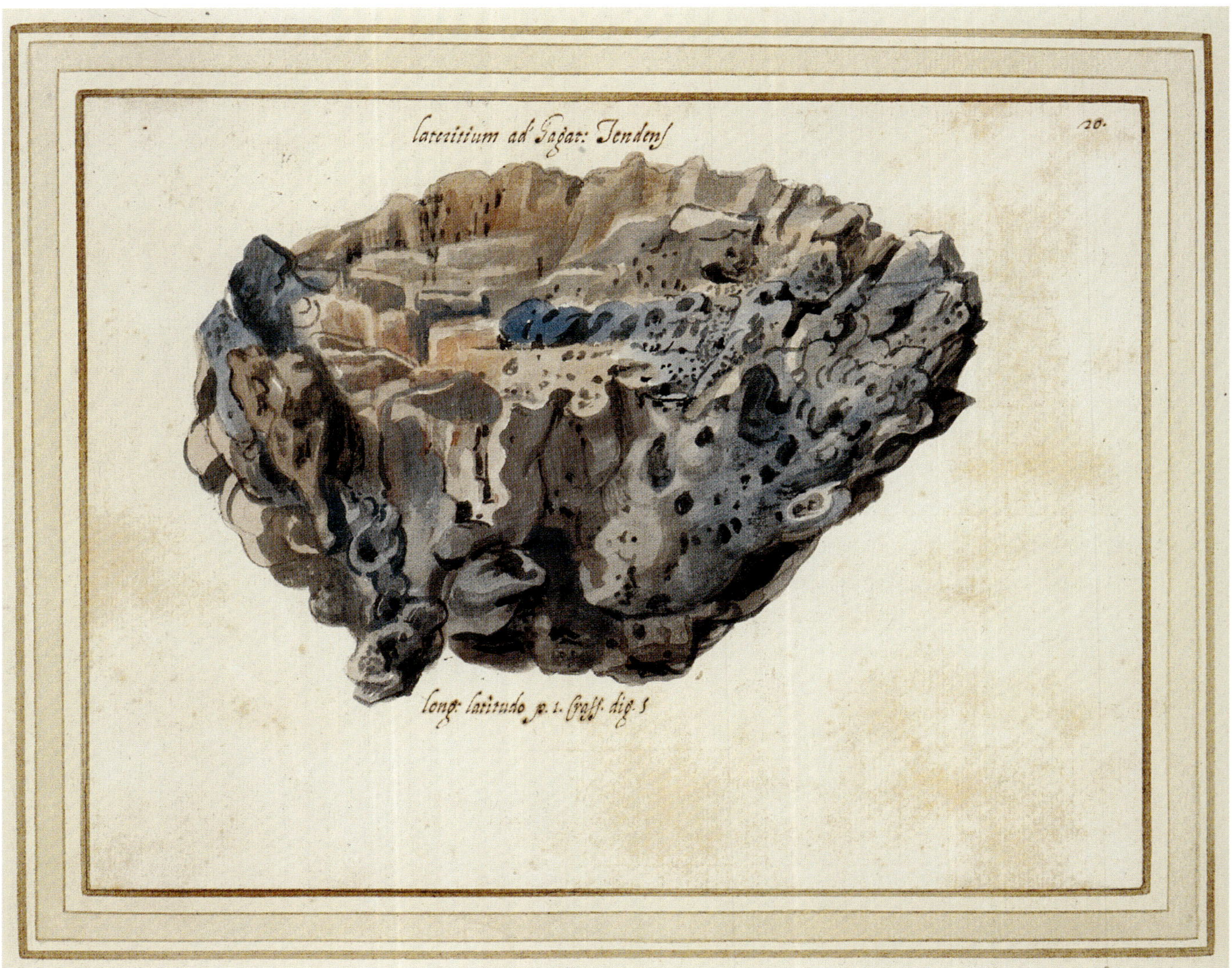

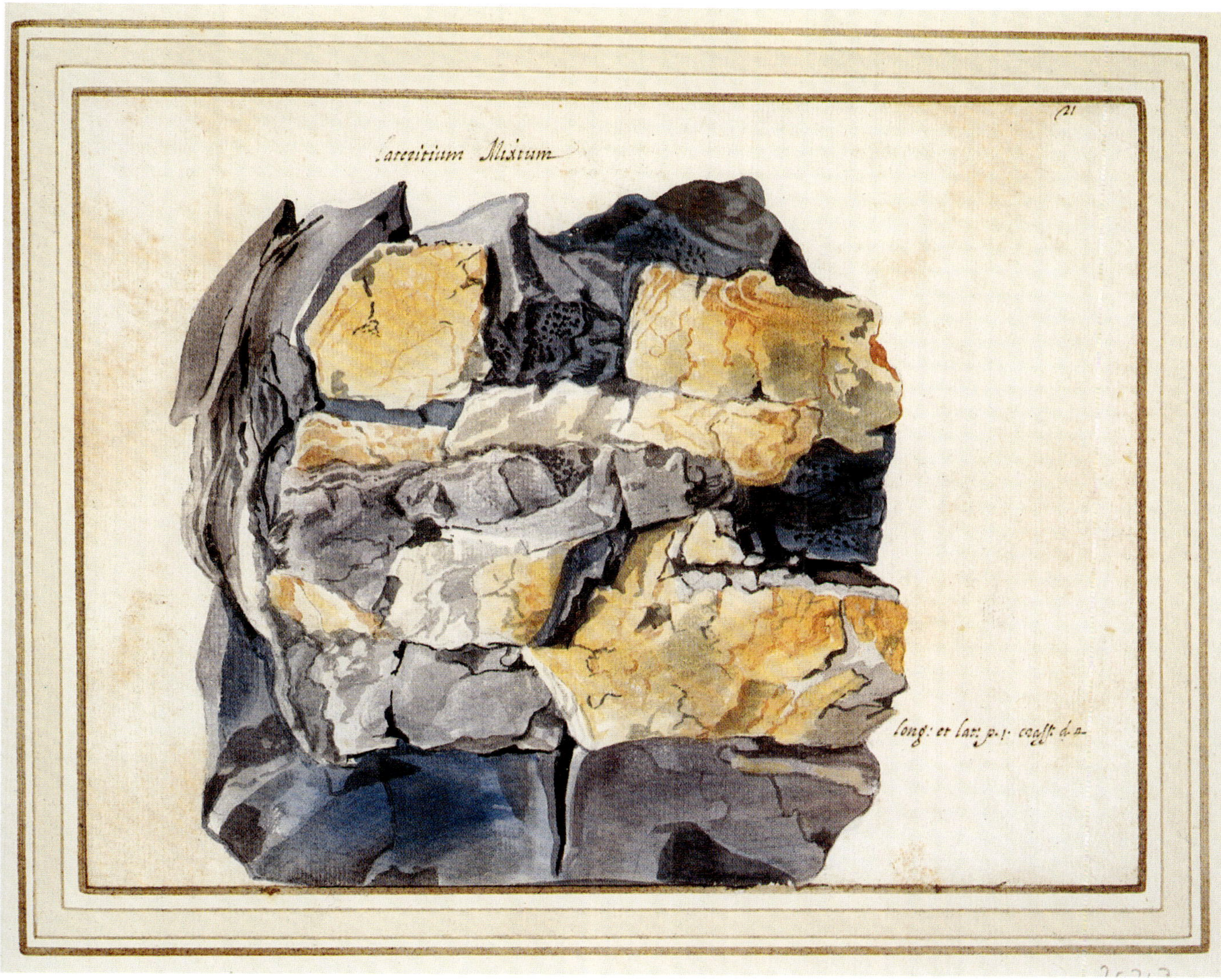

164. Baked clay 164

Windsor, RL 25717, *Fossils*, vol. XVII, fol. 22

Watercolour and bodycolour, over traces of black chalk

190 × 254 mm

NUMBERING: [ink, top right corner] *21*

ANNOTATION: [ink, top] *lateritium Mixtum*
[bottom] *long: et lat: p. 1. crass: d. 4*

The specimen is described as a 'mixed *lateritium*' and is another example of an organic-rich clay which has been partially baked. The outer surface of the specimen shows intense cracking and colour alteration.

REFERENCES: 17th-cent. Index, p. 371

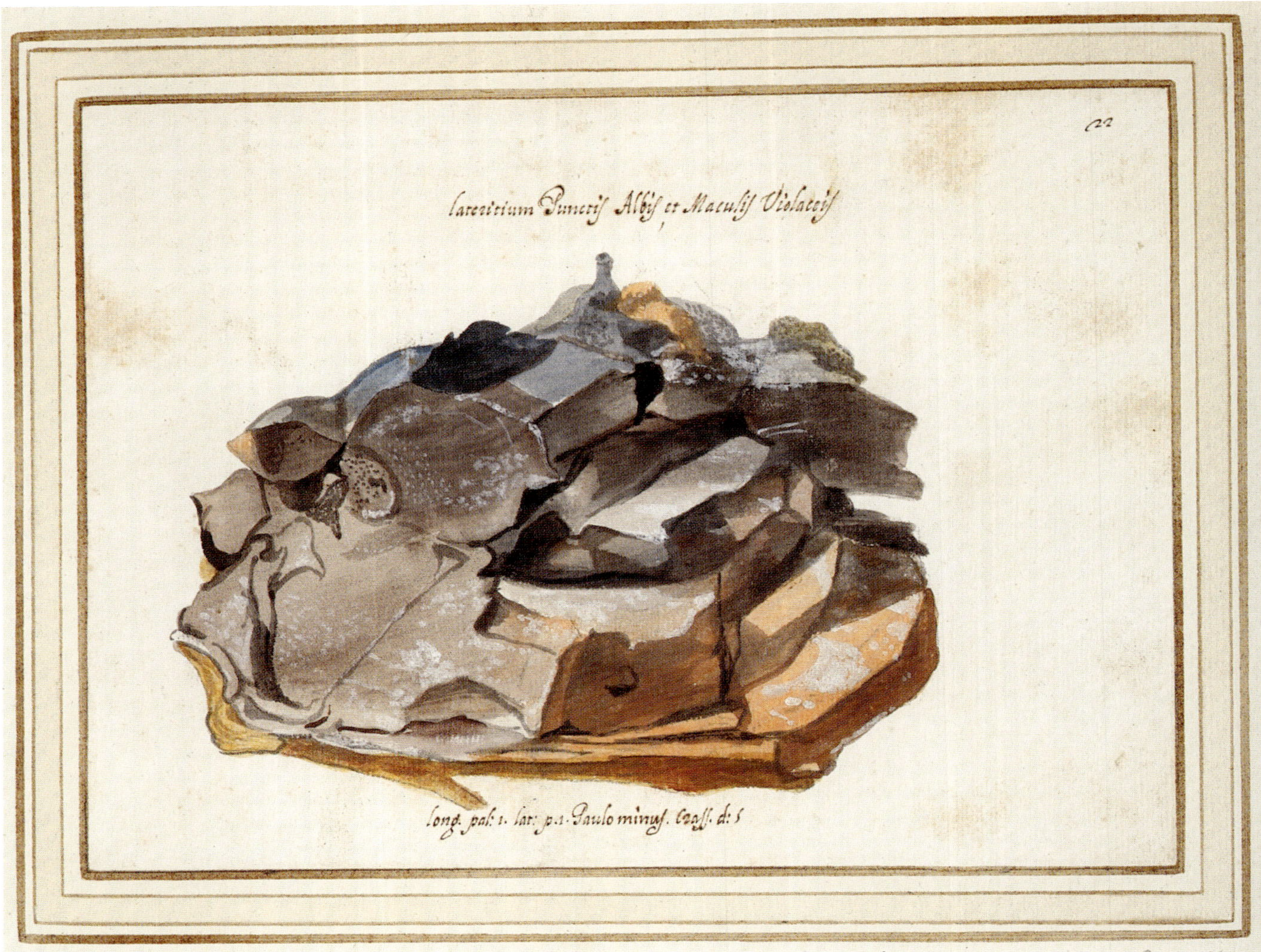

165

165. Baked clay

Windsor, RL 25718, *Fossils*, vol. XVII, fol. 23

Watercolour and bodycolour, over traces of black chalk

177 × 249 mm

NUMBERING: [ink, top right corner] 22

ANNOTATION: [ink, top] *lateritium Punctis Albis, et Maculis Violaceis* [bottom] *long. pal: 1. lat: p. ½. Paulo minus. crass. d: 5*

This specimen is described as having 'white specks and violet spots'. It is another example of a partially baked organic-rich clay.

REFERENCES: 17th-cent. Index, p. 371

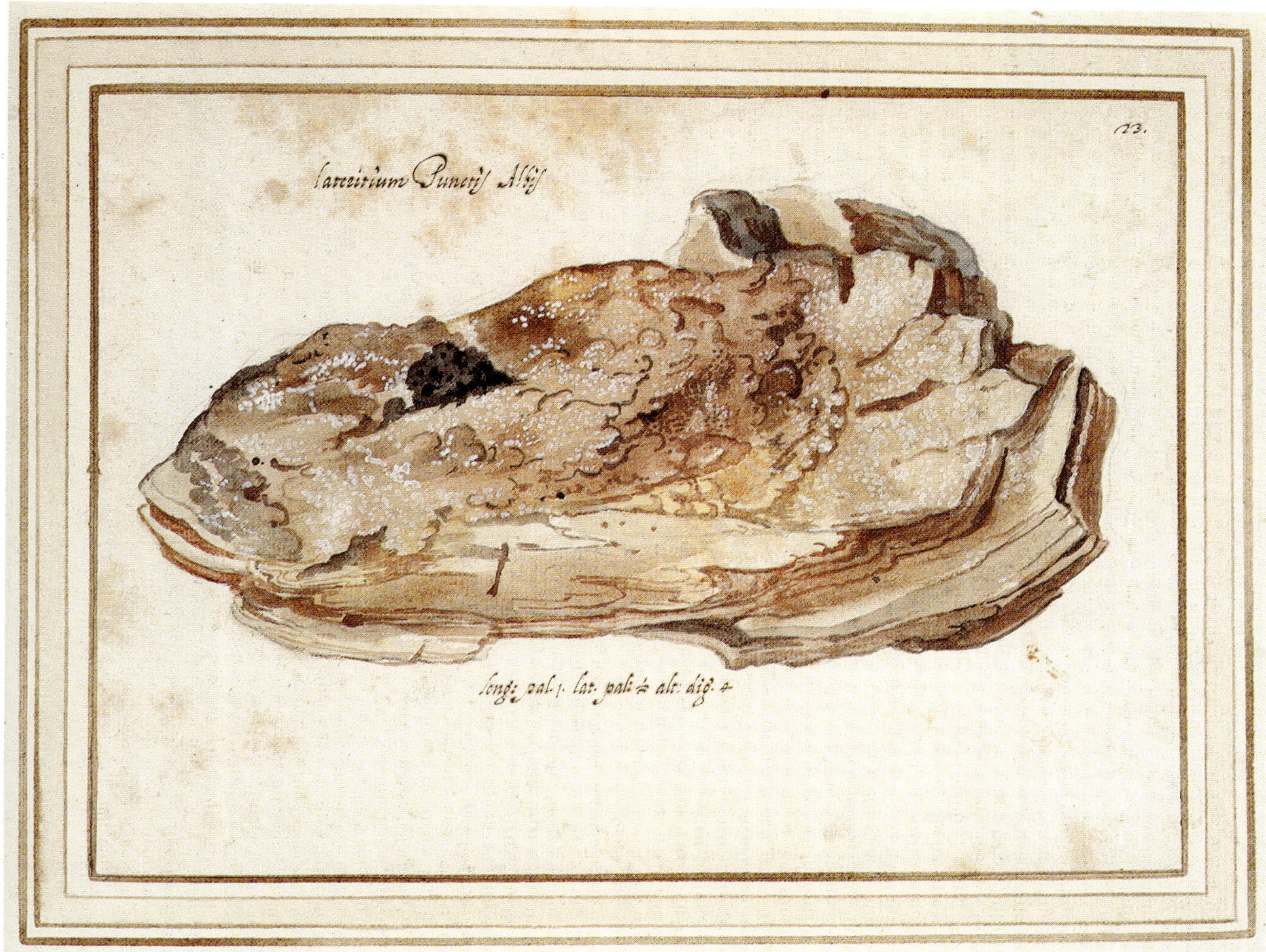

166

166. Baked clay

Windsor, RL 25719, *Fossils*, vol. XVII, fol. 24

Watercolour and bodycolour, over black chalk

178 × 248 mm

NUMBERING: [ink, top right corner] *23*

ANNOTATION: [ink, top] *lateritium Punctis Albis* [bottom] *long: pal. 1. lat. pal: ½ alt: dig. 4*

Another specimen described as having white specks. It shows laminations at its base. The upper part shows the 'flowery' (or 'frothy') formation characteristic of massive pyrite. The baking has turned the rock red–brown and caused the pyrite to decay, resulting in the white spots.

REFERENCES: 17th-cent. Index, p. 372

167. Baked clay

Windsor, RL 25720, *Fossils*, vol. XVII, fol. 25

Watercolour, over black chalk

178 × 257 mm

NUMBERING: [ink, top right] *24*

ANNOTATION: [ink, top] *lateritium Strijs Purpureis* [bottom] *long. pal. 1½ lat. pal. 1. alt: dig. 4*

Described as '*lateritium* with purple striations', this is a well-laminated clay. It has been baked thoroughly at the base, resulting in increased reddening.

REFERENCES: 17th-cent. Index, p. 372

167

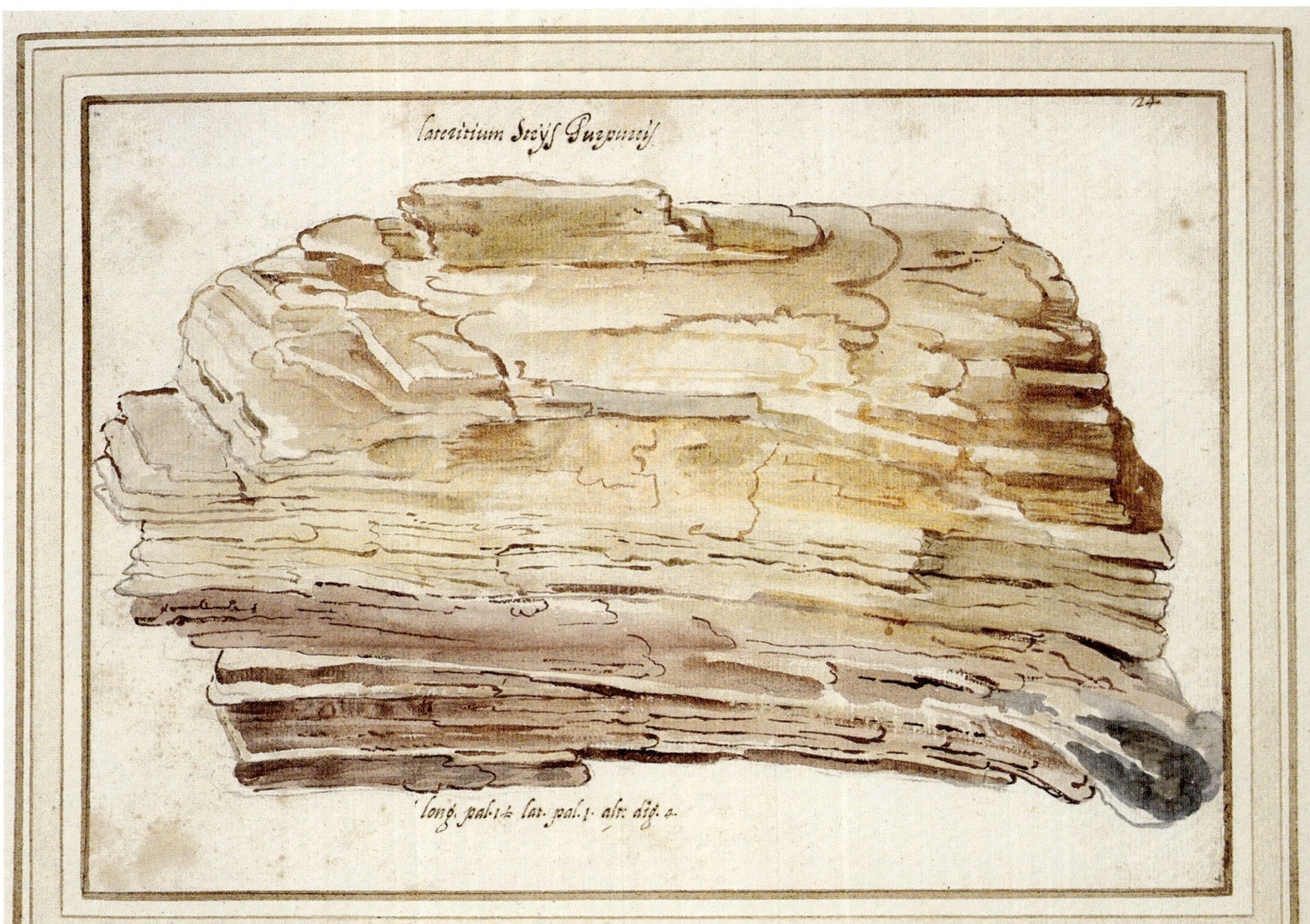

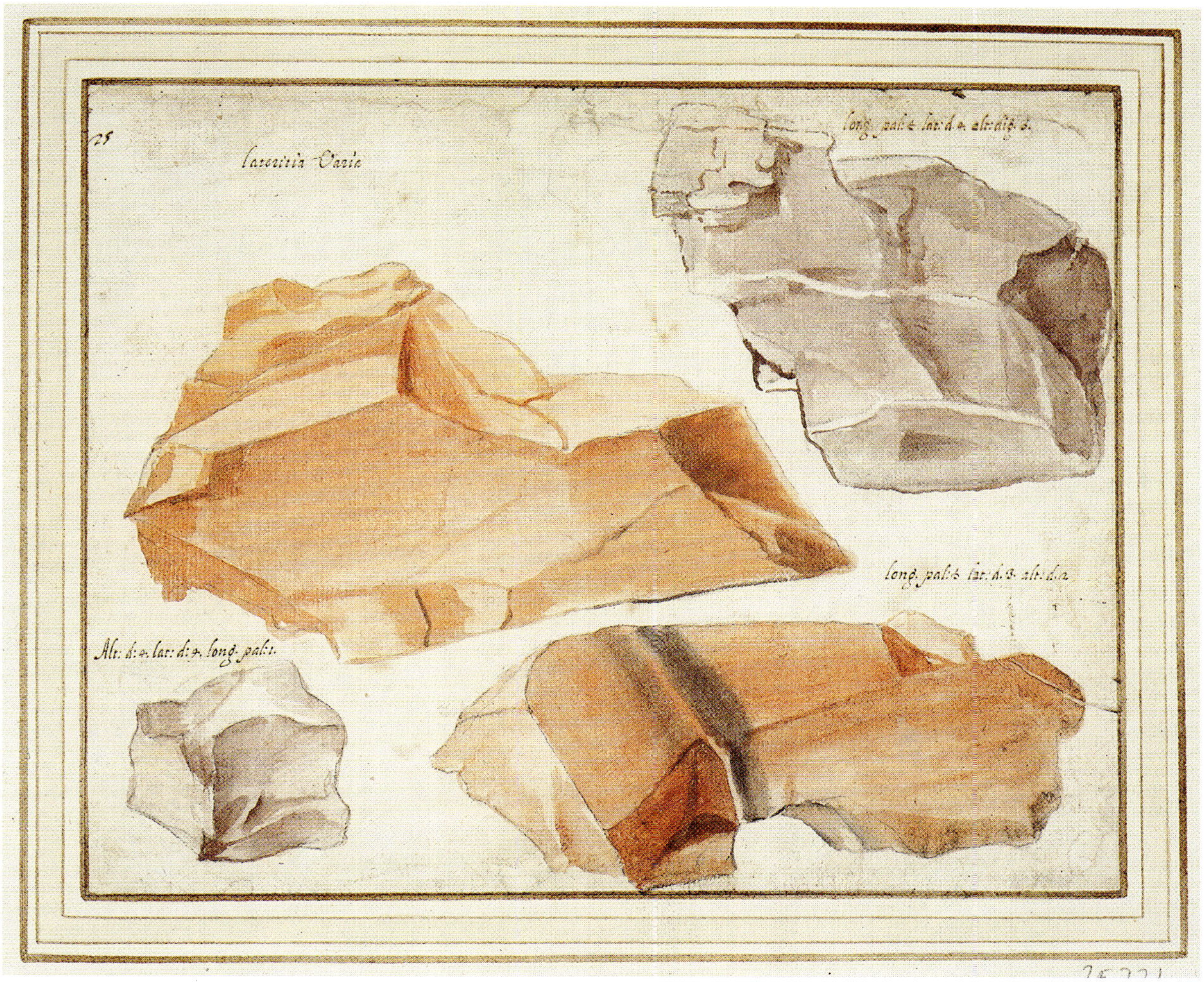

168

168. Three examples of baked clay

Windsor, RL 25721, *Fossils*, vol. XVII, fol. 26

Watercolour, over black chalk; measurements are given for only three specimens

205 × 261 mm

NUMBERING: [ink, top left] *25*

ANNOTATION: [ink, top left] *lateritia Varia*
[top right] *long. pal: ½. lat: d. 4. alt: dig. 3.*
[lower right] *long. pal: ½ lat: d. 3. alt: d. 2*
[lower left] *Alt: d: 4. lat: d: 4. long. pal: 1.*

A variety of *lateritia*. The main specimen of bright-red baked clay is shown in two views. The two other mauve-coloured specimens are less intensely baked.

These drawings appear less finished than the others in the series. The sheet has been damaged at some stage (possibly by water) and has been repaired along the upper and lower edges.

REFERENCES: 17th-cent. Index, p. 372

169. Baked clay

Windsor, RL 25722, *Fossils*, vol. XVII, fol. 27

Watercolour and bodycolour, over traces of black chalk

188 × 261 mm

NUMBERING: [ink, top right corner] *26*

ANNOTATION: [ink, top] *lateritium Violaceo Mixtum* [bottom] *long. lat: Crass. pal: 1.*

Described as a '*lateritium* mixed with violet', the specimen represents a well-laminated organic-rich clay partially baked by one of the underground fires in the region.

REFERENCES: 17th-cent. Index, p. 372

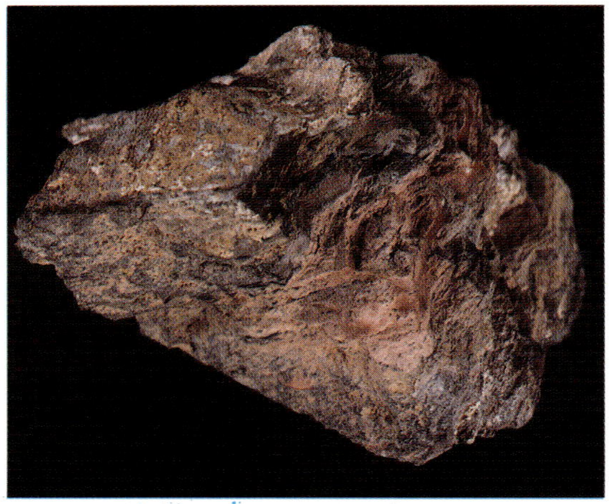

Comp. fig. 169. Baked clay from Podere Casanova

169

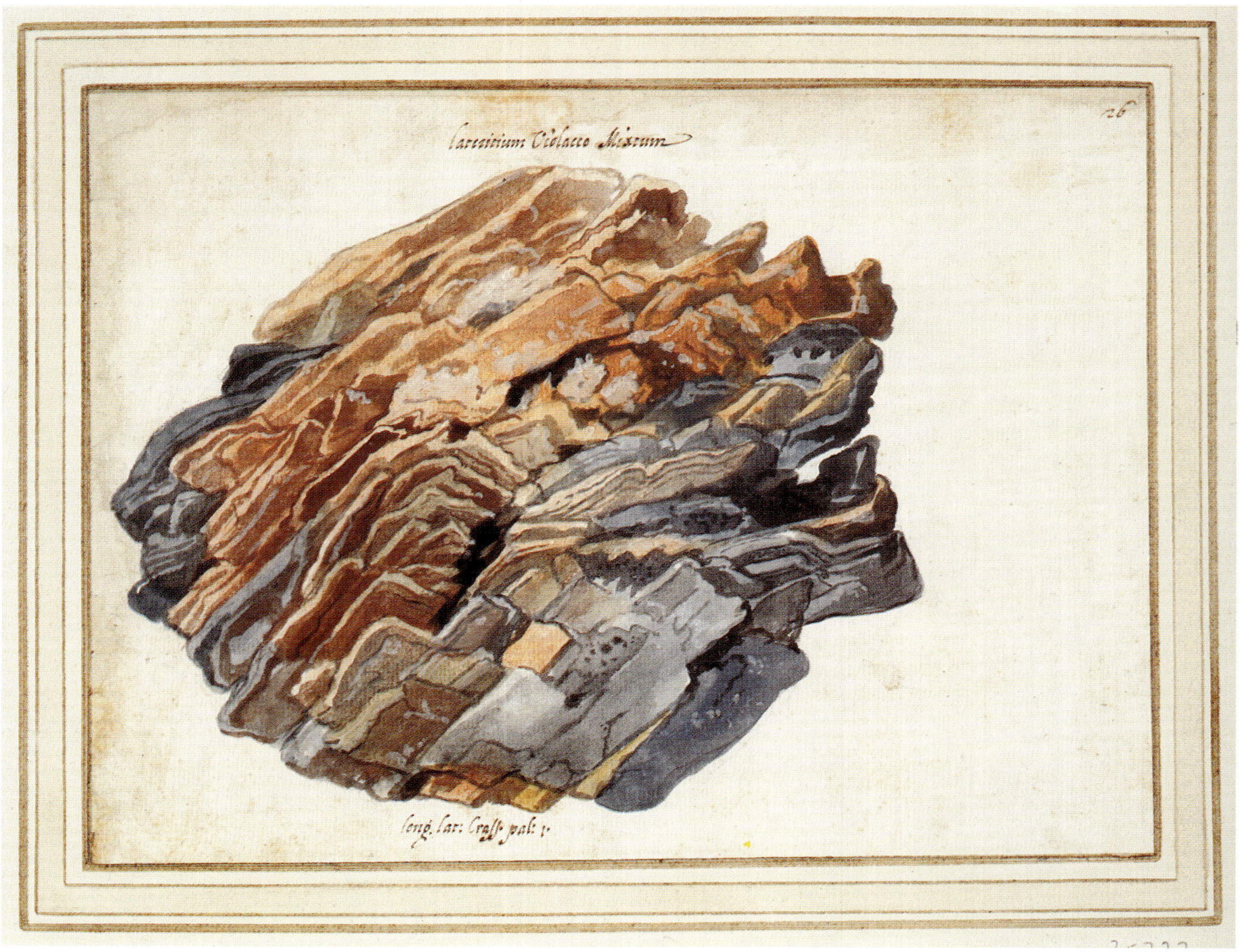

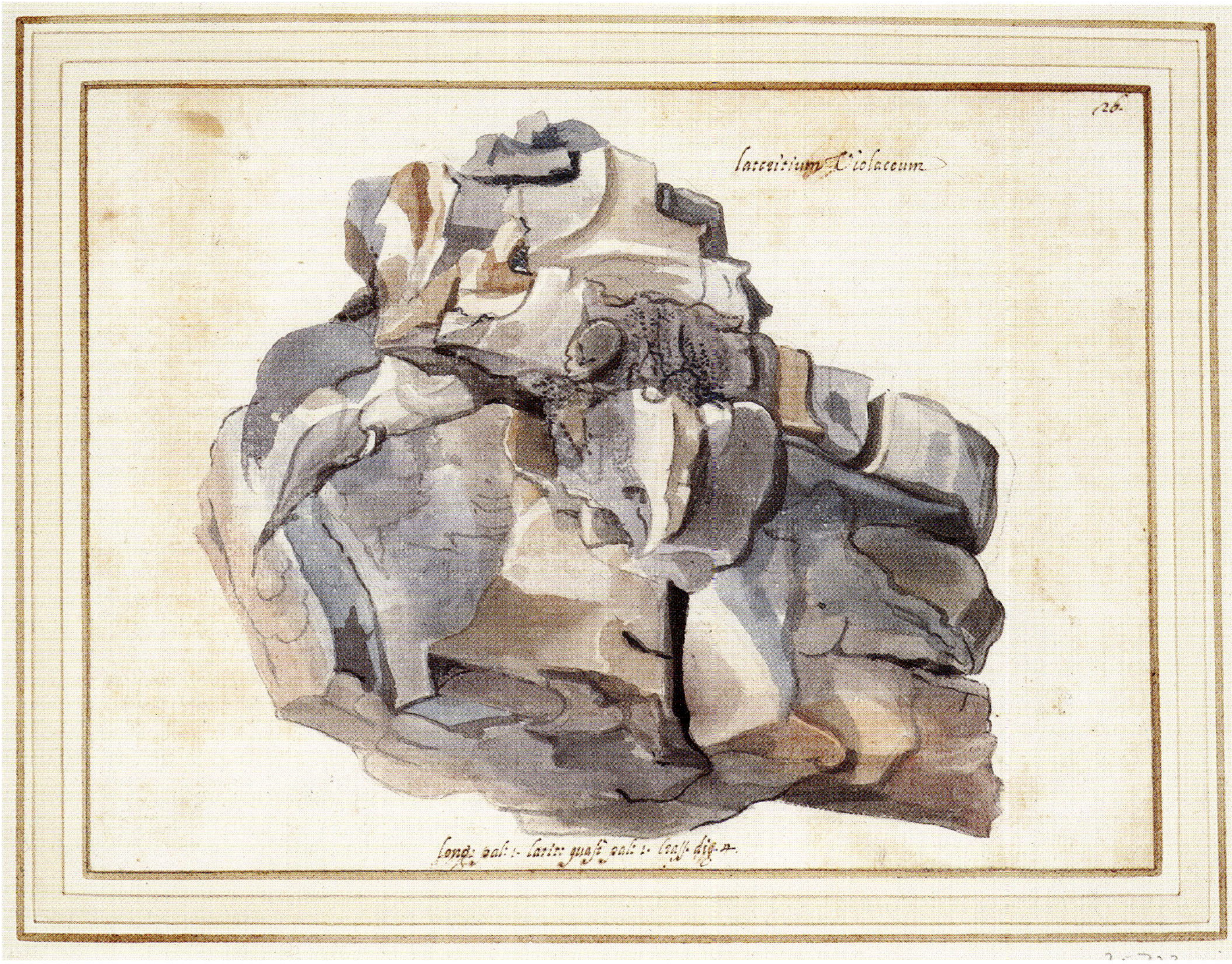

170

170. Baked clay

Windsor, RL 25723, *Fossils*, vol. XVII, fol. 28

Watercolour and bodycolour, over traces of black chalk

182 × 248 mm

NUMBERING: [ink, top right corner] *26*

ANNOTATION: [ink, top] *lateritium Violaceum*
[bottom] *long: pal: 1. latit: quasi pal: 1. Crass. dig. 4*

Described as a 'violet *lateritium*', the specimen probably represents a poorly baked organic-rich clay.

REFERENCES: 17th-cent. Index, p. 372

171. Baked pyritic clay

Windsor, RL 25724, *Fossils*, vol. XVII, fol. 29

Watercolour and bodycolour, over black chalk

179 × 240 mm

NUMBERING: [ink, top right] *28* over *27*

ANNOTATION: [ink, top] *Metallicum*
[bottom] *long: pal: 1. in cir.ª lat: p. ½ altitudo dig. 4*

Described as 'metallic', this specimen shows a central zone of massive iron pyrite. The specimen has been baked, and the clay adhering to the upper and lower surfaces of the pyrite has been reddened.

REFERENCES: 17th-cent. Index, p. 372

171

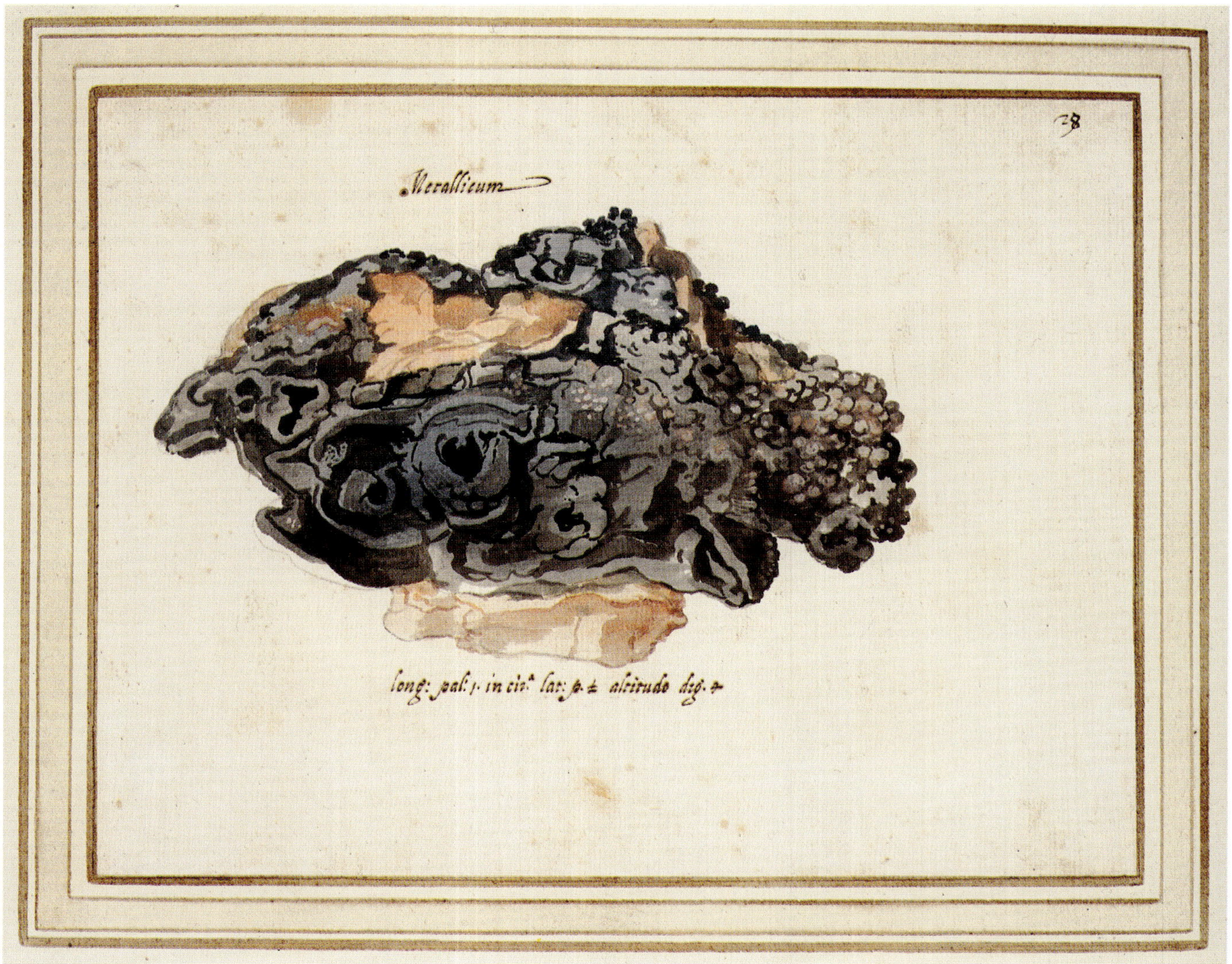

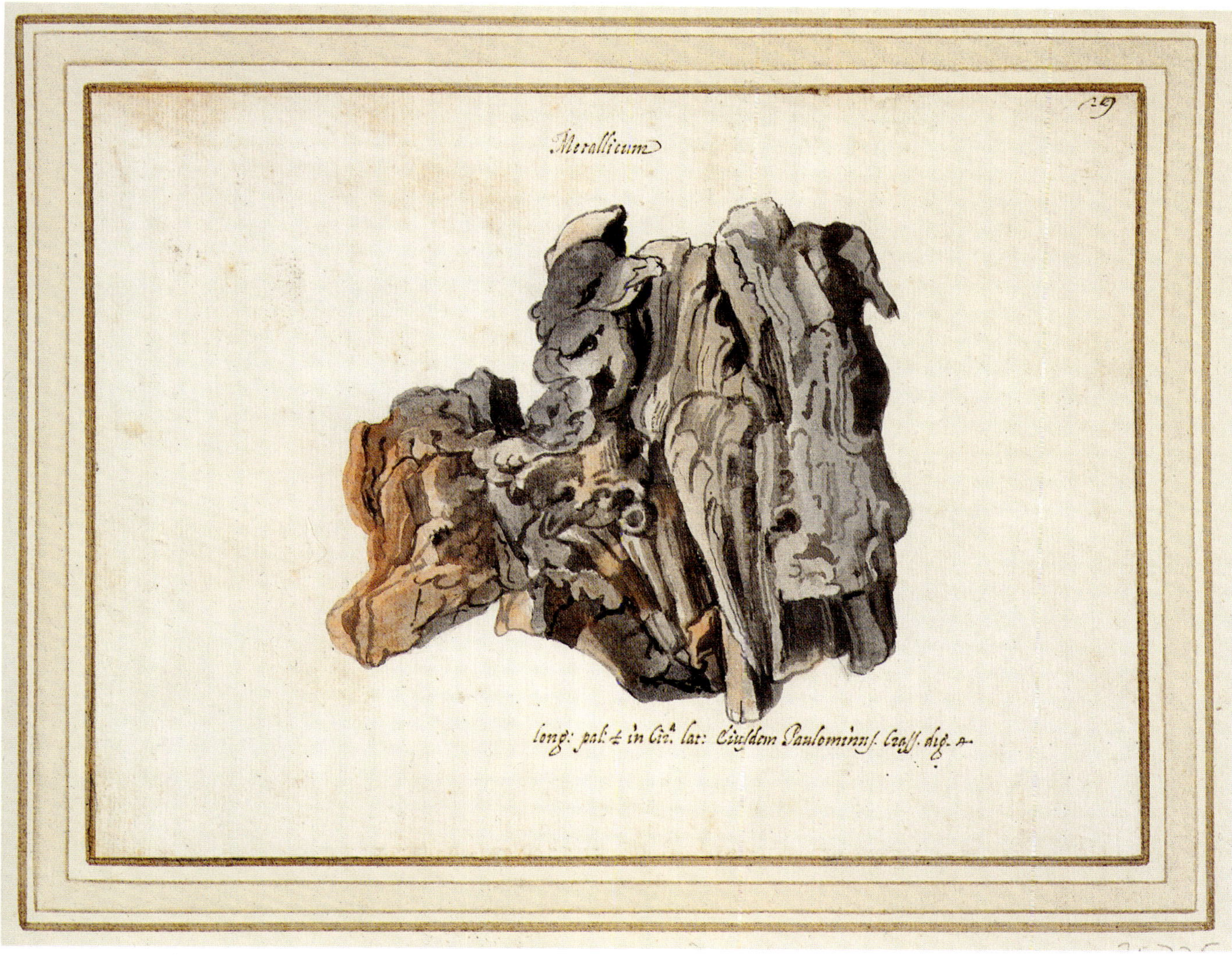

172

172. Baked lignite

Windsor, RL 25725, *Fossils*, vol. XVII, fol. 30

Watercolour and bodycolour, over traces of black chalk

179 × 245 mm

NUMBERING: [ink, top right] *29* over *28*

ANNOTATION: [ink, top] *Metallicum*
[bottom] *long: pal: ½ in Cir.ª lat: Eiusdem Paulominus. Crass. dig. 4*

Another specimen described as 'metallic', this is probably an iron-impregnated piece of wood. It has clay adhering to it, perhaps organic-rich, which has been partially baked.

REFERENCES: 17th-cent. Index, p. 372

173. Baked lignitic clay

Windsor, RL 25729, *Fossils*, vol. XVII, fol. 34

Watercolour and bodycolour, over traces of black chalk

179 × 242 mm

NUMBERING: [ink, top right corner] *33* over *32*

ANNOTATION: [ink, top] *Puniceum Varijs Maculis* [bottom] *long. pal: 1½ lat: pal: 1. Crass: dig. 4.*

Described as 'dark red with various spots', this specimen is probably a highly organic-rich baked clay. After baking, the lignite generates gas, which in turn causes vesiculation (bubbling). The outer surface of the specimen is a deep red colour, but where it is broken, the darker, unbaked centre is visible.

LITERATURE: Solinas 1989c, p. 85, fig. 62

REFERENCES: 17th-cent. Index, p. 372

173

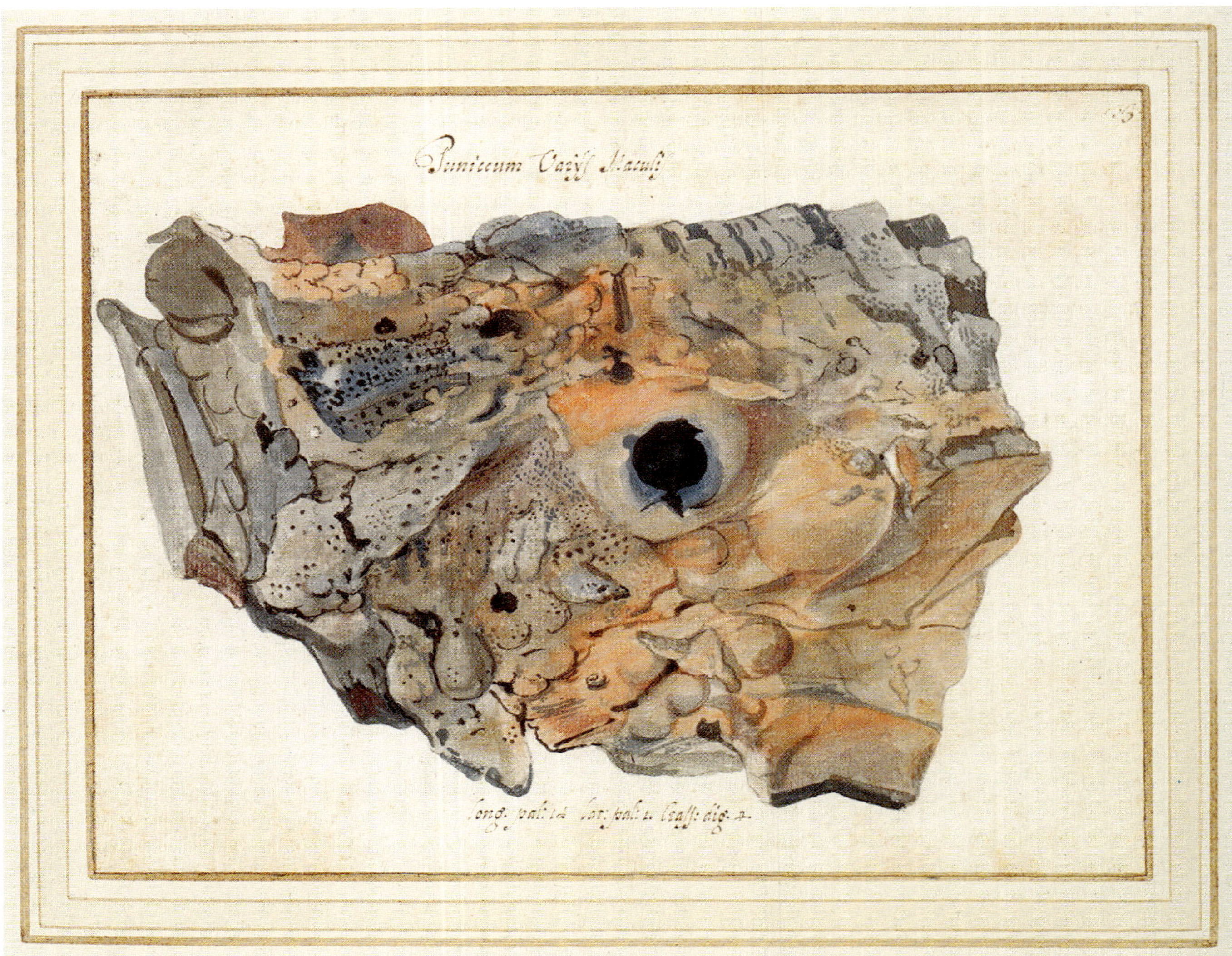

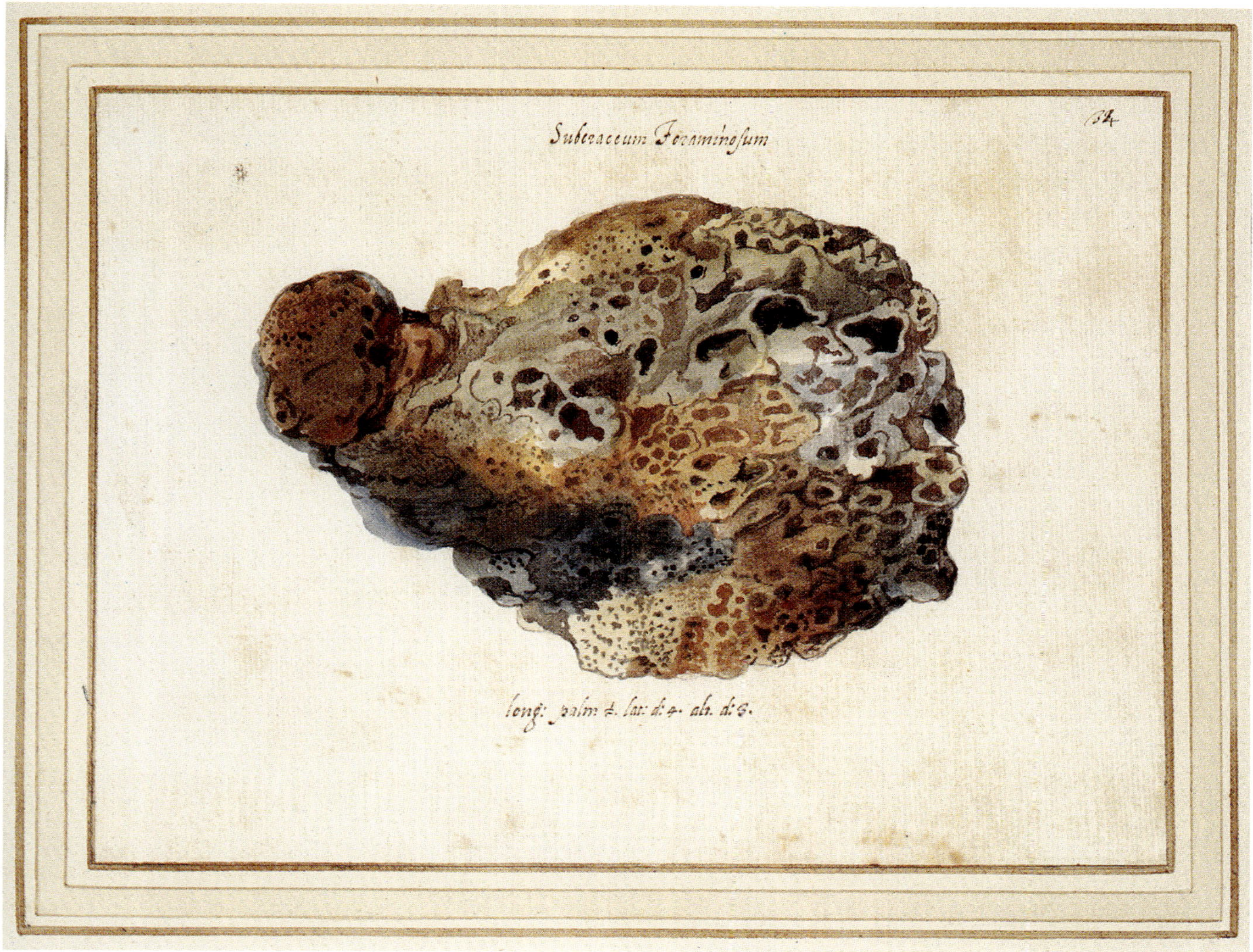

174

174. Baked lignite

Windsor, RL 25730, *Fossils*, vol. XVII, fol. 35

Watercolour and bodycolour, over traces of black chalk

174 × 242 mm

NUMBERING: [ink, top right corner] *34* over *35*

ANNOTATION: [ink, top] *Suberaceum Foraminosum* [bottom] *long: palm. ½. lat: d: 4. alt. d: 3.*

Described as 'cork-like with holes', this specimen may originally have been a lignite, which was later baked by an underground fire. The intense vesiculation gives it a sponge-like appearance.

REFERENCES: 17th-cent. Index, p. 372

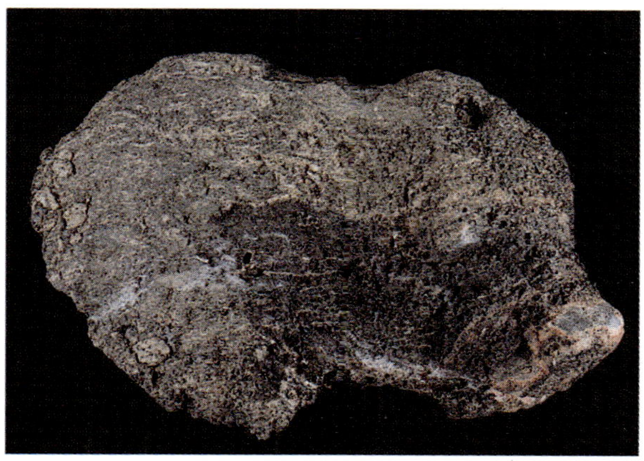

Comp. fig. 174. Baked lignite from underground fire showing spongy texture, found near Podere Casanova

175. Baked pyritic clay

Windsor, RL 25731, *Fossils*, vol. XVII, fol. 36

Watercolour and bodycolour, over black chalk

177 × 249 mm

NUMBERING: [ink, top right corner] *35* over *34*

ANNOTATION: [ink, top] *Sulphureum Ceruleo Mixtum* [bottom] *long: pal: 1: lat: d. 4. Crass. dig. 2.*

Described as a 'sulphurous [specimen] mixed with blue', this is possibly a pyrite-rich clay, partially baked and intensely altered chemically.

REFERENCES: 17th-cent. Index, p. 372

175

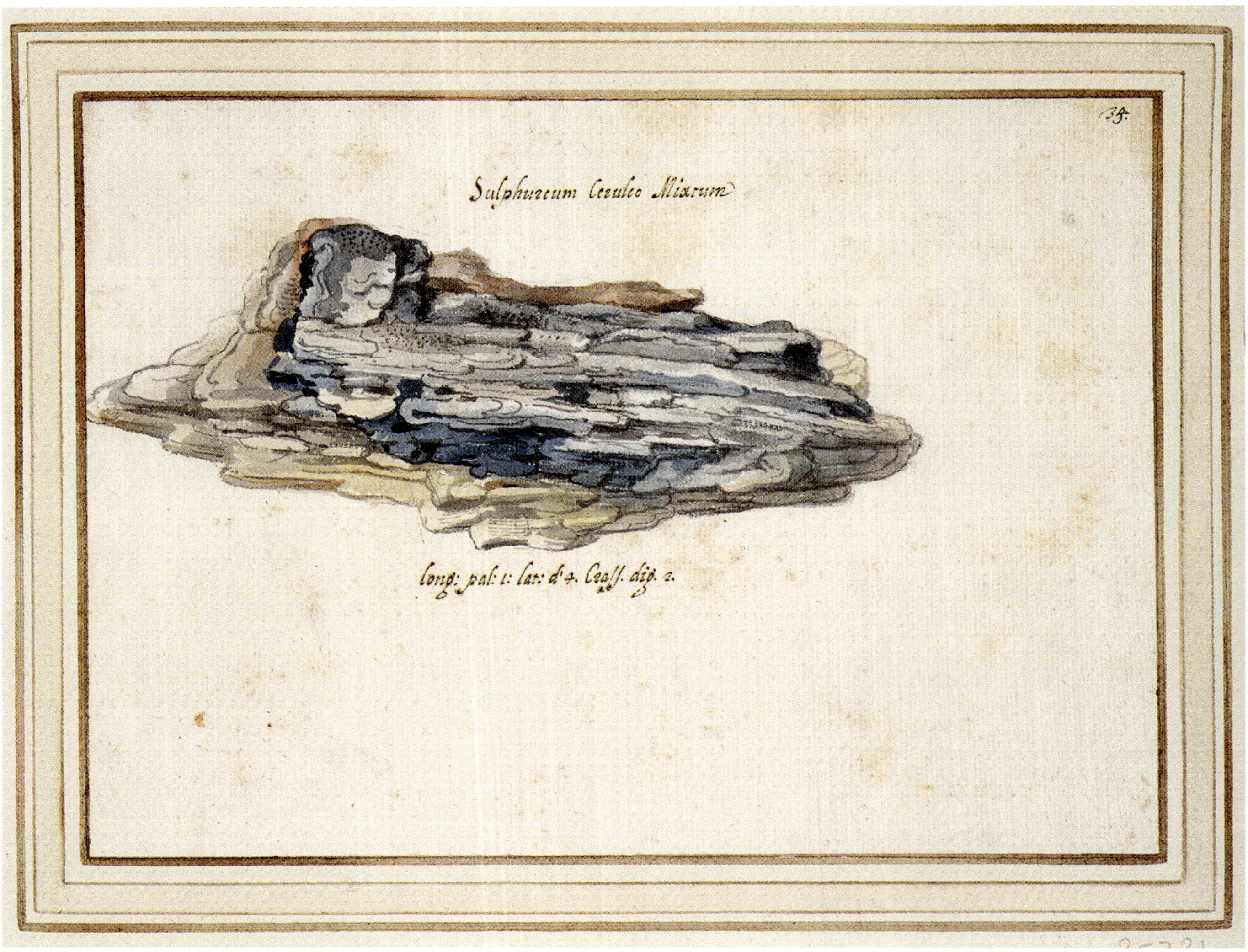

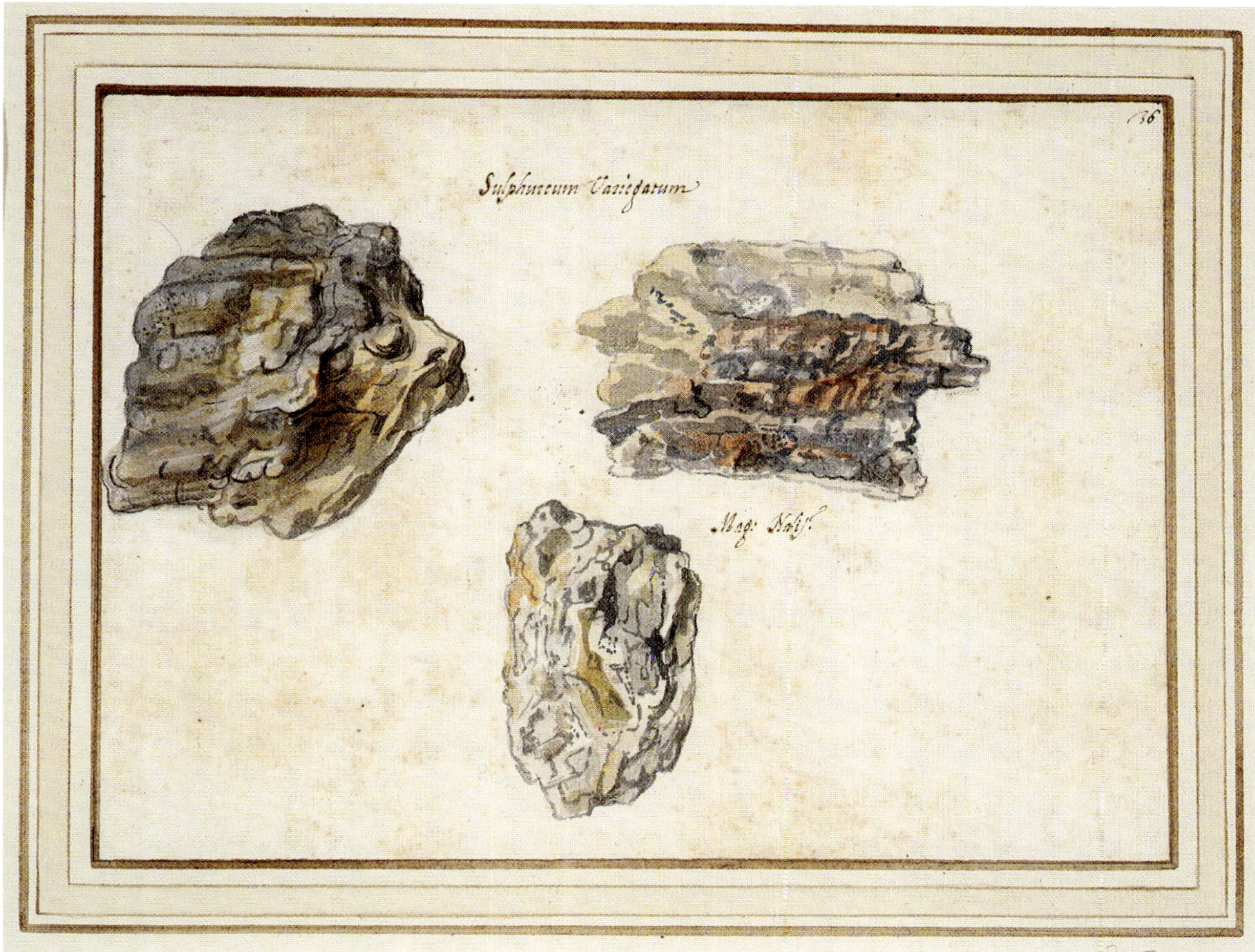

176

176. Three specimens of baked pyritic clay

Windsor, RL 25732, *Fossils*, vol. XVII, fol. 37

Watercolour and bodycolour, over traces of black chalk

171 × 240 mm

NUMBERING: [ink, top right corner] *36* over *35*

ANNOTATION: [ink, top] *Sulphureum Variegatum* [lower right] *Mag: Nalis~.*

These three specimens, described as having a 'variegated sulphurous colouring', are probably baked pyritic clays in various degrees of alteration. In particular, the top right-hand specimen shows intense alteration to a red colour, indicating that it has been burnt.

The specimens are drawn actual size.

REFERENCES: 17th-cent. Index, p. 372

177. Partially baked clay

Windsor, RL 25709, *Fossils*, vol. XVII, fol. 15

Watercolour and bodycolour, over traces of black chalk

184 × 252 mm

NUMBERING: [ink, top right corner] *13*

ANNOTATION: [ink, top] *Crustosum Multiplici Colore*
[bottom] *long: lat. pal: 1. Crass: d. 1.*

The specimen, described as 'crusty, of different colours', is probably a plant-rich clay, partially burnt by one of the underground fires. This interpretation is supported by the fact that the surface of the specimen is shown as highly cracked and fissured. The outside of the specimen is red, whereas the inside is blue and purple.

The lower and left edges of the drawing have been trimmed.

REFERENCES: 17th-cent. Index, p. 371

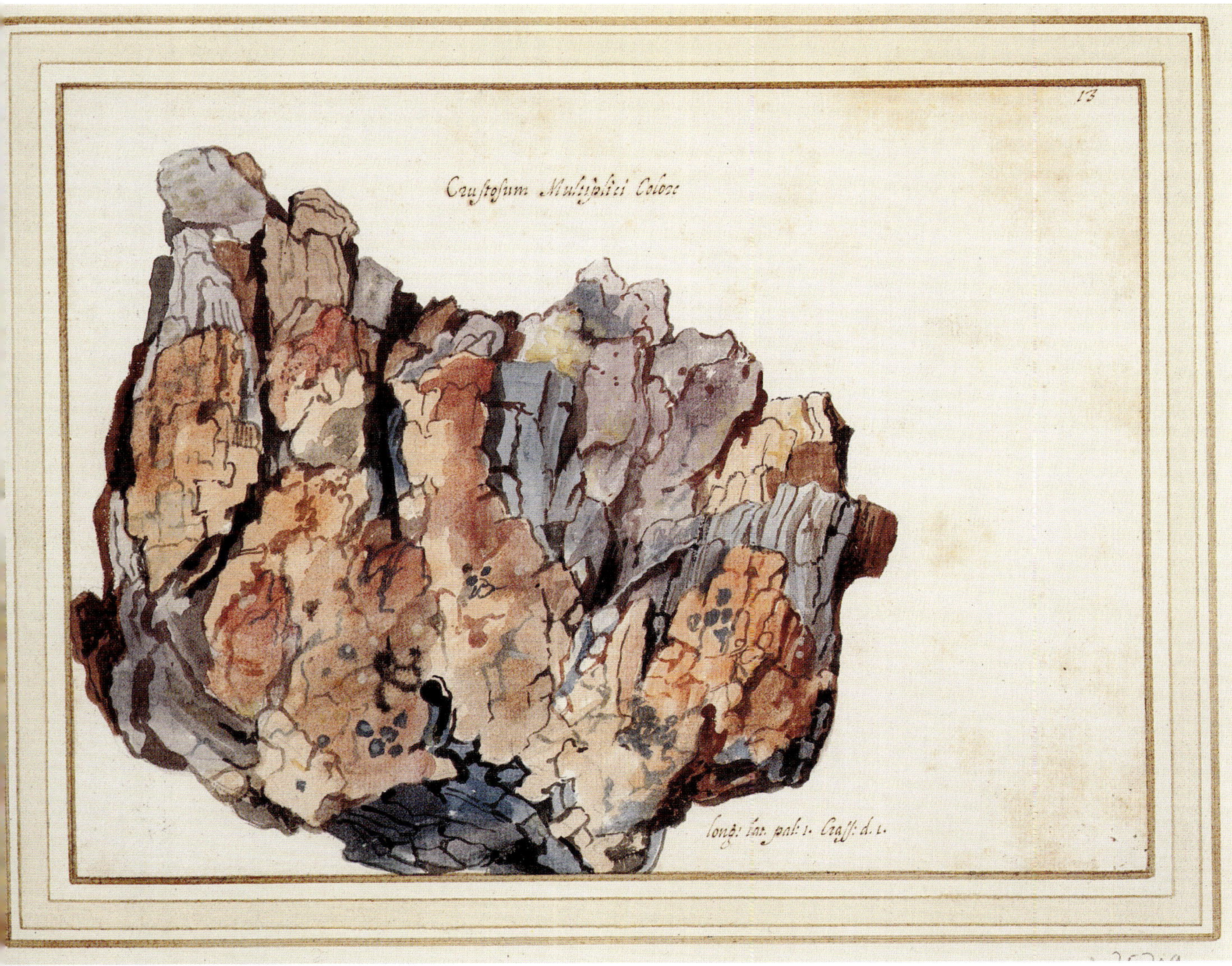

177

178. Possibly concretions, or baked clay

Windsor, RL 25593, *Fossils*, vol. XV, fol. 2

Pen and ink, over traces of black chalk; retouched in ink on mount sheet; measurements not given; ink frame line

181 × 242 mm; small holes caused by ink

NUMBERING: [ink, top left corner] *55*

ANNOTATION: [ink] *Ferruginosa Spumosa*

These two specimens are described as 'spongy and ferruginous'. They have the appearance of baked clay specimens, but it is possible they are concretions.

REFERENCES: 17th-cent. Index, p. 364

178

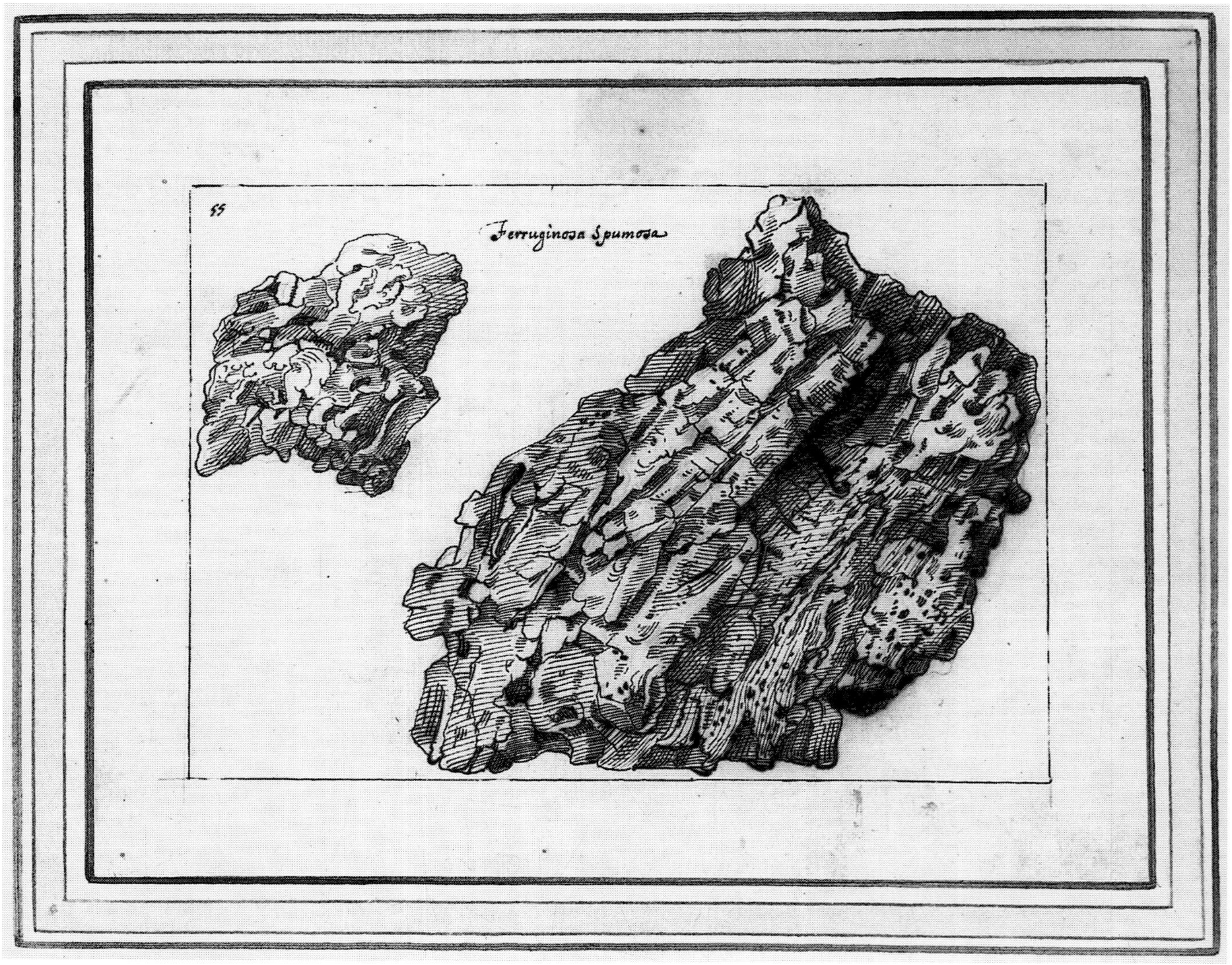

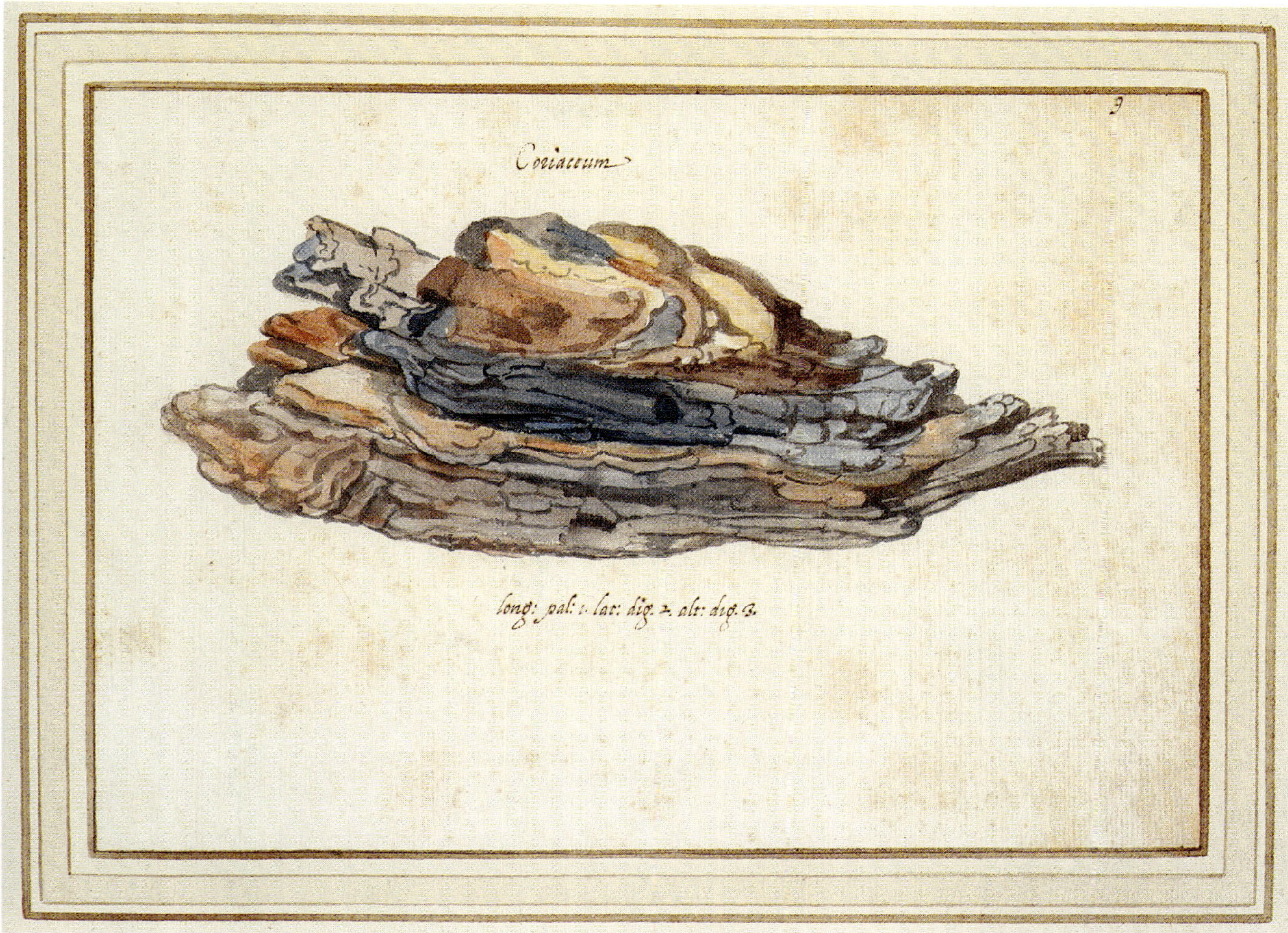

179

179. Baked or iron-rich clay

Windsor, RL 25705, *Fossils*, vol. XVII, fol. 11

Watercolour and bodycolour, over traces of black chalk

173 × 252 mm

NUMBERING: [ink, top right] *9*

ANNOTATION: [ink, top] *Coriaceum*
[bottom] *long: pal: 1. lat: dig. 4. alt: dig. 3.*

Coriaceum, 'leathery', may refer here to the flaky and layered texture of the specimen.

This appears to be a lump of rock showing good lamination (layering). The blue–grey colouring in the centre may possibly indicate an unaltered layer within the rock. Either side of this layer the rock is variously coloured from purplish-red through red to yellow.

One explanation is that the specimen has been iron- impregnated; it may also have been baked (see above, p. 292).

REFERENCES: 17th-cent. Index, p. 371

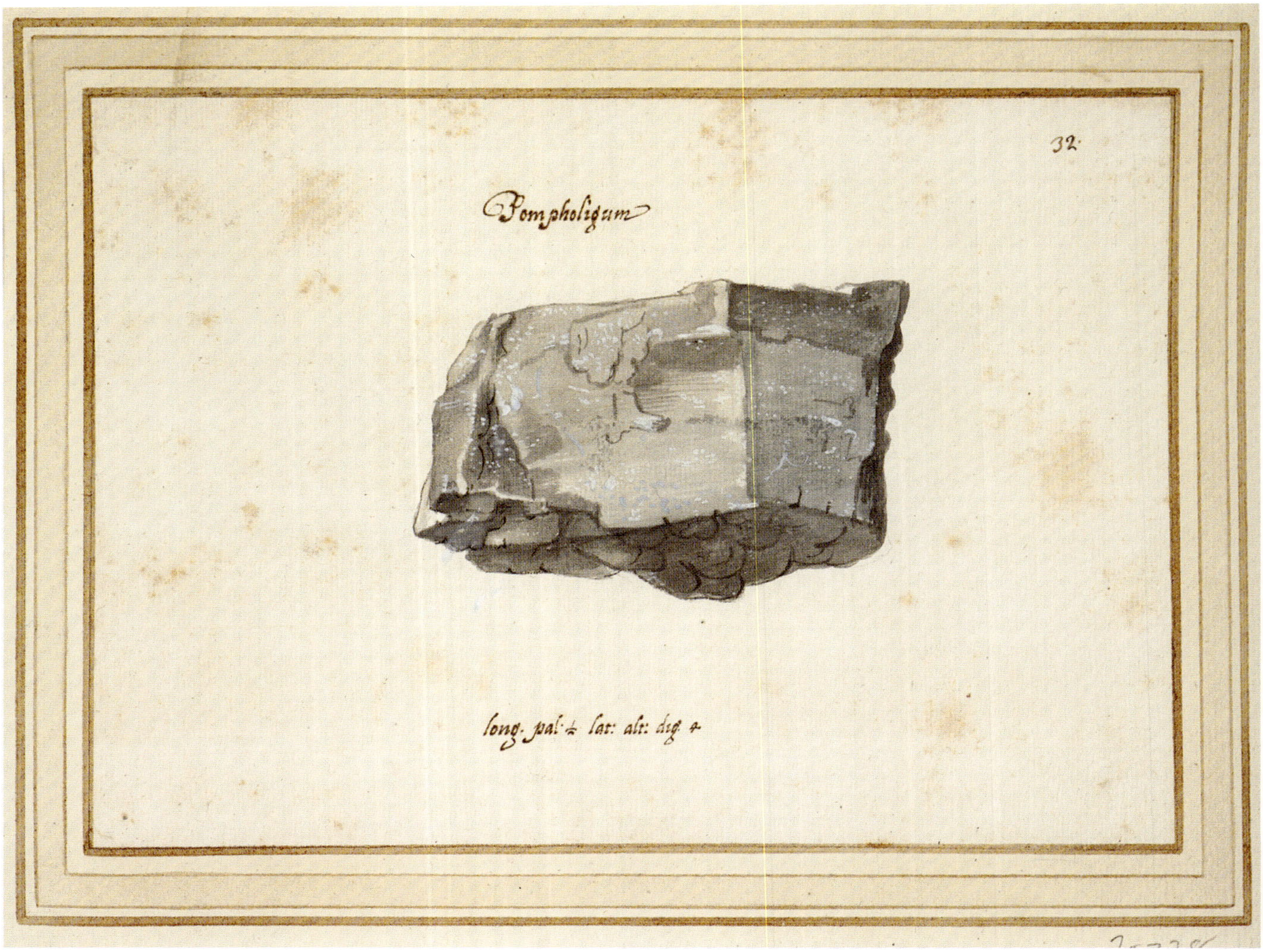

180

180. Pompholyx

Windsor, RL 25728, *Fossils*, vol. XVII, fol. 33

Watercolour and bodycolour, over traces of black chalk

169 × 244 mm

NUMBERING: [ink, top right corner] *32* over *31*

ANNOTATION: [ink, top] *Pompholigum*
[bottom] *long. pal. ½ lat: alt: dig. 4*

Given the difficulty of identifying this specimen precisely, the inscription *Pompholigum* is more than usually critical. *Pompholyx*, as both Pliny and Aldrovandi noted, means 'bubble'. It refers to the 'bubble and ash' that is a by-product from the smelting of copper, perhaps zinc oxide. It could be described as the lightest part of a sublimate, found normally in copper furnaces. Aside from the little pale grey and blue flecks on this specimen, there is little, however, that could be called 'bubble' here, though the connection with the ash produced by smelting may be seen here.

LITERATURE: Pliny *Natural History*, XXXIV, 128–30; Aldrovandi, 1648, p. 16

REFERENCES: 17th-cent. Index, p. 372

181. Possibly baked clay or a concretion

Windsor, RL 25549, *Fossils*, vol. XIV, fol. 20

Pen and ink, over traces of black chalk; retouched in ink on mount sheet; ink frame line

162 × 218 mm; small holes caused by ink

NUMBERED: [ink, upper right] *13*

ANNOTATION: [ink, upper right] *In latere est/ Suis nervis/ impressum folium* [lower right] *Creta excocta la-/teritia potius, ri/mosa, quam spumo/sa fulva sub fulva/ alicubi rubens ins/perso, et exundante/ quasi sale concreto/ præsertim in fissuris/ albo, nempe ex lapi resina/ illa preocupante coctione/ long: lat: pal: ½: alt. d. 4.*

This specimen could represent either baked clay or some form of concretion. The inscription notes that it is 'baked clay or rather a *lateritium* which is cracked and spongy, yellow or slightly yellow and red in parts and with a flowing white concretion almost like salt, especially in the fissures, a kind of stony resin (*lapiresina*), resulting from its baking'. The reference to a 'flowing white concretion' may indicate a secondary mineral precipitate. An interesting detail noted in the upper inscription, but not illustrated, is the presence on the side of the specimen of a 'leaf impressed with its ribs'.

REFERENCES: 17th-cent. Index, p. 361

OTHER DRAWINGS: Possibly same specimen as **182**

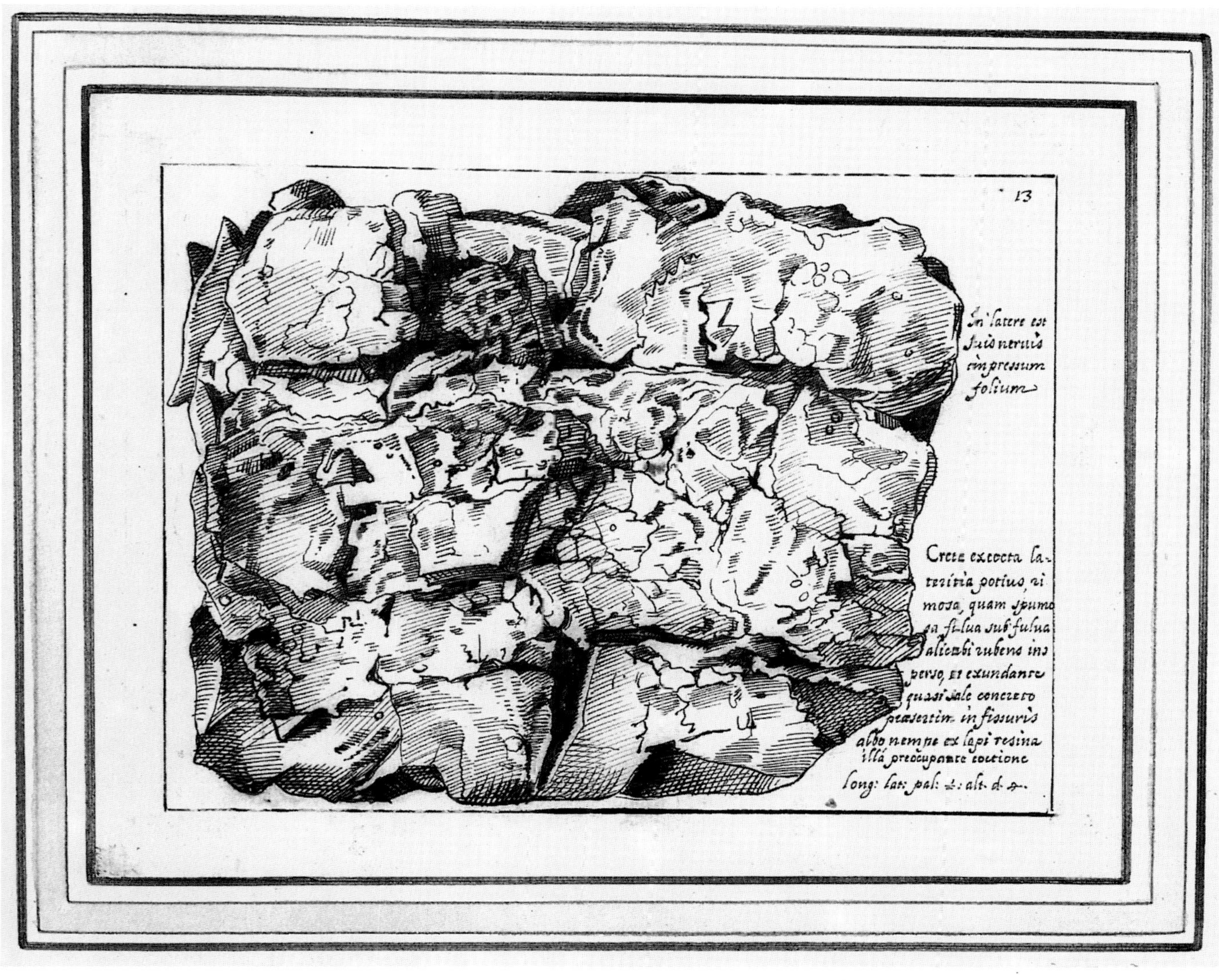

182. Possibly baked clay or a concretion

Windsor, RL 25550, *Fossils*, vol. XIV, fol. 21

Pen and ink, over traces of black chalk; retouched in ink on mount sheet; ink over black chalk frame line; no measurements given

164 × 234 mm; small holes caused by ink

NUMBERING: [ink, upper right] *14*

ANNOTATION: [ink] *Ab alia parte/ Crustosa diver/si coloris scilicet/ Ochreis crustis sub/ ceruleis et sub cinereis/ quibusdam partibus*

This seems to be a view from the other side of **181**. The inscription reads, 'On the other crusty part is a different colour, namely with ochre crusts and some parts that are somewhat blue and somewhat grey'. This description could refer to the fact that only the outer layer was affected by heat and as a result changed in colour from grey to ochre. Alternatively this could indicate iron impregnation.

REFERENCES: 17th-cent. Index, p. 361

OTHER DRAWINGS: **181**

182

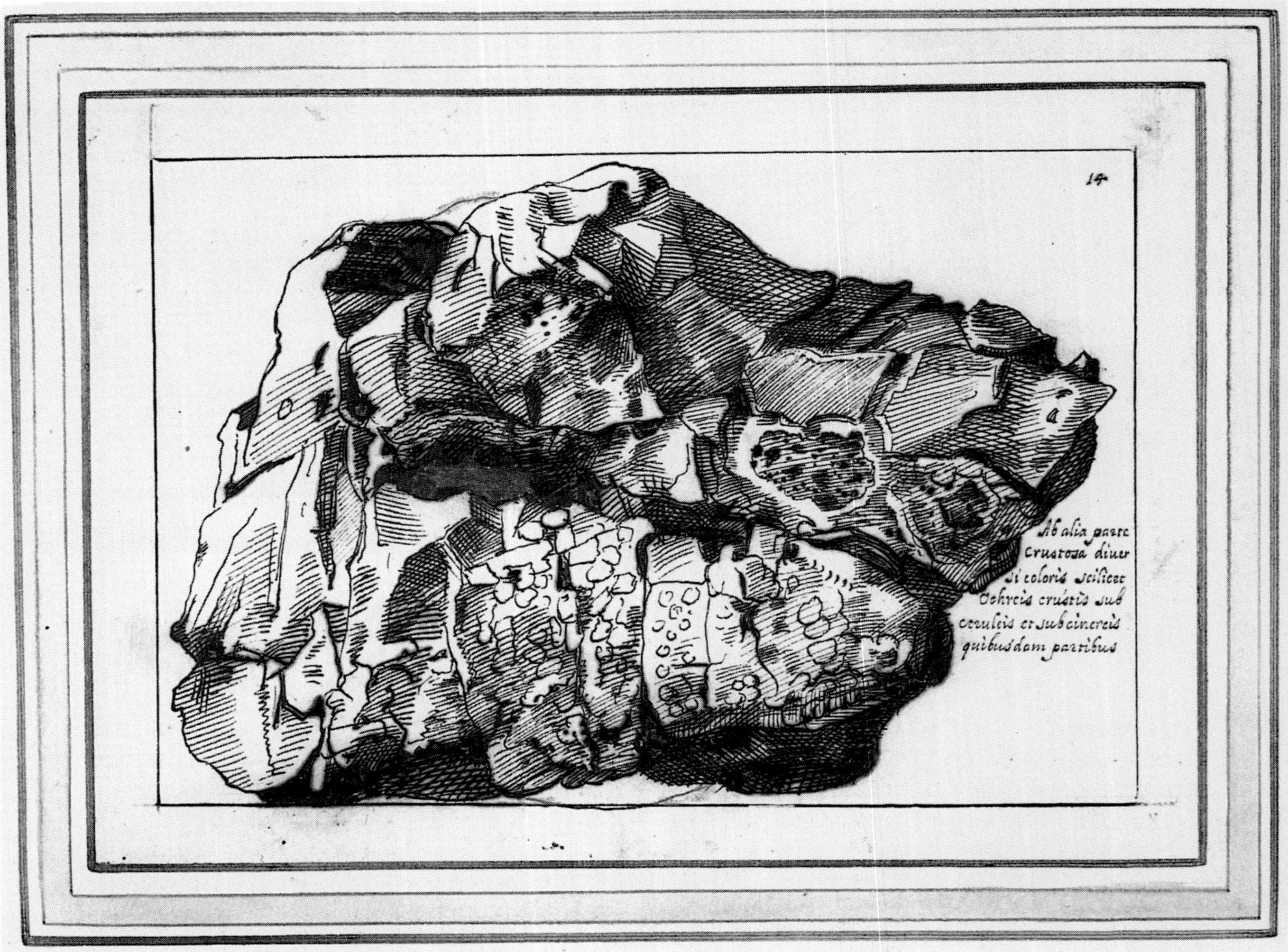

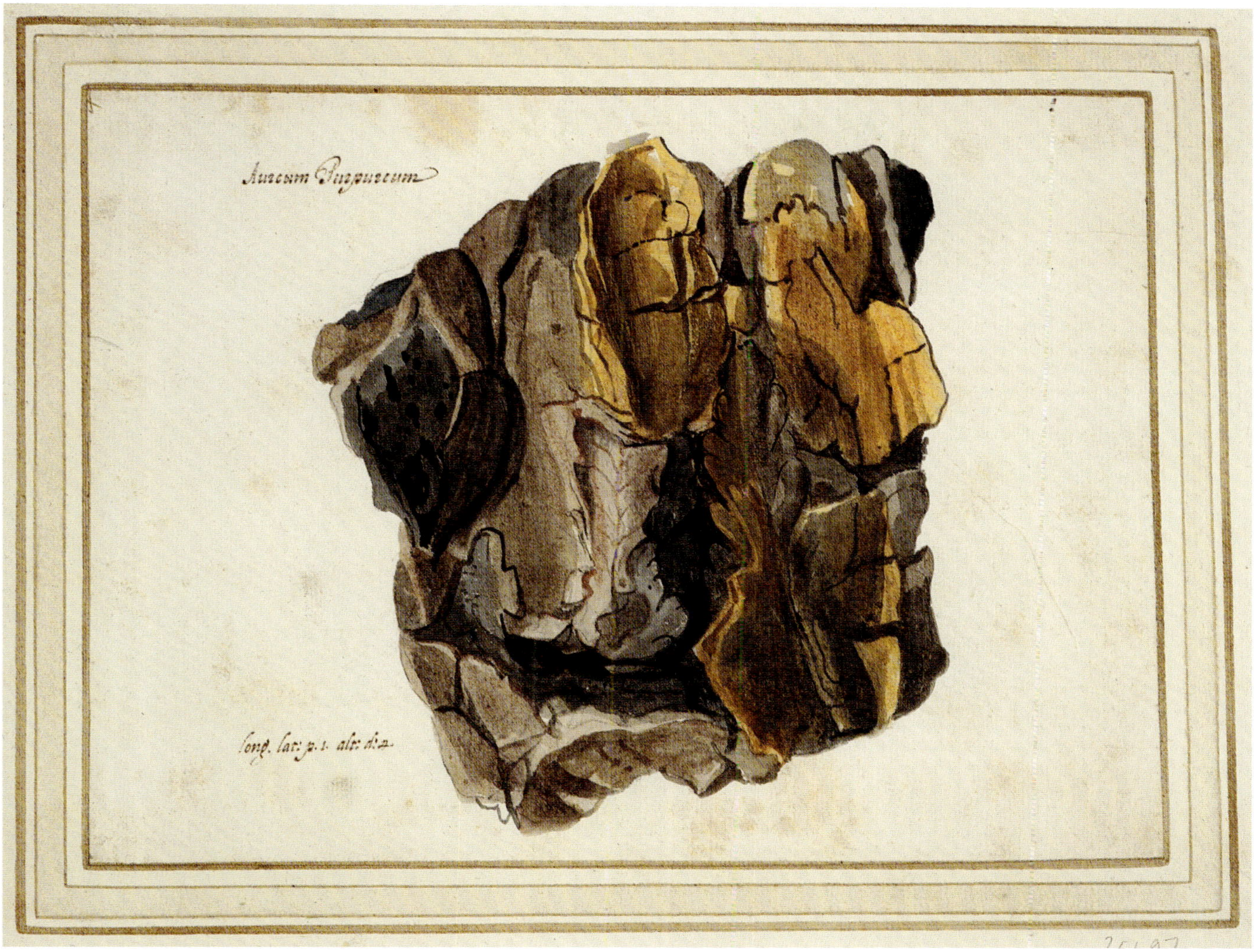

183

183. Fossil leaf in clay with ironstone

Windsor, RL 25697, *Fossils*, vol. XVII, fol. 3

Watercolour and bodycolour, over traces of black chalk; marked with a cross

181 × 253 mm

NUMBERING: [ink, top right] *1*

ANNOTATION: [ink, top] *Aureum Purpureum*
[bottom] *long. lat: p. 1. alt: d: 4.*

This specimen, 'purple [and] gold', is a complex mix of organic rich clay infiltrated by ironstone. In the centre there appears to be the impression of a fossil angiosperm leaf, as may be deduced from its venation (see **159** and **181**). The layers of ironstone appear to include both limonite (yellow) and siderite (red–brown).

REFERENCES: 17th-cent. Index, p. 370

184. Layered clay

Windsor, RL 25698, *Fossils*, vol. XVII, fol. 4

Watercolour, over traces of black chalk

179 × 251 mm

NUMBERING: [ink, top right] *2*

ANNOTATION: [ink, top] *Ceruleum lineis Varijs obductum* [bottom] *long: pal: 1½ lat: pal: ½ alt: dig: 8*

This specimen, described as 'blue with various lines running through it', appears to be a laminated, blue–grey clay with patches of clay–ironstone (light yellow–brown).

REFERENCES: 17th-cent. Index, p. 370

184

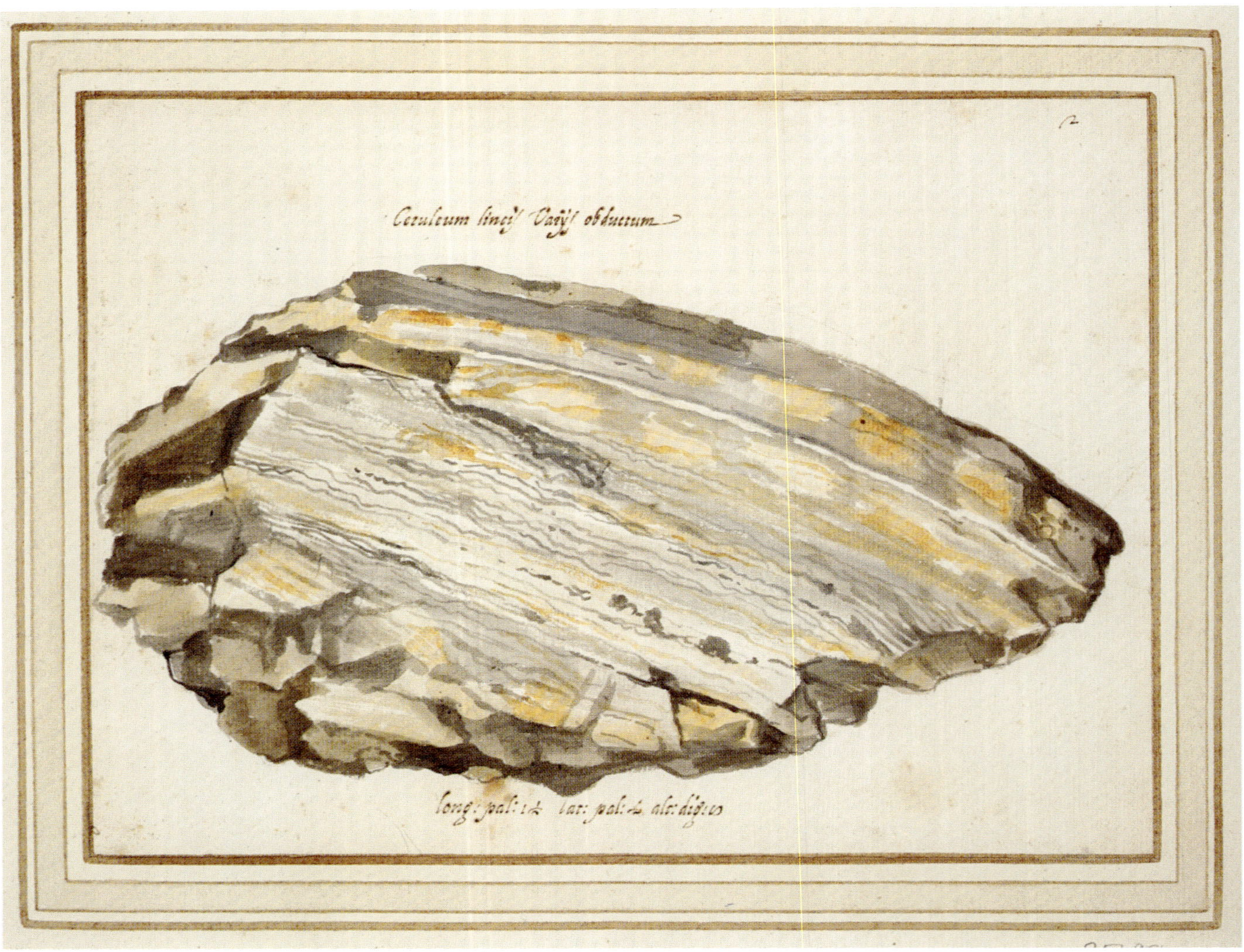

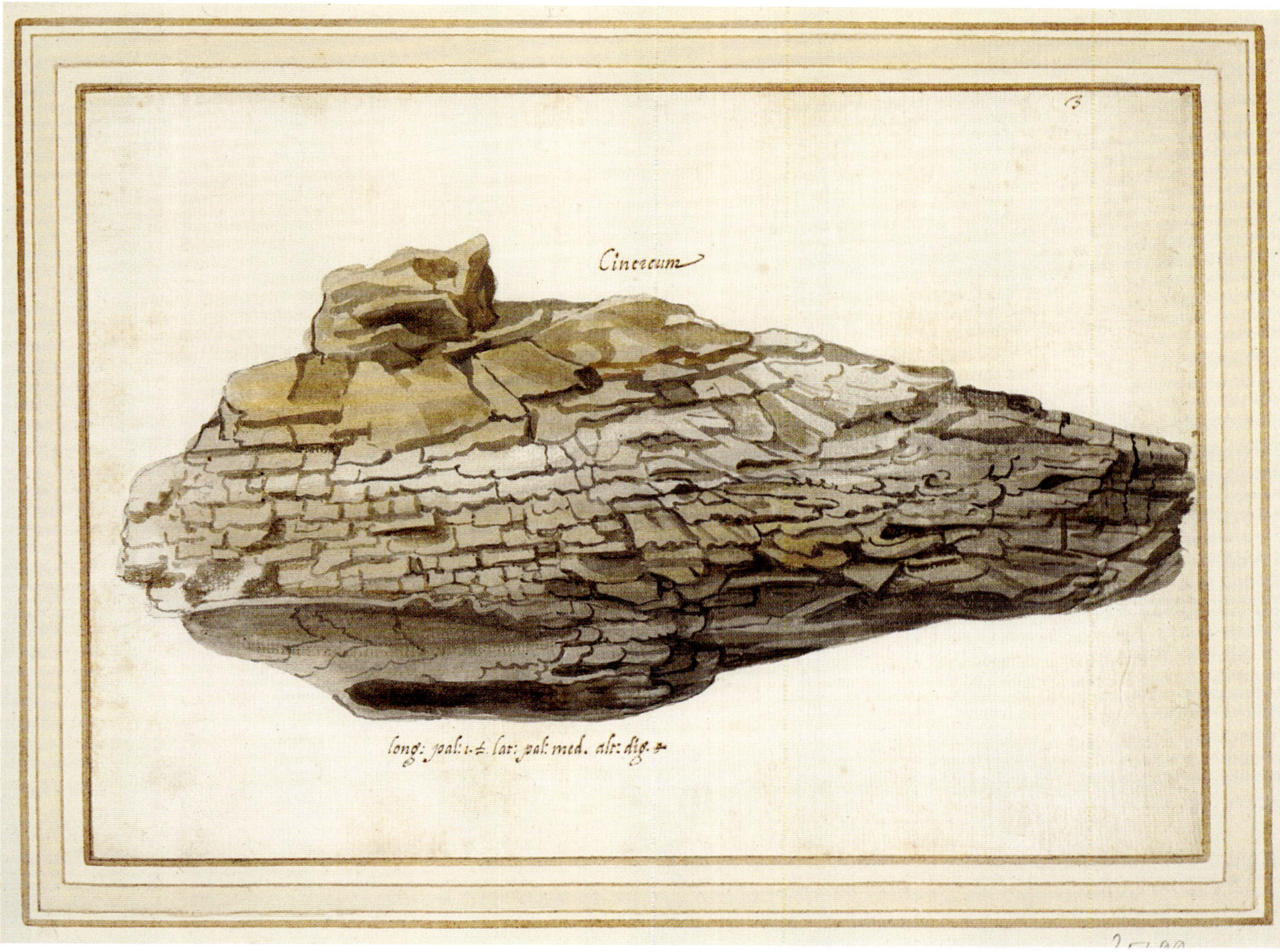

185

185. Lignite

Windsor, RL 25699, *Fossils*, vol. XVII, fol. 5

Watercolour, over traces of black chalk

182 × 263 mm

NUMBERING: [ink, top right corner] *3*

ANNOTATION: [ink, top] *Cinereum*
[bottom] *long: pal: 1.½. lat: pal: med. alt: dig. 4*

This cracked, grey rock, described as 'ashy', is probably an organic-rich clay. It may thus represent a piece of true lignite which has cracked upon drying (a feature well-shown in the drawing).

REFERENCES: 17th-cent. Index, p. 370

186. Pyrite-rich clay with wood fragments

Windsor, RL 25701, *Fossils*, vol. XVII, fol. 7

Watercolour and bodycolour, over traces of black chalk

189 × 247 mm

NUMBERING: [ink, top right corner] *5*

ANNOTATION: [ink, top] *Cinereum punctis Sulphureis, et lineis Nigris* [bottom] *long. lat: pal: ½ alt: d. 3.*

Described as 'ash-coloured with sulphurous specks and black lines', the specimen is probably an organic-rich clay with pyrite fromboids which have undergone decay. In addition, the blue–grey clay contains wood and other organic fragments.

REFERENCES: 17th-cent. Index, p. 371

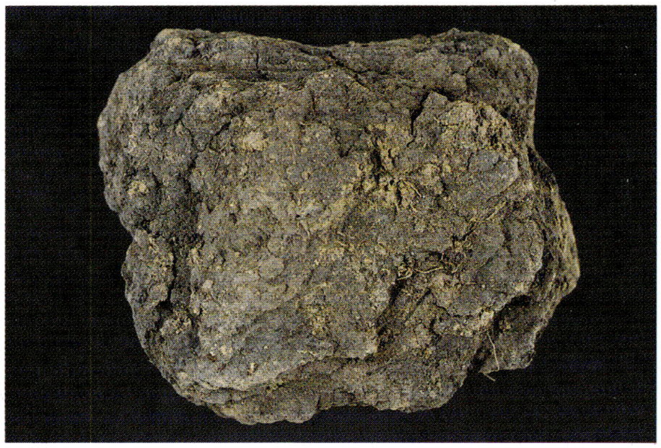

Comp. fig. 186. Dark grey-brown lignite from Podere Casanova

186

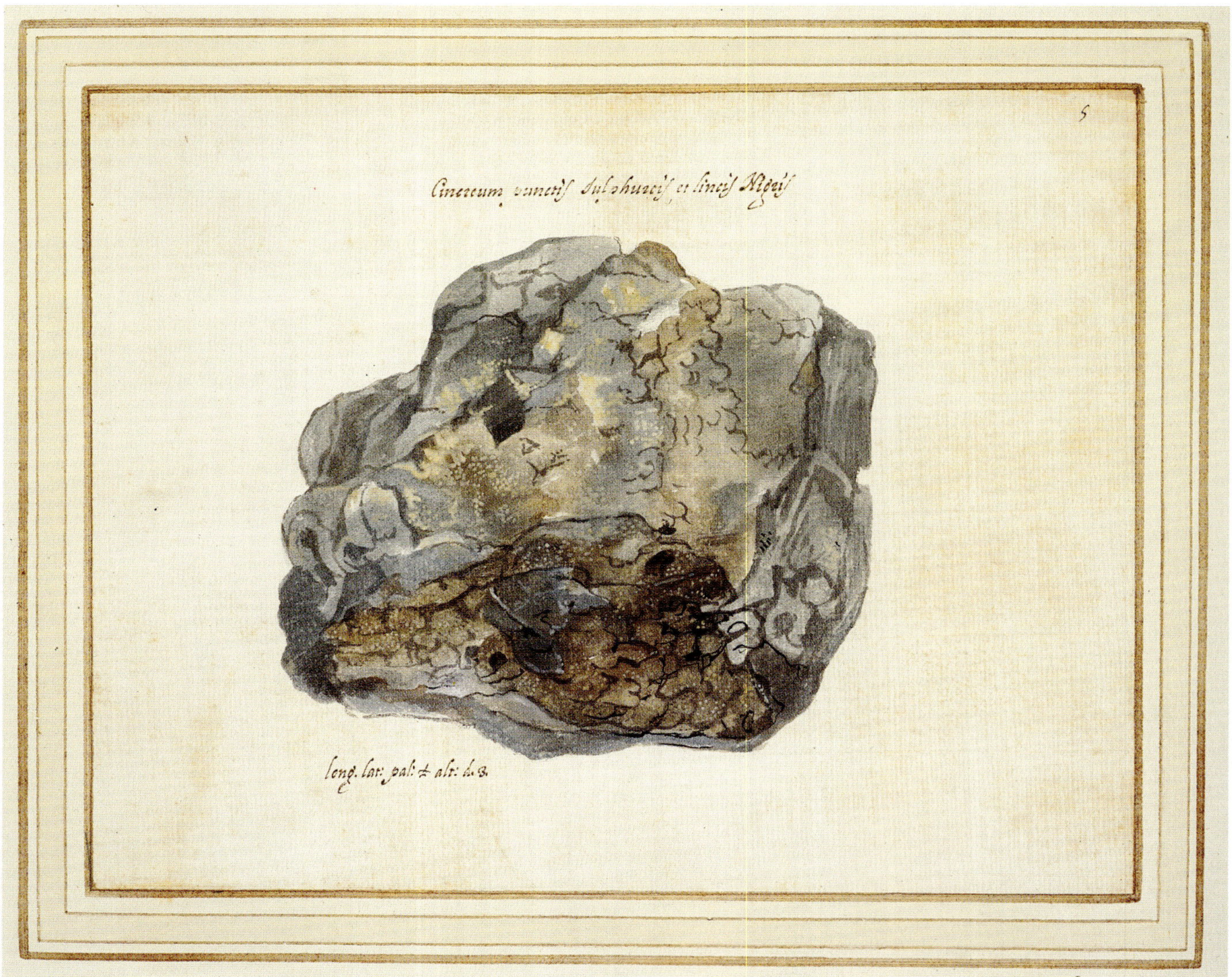

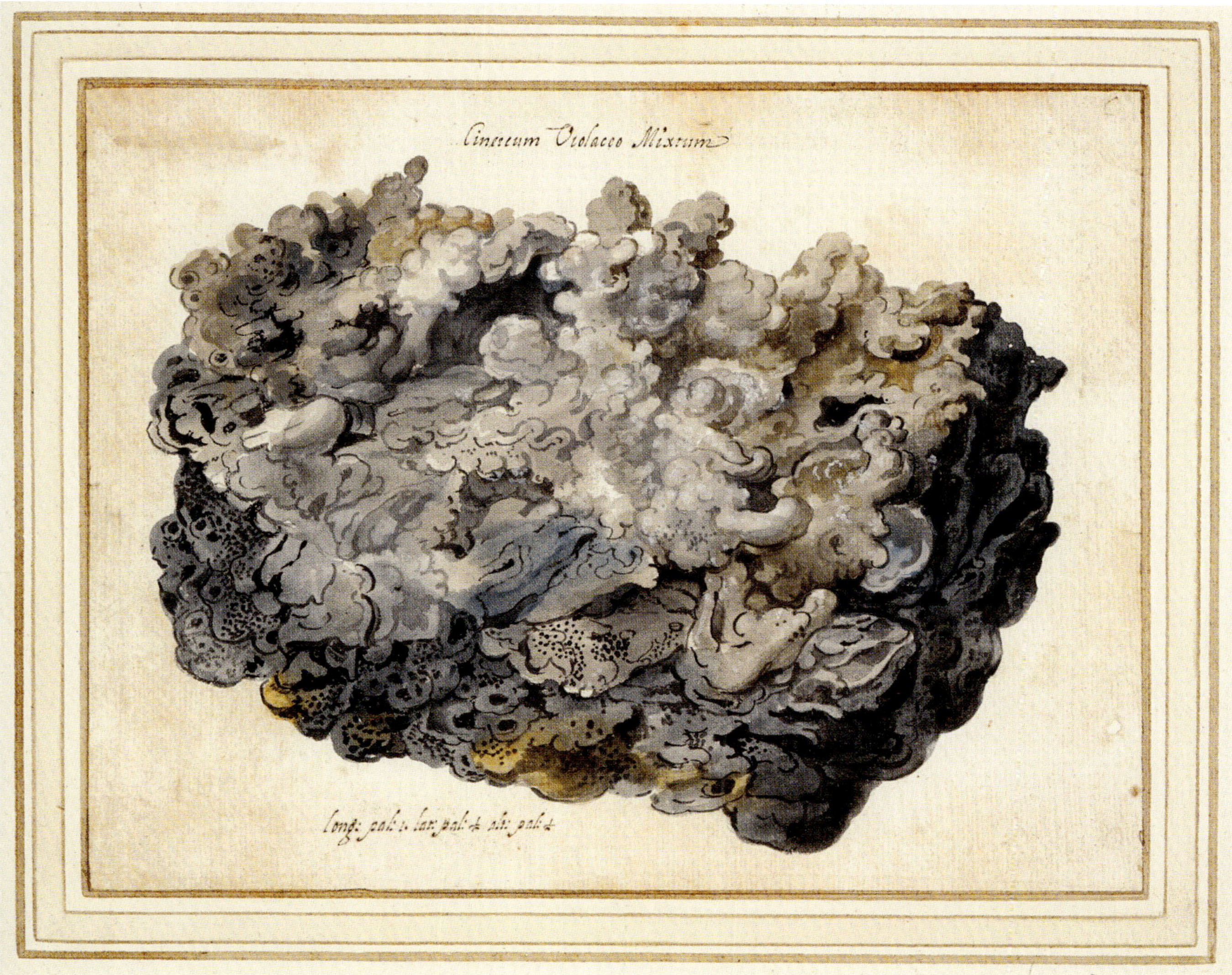

187

187. Iron pyrite in clay

Windsor, RL 25702, *Fossils*, vol. XVII, fol. 8

Watercolour and bodycolour, over traces of black chalk

187 × 244 mm

NUMBERING: [ink, top right corner] *6*

ANNOTATION: [ink, top] *Cinereum Violaceo Mixtum* [bottom] *long: pal: 1. lat: pal: ½ alt: pal: ½*

The specimen, described as 'ash-coloured mixed with violet', is probably a lump of massive iron pyrite in dark blue-grey clay. The curious cauliflower-like formation is a result of the accumulation and growth of iron pyrite in the clay.

REFERENCES: 17th-cent. Index, p. 371

188. Possibly a piece of lignite with iron-rich clay

Windsor, RL 25704, *Fossils*, vol. XVII, fol. 10

Watercolour and bodycolour, over traces of black chalk

182 × 249 mm

NUMBERING: [ink, top right corner] *8*

ANNOTATION: [ink, top] *Coriaceum Variegatum* [bottom] *long: pal: 1. lat: dig. 4 Crass. dig. 1.*

The specimen is described as 'variegated and leathery'. The grey–blue material is probably clay; the purple and yellow areas may reflect the impregnation of iron into a brown woody horizon.

REFERENCES: 17th-cent. Index, p. 371

188

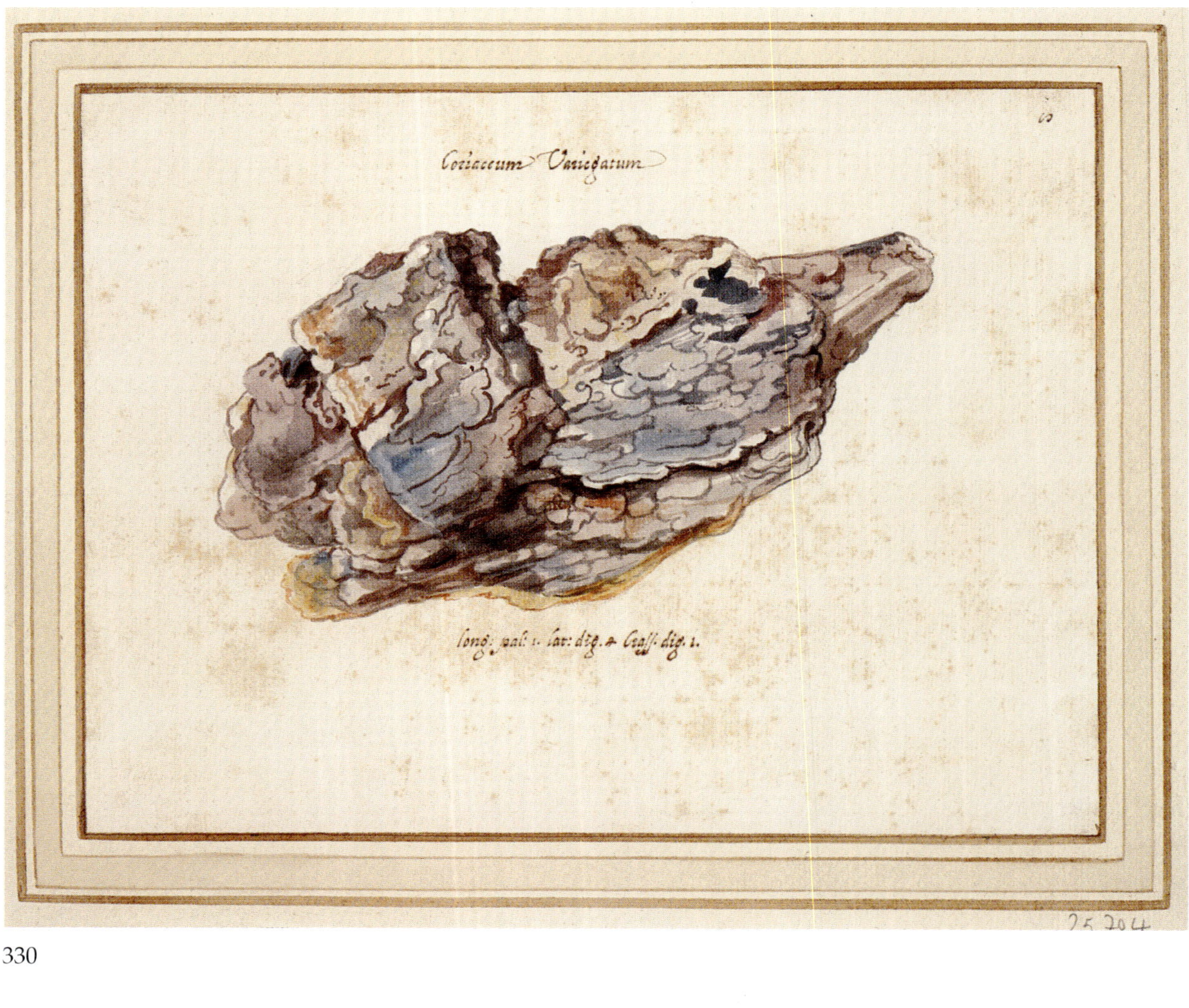

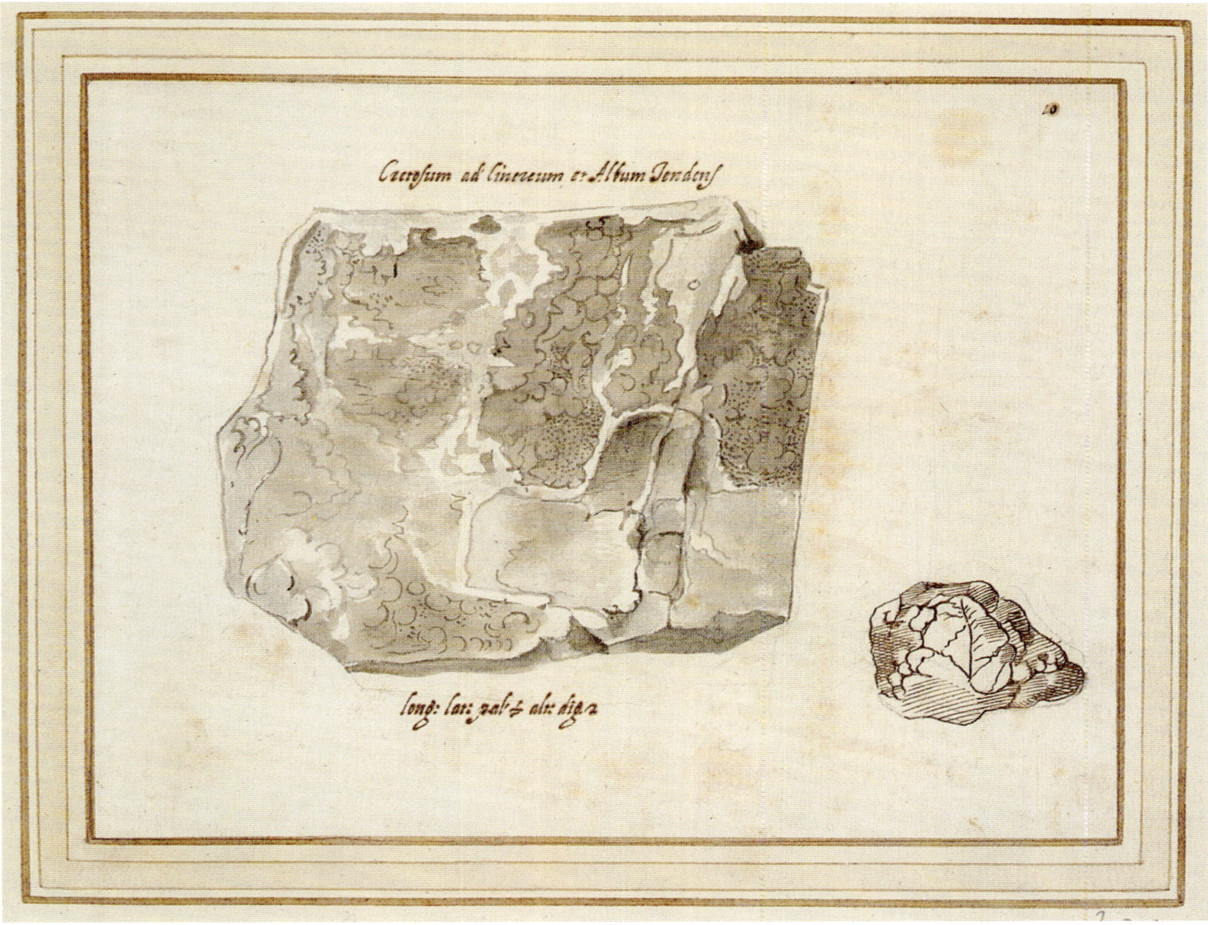

189

189. Specimen of clay and possibly rock fragment

Windsor, RL 25706, *Fossils*, vol. XVII, fol. 12

Watercolour and pen and ink, over black chalk with *pentimenti* in black chalk; measurements given for the left-hand specimen only

182 × 246 mm

NUMBERING: [ink, top right corner] *10*

ANNOTATION: [ink, top] *Cretosum ad Cinereum et Album Tendens* [bottom] *long: lat: pal: ½ alt: dig. 2*

The larger specimen, 'clayey, tending towards an ashy-white colour' represents a specimen of grey–white clay. The identity of the smaller specimen on the right is unclear; unusually for the drawings in Volume XVII, it is not drawn in colour.

An alternative outline for the larger specimen is visible in black chalk.

REFERENCES: 17th-cent. Index, p. 371

190. Iron-impregnated clay-rich lignite

Windsor, RL 25708, *Fossils*, vol. XVII, fol. 14

Watercolour and bodycolour, over traces of black chalk

185 × 247 mm

NUMBERING: [ink, top right corner] *12*

ANNOTATION: [ink, top] *Croceum Parvulis Maculis Interpositis* [bottom] *long. lat: alt: pal: ½ in Cir.*[a]

This yellow specimen 'interspersed with small spots', is possibly one of the clay-rich lignites which has been highly impregnated with iron (limonite and siderite).

REFERENCES: 17th-cent. Index, p. 371

191. Pyrite-rich clay

Windsor, RL 25712, *Fossils*, vol. XVII, fol. 18

Watercolour and bodycolour, over traces of black chalk

176 × 249 mm

NUMBERING: [ink, top right corner] *16*

ANNOTATION: [ink, top] *Fuscum Punctis Aureis* [bottom] *long: pal: ½ lat: dig: 4. alt: dig. 2*

A dark specimen with gold specks, of organic-rich clay, showing areas of iron pyrite (yellow), which has decayed in several places, shown by the small white areas.

REFERENCES: 17th-cent. Index, p. 371

190

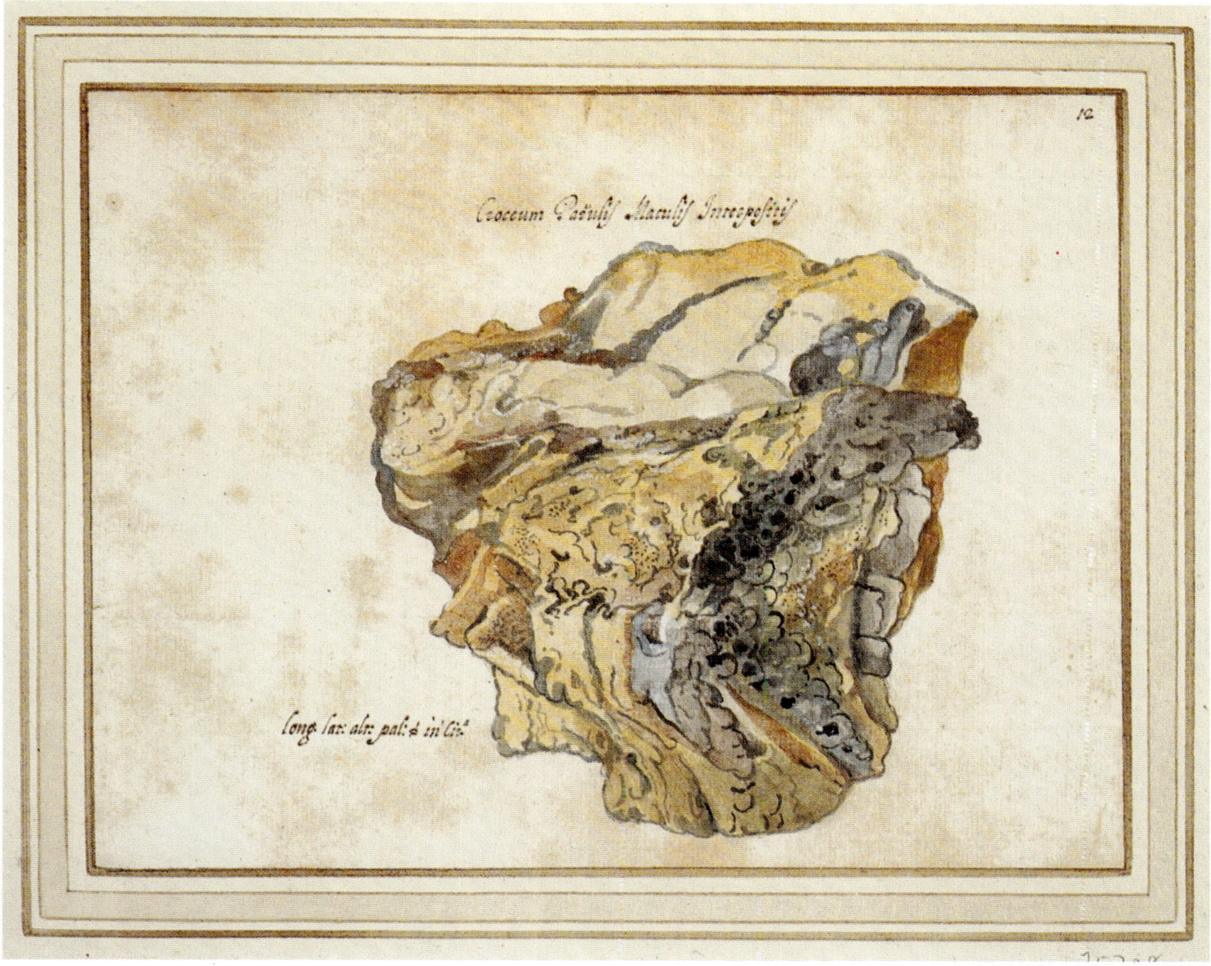

12
Croceum Paruis Maculis Interpositis
long. lat. alt. pal. 2 in Cir.

191

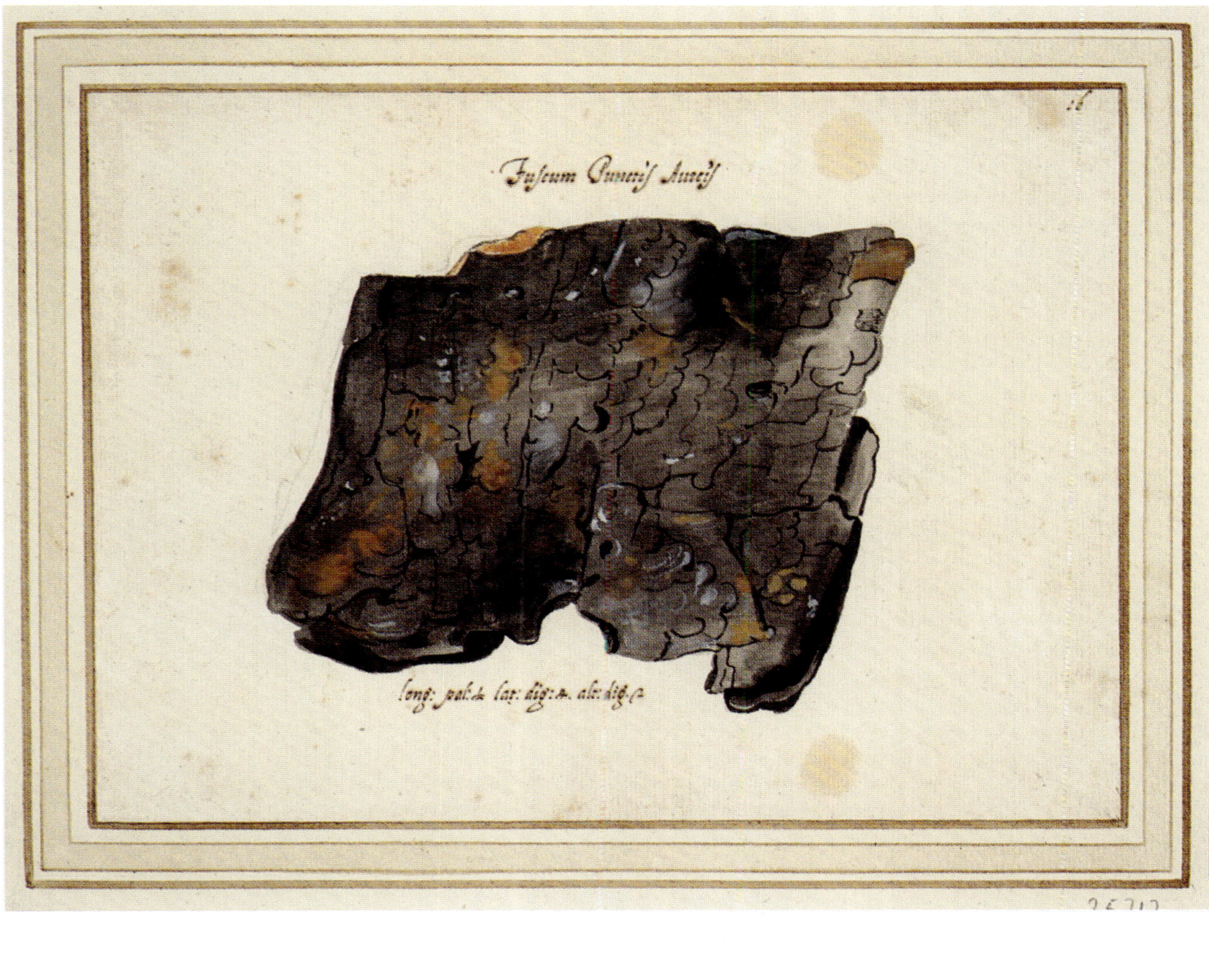

16
Fuscum Punctis Aureis
long: pal. 2 lat: dig: 2 alt: dig. 1

192. Ironstone

Windsor, RL 25713, *Fossils*, vol. XVII, fol. 19

Watercolour and bodycolour, over traces of black chalk

188 × 252 mm

NUMBERING: [ink, top right corner] *17*

ANNOTATION: [ink, top] *Gagat. non absolute Punctis Albis* [bottom] *long: et lat: p. 1 in Cir.ª Crass. d: 5*

This specimen of ironstone—which includes both limonite and siderite—is shown with an unusual area of white spots; it is not clear what the latter represents. The specimen is described as jet 'not entirely [covered] with white spots' (for further specimens described as jet or gagate, see **62** and **163**).

REFERENCES: 17th-cent. Index, p. 371

192

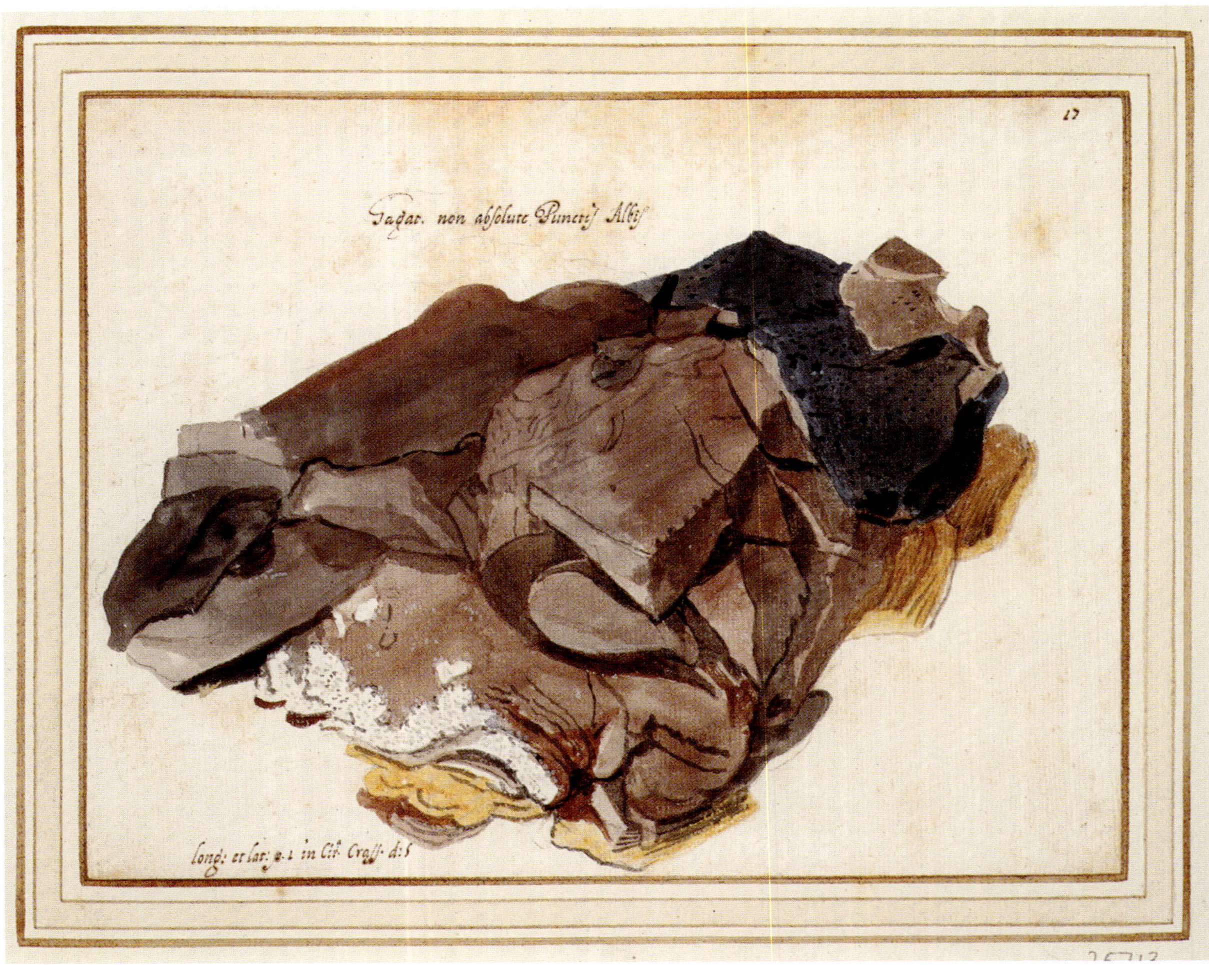

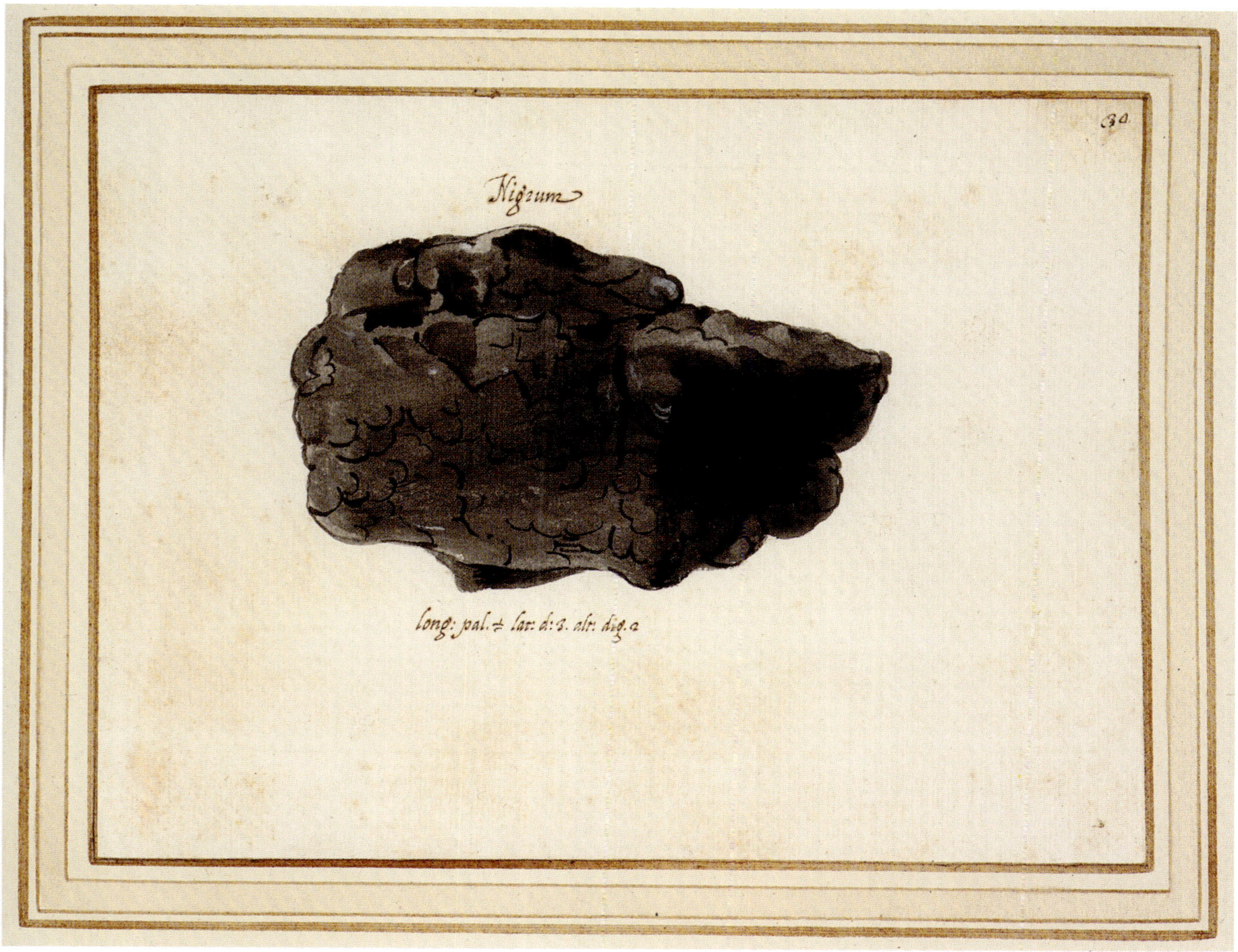

193

193. Black clay

Windsor, RL 25726, *Fossils*, vol. XVII, fol. 31

Watercolour and bodycolour, over traces of black chalk

177 × 241 mm

NUMBERING: [ink, top right] *30* over *29*

ANNOTATION: [ink, top] *Nigrum*
[bottom] *long: pal. ½ lat: d: 3. alt: dig. 2*

Described as 'black', the specimen appears to represent a highly organic-rich clay with little sign of lamination.

REFERENCES: 17th-cent. Index, p. 372

194. Black clay

Windsor, RL 25727, *Fossils*, vol. XVII, fol. 32

Watercolour and bodycolour, over traces of black chalk

183 × 245 mm

NUMBERING: [ink, top right corner] *31* over *30*

ANNOTATION: [ink, top] *Nigrum*
[bottom] *long. lat. pal: ½ alt. d: 2.*

Again described as 'black', this specimen probably represents a highly organic-rich clay. A piece of fossil wood is shown at its upper-left corner.

REFERENCES: 17th-cent. Index, p. 372

194

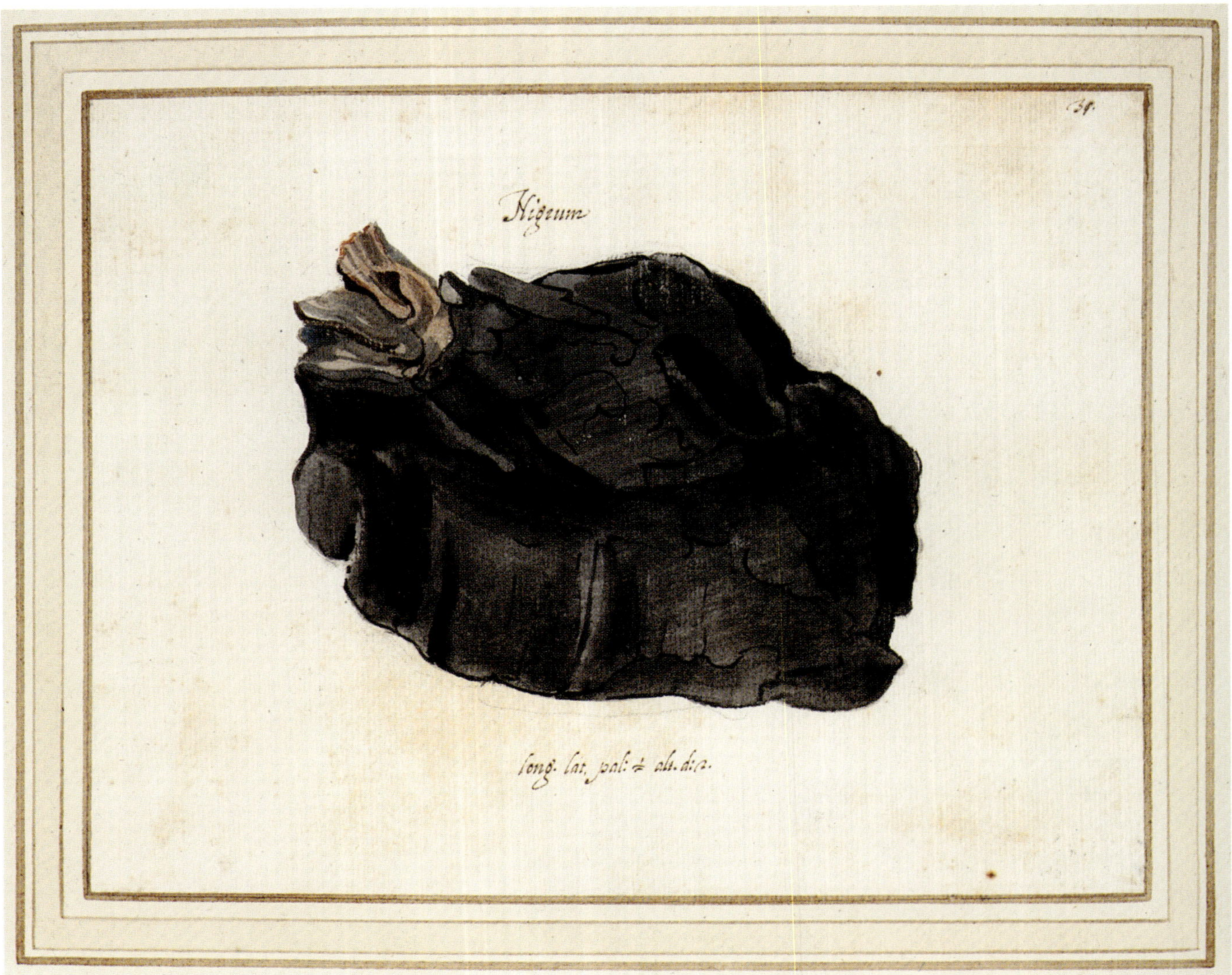

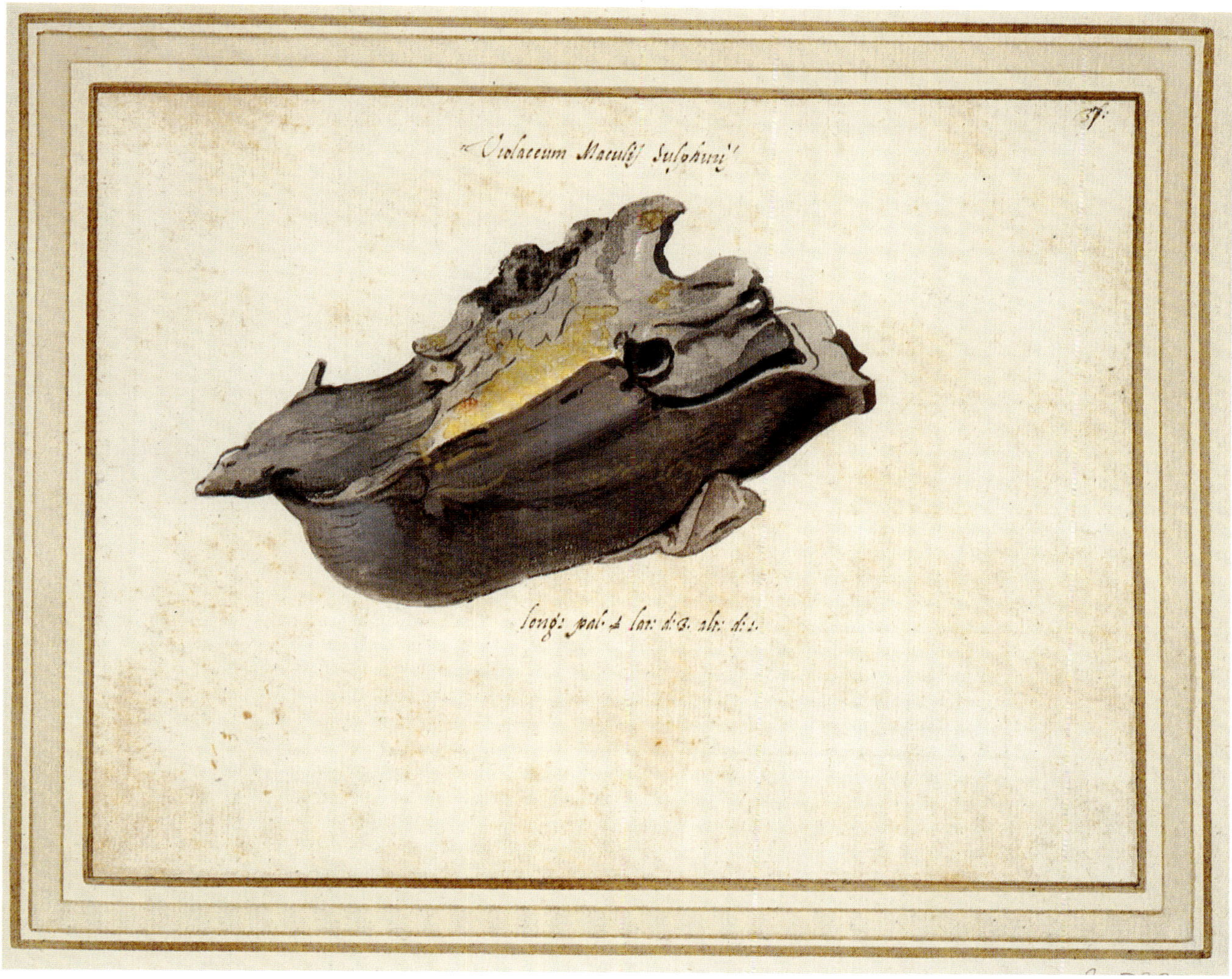

195

195. Organic-rich pyritic clay

Windsor, RL 25733, *Fossils*, vol. XVII, fol. 38

Watercolour and bodycolour, over traces of black chalk

178 × 236 mm

NUMBERING: [ink, top right corner] *37* over *36*

ANNOTATION: [ink, top] *Violaceum Maculis Sulphureis* [bottom] *long: pal. ½ lat: d: 3. alt: d: 1.*

This specimen is described as 'violet with sulphurous spots'. The violet colour appears to indicate an organic-rich clay; the sulphurous spots are evidence of pyrite decay.

REFERENCES: 17th-cent. Index, p. 372

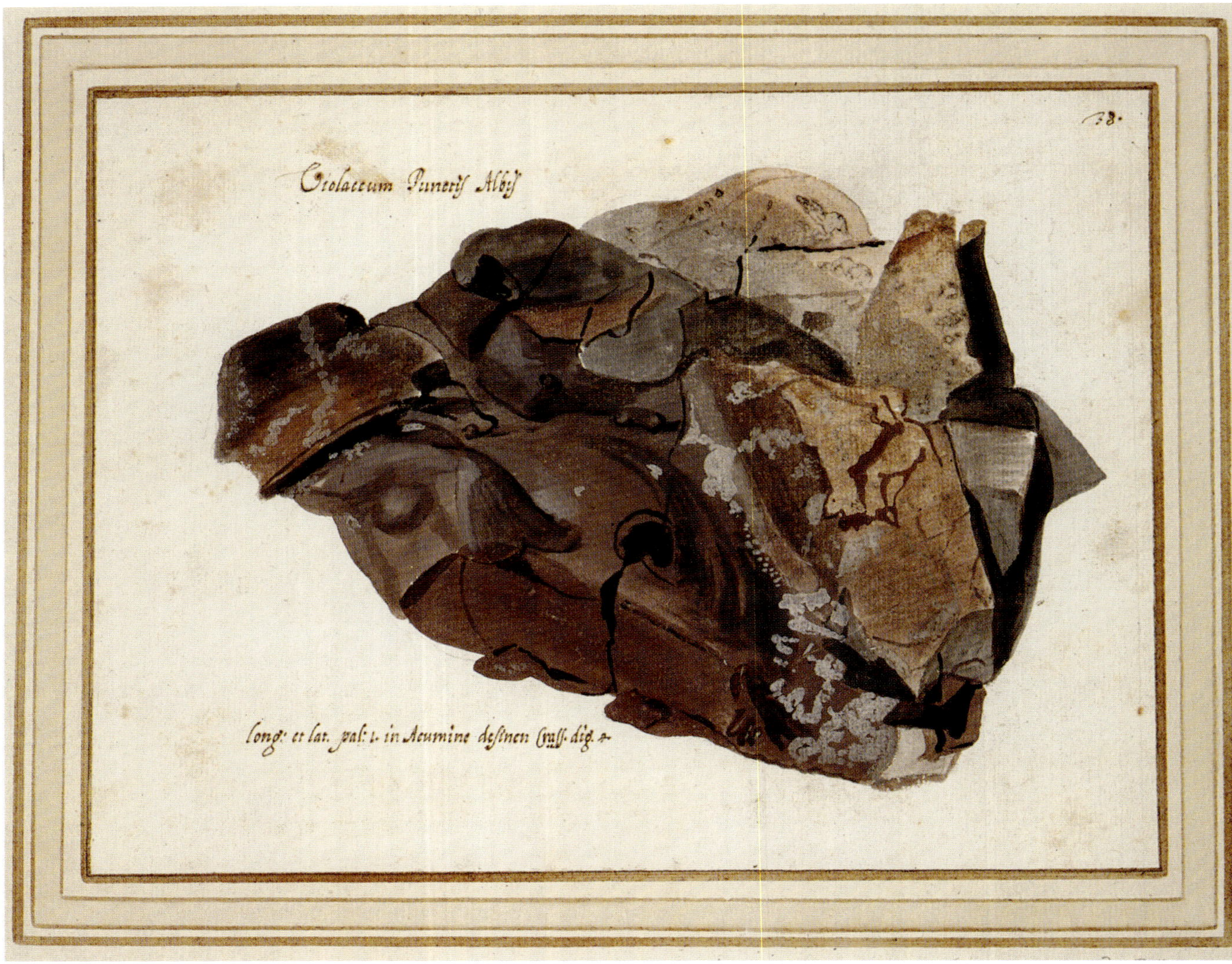

196

196. Black clay with clay ironstone

Windsor, RL 25734, *Fossils*, vol. XVII, fol. 40

Watercolour and bodycolour, over traces of black chalk

182 × 247 mm

NUMBERING: [ink, top right corner] *38* over *37*

ANNOTATION: [ink, top] *Violaceum Punctis Albis*
[bottom] *long: et lat. pal: 1. in Acumine desinen Crass. dig. 4*

This specimen, described as 'violet with white spots', is possibly a clay ironstone in which the siderite has only partially replaced the black clay.

Measurements are given; the thickness was measured to 'the sharp point'.

REFERENCES: 17th-cent. Index, p. 372

VI. AMMONITES [197–199]

Fossil ammonites were well known to naturalists of Cesi's time. They had been illustrated in numerous works including those of Conrad Gesner (1516–65) and Ulisse Aldrovandi (1522–1605).[1] They were called 'horns of Ammon' after the Egyptian deity, always represented with rams' horns. In some parts of Europe, they were known as snake stones. Their true origin—the shells of extinct cephalopods (a Class which includes the living nautilus, octopus, cuttlefish and squids)—was unknown to Cesi. Ten ammonites are illustrated and named as such among the drawings of fossils, but little information is given about their origin. They seem somewhat out of place in comparison with the other fossils from the young rocks east of Acquasparta. It is almost certain that they were included only because they had been found in the area around Acquasparta. The importance attached to them at the time is further demonstrated by their inclusion in Stelluti's *Trattato*, where they were described as:

> engraved stones in the shape of a snail and a snake coiled around itself; but on account of their even greater resemblance to the horn of the Ram, they are called the horn of Ammon, formed in this manner by Nature. And these spirals were made with such perfect proportions that I would always have believed them to be artificial, or else not natural, had I not discovered them like this, not only in the territory of Acquasparta, but also in my native territory of Fabriano; besides which, many naturalist authors also write of them; as may be seen in Gesner, in Pliny, in Agricola, Bellon, Imperato and others. And of their origins and nature His Highness Prince Cesi, mentioned above, had also in part written and was writing, when death prevented his finishing the treatise which dealt with this material.[2]

Ammonites are marine animals which became extinct some 65 million years ago at the end of the Cretaceous Period. Cretaceous and older Jurassic marine rocks are well exposed in the hills west of Acquasparta [Figs 6, 8]. Localities containing ammonites are rare. One of the few, west of Massa Martana near Acquasparta [Fig. 8], must have been known to Cesi, since specimens from that district were illustrated by Michele Mercati (1514–93).[3] This site, known as the Fosso di Pozzale, exposes fossiliferous Lower Jurassic rocks of the *Ammonitico Rosso*. Venturi[4] has described ammonites found there, and in 1995, he and the present author collected numerous specimens of the forms illustrated in the drawings (Fig. 43, Comp. figs 197a, 197b).

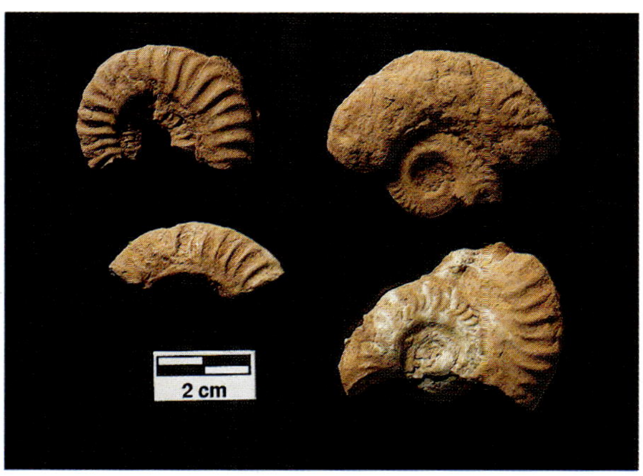

Fig. 43. Jurassic ammonites from Fosso di Pozzale, near Massa Martana

Comparisons between the specimens collected and these drawings suggest that the genera illustrated include *Hildoceras*, *Hildaites*, *Mercaticeras* and *Hammatoceras*. Interestingly, some of the ammonites are found in white limestones and others in the overlying red limestones of the *Ammonitico Rosso*,[5] but no colour is noted on the drawings of ammonites.

Little in the way of additional data is supplied by the inscriptions on the drawings. As these fossils are of quite a different character from the other material illustrated in the volumes presented here, it is not possible to comment further on their significance.

NOTES

1. See Rudwick 1985, pp. 26–9 for a review.
2. Stelluti 1637, p. 11.
3. Mercati 1719 (The mansucript was completed in 1576).
4. Venturi 1982, 1986.
5. Venturi 1996, pp. 32–3

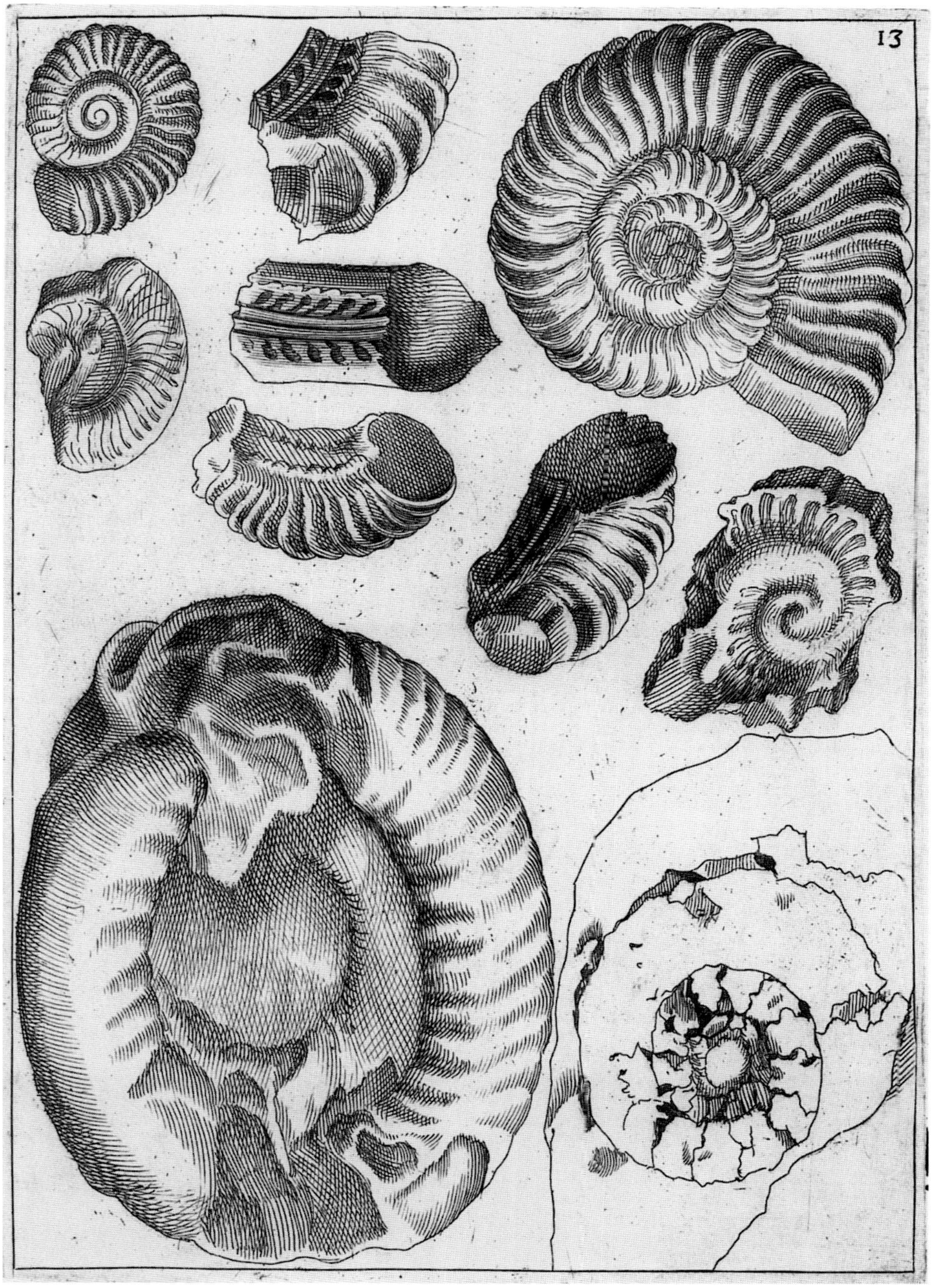

Comp. figs 197, 198 and 199. Stelluti 1637, Pl. 13

197. Ammonite, three fragments and a cross-section of ammonite in rock, probably from Carsulae

Windsor, RL 25536, *Fossils*, vol. XIV, fol. 7

Pen and ink, over traces of black chalk; retouched in ink on mount sheet; ink frame line; no measurements given

175 × 239 mm; small holes caused by ink

NUMBERING: [ink, upper right] *1*

ANNOTATION: [ink, top centre] *In lapide plano/ videtur quid/ ferreum cor/ticosum Spiriti/llis interceptum*
[right] *Ammonis Corsulan*
[upper left to right] *a9* [*8* crossed out], *a3*, *a10* [*9* crossed out], *a2*, *a1*

The specimens belong to the genus *Asteroceras* from the Lower Jurassic.

The main drawing shows a shell in three pieces. It probably came, as the inscription seems to suggest, from the ancient site of Carsulae, just south of Acquasparta. The upper-left drawing depicts a cross-section through an ammonite in a rock, showing the septal walls (Comp. fig. 197a).

The inscription notes the 'bark-like ferrous' qualities of the ammonite, indicating the presence of iron. This description may indicate that the specimens come from the *Ammonitico Rosso*, where the rock is a reddish brown.

The specimens, numbered *a1*, *a2*, *a3*, *a9* and *a10*, are part of a sequence continued in **198**: *a1* is probably *Hildaites*; *a2* is probably a piece broken off *a1*; *a3* and *a10* are probably *Mercaticeras*. This drawing was combined with the ammonites and fragments shown in **198-9** to form the last plate in Stelluti (Plate 13). For Stelluti's discussion of ammonites, see p. 339.

LITERATURE: Bromehead 1947a, fig. 2; Solinas 1989c, p. 91, fig. 69; *Cassiano cat.* 1993, no. 136; *Cassiano cat.* 1997, no. 74

REFERENCES: Stelluti 1637, p. 11; 17th-cent. Index, pp. 360, 365

ENGRAVINGS: Stelluti 1637, Pl. 13; for proof engravings in Montpellier, BEM, MS H.173, fol. 18, and Paris, BN, MS It. 1684, fol. 17, see p. 403

Comp. fig. 197a . Cross-section through ammonite, found near Fosso di Pozzale

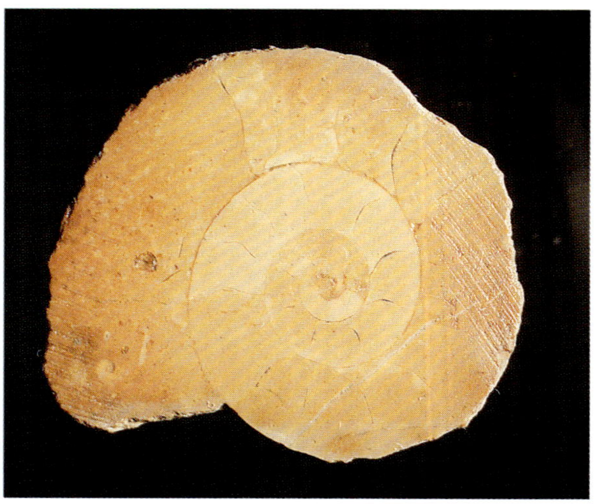

Comp. fig. 197b. Ammonite found in Ammonitico Rosso Formation, near Fosso di Pozzale

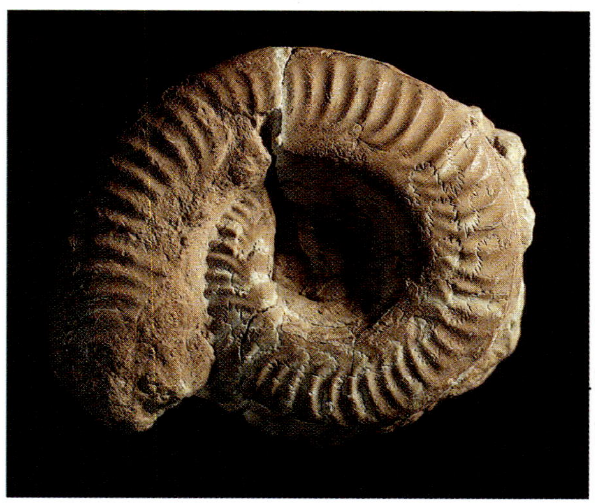

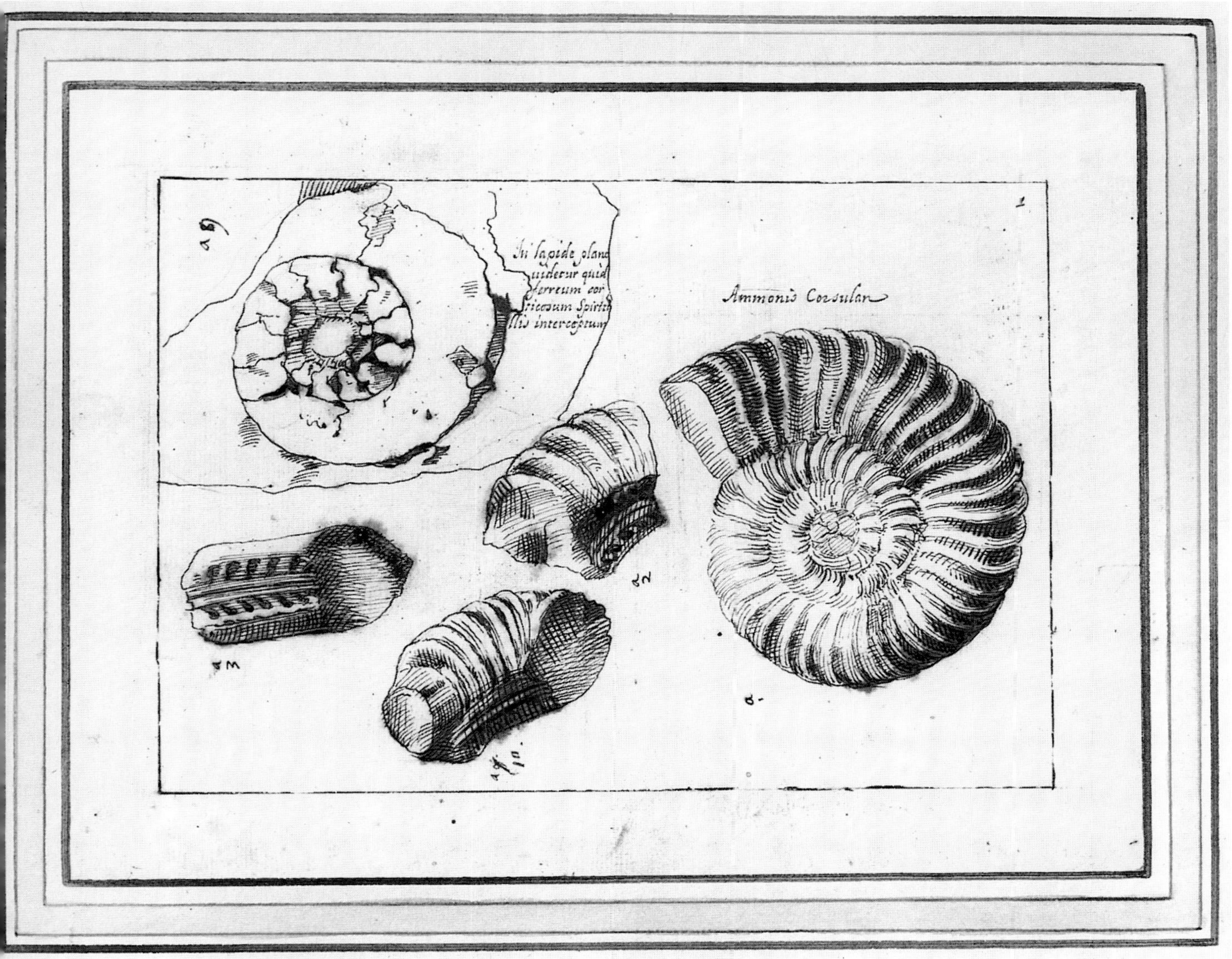
In lapide plan
uidetur quid
ferreum cor
ticosum Spirit
llis interceptum
Ammonis Corsular

197

198. Whole ammonite, two fragments and an impression, probably from Carsulae

Windsor, RL 25537, *Fossils*, vol. XIV, fol. 8

Pen and ink, over black chalk; ink frame line

183 × 246 mm. *Watermark:* Bird 41D

NUMBERING: [ink, upper right] 2

ANNOTATION: [ink, upper left] *Sedes C: ammonis eo impressu*
[upper right] *a4* [3 crossed out] *C: Ammonis profunde sectus*
[lower left] *C: Ammonis vix Signatus*
[lower right] *a6* [5 crossed out] *C: Ammonis sinuose pressus*
[Cesi's inscriptions, trimmed along lower edge]

The specimens appear to belong to the genus *Rarenodia*.

As the annotations indicate, the upper drawing on the left shows an impression in rock of an ammonite, while the upper-right drawing depicts an ammonite excavated from deep within the rock. The lower-left specimen, described as 'barely marked' (*vix Signatus*), appears to have been weathered and eroded.

The lower right specimen (numbered *a6*) belongs to the fragment shown in **197**, as the grouping of all four fragments in Stelluti's Plate 13 indicates. The top-right and bottom-left specimens may possibly be *Hildaites* and the bottom-right specimen may be a *Mercaticeras*.

LITERATURE: Bromehead 1947a, p. 68; Solinas 1989c, p. 91, fig. 70

REFERENCES: Stelluti 1637, p. 11; 17th-cent. Index, p. 360

ENGRAVINGS: Stelluti 1637, Pl. 13; for proof engravings in Montpellier, BEM, MS H.173, fol. 18, and Paris, BN, MS It. 1684, fol. 17, see p. 403

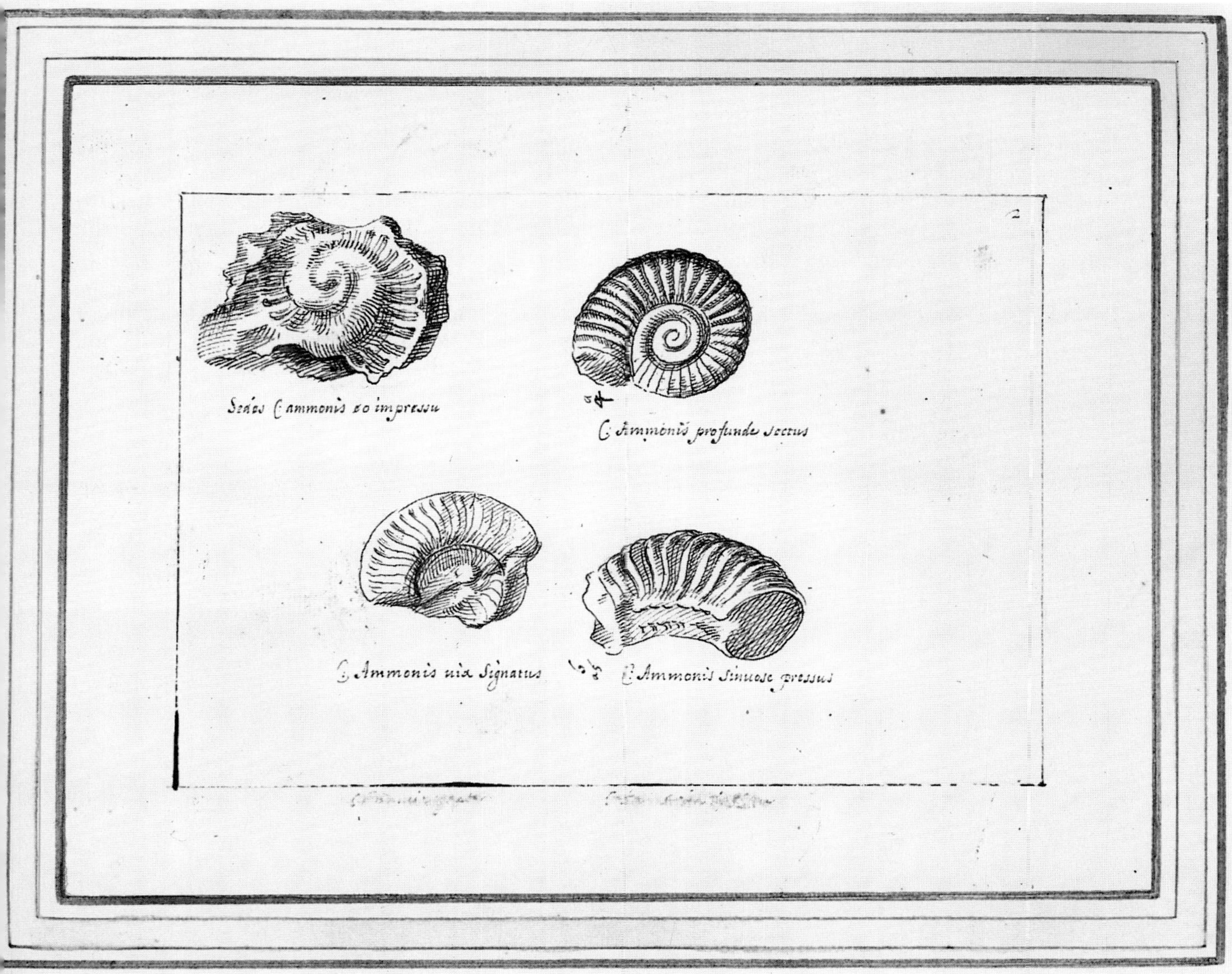
Sedes C. ammonis eo impressu
C. Ammonis profunde sectus
C. Ammonis uia Signatus
C. Ammonis sinuose pressus

198

199. Ammonite and fossil molluscs in rock

Windsor, RL 25547, *Fossils*, vol. XIV, fol. 18

Pen and ink, over traces of black chalk

183 × 251 mm

NUMBERING: [ink, upper left] *11*

ANNOTATION: [ink, top left] *Creta ex Mytulo concave*
[upper right] *Serpens lapideus an ex cornu/ Ammonis*
[centre left] *Creta cum cochleis minimis*

This ammonite possibly belongs to the Jurassic form *Hammatoceras*, but is too poorly preserved to make a secure identification. It is engraved in Stelluti's Plate 13. The upper-left drawing appears to be an impression of a bivalve shell, identified in the inscription as a *Mytulus* (cf. **151**). The lower-left drawing shows a piece of clay with very small shells—possibly gastropods. Both molluscs appear to belong to the Tertiary.

LITERATURE: Bromehead 1947a, p. 68; Solinas 1989c, p. 92, fig. 71

REFERENCES: Stelluti 1637, p. 11; 17th-cent. Index, pp. 367, 369

ENGRAVINGS: Stelluti 1637, Pl. 13; for proof engravings in Montpellier, BEM, MS H.173, fol. 18, and Paris, BN, MS It. 1684, fol. 17, see p. 403

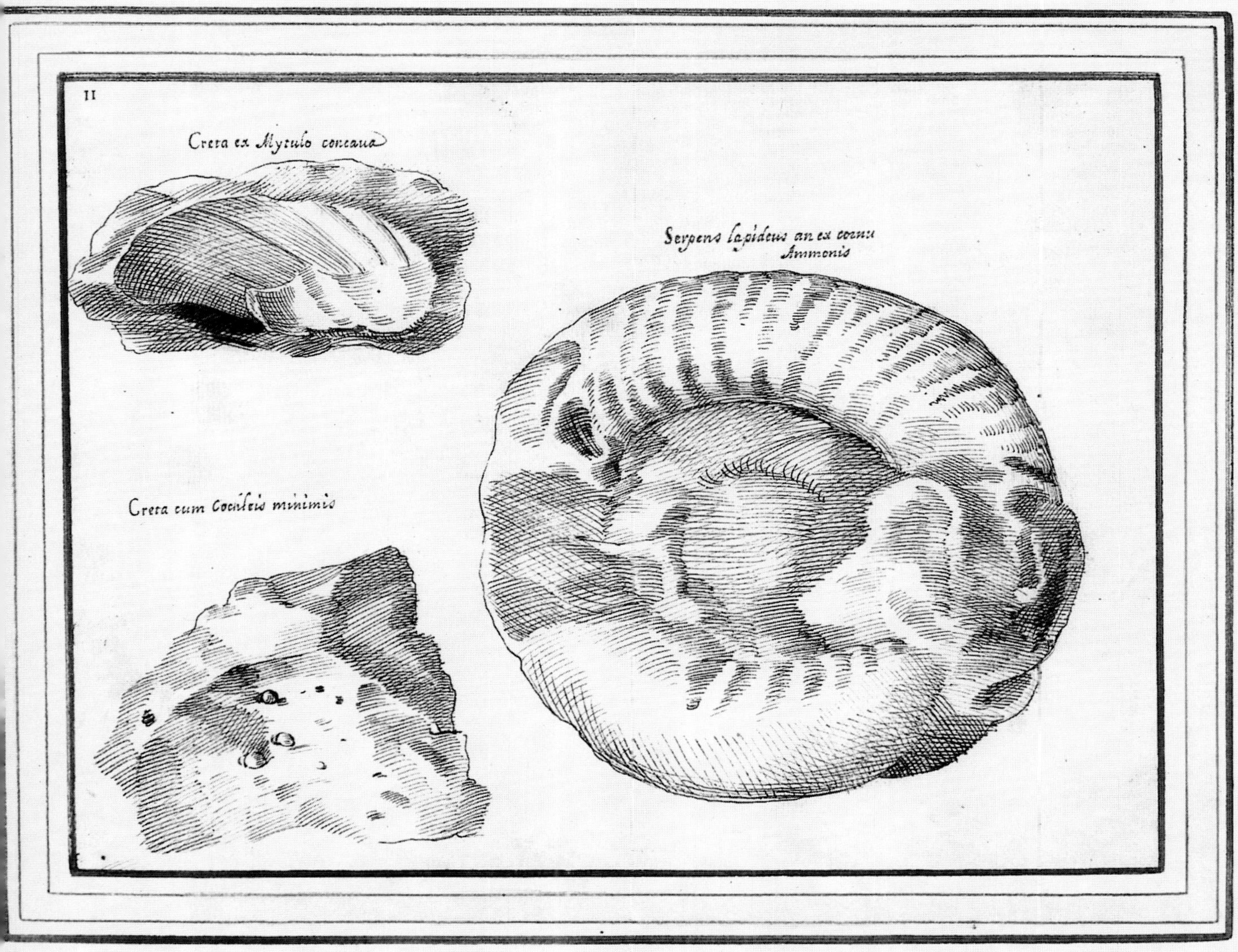
II
Creta ex Mytulo concaua
Serpens lapideus an ex cornu Ammonis
Creta cum Cochleis minimis

199

WATERMARK TYPES

Jo Taylor

THE manufacturing process of hand-made paper involves a pulpy mixture (made out of flax or cotton-rag fibres) being poured over a wire tray (or paper mould) and being left to set under gentle pressure from a felt mat which is laid over the pulp. The mould consists of a set of closely-spaced parallel wires held in place by a more widely spaced set of transversals: in other words, the wires in one direction are more narrowly spaced than those in the other direction. By the fifteenth century, paper-makers usually attached wire designs to the basic grid in order either to mark their own products or to differentiate particular sheet sizes. The paper would be slightly thinner (and thus more transparent) above each wire. Light transmitted through such a sheet of paper would reveal both the basic network of wires and the special design or watermark.

In 1993 all the watermarks found on the available dal Pozzo drawings and their mounts were photographed and classified by Martin Clayton, Assistant Curator of the Print Room, into 32 different types (or subject categories, such as Anchor, Bird or Figure). Within each type, watermarks were sorted by design into different groups: in a particular group the marks would all be identical in appearance and would have come from the same mould. Full details of all the watermarks found in the dal Pozzo drawings will appear as the 'Dictionary of Watermarks', to be appended to the multi-volume Catalogue Raisonné in due course.

A thorough analysis of watermarks in a particular series of drawings may reveal interesting patterns of paper use. However, the drawings discussed in this catalogue remain on the mount sheets and in the volumes where they were placed in the eighteenth century, and it is extremely difficult to see through the thick paper onto which the drawings were pasted. The watermarks were therefore neither studied nor photographed as part of the 1993 survey, and though fibre-optic light has been used to examine, trace and photograph the watermarks in a small number of cases, for the majority of the drawings catalogued no discernible watermark has been located.

Out of the 199 drawings included in the present catalogue, only 23 bear marks which were sufficiently discernible to be classified. All of these marks are unique to this catalogue and have not been found elsewhere in precisely this form. All but one of the marks show a Bird on hills within a circle surmounted by the letter 'G'. Seventeen of the sheets conform to a type now described as Bird 41B. The remaining five drawings have marks from the same group but with slight variants (Bird 41C and 41D). The only drawing here considered which does not to have this bird mark is

154 (a sheet showing fragments of bone with dendrites), which has a mark with an Anchor in a circle, found in this form on no other dal Pozzo sheet. Unfortunately such a small sample of watermarks provides few useful conclusions.

It is suggested above (on p. 34) that the drawings included in this catalogue were all produced during the same period, between about 1610 and Cesi's death in 1630. One of the drawings with a Bird 41B mark, **30**, bears an inscription stating that the specimen there depicted was found in 1611 and taken to Rome. Although this inscription does not date the drawing itself, it provides a *terminus post quem* for the sheet which fits readily into the period predicated.

Four of the manuscript pages which make up the Index section (see Appendix I) bear watermarks. The Index appears at the front of two of the bound volumes. Unlike the drawings, each of the Index pages has been inlaid into a larger sheet, thus rendering the watermarks easily legible. All four of the watermarked Index pages belong to the Bird type. The two marks identified (Bird 33 and Bird 45) are also variants of the Bird-on-hills (in circle) mark described above: one has a surmounting letter 'A', the other a pendant letter 'S'. To date, these two marks have not been found on any other dal Pozzo sheets.

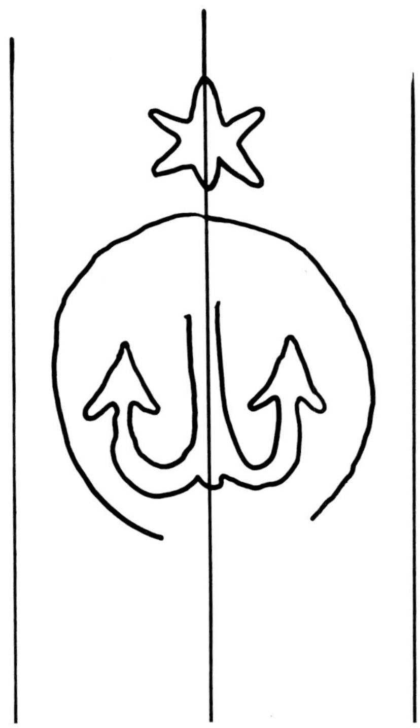

Fig. 44. Anchor 16B (tracing)

Watermark	**Drawing** (Cat. no.)	**Remarks**
Anchor 16B	**154**	Anchor in circle, with surrounding star
Bird 33	Index, Vol XIV, fol. 3	Bird on hills, in circle surmounted by 'A'
	Index, Vol XIV, fol. 6	
	Index, Vol XVII, fol. 1	
Bird 41B	**1**	Bird on hills, in circle surmounted by 'G'
	30	Inscription notes specimen found by Cesi in 1611 and taken to Rome
	54	
	51	
	54	
	61	
	69	
	70	
	72	
	76	
	79	
	90	
	94	
	137	
	147	
	152	
	162	
Bird 41C	**17**	Bird on hills, in circle surmounted by 'G'
	47	
	73	
Bird 41D	**64**	Bird on hills, in circle surmounted by 'G'
	198	
Bird 45	Index, Vol XIV, fol. 2	Bird on hills, in circle with pendant 'S'

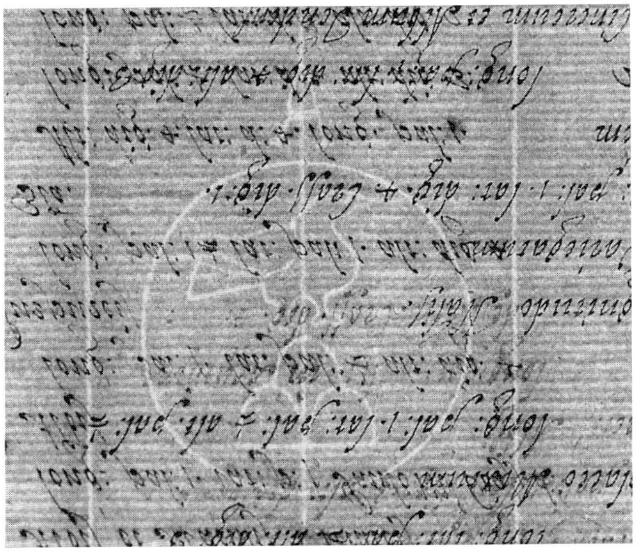

Fig. 45. Bird 33

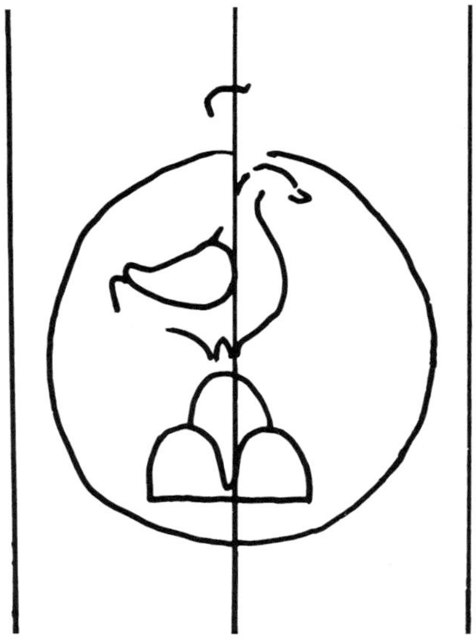

Fig. 46. Bird 41B (tracing)

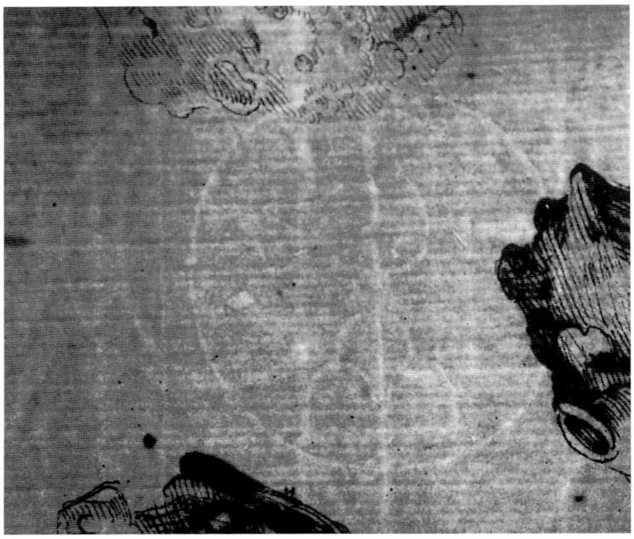

Fig. 47. Bird 41B

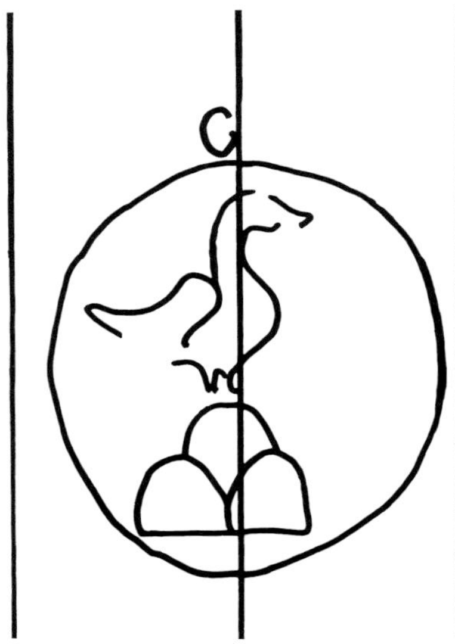

Fig. 48. Bird 41C (tracing)

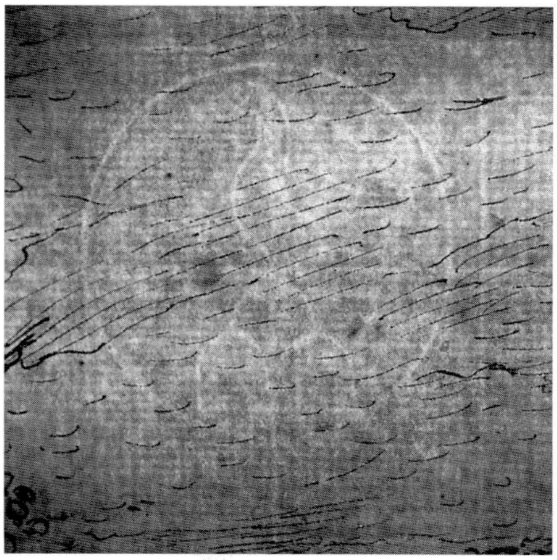

Fig. 49. Bird 41C

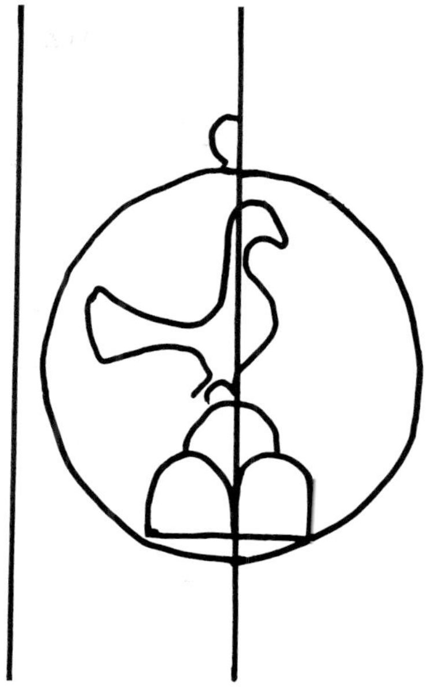

Fig. 50. Bird 41D

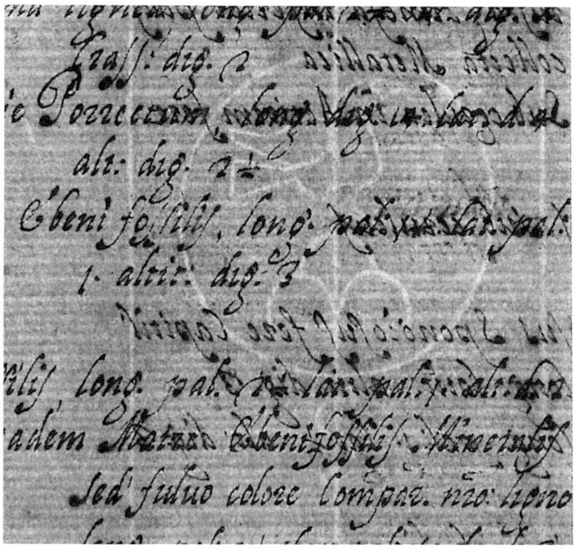

Fig. 51. Bird 45

GLOSSARY OF SCIENTIFIC TERMS

Andrew C. Scott

aetite: or eagle-stone: a clay or ironstone nodule; sometimes refers to some flint nodules.

ammonite: an extinct marine animal belonging to the cephalopod molluscs that have a chambered shell. Related to the nautiloid, squid and octopus.

angiosperm: a group of plants which have flowers.

anthracite: the final stage of coalification in the sequence peat—lignite—bituminous coal—anthracite.

bituminous coal: peat which has reached the black coal stage in the series peat—lignite— bituminous coal—anthracite.

bivalve: mollusc with two shells; [US] clam.

bp/ mybp: before present/ million years before present.

brown coal: a term used for lignite in the coal series.

calcite: a mineral composed of calcium carbonate.

cellulose: an organic compound which is found in plant cell walls.

charcoal: plant material which has been incompletely burnt.

coal: an organic rock (usually peat), containing more than 50% organic matter which has been altered by temperature and pressure with a loss of moisture and oxygen and an increase in carbon content.

coalification: the process by which plants and peats are altered by temperature and pressure during burial.

compaction: alteration by the application of pressure, usually by burial.

concretion: a cemented rock formed around a nucleus, often spherical in shape; usually caused by precipitation of minerals from solution.

conifer: a group of seed-bearing plants which include the firs, cedars and pines.

continental deposits: sediments deposited on land.

copperas: a name formerly used for all vitriols, a ferrous sulphate used as a mordant (fixative), dye pigment and coagulant in contemporary inks, medicine and tanning.

cortex: outer layer of specialized non-wood cells of a plant.

Cretaceous: a geological time–rock unit and the name of a Period of geological time (145.6–65.0 mybp).

dendrites: manganese oxide forming thin branching patterns on rock or fossil surfaces.

Eocene: a geological time–rock unit and the name of an Epoch of geological time (part of the Tertiary: 56.5-35.4 mybp).

'fool's gold': see pyrite.

Formation: a geologically mappable rock unit.

fossilization: the process which affects animals and plants to allow their preservation in rocks.

fromboids: minute crystal accumulations of iron pyrite; see pyrite fromboids.

fumarole: a sulphurous vent emitting vapours associated with volcanic activity.

fusain: fossil charcoal.

gastropod: an invertebrate mollusc, snail.

gelification: organic precipitation of cell walls during the peatification process.

Hettangian: a geological time–rock unit and the name for a stage of Geological time (208–203.5 mybp).

Horizon: a traceable or distinctive level in a rock sequence.

humic acid: organic compound resulting from the breakdown of cellulose in plant cell walls.

humins: organic compounds resulting from the breakdown of lignin in wood.

Jurassic sequence in the Fosso di Pozzale Monti Martani

Rock unit	Ammonite zone	Stage
Calcaia Posidonia	*Tmetoceras scissum*	Aalenian
Ammonico Rosso	*Dumortieria*	Toarcian
	Phymatoceras erbaense	
	Hildoceras bifrons	
	Hildaites serpentinus	
Marne del Serrone	*Dactylioceras mirabile*	Toarcian/Domerian
Corniola	*Emaciaticeras emaciatum*	Domerian

***in situ*:** in place, usually in position of growth.

Jurassic: a geological time–rock unit and the name of a Period of geological time (208–145.6 mybp).

lacustrine: living, or deposited, in a lake.

lamination: a thin layer of sediment.

lignin: an organic compound found in plants and giving rigidity, to wood in particular.

lignite: peat which has been slightly coalified as part of the peat —lignite—bituminous coal — anthracite series. Also known as brown coal.

limestone: a rock composed of calcium carbonate.

limonite: an iron oxide mineral with an earthy brown colour.

lithological: sedimentological composition of rocks.

marl: a mudstone which contains some calcium carbonate.

Miocene: a geological time–rock unit and the name for an Epoch of geological time (part of the Tertiary: 23.3–5.2 mybp).

mybp: million years before present

Oligocene: a geological time–rock unit and the name for an Epoch of geological time (part of the Tertiary: 35.4–23.3 mybp).

palaeobotany: the study of fossil plants.

palaeoclimatology: the study of ancient climates.

palaeoecology: the study of the way ancient animals and plants lived.

palaeontology: the study of fossils.

palynology: the study of fossil spores and pollen.

peat: an organic-rich deposit made from plants, containing more than 50% organic matter.

peatification: the process which converts organic matter to peat via biochemical processes.

permineralization: the process by which mineral matter precipitates within the cell spaces of plants.

petrification: a second phase of permineralization when the organic matter of the plant is replaced by mineral matter.

Pleinsbachian: a geological time–rock unit and the name for a stage of Geological time (194.0–187.0 mybp)

Pleistocene: a geological time–rock unit and the name of an Epoch of geological time (part of the Quaternary: 1.64-0.1 mybp).

Pliocene: a geological time–rock unit and the name of an Epoch of geological time (part of the Tertiary: 5.2–1.64 mybp).

pollen: microscopic propagules of seed and flowering plants which fertilize the ovule to produce the next generation of plants.

pyrite: or iron pyrites, a golden mineral made of iron sulphide, also known as 'fool's gold'.

pyrite fromboids: aggregates of small pyrite crystals.

pyritization: the process by which iron pyrite is precipitated from water or in mud, usually due to the decay of animals or plants. One of the minerals involved in permineralization and petrification.

rank: the position of a peat or coal in the coalification series.

Recent: the Period of geological time from 10,000 years bp to present(= Holocene).

relict population: a small population of plants or animals which at one time had a much larger geographical range.

resin: sticky organic polymer produced by some plants.

sedimentary rocks: rocks composed of particles of other rocks such as sand.

septal suture: the wall separating chambers in an ammonite.

siderite: an iron carbonate mineral usually yellowish or brown in colour.

silica: a mineral composed of silicon dioxide and often found in permineralizations and petrifications.

spore: the microscopic propagule of primitive plants such as ferns.

Stage: the name for a time–rock unit in geology.

stratum: a bedded rock unit.

System: the name for a time–rock unit in geology comprising several Stages.

taxonomy: a system of classification for the natural world.

Tertiary: a geological System and the name of a Period of geological time (65.0–1.64 mybp).

Time-scale: see Table below

Toarcian: a geological time–rock unit and the name of an Epoch of geological time (187.0–178.0 mybp).

ulmin: an organic precipitate formed by the mixing of humins and humic acid.

venation: slender bundle of tissues (usually tracheids) forming part of the framework of a leaf.

vesicular: full of holes (often formed by escaping gas).

vitriol: sulphuric acid or a sulphate resulting from the oxidation of sulphide minerals.

Tertiary sequence of the Tiberino Basin

Formation	Age
Acquasparta Formation	Early Pleistocene (Quaternary)
Santa Maria di Ciciliano Formation	Early Pleistocene
Ponte Naja Formation	Late Pliocene
Fosso Bianco Formation	Mid–late Pliocene

Time-scale

Geological Period/Epoch	Age (million years before present)
Quaternary	
Holocene (=Recent)	10,000 years bp
Pleistocene	1.64–10,000
Tertiary	
Pliocene	5.2–1.64
Miocene	23.3–5.2
Oligocene	35.4–23.3
Eocene	56.5–35.4
Palaeocene	65.0-56.5
Cretaceous	145.6-65.0
Jurassic	208.0-145.6

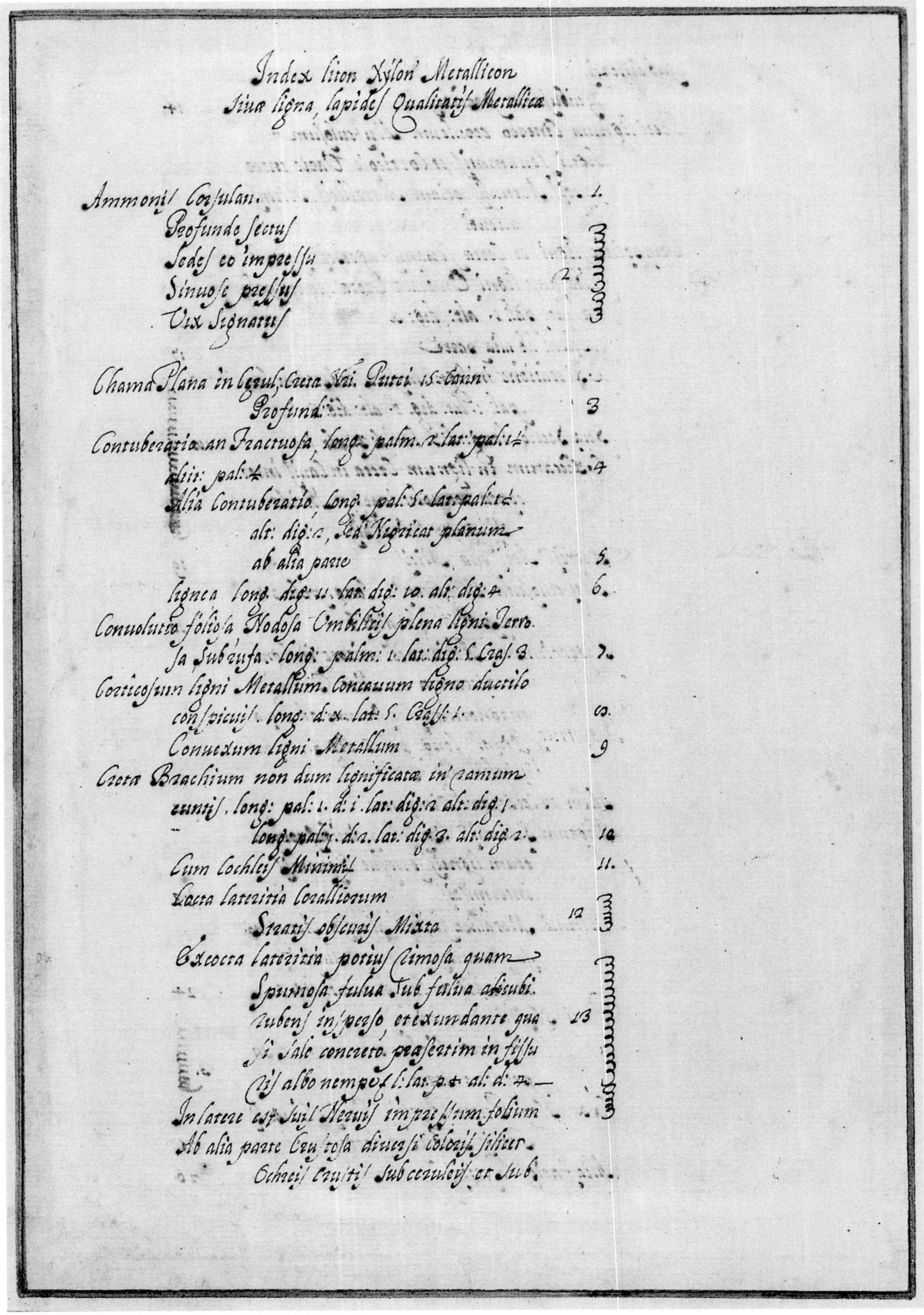

Index liton Xylon Metalliton
Siue ligna, lapides Qualitatis Metallicæ

Ammonis Cornu . . . . 1.
Profunde sectus
Sedes eo impressu
Sinuose pressus . . 2.
Vix signatus

Chama Plana in lapis Creta Vai. Putei 15 . . .
Profund. 3
Contuberatio an Fractuosa long. palm. 2 lat. pal. 1½
alti: pal. 1 . . . 4
Alia Contuberatio long. pal. 5 lat. pal. 1½
alt: dig. 2, sed Nigricat planum
ab alia parte . . 5.
lignea long. dig. 11. lat. dig. 10. alt. dig. 4 . 6.
Conuolutio foliosa Nodosa Umbilicis plena ligni ferro-
sa subrufa. long. palm. 1 lat. dig. 5 Crass. 3. 7.
Corticosum ligni Metallum. Concauum ligno ductile
conspicuis. long. d. 4 lat. 5. Crass. 1. 8.
Conuexum ligni Metallum 9
Creta Brachium non dum lignificata in ramum
euntis. long. pal. 1. d. 1 lat. dig. 2 alt. dig. 1 .
long. pal. 1. d. 2. lat. dig. 3. alt. dig. 2. 10
Cum Cochleis Minimis 11.
Creta lateritia Corallioidum
Steatis obscuris Mixta . 12
Excocta lateritia potius Limosa quam
Spumosa fulua sub fulua alicubi
rubens inspersa, et exundante qua . 13
si sale concreto præsertim in fissu
ris albo nempe s. l. lat. p. 4 al. d. 4
In latere est sius Neruis impressum folium .
Ab alia parte Crustosa diversis coloris silicet .
Ochreis Crustis subcerulesis et sub.

Fig. 52. A page from *Natural History: Fossils,* Vol. XIV, fol. 1r: Index

APPENDICES

APPENDIX I

Seventeenth-century Indices to the Fossils Drawings

Two seventeenth-century indices to the drawings exist. The first is mounted in the beginning of Volume XIV and covers the contents of Volumes XIV, XV and XVI. The second index, mounted in the beginning of Volume XVII, covers the coloured drawings contained in that volume. The handwriting of both indices is that of the scribe normally employed by Cassiano (see above p. 38). The indices appear to be based on the classificatory descriptions (derived from Cesi's annotations) on the drawings; they may well have been compiled during the 1630s, simultaneously with the preparatory work for the *Trattato*. The wording is similar to, but not identical with, that of the inscriptions of the drawings.

In the left-hand column below the numbers are those included in the manuscript indices. These numbers correspond to those still found on the drawings (for which see Concordance C). With the help of these indices it has been possible to identify the numbers originally inscribed on three currently unnumbered drawings (**34**, **57**, **66**). The right-hand column gives the catalogue numbers that relate to the index entries. Where these are given in parenthesis, the descriptions correspond but the inscribed numbers do not. Only six catalogue numbers are not included in this chart: **16**, **26**, **51**, **81**, **83** and **99**.

The title of the Index in Volume XIV may be translated as: 'Index of stone wood metal [specimens] or wood with the qualities of stones and metals'. The title of the Index in Volume XVII may be translated as: 'Other coloured woods follow, as found in mines'.

[*Fossils, Vol. XIV, fol. 1*: see Fig. 52]

Index liton Xylon Metallicon

Sivae ligna, lapides Qualitatis Metallicae

Ammonis Corsulan.	1	**197**
Profunde Sectus		
Sedes eo impressu		
Sinuose pressus	2	**198**
Vix Signatus		
Chama Plana in Caerul, Creta N[ost]ri Putei 15. Cann		
Profund:	3	**151**

Contuberatio an Fractuosa, long: palm. 2. lat: pal: 1½		
altit: pal: ½	4	**85**
Alia Contuberatio, long. pal: 5: lat: pal: 1½.		
alt: dig: 2. Sed Nigricat planum		
ab alia parte	5	**84**
Lignea long. dig: 11. lat: dig: 10. alt: dig: 4	6	**106**
Convolutio foliosa, Nodosa Umbilicis plena ligni Terrosa		
Subrufa. long: palm: 1. lat: dig: 5. Cras. 3.	7	**109**
Corticosum ligni Metallum. Concavum ligno ductilo		
conspicuis. long: d: x, lat: 5. Crass: 1.	8	**120**
Convexum ligni Metallum	9	**121**
Creta Brachium non dum lignificatae in ramum		
cuntis, long: pal: 1. d: 1. lat: dig: 2 alt: dig. 1.		
long: pal: 1. d: 2. lat: dig: 3. alt: dig: 2.	10	**53**
Cum Cochleis Minimis	11	**199**
Cocta lateritia Coralliorum		
Stratis obscuris Mixta	12	**157**
Excocta lateritia potius rimosa quam		
Spumosa fulva sub fulva alicubi		
rubens insperso, et exundante quasi	13	**181**
sale concreto. praesertim in fissuris		
albo nempe fl: lat. p. ½. al: d: 4. -		
In latere est suis Nervis impressum folium		
Ab alia parte Crustosa diversi coloris silicet.		
Ochreis crustis sub ceruleis et sub		

[*fol. 1v*]

sub cinereis		
quibusdam partibus	14	**182**
Creti lignum Cinereo ceruleum Musculosum ligni squammis,		
et Corticose Varie interceptis Arenulis etiam		
Metallicis, Nitens alicubi	15	**113**
Compositio ligni et Cretae Squammis agrestis simul'		
alba sunt ligni Umbratae Cretae. long. pal 2.		
lat: pal: 1. alt: dig: x	16	**123**
Idem ab alia parte	17	**124**
Exiccatione in lignum fere redactum lon.		
pal: 1. lat: dig. 6. alt: dig. 3		
Sua Sectio long: d: 7. lat: d: 3. alt. 1½.	18	**66**
Exiccatum in lignum Creta in Cavis intercepta		
convolutiones, et Nitida...		
Magnit: Cir:[a] Nalis.		**95**
Ferreis Aerisquae Mitis	19	**133**

Musculosum Nigricans, long. pal: 2½ lat: pal: 2. alt: dig: 5	20	**107**
Nigerrimum Acciavacci nitore in lignum exiccatum ex M[at]re 3ium magnorum frustorum ad Rosarium. long: pal. 3. dig. 3. lat. pal. 1. altit dig. 3.	21	**58**
Idem exiccatione rimosum	22	**59**
Nigrum squamis Varijs Crassioribus compactum ligneis quoque Squamis intermistis	23	**115**
Nitenti Metallica scabie aspersa Aeruginosa squamosa nigra fulvosa long: dig. 7. lat: d: 11. alt: dig. 3	24	**128**
Putei Nostri cum Cochleis Coniunct: Valu. XIV. Cann. a superficu Terrae	3	**151**
Ex N[ost]ro. Puteo Pyritis Venis Virgulis Granis Scabiequae freta	25	**144**
Alia facies Pyritis	26	**145**

[*fol. 2*]

Sectio a presso latere ipsa facies long. pal. 1. lat. dig. 3.	27	**57**
Spumosa Ustae, long. pal: 2. altit: et lat p. 1.		**158**
Squamis ligneis mixta Creta, long. pal: 1½ lat: pal: 1. altit. dig: 4.	28	**116**
Squamosum Crassioribus Squamis.	29	**122**
In textum Cocta rubens fissilis. long. dig: 8. lat: dig: 4. alt: dig. 3.	30	**155**
Alia long: pal: 1. lat. dig. 4. lat: dig. 3.		
In Textum rubens cocta, long. dig: 10. lat dig: 8.. alt: dig. 3.	31	**156**
Crustosa Corticosa ligni Terrea long. dig. 14. lat dig. 8: Crass: dig. 1½	32	**61**
Similis eiusdem Magnitudinis lapidea		
Crusto squama lignea long. pal: 2½ lat: dig: x Crass: dig. 2	33	**110**
Varie Porrectum, long. dig. 14 lat: d. 4 alt: dig. 2½	34	**60**
Creta Mater Ebeni fossilis, long. pal: 1½: lat: pal: 1. altit: dig. 3	35	126
Ebenus fossilis, long. pal: 2½ lat: pal: 1. alt: d: 2	36	**63**

Ex eadem Matre Ebeni fossilis Mineralis sed' fulvo colore compar[e]. n[ost]ro: ligno long. pal: 2 lat: pal: ½ alt: d: 3	37	**35**
Ex P[onte]: Card[ano]: cum Virgulis Metallicis interceptis, Niger medio Fulvus. long. pal: 2. lat: pal: 1. alt: palm. ½	38	**36**
Ex P[onte]. Cardano frangibilis Gummosus ad. Dini Accedens, long. pal. 1. lat: dig. 4. altitudo dig. 2.		
Nigrum Nitidum Gagat[e] Frangibilis Gummosus Magnitudo Na[tura]lis.	39	**62**
Ex Ponte Cardano long. palm 1. dig: 8 lat: dig. 7. altit: dig. 3.		**34**

[fol. 2v]

Aetitis affinia frusta Crispa	12	**157**
Arte angulatus	40	**134**
Bombicyophiton	41, 42	**135**
Christallus Intestina	41	**135**
Collectitium Saxum Rimosum	43	**140**
Cortices	46	**146**
Crusta Cochleis minimis intercepta	47	**149**
ex Duna robba	45	**148**
Fracti lapi Resinae	44	**141**
Globulosus contortus lapis ex Aetitis. Contubernio ipsi affinis	43	**140**
Globus ferrae compactionis Scissus	46	**146**
Integer	46	**146**
Intestina cretosa Rubentia	42	**135**
lateritia Crusta rimosa inferior pars collecta Metallica	48	**136**
Sub caerulea in Aetitem fere	48	**136**
Magnus	42	**135**
Metallicus sub ceruleus	48	**136**
Minor	41	**135**
Monstruosus spongiosus fere Capitil long. latit. altit. dig. 3 long: pal: 1. lat ½ alt: dig: 3	49	**142**
Pars lateritia ab alia parte	50	**129**
Rubra facie cretosa Tingens	46	**146**
Scabro arenoso cortice	40	**134**
Solidus Macula Aetitis Vacuus Aemulatus	48	**136**
Squamis multis	46	**146**
Vacuis septa	40	**134**
Variae crustosi	51	**137**
Varii	52, 53	**138, 139**

Fasciae Tenue Variae quasi coriacae Sub nigrae	54	**67**
Crassiuscula Crispata Tortuosa quasi cocto		
[*fol.* 3]		
Corio alicudi sub rubra	54	**67**
Corticosa		
Quasi cinturae Anulus	54	**67**
Tenuissimae quasi gladius plana prorsus		
Ferruginea Spumosa	55	**178**
Spumosa exprimens quasi ligni cortice colore		
Suberaceo long: d. 14. lat. d: 8. crass.		
ferae 2, et long: d: 11. lat: d: 7 alt: d. 3½	56	**127**
Squamosa coniungatur cum Superiori	57	**108**
In frustis Cornu Monocerotis Sivae Terrae Eretriae		
Albae squamosae dentroites	58	**154**
Frustis Metalliginis	59	**119**
Alia facies frustis Metalliginis media lateritia		
parte, long. d: 4. lat. 3. crass. c.ª 3.	60	**(118)**
Frustum Anceps ligni Terreum ab una parte lignum	61	**(117)**
ab' alia creta		
Prope Magna frusta Rosarii	62	**15**
Tria ligni frusta cum sectionibus ad Scismanum	63	**8**
2m. Frustum ligni	64	**7**
3m. Frustum ligni	65	**11**
Ager, seu Vallis ad Tria illa frusta ingentia	66	**6**
Frustum ligni sectio	67	**9**
Prope contra frusta Ingentia ad Scismanum	68	**5, 10**
Nodo in medio exuberante	69, 115	**47, 24**
Tribus volutis nodo et plexu Varium,		
long. pal. 7. lat: palm: 1½ alt: pal. 1.		
2m long: palm. 6. d. 2 lat: palm. 1½ alt dig: 9.	70	**52**
Undatum	157	**76**
Lapideum foliosum sub rufum long: pal: 1½. lat: d. 3		
crass: dig: 1½	75	**55**
Folijs compactum sub rufum, long. pal. 1½	76	**40**
Nodosa Tuberatio, long: d: 8. lat: d: 6. crass. d. 5		
Rubens extra Veste lignea, Nigra Rugosa	77	**19**
long: pal: 6. lat: pal. 1 crass. dig: 4		

[*fol. 3v*]		
Veste lignea nigra, long: palm: 2½. lat: pal: 2 alt: pal: 1	78	**28**
Lapis Conchites Firmanus, long. pal: 1. lat: p: ½ a:t p ½	79	**153**
Lapila Terri lignum homogeneum sub ochreus tingens		
long. pal: 2½ lat: d: 4. alt: d. 2		
long. pal: 1. alt: d: 2	80	**38**
Idem. long: pal: 2: lat: p.1. alt: d: 1.	81	**71**
Lapi factum ex Albo et Fusco Varium, long: pal: 3.		
lat. pal: 1. alt: palm: ½		
long: pal: 2.½ lat. pal. 1. alt. c.ª pal. 1.	82	**73**
Lapi lignum ex Duna robba		
long. pal: 3. lat: p. 1. d .2. alt: d. 10	83	43
Eius Sectio ex toto lapide		
Idem et Icones rudiores cum Sectione		
long: pal: 2 ½. lat: d: 10. alt: d: 6.	84	**69**
Crustis diversis lapi Ter.r. lig., long. pal 2 lat: palm: 1. alt: dig. 1.	85	**91**
Fere tota petrefacta Album ferens Squamis crassioribus, long. d. x. alt: et Crass. dig: 5	86	**105**
In lapide plano Videtur quid' ferreum Corticosum Spiritillis interceptum	1	**197**
Intus lapideum extra ligneum Nigro Fulvum, long: pal. 3. lat: pal: 3 alt: palm. 1½	87	**98**
Lapi lignum Mixtum Contuberatum long: pal: 2. lat: palm. 1½ alt pl. 1.	88	**29**
Lateritium ligni Musculis rubens cavernosum Nervosum.	89	**114**
long: pal: 2. alt: d: 3. crass. d: 6.		
Idem ab alia parte	90	**97**
Ligneum, long. pal: 2. lat: pal: 1.½ alt: dig. x	91	**75**
lignea long: pal: 2½. lat: d: 6. alt. d. 1.	92	**56**
[*fol. 4*]		
Lignea Capita Nodo voluta, aut Nodo fasciata		
long: pal. 5. lat. pal: 1. d: 8. alt. dig: 9.		
Eius Sectio	93	**77**
long: pal: 1. d: 8 lat: pl: 1. d 2. al. d. 5		

Ligneum Caput cum Intruso ramo in fractione, et		
Creta Mixta in Cavitate fractionis.	94	**54**
long: pal: 1. lat: pal, med. alt: p ½		
Corticosum suberaceum Crispum levius		
long: pal: 1. lat: pal: 1. d: 2 alt: d. 2½	95	**21**
Costae erectae procurs. Ambo praessi		
long: pal. 7. lat: pal: 1. alt: d. 3.	96	**45**
long: palm. 4.½ lat: d: 8. alt: d: 2.		
Costae pressae		
long: pal: 6. lat: p. 1½ alt: d: 4.	97	**46**
long: pal: 3. lat. dig: 7. alt: d: 3.		
An Fractuosum		
long: dig: 11. lat: d: 10 alt: d. 4.	29	**122**
Lapiferreum		
long: pal: 2. lat: pal: 1½ alt: d: 2	98	50
Metallum e ferrugine Vergens ad. Colorem		
Ferretti di Spagna dicti eo Tingens		
long: dig: 14 lat. pal: ½ Crass. dig: 5.	99	96
Metallum cum Corticibus lateritijs		
long: dig: 14 lat: dig: 8. crass fere. 2	56	**127**
Metallum ductibus conspicuis et crusti lat[er]ijs		
long: palm: 1. alt: pal: ½ Crass: d: 1.	100	**112**
Metallum ductibus ligni, et Crustosa parte coniunctis		
long: dig: x. lat: dig. 3 Crass: dig: 2	101	**64**
Nigrum cum Ramo Ponderosum intus lapideum		
long: pal: 2. lat: pal: 1½ alt: p.½	102	**65**
Nigra		
long: pal. 2½ lat: p. 1. d. 8. crass. p. 1	88	**29**
Nodo Volutum		

[*fol. 4v*]

long. pal. 2½, lat. p. 1½, crass. pal. 1	103	**74**
Nodo Volutum praessum Navicula facie		
Turbine acuto emisso		
long. pal. 2½ et Turb crass. circa		
pal: 1.	104	**88**
Nodo volutum productum caudo culatum		
long. pal: 2½ lat: pal. 2½ alt: p.1.	105	**90**
Nudi Metallificati ligni Metallum Superius		
quadam Tenus Spumosum	106	**104**
Laxius aliud solutius exoticus.		
long. d: 8. lat: d: 6. alt: d: 4	106	**104**
Ponderosi ligni Metall. nodo		
long: faere. dig: 8		
Sulphureum Terrosum floribus Sulphureis		
Scabrosis		
long: palm. 1½: lat: d: 6. alt: dig. 3	107	**41**

Terr. Nigr:		
long: pal: 3. d: 3. lat: d: 8. alt: d: 4	108	**42**
Terr. Nervosum fulvo fuscum		
long: pal: 1½. alt: d: 3. crass: d: 2	76	**40**
Terrosa	75	**55**
long. pal. 1. lat. d: 8. crass: d: 4		
Tuberatio nodosa ligneo colore		
long: palm. 5. Lat. 3. Crass: 1.	109	**101**
Undatum		
long: pal: 1½ lat: d: 10, alt: d: 3.		
In secta ligni crassi parte media	110	**44**
Sectio, Naturalis Magnit:		
Undatum Supina parte		
long: pal: 1½ lat: dig: 10: alt: d: 3.		
long: pal: ½. d: 4. lat: d: 7. alt: d: 5.	111	**79**
Lito Xila intus Lapidea extra lignea ex Duna Robba		
long: pal: 6. lat: pal: 1. d: 3. alt: d: 6	114	**20**
Sectio, lat pal. 1 d: 4. alt: pal: 1½	83	**43**

[*fol. 5*]

Lito Xilum ex Duna robba		
long: palm. 4. d: 4. lat: p: 2. alt: pal: 1.		
Idem	115	**24**
long: pal: 3. d: 4. lat: pal: 1. alt: d: 4.		
Metalligo convoluta	118	**111**
long: pal: ½ lat: dig: 4. alt: dig: 4		
Floribus erectio quasi Corallis cullis speciosa	117	**(132)**
Metallophiti ligni Secta Sponte capita cum lapi resina ad' creta	116	**125**
Metallophiti ligni Squama pressa et Volutata reflexo Simplici Anno 1611. a.' nobis inventum, et Romae allat.		
Magn: N[atura]lis. Crass: dig: ½	119	**(30)**
Ex Mytulo Creta Concava	11	**199**
Mytuli cretae cerulae N[ost]ri. Putei intercepti Apertis Valvis fractus 12 cann. a' Superficie Terrae		
Coniunct: Valv. 14. Cann. a' Sup Terr.	3	**151**
Ex Monte Nigro Musculis intus interiecto Creti ligni in gyrum dispositis		
long: pal. sivae. d: 10. alt: d. 5. lat: d: 7.	120	**37**

Ex Picchia Niger Usto colore rimosus Terrosus floribus		
Vitrioli		
long. pal: 3. lat: pal. 1. alt: d: 6.	122	**48**
Nigra ligna Crassa Veste intus aliquando lapidea		
long: palm. 3. lat: p. 2. crass: pal: 1.	121	**100**
Nigrescens cum Interiori exiguo lapide lignum		
nodi Tuberati		
Alt: pal: 2. ½ lat: 2 Crass. 2½	112	**80**
Nigrescens nodo Volutum lapi lignum Dunarobbae		
forte aliquantulum Metallico pondere		
long: pal: 3. lat: p. 1. alt: palm: 1.	123	**78**
Nigricans extra lapideum Intus Lapideum et Varius		
[*fol. 5v*]		
Varium ex Duna robba		
long: palm 2.½ lat: pal' 1½ alt: d.x	136	**25**
Nigrum ex Scismano		
long pal: 4. lat: p. 1. alt: p. ½	124	**82**
Nodosa Ligna		
Palm 1. per ogni Verso	125	**33**
Ramos promittens		
long. pal: 1½ lat: pal: 1. crass. d: 4	126	**86**
Idem	127	**87**
Medio Terrosa		
long: pal: 1. lat: p. 1. d: 3. alt: d: 8	128	**32**
Nodo Volutio Turbinata nigra		
long: pal: 3. lat: pal: 2. alt: d: 4.		
long: palm. 3. lat: d: 8. et crass.	129	**72**
Nudosa ligna		
Voluta Utrinque crustosa		
long: pal: 2. lat: d: 10. alt: d: 6	130	**92**
Idem.	131	**89**
Petrefactum aliqua medio in nodo ligni particula		
Album		
long: palm. 2½ lat: pal: 1. alt: p.½	132	**70**
Cavum veluti ad' Corticem, Musculus		
ob rotundus Albus		
long. pal: 1. alt: pal: 1. lat: rotund.	133	**49**

Ex Duna robba in Corde parum ligneum		
long. pal: 2½ lat. p: 1½ crass: p. 1.	134	**39**
An Fractuosum		
Petri lignum extra et pro Maiori parte lapideum		
intus alicubi ligneum		
long: pal. 5 lat: pal: 2. alt: pal: 1½	135	**25**
Ex Duna robba		
long: pal: 5. lat: pal: 2.	84	**69**
Pendroites Breviosa Christallina Tuberculosa	136	**143**
[*fol. 6*]		
Ramus ligni Cretosus ad pact~ in Ramum A' Creti		
lignum et in' Massa B' Sub Ala et A'		
Sciunctum Ramum	137	**68**
long: pal. 1. d. 5. lat: d: 5. alt: d: 2.		
Multi nodosus ab Utraque parte		
long: pal: 3. lat: d: 3. alt: d: 3.	138	**18**
Ramulos Tuberculose prominens		
long: pal. 1. d: 3. lat: d: 3. alt: d 1½	139	**22**
Idem, long: pal: 7, d: 4. lat: pal: 2. alt: d. 2.	95	**(23)**
Ex Visceribus crassi[ssi]me molis ad' Rosarium		
Secta Superficies Transversis rectisq		
multis confertis Nodis comprehensis		
ex' Ramulis	140	**23**
In Ramum Nitidum Acciavacci fere		
long: pal. 1½ lat: d: 5. alt: d: 2.	34	**60**
Serpens lapideus	11	**199**
Squama Ampla	5	**84**
Stratum ad' Duna robba Supra cretam	141	**2**
A' Duna robba, sed' Torrens Medio confundit	142	**1**
ligneum Crassum Super Ingentia frusta ad' Scismanum	145	**3**
Ad' Scismanum prope frusta ingenti	143	**4**
Inter Scismanum et Rosarium	144	**14**
Tartar. ex N[ost]ro. Puteo in Profundum 18 Cann. cum		
Turbinibus, et Cochleis rubens	153	**152**
Terri ligni 3bus. Nodis ab alia parte		
long: pal: 6½ lat: pal: 1. d: 4. alt: d: 7.	140	**23**
Tiburtinum cochleis, Marmor Aqua Spartae	146	**147**

Tractus prope Ignem Cretosus praefractus	149	**13**
Peroscioli Strato Secante	148	**17**
Sive Ora Fatiscens intra schismanum, et Rosarium	147	**12**
Tuberatio ligni		
Magnitudo Na[tura]lis:	150	**27**
Turbinatum pressum ligneum		
[*fol. 6v*]		
long. pal: 2: lat: pal: 1. lat: d: 3	152	**93**
Pressum Sectio media	151	**94**
Vitrioli flos Capillaris è ligno Capitonij Virgulatum		
Metallicarum		
long: pal. 3. lat. p: 1. d: 3. alt: d: 8	154	**130**
Voluta lignea Sulphurea Scabie conspersa		
long: pal: 1½ lat: p. 1. alt: dig. 4		
Eius Pes.	155	**102, (103)**
Pressa Obrotunda		
long: pal: d: 8. lat: d: 10. alt: dig: 9	156	**(76)**
Pressa Pyramidalis		
long: pal: 2½ lat: pal: 1½ alt: dig: 7	157	**76**

[*Fossils XVII, fol. 1*]

Sequuntur alia ligna Colorata

Ut ex Fodinis Eructa

Aureum Purpureum		
long. lat: palm: 1. alt: dig: 4	1	**183**
Ceruleum lineis Varijs obductum		
long: pal: 1½. lat: pal: ½: alt: dig: 8	2	**184**
Cinereum	3	**185**
long: pal: 1½ lat: pal: ½ alt: dig: 4.		
Cinereo Ceruleum Pallidum cum Cochleis		
long: pal: 2. lat: pal: 1½ alt: dig: 6	4	**150**

Punctis Sulphureis et lineis Nigris
long: lat: pal: ½. alt: dig: 3. 5 **186**
Violaceo Mixtum
long: pal: 1. lat: pal: ½ alt. pal: ½ 6 **187**

Coralloides
Magnitudo Na[tura]lis. 7 **159**

Coriaceum Variegatum
long: pal: 1. lat: dig. 4. Crass. dig: 1. 8 **188**
Idem 9 **179**
long: pal: 1. lat: dig. 4 alt: dig. 3.

Cretosum ad Cinereum et Album Tendens
long: lat: pal: ½ alt: dig. 2 10 **189**

Croceum Punctis Albis
long: pal: 1. lat: pal: ½ alt: pal: 1. 11 **160**
Parvulis Maculis interpositis
long. lat: alt: pal. 2 in Cir[a] 12 **190**

Crustosum Multiplici colore
long. lat. pal: 1. Crass. d: 1. 13 **177**

Fuscum, intus Aureum Punctis Argenteis foris
long: pal: 1. dig. 3, Alt: et lat: dig.،8. 14 **131**

Fuscum in Medio lateritium

[fol. 1v]
long. pal. 2. lat: pal: 1. d: 2. Alt: pal: ½ 15 **31**
Punctis Aureis.
long. pal: ½ lat: dig. 4. alt: dig. 2. 16 **191**

Gagat[e] non absolute Punctis Albis
long: et lat: pal: 1 in Cir[a]. Crass: dig. 5 17 **192**

Lateritium Foraminosum
long: lat: pal: 1. Crass. dig. 4 18 **161**
Fulvo, et Albo Mixtum
long. lat: pal: 1. Crass. dig: 1 19 **162**
Ad Gagat.[e] Tendens
long. lat: pal: 1. Crass. dig: 5 20 **163**
Mixtum
long: et lat. pal: 1. Crass. dig. 4 21 **164**
Punctis Albis, et Maculis Violaceis
long: pal: 1. lat: p: 1 Paulo M[inu]s: Crass. d: 22 **165**

Punctis Albis		
long: pal: 1. lat: pal: ½ alt: dig: 4	23	**166**
Strijs Purpureis		
long: pal: 1½ lat: pal: 1. alt: dig. 4	24	**167**
Varia 3ia		
Alt: dig: 4. lat: d: 4. long. pal: 1.		
long: pal: ½ . lat: d: 4 alt: dig. 3.	25	**168**
long: pal: ½ lat: d: 3. alt: dig. 2.		
Violaceum		
long: pal: 1. lat: quasi pal: 1. crass. d. 4	26	**170**
Mixtum		
long: lat: Crass. pal: 1.	27	**(169)**
Metallicum		
long: pal: 1 in Cir^{a}. lat: p: ½ altit: dig: 4	28	**171**
Idem		
long: pal: ½ in Cir^{a}. lat eiusdem Paulo-minus Crass. dig. 4	29	**172**
Nigrum		
long: pal.½ lat: dig. 3. altit: dig. 2	30	**193**
[*fol.* 2]		
Nigrum ut Superius		
long: lat: pal: ½ . altit. dig. 2	31	**194**
Pompholigum	32	**180**
long: pal. ½ lat: alt: dig: 4		
Puniceum Varijs Maculis		
long: pal: 1½ lat: pal: 1. Crass. dig: 4	33	**173**
Suberaceum Foraminosum		
long: pal: ½ lat: dig: 4. alt. dig: 3.	34	**174**
Sulphureum Ceruleo Mixtum		
long: pal: 1. lat: dig: 4. Crass. dig. 2.	35	**175**
Variegatum		
Mag[nitudo]. na[tura]lis	36	**176**
Violaceum Maculis Sulphureis		
long: pal: ½ . lat: dig: 3. alt: dig. 1.	37	**195**
Punctis Albis		
long. lat: pal: 1. in' Acumine desinen. Crass. dig. 4.	38	**196**

APPENDIX II a

Transcription of Stelluti's *Trattato*

[*Stelluti 1637, title page*: see Fig. 53]

TRATTATO
DEL LEGNO FOSSILE MINERALE NUOVAMENTE SCOPERTO
NEL QUALE BREVEMENTE SI ACCENNA LA VARIA, & mutabil natura di detto Legno, rappresentatovi con alcune figure, che mostrano il luogo dove nasce, la diversità dell'onde, che in esso si vedono, e le sue così varie, e maravigliose forme.

DI FRANCESCO STELLUTI ACCAD. LINCEO DA FABRIANO

All'Emin.mo & Rever.mo Sig. Card. FRANCESCO BARBERINO

IN ROMA, appresso Vitale Mascardi, MDCXXXVII.
CON LICENZA DE'SUPERIORI
Bibliothecæ Domus Romanæ Pauperum Matris Dei Scholarum Piarum.

[*Stelluti 1637, p. 3*]

ALL'EMIN.MO ET REV.MO SIGNORE E PADRON MIO COLENDISSIMO, IL SIGNOR CARDINALE FRANCESCO BARBERINO.

DIEDI l'anno passato a V. Eminenza un mio breve trattatello a penna di quel legno fossile, che nasce in queste nostre parti, il quale in quella guisa distesi per curiosità di molti, che di legno così strano desideravano haverne una vera, e fedele relazione: e perche dopo me ne son state dimandate molte copie, non potendo con esse a tutti soddisfare; ho per ciò risoluto darlo alle stampe, con aggiunta d'altre figure assai curiose, per maggior soddisfazione de'lettori: acciò che col mezzo della vista, restino maggiormente appagati, & persuasi di quanto in questo trattato saranno per udire, il quale dovendo io publicare con le stampe, ho voluto a V. Eminenza dedicarlo, sì per il buon gusto, ch'ella suol prendere della contemplazione degli occulti parti della natura; si anche perche porti in fronte un / [*p.* 4] testimonio così degno, & eminente, per render'altri sicuri, che non propongo loro cose finte, e favolose.

Porto avanti a V. Eminenza una materia in vero totalmente nouva, & ambigua; ma verrà ciò contraposto all'antica servitù, che le professo, & alla mia indubitata osservanza, & divozione, a che s'haverà ella risguardo, potrò sperare da lei l'aggradimento di questa ancor che piccola offerta; e tanto più, per esser'io certo, che il non dispregiare le cose humili è proprio dell'alta, & innata generosità dell'Eminenza V. alla quale riverentemente m'inchino. Di Roma li 8. di Maggio 1637.

Di V. Eminenza Humiliss. e devotiss. servitore

Francesco Stelluti.

[*Stelluti 1637, p. 5*]

TRATTATO Del Legno fossile minerale nuovamente discoperto DI FRANCESCO STELLUTI LINCEO.

I PARTI della Natura non si può negare, che non siano tutti maravigliosi, ma molto più mirabili si rendono quelli, che di rado suol ella partorire, perche son men da noi conosciuti; come avviene di questo legno fossile, o sotteraneo, che vogliamo nominarlo; detto dal nostro Signore D. Federico Cesi Principe di S. Angelo, e Duca secondo di Acquasparta, da cui fù gli anni passati discoperto, Metallofito per la natura, c'ha della pianta, e del metallo: se bene considerato in se tutto, molto più participa della pianta, che del metallo.

Questo legno per la sua rarità è degno di maraviglia; non sapendosi fin hora che nasca in altre parti, & che n'habbia Autore alcuno scritto. E se bene Teofrasto (come scrive Plinio nel lib. 36. cap. 18. della sua Istoria naturale) dice trovasi l'Ebano fossile di bianco, e nero colore; e che di terra si generano ancora l'ossa: non perciò fa mentione alcuna di questo legno ondato, che nasce in queste nostre parti. E non solo è mirabile per la rarità, ma ancora per la varietà dell'onde, e sì belle, che in esso si scuoprono; e per le sue tante, e così diverse forme, con le quali si fa vedere, come appresso diremo, e si vedrà nelle seguenti figure dal natruale ritratte, & intagliate, E se questo legno con altro si potesse assimigliare, crederei, che gran somiglianza havesse col cedro, che nasce nella Mauritania nel monte Atlante, del quale facevano gli antichi i letti, e le mense così belle, e di tanta stima, come nella traduttione delle satire di Persio da noi fatta habbiamo ancora accennato, e come si può vedere in Plinio al cap. 15. del libro xiij.

Hora prima che veniamo ad altro particolare, diremo il luogo dove questo legno nasce, che è nella Provincia dell'Umbria della nostra Italia, nel territorio di Todi in diverse parti, come si potrà vedere nella prima figura qui alligata, che rappresenta tutto il detto territorio, essendovi intagliati dove questo legno nasce alcuni fumi col nome di sotto di metallofiti. Ma fra li due Castelli di Collesecco, e Rosaro (che non molto /[*p. 6*] son distanti dalla strada Romana) se ne trova maggior quantità, che atrove, & anche di più varietà, che ne gli altri loughi notati in detta figura.

La generatione poi di questo legno, per quanto ho potuto vedere, & osservare non procede da seme, ne da radice di pianta alcuna, ma solo da una spezie di terra, c'ha assai del cretoso, la quale a poco a poco si va trasmutando in legno; così operando la natura fin che resta tutta in detto Legno convertita; e questo credo con l'aiuto di alcuni calori di fuochi sotteranei, che in quei loughi sono, li quali vanno serpendo sotterra, e mandando fuori del continuo un fume assai spesso, & alle volte fiamme, e particolarmente ne'tempi piovosi, e con l'aiuto ancora di acque solfuree, & minerali.

E se il calore sarà assai, il legno s'abbronza, o leggiermente abbrucia, e resta come carbone. Se poi la materia terrestre non è per ancora trasmutata in legno, quel fuoco la cuoce, & rimane come quei vasi di terra cotta nella fornace, o come mattoni. E da questo ancora si manifesta la materia di detto legno non esser altro, che terra, perche n'ho veduto io pezzi, una parte de quali era creta dura, l'altra legno, & il resto come carbone.

Che poi non nasca da seme alcuno, ne da radice, o rami come l'altre piante è chiaro, per non trovarsi pezzo alcuno di questo legno con radiche, ne con rami, ne con nervi come l'altro legno, & alberi, ma solamente tronchi semplici di diverse forme, e non eretti come nascono tutte le piante, ma in terra distesi, o vogliamo dire a giacere; e l'onde, e vene, che esso ha, non sono continuate in una medesima forma per tutto il legno, ma diversamente figurate, hora lunghe, e diritte, hora strette hora larghe, hor in giro, hor serpeggianti, & in altre mille varie guise; dove che nell'altro legno, l'onde si vedono per lo più ne'nodi, e nelle parti vicine alle radiche, e le lor vene sono distese per la lunghezza, o altezza dell'albero, il quale dalle radici ha il suo nutrimento, e però i nervi, e le vene crescono per lo lungo della pianta: ma questo legno minerale da ogni parte della terra piglia il suo nutrimento, e però ha le sue onde così varie.

Ne meno si può credere, che questi legni siano tronchi, o frusti d'altri alberi sotterrati in quei luoghi, o caduti, e dalla terra ricoperti, e formati poi con quell'onde da quell'acque minerali, che ivi scaturiscono, e da fuochi sotterranei, com'io nel principio mi persuasi, per haver trovato alcuni olmi ricoperti dalla terra in quei luoghi, dove detto legno si trova, perche la sua forma si varia, e la mole si grande mi fa credere il contrario, non trovandosi alberi mai dalla natura formati come nelle seguenti /[*p. 7*] figure si vedra, & in particolare nella figura settima in ordine, dove n'è disegnato un pezzo grandissimo, ch'io feci cavare dalla terra di figura ovata; si come anche li due pezzi della figura seguente: e sono (credo io) in quella forma ovata da i lati destro, e sinistro, perche havendo di sopra il peso della terra non han potuto crescere verso quella parte che gli sovrasta per ridursi in forma circolare, o per dir meglio cilindrica, come sono ordinariamente tutti li tronchi degli alberi: di modo che potemo sicuramente affermare, la prima materia di questo legno non esser'altro, che terra di sostanza cretosa, & in confirmatione di questo vi si può aggiungere la molta sua gravezza, e l'andare al fondo nell'acqua per ben piccola scheggia, che sia.

L'altra maraviglia c'ha in se, è che l'onde non sono continuate uniformamente da una estremità all'altra del tronco, ma un tantino di legno che se ne levi, quanto sarebbe la grossezza di un giulio, o poco più, muta la sua figura, apparendo l'onda d'altra forma; e così va seguitando questa mutatione sino all'altra parte del legno. E di più si deve notare che il fianco di questo legno non ha onda, ne vena di forta alcuna, essendo tutto uniforme d'un colore leonato oscuro; ne vi si vede di onde pur un minimo segno, salvo che in alcuni luoghi certi punti bianchicci, dove più, e dove meno, li quali altro non sono, che granelli di terra non convertita per ancora in legno, per esser forse d'altra natura.

L'altra sua proprietà mirabile è, che mentre si lavora fresco, e non per anni stagionato, si piega a guisa d'arco sempre dalla parte del fianco, dove non appariscono l'onde, mentre però sia il lavoro stretto, e non tavola larga, dico stretto due o tre dita, come più volte ho osservato nelle croci, che sono state fatte di questo legno, di poco tempo levato dalla terra, che è cosa strana a vedere.

Se si mette al fuoco mentr'è stato cavato di fresco dalla terra, s'abbrucia, ma lentamente con gran fumo, e con odore spiacevole. quando poi il legno è secco, l'odore è più grato, e fa un fuoco di calore così intenso, che non gli si può stare da vicino, come al fuoco fatto d'altro legno, ne si consuma cosi presto, conservando il fuoco assai più tempo di qualsivoglia altra materia combustibile; e quando il legno è cominciato ad impetrirsi resta al fuoco due, e tre giorni senza consumarsi.

E perche di sopra si è accennato trovarsi questo legno ancora impetrito, si deve sapere, che pezzi grandi, e piccoli di varie figure sono stati trovati impetriti, alcuni de quali non si potevano discernere s'erano legno, o pietra, restando così d'una mezzana natura, che non potevano dirsi esser'altro che legno pietra, non essendosi per ancora tutto il legno trasmutato in pietra. Altri tronchi havevano la sembianza del legno, ma la / [*p. 8*] sostanza tutta di pietra; & alcuni in particolare n'habbiamo trovati, la parte interiore de'quali era legno, e l'esteriore pietra, & altri che di dientro erano pietra, e di fuori legno; scherzando cosi la natura per farci forse maravigliare dell'opere sue così rare, come sono stati altri pezzi di legno, ritrovati in detti luoghi con vari schegge di metallo, c'havevano qualche somiglianza con la marchesita, fra quali alcuni ve n'erano pieni di pezzetti di detto metallo, lunghi come ferro filato, mà però fragili, situati per la lunghezza del legno, e fatti cosi tondi, e politi, come se fussero passati per quei fori, per il quali fanno passare l'oro, el'argento quelli, che tirano questi metalli, che ci reco non poca maraviglia.

In alcune ripe de fossi gia dell'acque discoperte si vede un lungo strato di questo legno seguitamente per lo lungo della ripa di lunghezza di quindici, e più canne, parte del quale è pura creta, parte creta, e legno e parte tutto legno; e la scorza, c'ha d'intorno è per lo piu dell'una, e dell'altra natura, che potremo chiamarla creta legno.

In oltre vi si vedono moltissimi tronchi di strane, e diverse figure; essendo alcuni di essi di forma lunga, e di circonferenza assai varia, altri piramidali, altri turbinati, altri ovati, altri tondi, e globosi, altri cilindrici, altri ondosi, altri angolari, e semicircolari insieme, altri come colonne scannellate, & altri di forme così varie, e stravaganti che difficilmente si possono descrivere. In alcuni poi di questi tronchi la parte esteriore fu trovata tutta stopposa, e fragile; in altre durissima, crespa, & aspra con l'anima di dentro come osso dura, & in altri nera come carbone, e di tutte queste varietà se n'è preso il disegno, che si dovevano intagliare in rame, come già s'e fatto di alcune di esse, che si vedranno ne'seguenti fogli.

Resta che veniamo alla dichiaratione delle figure, che seguono; ma prima diremo, che quando altro testimonio non vi sia della generatione di questo legno, ne altra prova, basti questa; & è che essendo stata levata una parte di terra humida, ch'era intorno ad un pezzo di detto legno, e riposta in una stanza del palazzo di Acquasparta del Signore Duca Cesi, dopo alcuni mesi fu trovata tutta convertita in legno, non senza maravigna del sopradetto Signore, e di altri, che ciò videro: onde non ha dubio alcuno, che la terra è seme, e madre di questo legno, atta nata in quelle parti alla generatione di esso.

E per venire alla dechiaratione delle quì impresse figure, diremo, che la prima è il territorio di Todi con quello di Acquasparta, con i suoi confini, nel quale son notati quattro luoghi in particolare dove nasce questa spezie di legno, come in detta figura si vede, notativi col nome di metallofiti. / [*p. 9*] La seconda figura è un luogo particolare non molto distante dal Castello di Rosaro, dove è stata trovata gran quantità di detto legno di varie grandezze, come si accenna nel disegno; e quei fumi, che quivi si vedono dalla terra uscire, sono stati osservati per dieci anni continui da uno di detto Castello senza mai cessare; ma però detti fumi non sempre uscivano da un medesimo luogo, ma andavano cangiando sito, secondo forse ch'il fuoco trovava materia da bruciare.

L'altre quattordici figure, che sono nelle quatro carte seguenti, cioè 3. 4. 5. & 6. mostrano la varietà delle vene, o vogliamo dire onde, che si son trovate in diversi pezzi di questo legno, fatti segare, e pulire, così ritratti dal naturale; essendosi ciò fatto per mostrare la varietà di esse onde, e loro bellezza.

La figura poi del foglio di num. 7. mostra quel pezzo di legno così grande, di cui n'habbiamo fatto di sopra menzione, ritrovato non in quel modo, che sta intagliato, cioè col fianco in sù, ma col ventre, per così dire, nel piano della terra, e schiena rivolta verso il Cielo, nel modo, che stanno gli altri due pezzi del foglio seguente; e perche la misura della sua grandezza l'habbiamo notata in piè di detta figura, quivi si potrà vedere: soggiungendo, che fu prima scoperta quella parte, che sembra segata, dove son notate le lettere A B C D; e fatto levare la terra che di sopra haveva fino ad F, fu in quella parte trovato il legno disgiunto, come se fusse stato segato; e seguendo a levare la terra, giunti dove è la lettera G si trovò un'altra separatione, come s'accenna in detta figura; e di nuovo levandosi altra terra per discoprire tutto il legno, e per sapere la sua lunghezza, per la gran cava fatta, dopo che ne fu levata tanta, che si scoprì un braccio di esso legno, gli cadde la terra sopra, e restò ricoperto; ne vi si usò altra diligenza per finire di scoprirlo, perche il padrone di quel terreno non volle, che si guastassse con maggior cava il suo campo; e perciò si lasciò stare il rimanente di esso dalla terra ricoperto: ma hora che sto queste cose scrivendo, ho inteso, che l'acque l'hanno di nuovo discoperto.

Li due pezzi della seguente figura di num. 8. così trovati, come si vedono, sembravano ancor essi dalla parte dinanzi segati, & erano alquanto più grossi del pezzo sopradetto ma più corti, e mostravano la scorza come li tronchi degli altri alberi, havendo qualche segno ancora dell'anima di dentro; sforzandosi qui la Natura di fare assomigliarequesti suoi parti imperfetti quanto più può a quelli perfetti. In quella parte poi dove il legno mostra esser segato, haveva una certa resina di color bianco, come mastice, & incenso bianco, risudando in detto luogo il legno come appunto fanno quei muri incollati di fresco in luoghi humidi. E con / [*p. 10*] questa resina fu trovato parimente l'altro pezzo grosso detto sopra, si nella faccia anteriore, si anche in ciascuna delle dette separationi; non essendo quelle altro, che crepature della terra, mentre ancor creta non s'era trasmutata in legno, come s'è osservato in altri pezzi in più luoghi. Di sopra poi, e d'ogn'intorno a questi due pezzi così grossi ve n'erano alcuni altri di varie grandezze, come nell'istessa figura si vede; parte de quali non s'era interamente lignificata, parte era tutto legno, e parte come scorza di legno, e nera come carbone.

Nel foglio nono vi sono due altre figure, la superiore delle quali è di quel legno dove sono stati trovati quei pezzetti di metallo, rappresentati da quei tondini, che quivi si vedono. L'altra figura inferiore mostra l'onde assai grosse, e qualche pezzetto ancora di detto metallo.

Segue la decima figura, nella quale vi sono intagliati tre pezzi di detto legno: quello notato con lettera A è di quella spezie di legno dove nascono quelle lunghe fila di metallo, c'habbiamo già detto di sopra, grosse nel modo, che quivi si vedono figurate; & è quello un frusto di un pezzo maggiore, il quale perch'era fragile assai fu facilmente diviso, e restò poi in quella forma che si vede, scorgendosi la parte di fuori nera come s'havesse havuto il fuoco.

Gli altri due pezzi, o tronchi B, & C erano tutti stopposi, li quali potremo appellarli capillari, per la gran somiglianza, c'hanno quelle tante lor fila co'capelli: tronchi in vero assai stravaganti; e questi parimente erano assai fragili, e con la scorza nera, & abbronzata.

Il tronco D, ch'è fra li due capillari sopradetti, era ancor esso mirabile: poiche la parte interiore era già tutta impetrita, e d'un colore rossiccio, e quella veste di fuori era di legno scorzoso, e nodoso, come appunto ivi è disegnato. Era la lunghezza di tutto il tronco palmi sei, e di grossezza haveva un palmo in circa di diametro. Vi sono stati trovati altri tronchi grossi simili; cioè con la parte interiore impetrita, e di fuori vestiti di legno, de quali già vi sono li disegni.

Nell'undecimo foglio vi si vedono quattro figure del medesimo legno: quella superiore, cioè E è un frusto, che non si può dir legno assolutamente, ma ligniterreo, perche da una parte, che è quella dinanzi, che già si vede era tutto legno, conoscendosi benissimo alli segni della scorza, e dall'altra parte di dietro, tutto creta dura, non per ancora in legno trasmutata.

Il pezzetto H, fu in quella forma ritrovato fatto a onde come mostra il suo disegno.

L'altro pezzo F è di figura turbinato, assai aspro, e scorzoso, la sua lunghezza e di piedi due, & un piede largo. fu segato per lo mezzo per / [*p. 11*] vedere la parte interiore: onde l'altro pezzo G, rappresenta come era di dentro. Molti frusti di legno di questa forma turbinata sono stati in quei luoghi ritrovati; ma basterà d'haver fatto mostra di questo per non esser gli altri molto differenti; eccetto che ven'erano alcuni dalla parte di dentro duri come osso, che col tempo (credo io) si sariano impetriti.

Li cinque pezzi del 12. foglio, sicome sono fra loro diversissimi di forma, così sono parimente di sostanza. Il pezzo H, in quella parte che si vede piu oscura era legno, e metallo insieme, e quella crosta di fuori più chiara era di sostanza lateritia, cioè di terra cotta come sono li mattoni.

Il pezzo I, fu ritrovato insieme con altri pezzi simili in un fosso non molto distante dal Castello di monte nero; ed era tutto legno, ma di forma somigliante ad una colonna scannellata, e in quella parte, dove mostra esser segato v'era quella resina, c'habbiamo detto di sopra, così ancora nella parte a quella opposta.

L'altro pezzo L, pareva tutto scorza di legno, come già all'occhio si dimostra, ma participava della natura del metallo, & alcuni nodi ch'intorno haveva s'erano già totalmente metallificati, che perciò era di peso assai grave.

Il frusto M, era di sostanza cretilignea, cioè creta, e legno, non però molto duro, e dove si vede più oscuro, era lucido, e risplendente simile all'accia vaccio, ma assai frangibile.

Il pezzo poi inferiore N, haveva la forma d'un pesce, & era di materia legnosa, e ferruginea insieme, e di colore come il ferretto di Spagna; e questo ancora era più dell'altro legno pesante. Era lungo un palmo, e due dita, largo mezzo palmo, e grosso cinque dita.

Nel seguente foglio xiij. vi sono intagliate alcune pietre di forma simili alla Chiocciola, & alla serpe in se stessa ravvolta; ma per la gran somiglianza ancora, c'ha col corno dell'Ariete, son chiamate corno d'Ammone, in quella guisa formate dalla Natura: e son fatte con tanta proporzione quelle circolationi spirali, ch'io l'haverei sempre stimate artificiose, e non altrimenti naturali, se non l'havessi così ritrovate, non solo nel territorio di Acquasparta, ma ancora nel territorio di Fabriano mia patria; oltre che ne scrivono anche molti Autori naturalisti; si come si può vedere nel Gesnero, in Plinio, nell'Agricola, Bellonio, Imperato, & altri. E della loro generatione, e natura n'ha in parte scritto, e ne stava scrivendo il nostro Signor Principe Cesi sopradetto, il quale dalla morte prevenuto, non potè finire il trattato, che di queste materie faceva, che per la novità sarebbe stata una lettura gustosissima, & curiosissima: poiche non solo scriveva della generatione di dette pietre, e legno, e delle pietre Aquiline, che pure in detti luoghi se ne genera gran quantità, ma di tutte / [*p. 12*] l'altre pietre note sin quì, e di altre ancora non più osservate, ne descritte da altri Autori; havendo egli con lunga, e diligente osservatione trovato come si generi la pregnante pietra Aquilina; come si formino nelle pietre Arboree, che l'Imperato chiama imboscate, quegli alberetti con rami, e frondi, tanto alli naturali somiglianti, che sembrano esservi artificiosamente dipinti. Come nelle pietre Astroite, o Stellarie nascano quelle stelle; le croci nelle crocigere; quelle righe, o solchi nelle pietre Giudanche; quelle strie ne'fonghi, che chiamamo petrificati; e quei tanti, e cosi varij colori ne'marmi mischi: e parimente come prendano quella lor forma la pietra Bucardia, il Lincurio, l'ostracite, la vermiculare, la millepora, retepora, e frondipora, overo escara marina & altre molte, che per brevità tralascio.

Ma pare che la Natura invidiasse il gran sapere di questo Signore; e che non per altro habbia voluto abbreviare li giorni della sua vita, se non perche egli non venisse a discoprire il modo, ch'ella tiene nell'occulte operationi sue, mentre va producendo tante varietà d'oggetti, e di qualità si diverse, e sì mirabili, e con tanto artificio, senza, che v'adopri ne scalpello, ne pennello, ne strumento di sorta alcuna; che perciò tanto maggiormente son l'opere sue degne di maraviglia.

E questo è quanto m'è sovvenuto dire intorno à questa materia con quella maggior brevità, c'ho potuto, per quelche più volte ho visto, & osservato in quei luoghi, dove nasce questo nuovo, raro, e mirabil parto della Natura.

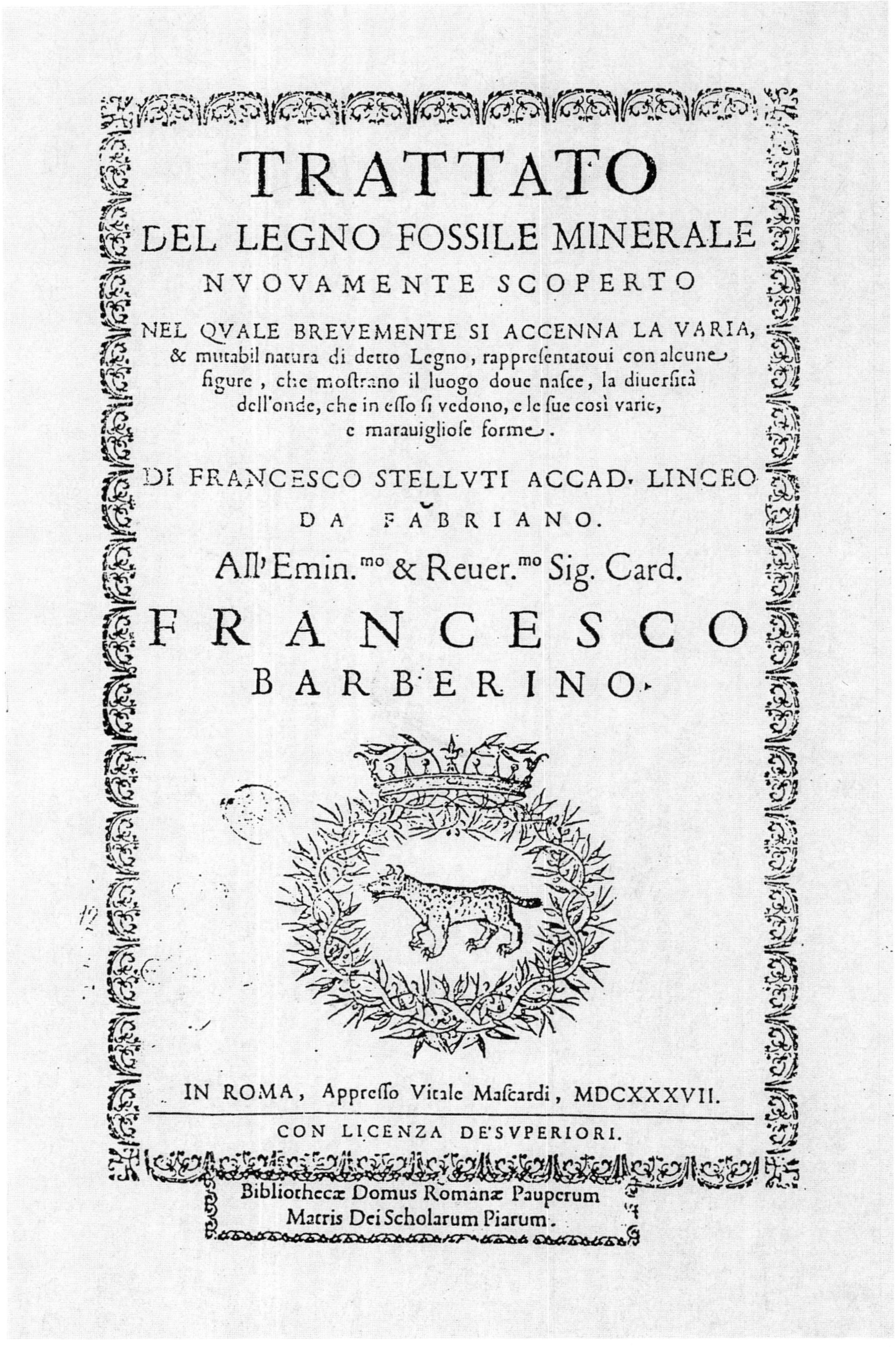
TRATTATO
DEL LEGNO FOSSILE MINERALE
NVOVAMENTE SCOPERTO
NEL QVALE BREVEMENTE SI ACCENNA LA VARIA,
& mutabil natura di detto Legno, rappreſentatoui con alcune
figure, che moſtrano il luogo doue naſce, la diuerſità
dell'onde, che in eſſo ſi vedono, e le ſue così varie,
e marauiglioſe forme.
DI FRANCESCO STELLVTI ACCAD. LINCEO
DA FABRIANO.
All'Emin.mo & Reuer.mo Sig. Card.
FRANCESCO
BARBERINO.
IN ROMA, Appreſſo Vitale Maſcardi, MDCXXXVII.
CON LICENZA DE'SVPERIORI.
Bibliothecæ Domus Romanæ Pauperum
Matris Dei Scholarum Piarum.

Fig. 53. Title page of Stelluti's *Trattato*, 1637

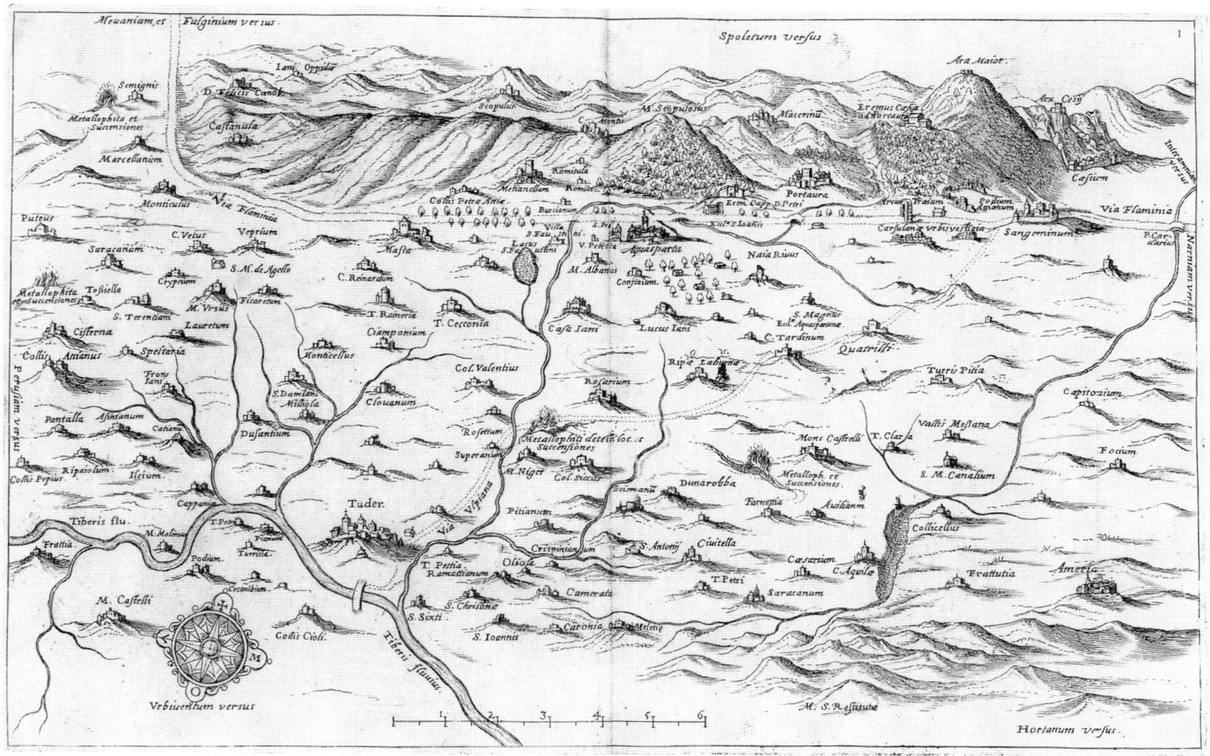

Fig. 54. The district around Todi and Acquasparta, showing sites with *metallophytes* and underground fires (Pl. 1)

Fig. 55. A site of underground fires near the Castle of Rosaro (Pl. 2)

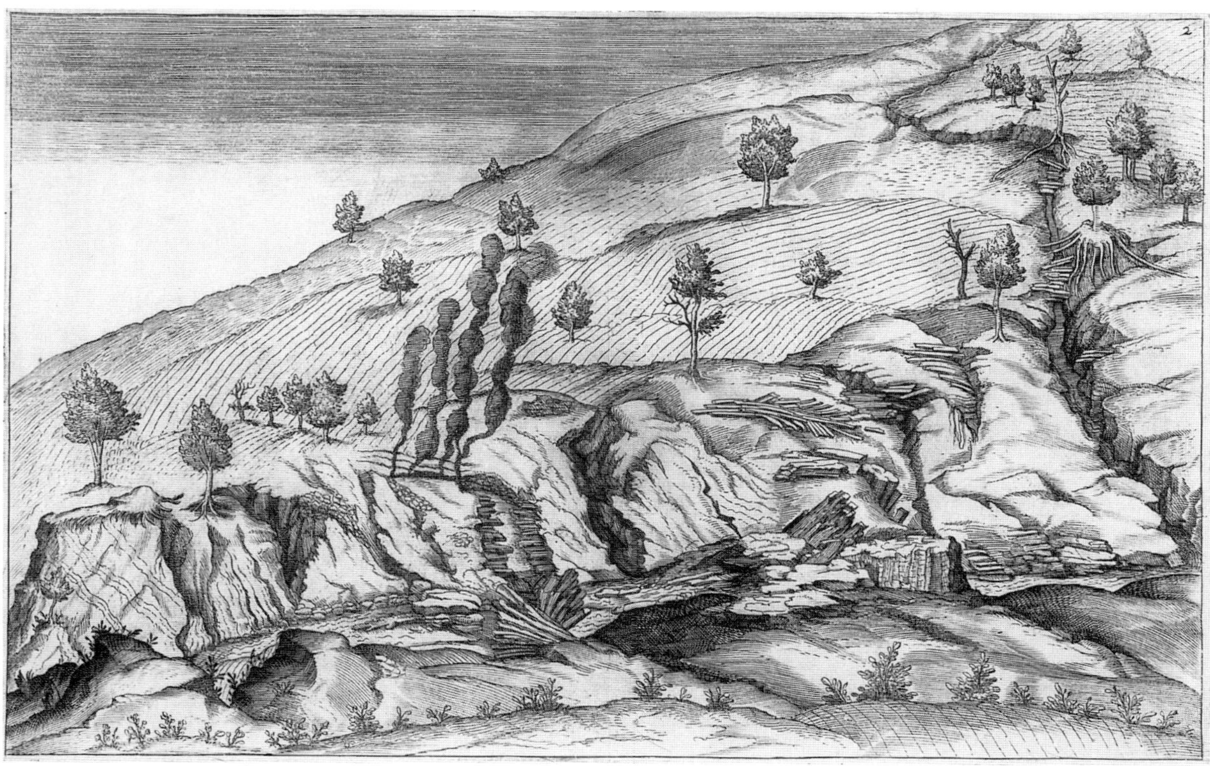

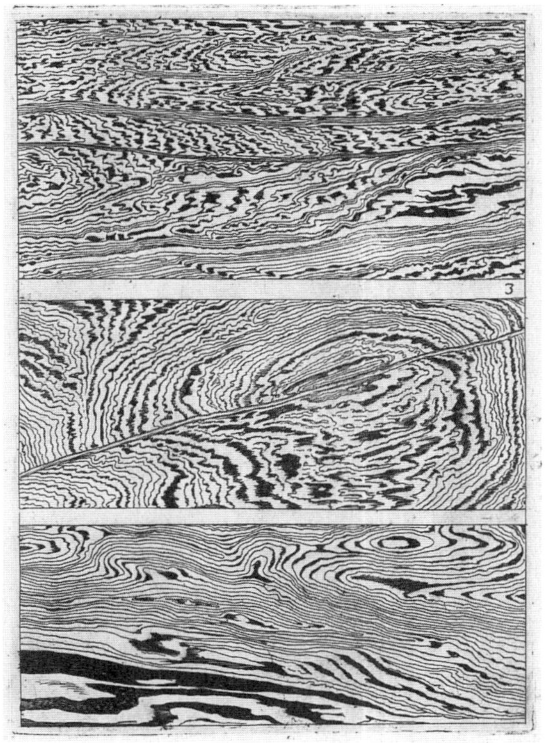

Fig. 56. Three sections of fossil wood (Pl. 3)

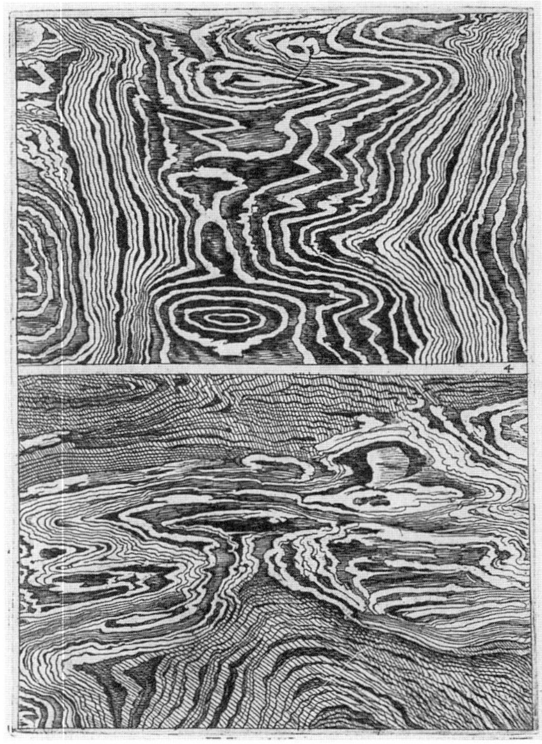

Fig. 57. Two sections of fossil wood (Pl. 4)

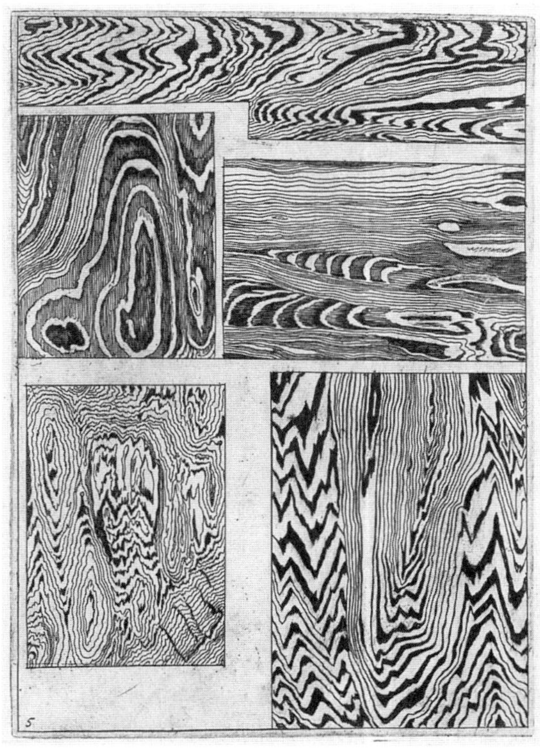

Fig. 58. Five sections of fossil wood (Pl. 5)

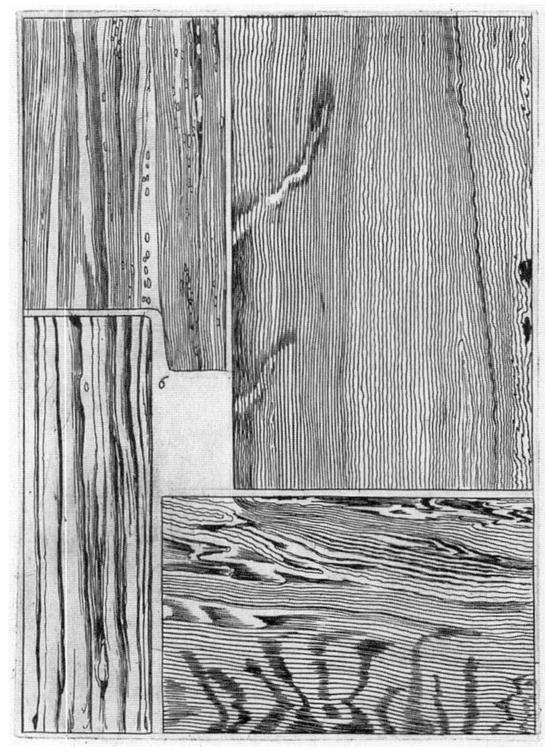

Fig. 59. Four sections of fossil wood (Pl. 6)

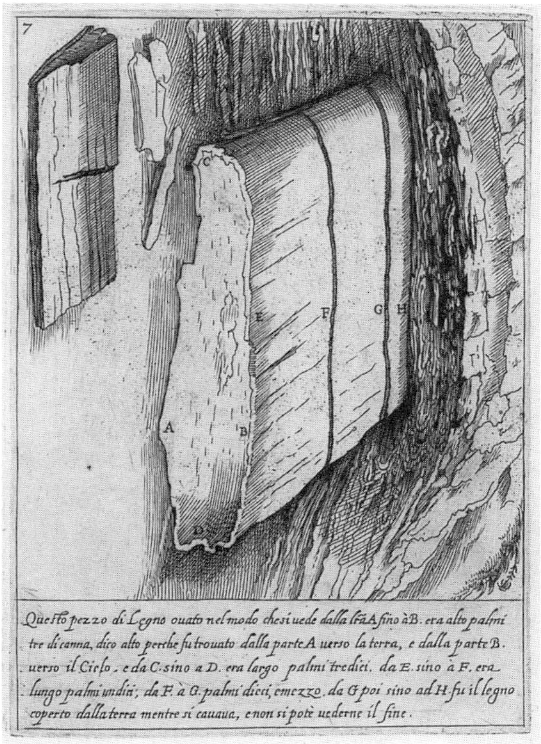

Fig. 60. An exceptionally large specimen of fossil wood (Pl. 7)

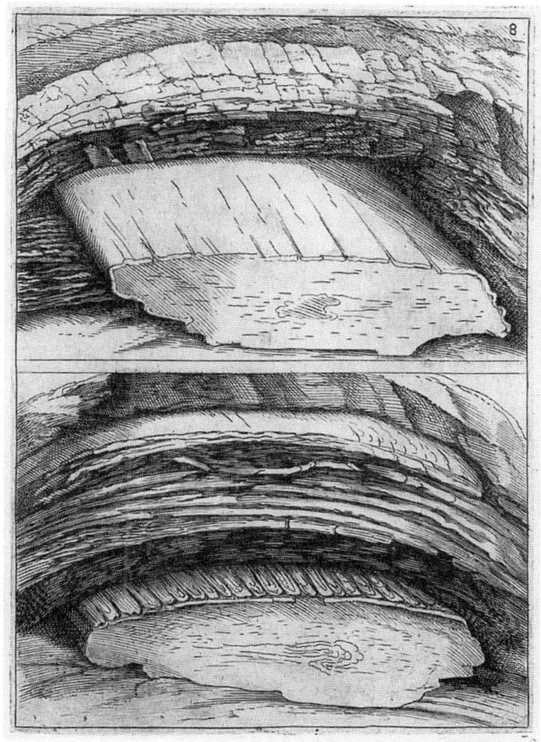

Fig. 61. Two specimens of resinous fossil wood, as found (Pl. 8)

Fig. 62. Fossil wood showing 'metal' and 'wavy' grain (Pl. 9)

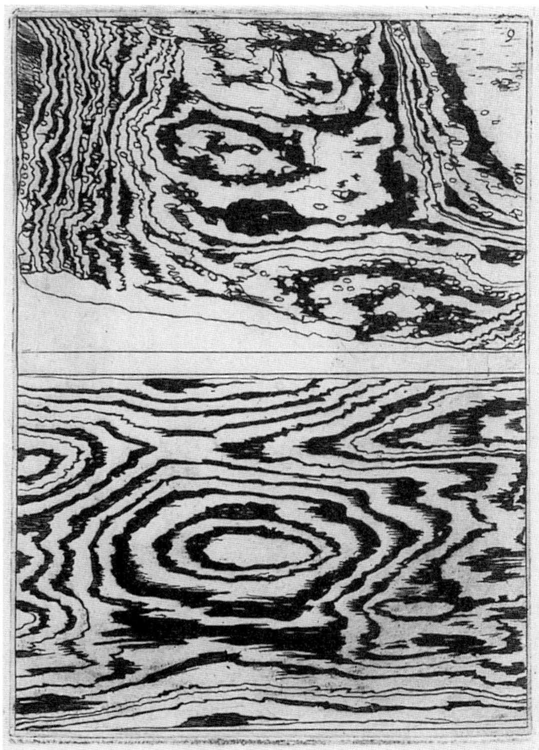

Fig. 63. Three pieces of 'thread-like' fossil wood and one 'half-petrified' (Pl. 10)

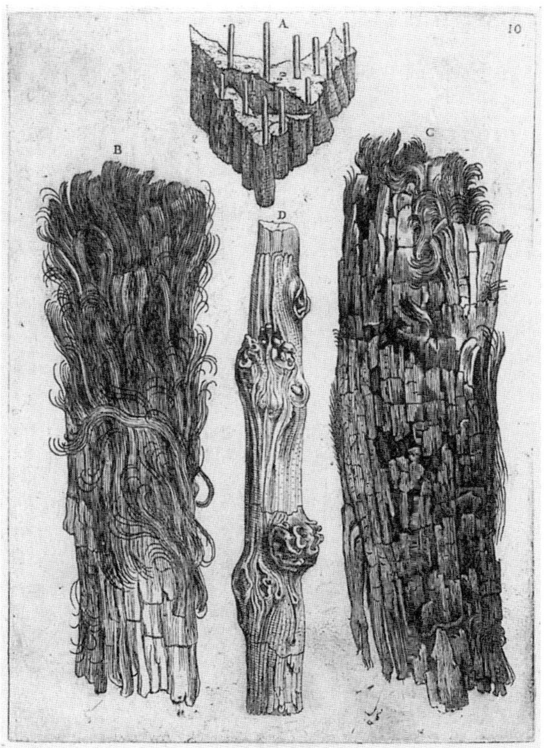

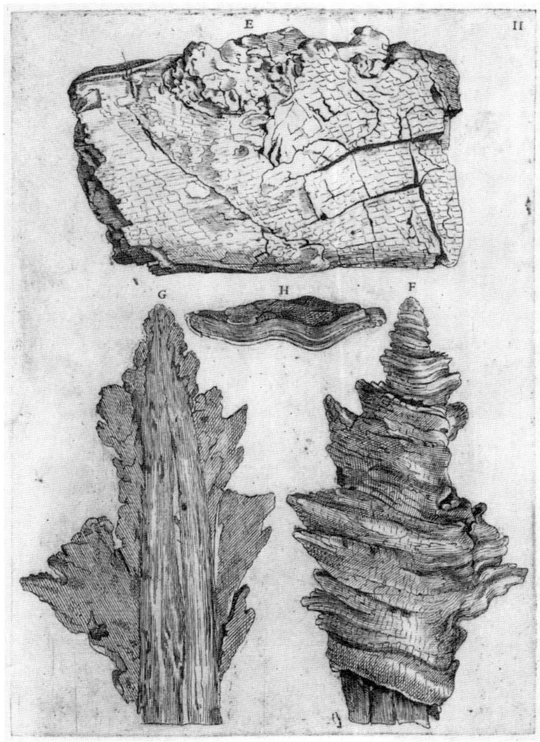

Fig. 64. A fragment of 'ligneous earth' and two specimens of 'swirling' shape, with a cross-section (Pl. 11)

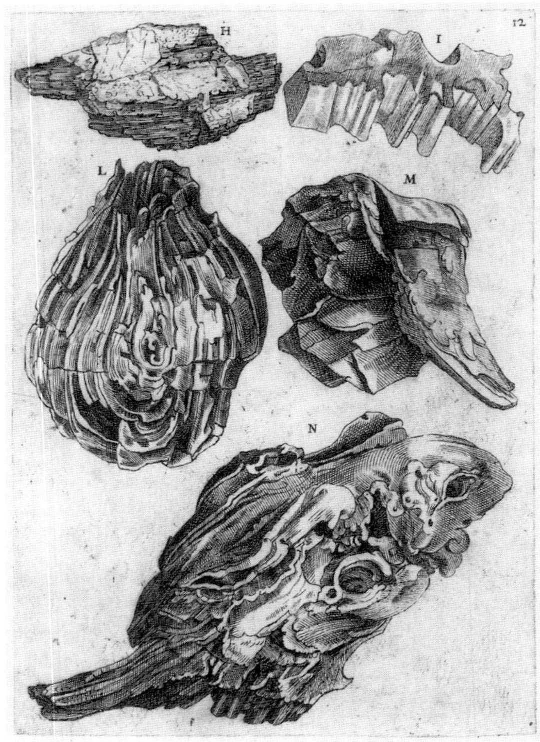

Fig. 65. Five specimens that seemed variously compounded of wood, metal, clay and stone (Pl. 12)

Fig. 66. Stones in the shapes of 'snail or snake' (Pl. 13)

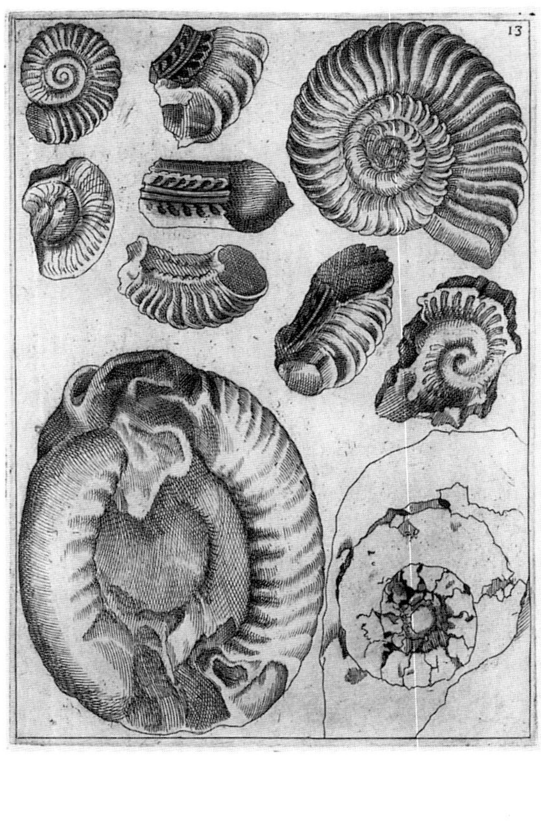

APPENDIX II b

Translation of Stelluti's *Trattato*

with notes by Andrew C. Scott

[*Stelluti 1637, title page:* see Fig. 53]

TREATISE

ON THE NEWLY DISCOVERED FOSSILIZED MINERALIZED WOOD

IN WHICH IS BRIEFLY SUMMARIZED
the varied and mutable nature of the said wood, represented here with some illustrations which show its place of origin, the diversity of the undulations[1] that are seen within it, and its varied and wondrous forms.

BY FRANCESCO STELLUTI, MEMBER OF THE ACADEMY OF LINCEANS AND NATIVE OF FABRIANO

To His Most Eminent and Most Reverend Lord Cardinal
FRANCESCO
BARBERINI

ROME, Vitale Mascardi Printer, 1637.
BY PERMISSION OF THE AUTHORITIES
Bibliotecæ Domus Romanæ Pauperum
Matris Dei Scholarum Piarum.

[*Stelluti 1637, p. 3*]

TO THE MOST EMINENT AND REVEREND LORD AND MY MOST WORTHY PATRON, LORD CARDINAL FRANCESCO BARBERINI

Last year I gave Your Eminence my hand-written treatise on the fossil wood which is indigenous to our region, which I wrote about in this way to satisfy the curiosity of many who wished to have a true and faithful account of such strange wood. And because afterwards many copies were requested of me, being unable to satisfy everyone, I decided for this reason to put the work into print, with the addition of other, rather curious figures, for the greater satisfaction of my readers: so that through visual means they might be more satisfied and convinced of all they will learn in the treatise, which I being obliged to publish with engravings have wished to dedicate to Your Eminence, both for the pleasure which you are wont to take from the contemplation of the secret parts of Nature; and also that it may bring to the fore a/[*p. 4*] testimony so worthy and so eminent as to assure others that I present nothing that is false or fantastic.

I bring before Your Eminence a matter which is in truth wholly new and ambiguous. But you will see it set against the long-standing subservience which I profess for you as well as against my unquestionable respect and devotion. If you will acknowledge these [professions], I will be able to hope that my modest offering is pleasing to you; all the more so because I am certain that it is in the nature of the lofty and innate munificence of Your Eminence, to whom I reverently bow, not to despise humble things. Rome, 8th of May 1637.

Your Eminence's most humble and devoted servant,

Francesco Stelluti.

[*Stelluti 1637, p. 5*]

TREATISE On the Recently Discovered Mineralized Fossil Wood BY FRANCESCO STELLUTI, ACADEMY OF LINCEANS

That all the products of Nature are marvellous cannot be denied, but infinitely more marvellous are those which she produces rarely, because they are less known to us. Such is the case with this fossil, or subterranean wood, which we wish to call it; for thus it was named by His Lordship D. Federico Cesi, Prince of S. Angelo and second Duke of Acquasparta, who in recent years made its discovery. [It is] by nature a metallophyte, being part plant, part metal:[2] if well considered in its entirety, it partakes more of the plant than of metal.

This wood is worthy of wonder for its rarity. Until now it was not known to exist in other regions and no author has ever written about it. And though Theophrastus (as Pliny writes in Book 36, Chapter 18 of his *Natural History*) affirms that fossil ebony[3] is found both white and black in colour; and that bones are still born of the earth: he makes no mention of this wave-patterned wood, which is produced in our regions. And not only is it wondrous on account of its rarity, but still more so on account of the variety and beauty of the waves that are discovered within it, also for its many and diverse forms with which it manifests itself, as we shall shortly explain, and this will be seen in the following figures drawn from nature, and engraved. And if this said wood may be compared to another, I believe it greatly to resemble the cedar,[4] which grows in Mauretania, on Mount Atlas, [a wood] from which the ancients fashioned beds, and most beautiful tables, and so greatly esteemed, as we noted in our translations of the satires of Persius,[5] and as may be seen in Pliny, Book 13, Chapter 15.

Now before we speak of any other particular, we shall speak of the place where this wood is generated, which is in the Province of Umbria in our Italy, in various parts of the territory of Todi, as may be perceived in the first illustration appended here, representing the whole of the said territory, there being engraved there, where this wood is generated, some fumaroles[6] with the name 'metallophyte' beneath. However, between the two castles of Collesecco and Rosaro (not far / [*p. 6*] distant from the Roman road) a greater quantity of them is found than elsewhere, and of greater variety, than in the other places noted in the illustration.

From what I have been able to see and observe, the wood is not generated from the seed or root of any plant whatsoever, but only from a type of earth, containing much clay, which is slowly transformed into wood, nature thus operating until all the earth is converted into the said wood. This I believe to take place with the assistance of the heat of subterranean fires which exist in these areas, and which, snaking underground, continuously emit very thick smoke, and sometimes flames, and especially when the weather is wet, and with the further assistance of sulphurous and mineral waters.

And when the heat is sufficient, the wood is browned, or lightly scorched, and remains similar to coal. If then the earthen material is not yet converted into wood, this fire bakes it and it ends up similar to vases of terracotta baked in a kiln, or to bricks.[7] And from this, too, it is clear that the essence of this wood is nothing other than earth, because I myself have seen pieces, one part of which was hard clay, the other wood, and the rest similar to coal.

It is clear that it is not born of any seed or root or branch as are other plants, for there exists no piece of the wood bearing roots or branches,[8] or with nerves as other wood and trees,[9] but merely simple trunks of diverse forms, and not erect, as all plants are born, but recumbent or, as it were, lying on the earth. And the waves and veins it possesses do not extend in the same form through the whole of the wood, but are variously fashioned, now long and straight, now narrow, now wide, now spiralling, now undulating, and in a thousand different guises; while in other kinds of wood, the waves are visible principally in the knots, and in those parts which are close to the roots, and their veins are extended along the length and height of the tree, which receives its nutriment from its roots, but the nerves and veins grow along the whole length of the plant: but this mineral wood takes its nutriment from every part of the earth and still has such varying waves.

Neither can these pieces of wood be considered trunks, nor [are they] fragments of other trees buried in this area, or which have fallen and been covered by earth and then given wave-like form by the mineral waters that have their springs there. Nor have they been created by subterranean fires, as was my initial persuasion, having found some elms covered by earth in the area where this wood is to be found, since its varied form and its great bulk makes me believe the contrary and no trees ever having been found formed by Nature like those in the following / [*p. 7*] illustrations, and especially in the seventh one, showing a piece of considerable dimensions[10] which I had dug out of the earth. [In section, the piece is] ovate in shape;[11] as are also the two pieces in the following illustration. And they are (I believe) ovate in form on both the left and right sides because, bearing the weight of the earth above them, they were unable to grow towards the part that hangs over them and to thus reduce themselves to a circular, or rather, cylindrical shape, as all tree trunks normally are. So that we may safely affirm that the raw material of this wood is nothing other than clay-rich earth, and in confirmation of which one may add its great weight and the fact that even a very small flake sinks to the bottom of water, which it does.

Another marvellous quality of this wood is that the waves do not uniformly spread from one end of the trunk to the other, but the removal of the smallest quantity of wood, the size of a Julius [that is, a small coin] or a bit larger, alters its shape, revealing a different form of wave; this mutation then comes to the other part of the wood. Furthermore it should be noted that the side of the wood possesses neither wave nor vein of any kind but is uniform throughout of a dark, tawny colour. Nor is there the least trace of a wave, except that in some places there are certain whitish points,[12] of uneven distribution, which are nothing other than grains of earth not yet converted into wood, because they are perhaps of a different substance.

Another wondrous property [this wood] possesses is that if [it is] worked when fresh and not seasoned for some years, it bends like a bow always on the side where no waves appear, because the piece of work [that is, the wood] is narrow and is not a wide baulk—as narrow as the breadth of two or three fingers perhaps—as I have frequently observed in crosses fashioned from this wood which has been lifted from the earth, and which is a marvellous thing to see.[13]

If set on fire when freshly dug from the earth, it will scorch, but slowly with great quantities of smoke, and with a noxious odour. When the wood is then dry, the odour is more pleasing, and so intense a heat is produced that it is impossible to stand nearby, as one can with the fire made by other woods, nor does it soon consume itself, but retains the fire for a much longer time than does any other combustible material; and when the wood has begun to petrify, it may stay on fire for two or three days without being consumed.

And having stated above that the wood may be found already 'petrified', it should be known that large and small pieces, of diverse shape, have been found petrified. It was impossible to tell whether some of these pieces were of wood or stone, being thus of a middle nature which may only be named 'wood-stone', not all the wood being yet converted into stone. Other trunks bore the semblance of wood, but the substance / [*p. 8*] was entirely of stone; and we found some especially which had an interior of wood and an exterior of stone, and others which were stone within and wood without,[14] Nature thus joking perhaps to make us marvel at her rarities, such as other pieces of wood found in these parts with various splinters of metal, bearing some resemblance to marcasite,[15] among which there were some full of pieces of the said metal, as long as iron wire, but fragile nevertheless, distributed through the length of the wood, and as round and polished as if they had passed through the holes which gold and silver workers make those metals pass through, which caused us greatly to marvel.

On the banks of some ditches already uncovered by the water, a long layer of this wood may be seen along the whole length of the bank, of a length of fifteen or more *canne*, part of which is pure clay, part clay and wood, and part entirely of wood, and the bark around it consists for the most part of one and the other nature, which we will call clay–wood.

In addition, many trunks of strange and diverse shape are seen there; some long, and of greatly varying circumference; others shaped like a pyramid, others like a spiral, or ovate, others round, or spherical, others cylindrical, others undulating, others angular and semi-circular at the same time, others fluted like a column, and others of shapes so varied and extravagant that they can barely be described. In some of these trunks the outside is completely string-like and fragile; in others very hard, puckered and sharp, with a core like [that] inside hard bone, and in others black as coal; all the various kinds have been sketched which must be engraved in copper, as has already been done with some of them, which may be seen in the following pages.

We have now only to pass to an explanation of the illustrations which follow. But first we will say that, when there is no other evidence or proof of the origins of this wood, let this suffice: that a quantity of damp earth, which was around a piece of this wood, having been removed and placed in a room in the Palace of Acquasparta, belonging to the Lord

Duke Cesi,[16] it was found after some months to be wholly converted into wood,[17] to the great wonder of the aforementioned lord, and of all the others who saw it. Whence he has no doubt whatsoever that the earth itself is the seed and mother of this wood, the earth of these parts being most suitable for its generation.

And to come now to the explanation of the illustrations printed within, we will say that the first shows the territory of Todi and that of Acquasparta, with its boundaries, in which four places in particular are marked where this type of wood grows, as is seen in the said illustration, marked there with the name of metallophytes.

[*p. 9*] The second illustration is of a particular place not far distant from the Castle of Rosaro, where a large quantity of the said wood of various sizes has been found, as is indicated in the drawing; and those fumaroles which are seen to issue from the earth have been observed by someone of the castle for ten consecutive years, and they have never ceased;[18] nevertheless these fumes did not always issue from the same spot, but kept on changing their site depending, perhaps, on whether or not the fire found material to burn.

Fourteen other figures in the next four pages, that is [pages] numbers *3*, *4*, *5* and *6*, show the variety of veins, or we would say rather, [those] that have been found in diverse parts of the said wood, [having been] sawn off, cleaned, [and] thus drawn from nature, this being done to demonstrate the variety of these waves, and their beauty.

The illustration of page number *7* shows the piece of wood of vast proportions mentioned above, discovered not as it is shown engraved, with its side uppermost, but with its belly, if I may so term it, on the ground, and its crest turned towards the sky, as are the other two pieces in the following sheet; and for this reason we have noted the dimensions of its size at the foot of the illustration, where it can be seen: adding that the part which appears to have been sawn off, bearing the letters *A*, *B*, *C*, *D*, was the first to be discovered; and the earth which covered it having been removed as far as point *F*, it was here that the severed piece was found, as if it had been sawn off; and more earth having been removed, another juncture was found at the part indicated by the letter *G*, as illustrated in the said figure; and more earth having been removed to make a large hole in order to uncover the entire piece, and to ascertain its length, after so much had been removed that an arm's length of this wood was discovered, the earth fell over it, and it was covered once again; nor were more attempts made to finish uncovering it, because the owner of the land did not want his field to be ruined by further holes; so that the remainder was left covered by the earth; but now, while writing this account, I have heard that the waters have once more uncovered it.

The two pieces of the following figures, number *8*, found just as they are illustrated here, also appeared sawn off at the front, and were somewhat thicker than the aforementioned piece, but shorter, and having bark like the trunks of the other trees, still having some sign of their inner core, Nature here straining to make these her imperfect parts resemble as closely as possible those which are perfect. In that part where the wood appears sawn off, a resin was found, white in colour, like gum, and white incense, the wood here sweating just as those walls do which have been freshly plastered in damp places.[19] / [*p. 10*] The other large piece, mentioned above, was also found with this resin, both on the front and in each of the said divisions, these being nothing other than cracks in the earth, while the clay had not yet lignified, as has been observed in other pieces in several places. Then above, all around these two large pieces, were others of varying size, as is seen in the same illustration, some of which were not yet wholly turned into wood, some were all wood, and some like the bark of trees and black as coal.

On the ninth sheet there are two more figures; the top one is that wood in which were discovered pieces of metal, here represented by circles. The figure below displays very large waves and again some little piece of the said metal.

The tenth figure follows in which are engraved three pieces of this wood: that indicated by the letter *A* is of the kind of wood in which grow those long threads of metal of which we already spoke earlier,[20] large in size and which are represented here; and this one is part of a larger piece which, being fragile, was easily separated, and remained in the shape in which it is seen, the outer part appearing as black as if it had been burned.

The other two pieces, or trunks, *B* and *C*, were stringy throughout, and we could call them capillaries on account of the great similarity which their abundance of threads has to hair: most fantastic trunks, indeed; and these too were most fragile, with a black and tanned bark.

Trunk *D*, in between the two capillaries mentioned above, also excited great wonder: because its inner part was already totally petrified, and reddish in colour, and the outside covering was of a bark-like wood and knotty, exactly as it has been drawn. The length of the entire trunk was six palms, and its width approximately one palm in diameter. Other similarly large trunks have been found there, that is with their centre petrified and their outside covered with wood, of which drawings have already been made.

Four figures of this same wood are shown in the eleventh sheet; the top one, *E*, is a fragment, which cannot quite be defined as wood, but as ligneous earth, since one part, the foremost, can be seen once to have been all of wood, traces of bark being clearly distinguishable, and the other, the hindmost, all of hard clay not yet transformed into wood.

Piece *H* was found in that form patterned with waves as its drawing illustrates.

The other piece is swirling in form (*F*), rather rough and bark-like, two feet long, and one foot wide. It was sawn in half / [*p. 11*] to reveal the interior: hence piece *G* shows the appearance within. Many pieces of wood of this swirling shape have been found in this area; but it will suffice to have shown this one, the others not differing greatly, except that some were hard as bone inside, which in time (I believe) would petrify.

The five pieces on sheet *12* differ as much from one another in form as in substance. Piece *H* in that part which appears darker, was both wood and metal, and the outer, lighter crust was of laterite,[21] that is, of baked earth, as bricks are.

Piece *I* was found together with other similar pieces in a ditch not very far from the Castle of Montenero; and was entirely of wood, but of a shape resembling a fluted column, and in the part where it seems to have been sawn off there was that resin spoken of above,[22] also again in the part facing it.

The other piece, *L*, seemed to be entirely of bark, as it appears to the eye, but partook of the nature of metal, and some knots around it had already [become] completely metallized, its weight therefore being considerable.[23]

Fragment *M* was cretaceous-wood in substance, that is, clay and wood, yet not exceedingly hard, and where it appears darker it was shiny and resplendent, like anthracite,[24] but most brittle.

The piece below, *N*, had the shape of a fish, and was of a substance both ligneous and ferruginous, and in colour like *ferretto di Spagna*; and this wood was even heavier than the other. It was one palm and two fingers in length, half a palm in width, and five fingers thick.

In the following sheet, *13*, are engraved stones in the shape of a snail and a snake coiled around itself; but on account of their even greater resemblance to the horn of the Ram, they are called the horn of Ammon, formed in this manner by Nature. And these spirals were made with such perfect proportions that I would always have believed them to be artificial, or else not natural, had I not discovered them like this, not only in the territory of Acquasparta, but also in my native territory of Fabriano; besides which, many naturalist authors also write of them; as may be seen in Gesner, in Pliny, in Agricola, Bellon, Imperato, and others. And of their origins and nature His Highness Prince Cesi, mentioned above, had also in part written and was writing, when death prevented his finishing the treatise which dealt with this material, the novelty of which would have made it most delightful and curious to read: because not only was he writing of the generation of these stones, and wood, and eagle-stones, which were generated in such quantities in these places, but of all / [*p. 12*] other stones known up until now, and even of others not yet observed nor described by other authors; he having through long and diligent observation discovered how the pregnant eagle-stone is generated, how in arboreal stones, which Imperato calls 'imbosked', are formed those small trees with branches and fronds so resembling their natural counterparts as to appear artificially painted. How stars are formed in astroite or stellate stones; crosses in cruciferous [stones]; those lines or furrows in Judaic stones;[25] those stripes in mushrooms that we call petrified; and the many and diverse colours in variegated marbles: and likewise how are formed *bucardia* stone,[26] *lincurio*,[27] ostracite, vermiculite, milleporite, reteporic and frondeporic stones, or sea eschar,[28] and many others which for the sake of brevity I pass over.

But it seems that Nature envied His Lordship's great knowledge; and wished to shorten the days of his life, for no other reason than that he should not come to discover the manner of her secret operations, while she continued producing such a variety of objects, and of such diverse quality, and so wondrous, and with such artifice, with the assistance of neither scalpel, nor brush, nor any instrument whatsoever; which works of hers are thus the more worthy of wonder.

And this is all that comes to my mind to say about this subject and with the greatest brevity that I can from those several times that I saw and observed it in those places, where Nature's new, rare and wondrous offspring is born.

NOTES

1. By 'undulations' (*onde*) Stelluti means the distinctive grain that is shown when the wood is cut.
2. Specimens which Cesi and Stelluti called 'metallophyte', we would probably describe as pyritized. Throughout, 'metal' probably refers to iron pyrite or other iron minerals.
3. 'Ebony' refers to the wood rather than to the ebony tree.
4. 'The cedar, which grows in Mauretania' is a conifer.
5. The reference is to Stelluti's *Persius, tradotto in verso sciolto* ... (in this volume referred to as Stelluti 1630).
6. 'Fumarole' is a modern term associated with volcanic activity, in which sulphurous gases are emitted from the earth. The metallophyte or pyritized wood is composed of iron sulphide, but it is not certain whether it was then appreciated that sulphur was involved in both.
7. This observation—that subterranean fires can produce material of the character of terracotta or brick—is quite correct.
8. This absence of any specimens of roots or branches is indeed the case in many fossil deposits, where the trunks have been washed into a lake or river; it probably represents a fossil 'log-jam'.
9. The term 'nerve' may refer to the elongated cells of the wood known as tracheids. Plant anatomy at this time was poorly known.
10. This important observation—that fossil trees may be larger than anything growing—may originally be attributed to Cesi. The fossils are of large conifers no longer growing in Italy.
11. Some of the trunks may have been partly compacted, resulting in their ovate shape.
12. The white substance may have been decayed pyrite, which is characteristically powdery grey-white and smells of sulphur (see Comp. fig. **36**).
13. Fossil wood is saturated with water; when exposed to the air it dries out, and this may cause cracking and bending of the wood.
14. Woods which show variation in their type of fossilization are common in many places.
15. Marcasite and pyrite are both forms of iron sulphide, which is yellow in colour. Pyrite may fill single long tracheids of vessels in the wood and could easily appear as thin strands.
16. It is possible (see above p. 66) that Cesi believed the fossil wood to be from once-living trees. This account by Stelluti cannot be taken as reliable evidence of Cesi's views.
17. Clay-encrusted wood is not uncommon. After drying out, the clay would crack, separate from the wood inside and fall off, as he observed.
18. In La Ferrière's letter to Peiresc a different account is given: 'My guide did show me a place where, he said, a fire had burnt for seven years and shepherds had set fire to the soil'. La Ferrière also reports that Cesi had mentioned something burning for ten years. (See Appendix III b.)
19. This observation—of the wood 'sweating' and producing a white deposit—could indicate the occurrence of pyrite decay, which takes place under moist conditions. The reaction results in the formation of a white powder.
20. A different version of Pl. 10, fig. A, occurs in the early manuscript version in Montpellier (see Fig. 67, p. 401).
21. Laterite is a red earth formed by the concentration of iron in the soil of hot, humid climates.
22. 'Resin' here is probably used simply as a descriptive term.
23. Pyritized wood can be very dense and heavy.
24. Anthracite is coal of high rank, and is black and shiny.
25. Judaic stone—*Pietra Giudaica*, also defined as 'Phoenician' or 'Syrian' stone (Battaglia 1967)—was considered to be of meteoric origin and to have therapeutic properties.
26. *Bucardia* is a sort of shell, from the Latin *bos cardia*.
27. *Lincurio*—often illustrated in the Cesian corpus—was Pliny's *lyncurium*, a kind of stone-wood with therapeutic properties, fabled as being produced from the urine of the lynx.
28. The sea eschar is the species *Escara marina*.

APPENDIX III a

Letter to Cardinal Francesco Barberini from Duke Federico Cesi Acquasparta, 1 December 1624

Original Italian copy in the author's hand, in the Vatican, BAV, MS Barb. Lat. 6461 (published in Gabrieli 1959, III, no. 798, pp. 956–7 with preceding bibliography), fols. 148–9; secretary's copy on fols. 157–8; author's handwritten draft in Rome, BANL, Archivio Linceo 2, fols. 57–60, 62.

Most Illustrious and Most Revered Lord,
my Most Esteemed Patron

In the little free time my troublesome affairs have left me, I have pursued my physical observations, and particularly on the new form of middle nature between plants and metals, on the veined and varied mineralized wood discovered by me some years ago in this region, and which I have already brought to you. And after having studied and seen many differences in these things, which are so full of curiosities and such immeasurable excesses, that within the species of similar composite things they can even exceed the proportion the cetaceans have among sensate creatures, to whom, it seems that, even the sacred texts attribute the notion of largeness.

I have found a great variety of petrified objects, that is, some completely petrified, of the consistency of stone or brick; or mixed—part wood, part stone—or a similar condition, with internal and external combinations of many forms.

That which gave me the greatest cause first for wonderment, then for reflection, was the discovery of metallized woods, and those which were similarly either completely or partially transformed into metal, and strangely altered. I believe this phenomen to be something completely unexpected by natural historians, as the mixtures described above are most pleasing and at various stages of petrification. But for many the matter has remained very doubtful and difficult, being more of a natural problem than a scholastic one.

Moreover I believe that the relationship between these terrestrial natures will open up to me the road to a full understanding of them, which until now has been poor and confused. Because apart from the aforesaid intermediate state, I have found many ambiguities among them, which display mixtures in various ways: of wood and soil, of wood and stone, of wood, metal and stone together, and wood and mineral fluids, brought together and participating equally in the substance, in an almost neutral body, or rather these things are reduced to quite apparent unity; from which compound the very obscure nature of those terrestrial liquids we call bitumens may receive quite some light.

Among these, I have seen the generation of jet, even though it was a long way from the Ganges, and of aetite which is similar, in which Pliny is to be praised and freed of the calumny which is commonly bestowed on him, for having united them. Similarly [I perceived] the nature of anthracite which they call fossilized ebony, lithanthrax and such like. Nor perhaps is agalloch so different from this wonderful new species of metallophite (which is as famous and prized for its fragrance as it is obscure in its origin; as though it were rivalling the Nile, since its birthplace is unknown, [and] only its transport by floods is related), whose nature I have been and am still discovering with the above matters. [I have] also [been] obtaining, as much as it is permitted to me to skim over them in writing, not only the full historical point of view, but also every natural consideration and discussion that I have been able to [contemplate], in searching for causes and examining how much can be found out about these subjects or concluded from them.

On such occasions, I have also observed furnaces kindled by nature, their manner and their fumes. I also saw in this new class of things, the sulphurous efflorescence as well as the saline, vitreous and crystalline ones; andI confirmed with particular satisfaction many of my theories, made over many years about the essence and causes of these fires which although very little known and hidden to human intellects, always seemed to me for the most part to be full of light.

Of all the above I hope and desire an occasion to give Your Most Illustrious Lordship a full account in person next week, unable to do so earlier on account of my family duties. I also wished, in accordance with my obligation to you, to send in advance this little token: a petrified trunk in which both substances are very clearly evident, a piece of simple fossilized ebony, and a more solid and variegated fragment of the type of the same substance, in which are interposed metal veins similar to those which appear much more in some additional pieces of a much more fragile and dissolved nature similar to lithanthrax, in the hope that these will be agreeable to the great virtue of your Most Illustrious Lordship, to the delight which you take in the natural disciplines which you so fully master, and above all to the most human affection by which you see fit to protect me and favour me with the most abundant favours.

And thus I humbly revere Your Most Illustrious Lordship, praying that God will give you every contentment.

From Acquasparta, on the first of December 1624.
The most humble, devoted and obliged servant of
Your Most Illustrious and Reverend Lordship

Federico Cesi
member of the Accademia dei Lincei

APPENDIX III b

Letter to Nicolas Claude Fabri de Peiresc from Jacques de la Ferrière
Rome, November 1635.

The original in French is in Paris, BN, MS It. 1684. The only known copy was probably written by Jacques de la Ferrière himself and left with Cassiano dal Pozzo in Rome. It is now in Montpellier, BEM, MS H.170, fols. 11–12 (published with a bibliography in Alessandrini 1978, pp. 214–17). The translation here is based on the original.

The notes on the top and top-left margins of the first page are in Cassiano's hand: 'Concerning the fossilized or mineral wood which is found in Acquasparta / By Monsieur de la Ferrière, physician to the Lord Cardinal of Lyon, brother of Cardinal Richelieu.

Sir,

Having promised you in all my previous letters to go to Acquasparta so as to acquaint you with the peculiarities of the fossil woods which are to be found near there, I was obliged to allow the great heat of summer and of the early autumn to pass, so that I left Caprarola only on 22 October, having first asked the Cavaliere dal Pozzo to write on my behalf to someone there of his acquaintance, which he did. I was taken to a place three miles [5 km] from the aforementioned Acquasparta by someone who lives below the Castle of Rosaro, close to which these woods are found. I will therefore begin with the site of the place, as you have asked me to do in all your letters, and you will say that the main road from Narni to San Gemini and from there to Todi passes northwards beneath the said Castle of Rosaro, leaving it to the right, and the terrain or land where this wood is found an equal distance to the left, it being no more than a musket-shot from Rosaro to where the fossils are found, as the slope of the said hill faces the east and that of the castle the west. The discovery of these fossils was made on account of the torrents of water, at the time of heavy rains which have eroded the earth on the slope of the hill over a period of time, this aided by constant ploughing which has made the earth more liable to be washed away. The soil, having thus been lowered by two *piques* [unit of measurement equal to the length of a pike], reveals daily new woods which resemble that wood which is crudely split to make stave-wood for fashioning wine barrels: being smoothed and cut at both ends, some more or less half a foot wide [15 cm] and five or six feet [1.5–1.8 m] long. Aside from these, one finds tree trunks which seem to have been sawn in former times into large logs eight to ten feet [2.5–3.0 m] in length in order to be split for the same use rather than to make joists or beams. In one of these trunks, I could see wedges of wood which my guide told me that his Eminence my Lord Cardinal Baigny had had hammered into it for fifteen days before, with the intention of splitting it, which he had not been able to do as this wood is very strong and hard, having no crack at all to allow it to break easily [across the grain], though it splits easily lengthwise. I, having particularly asked the Lord of Acquasparta what was the largest piece that had been found, he told me that the late Prince Federico Cesi had had a table made of a single piece which he had measured to be nine hand spans [2.1 m] wide and ten [2.3 m] long. I also asked the aforementioned Lord and particularly my guide about that vein [of rock] you had mentioned in one of your letters and which the Cavaliere dal Pozzo told me had been recently discovered which was forty or fifty feet [12–15 m] long but no one could give me any news and they had never even heard anything of it. My guide did show me a place where, he said, a fire had lasted for seven years, and the shepherds, warming themselves, had inadvertently set fire to the soil in this place which I found to be a perpendicular drop, of the depth of a *pique* or more, and a bit to the right of the trench or ravine which the water has carved into the slope of the above-mentioned terrain. I suddenly remembered the treatise, which the late Prince Cesi had written about this fossil wood, of which you have a copy, which states that there was a place where, the earth being naturally sulphurous, it had burnt for ten years and without doubt this was the place my guide had shown me, in which I found that the soil broke off in what looked like terracotta scales resembling earthen pots or broken bricks, and there were many of them even at the bottom of the escarpment that had fallen and which cracked and splintered under my horse's hooves like broken pots. I also saw large lumps of this baked brick-coloured clay, but of rounded shape of about the size of an ox head or larger, which the water had caused to roll down quite a distance from their starting place, as well as pieces of the same size and shape as slag [*scoria ferri*], which is the waste of unusable iron that is thrown into forges in large quantities and which, when broken up with hammers, reveals the iron inside. I must confess to you that having wished to look for some piece in which one could see the chalk or clay turning into something akin to wood, and then into wood itself, I could find none. But I did unearth more than a dozen well-formed pieces of wood, separated as if they had been cut, of which only the ends were visible, the remainder being buried in the earth. I was told that the late Prince Cesi had once harnessed eight pairs of oxen to a trunk to uproot or exhume it, but that was because of its bulk, as none have been found that were longer than ten or twelve feet [3.0–3.6 m] at the most. To get to the examination of this wood I must tell you that the soil below is surely sulphurous for I noticed that, although it had been cultivated it was split and it had cracks in it that were larger than those produced by the summer drought; and they could not be seen at all even in the cultivated land, but in this terrain, they were so deep that my horse, which I was having led by the bridle, sank his legs into them to a depth of a hand's span [about 23 cm] with the risk of breaking them. This makes me think that the inner aridity of the soil caused by underground fires made it crack like this, though I saw no smoke passing through, and I believe that the great drought of the season caused it and that it must rain for the fires to be rekindled which even the locals rarely see.

As for this wood, educated inhabitants of these parts have told me that they think that it was wood which had been buried since the Flood or by an earthquake, and it looks as if it had been cut and sawn in order to be fashioned into barrels or other wine containers, of which parts had been split to a thickness of two to three fingers, and that the remaining trunks had been sawn but that there had not been time to split them. However, here is a conjecture which persuades me to the contrary view, namely that to split trunks after they have been sawn they are placed upright, since one cannot split them properly otherwise, and none of these trunks has ever been found standing or even inclined but always lying on the ground. Secondly, the trunks which I have carefully observed are all rotten on the outside to such a point that not only the bark but all of the tree's albumen is destroyed and unusable. Yet Prince Cesi had a table made nine hand spans [2.1 m] wide which was made of the core of such a tree. One must therefore infer that this is no ordinary tree for in this part of the world we don't have trees of this size and diameter. And as for what Prince Cesi says to prove the generation, that he had seen pieces which at one end were like hard clay and at the other end like wood, I think he is wrong and that he has mistaken the beginning of petrification for the beginning of the generation of (real) wood. Indeed, I have a piece given to me by the Cavaliere dal Pozzo which is like wood at one end and like a half-baked brick at the other. But it is easier for nature to turn wood into terracotta or stone than to turn earth into wood. We have even seen flesh which has been similarly petrified, and Dumoustier in Paris has a petrified trout pâté where the crust and the trout have lost very little of their natural colour. Therefore I don't think that one can rely on Cesi's report as to how this came to be formed, having found two points which cast doubt upon it, namely the recumbent positions of the trunks and their extraordinary size.

I leave the resolution of this difficulty to your judgement—to which I will always yield.

CONCORDANCES

CONCORDANCE A

Royal Library inventory numbers and Catalogue numbers in this volume (B.III)

RL no.	Volume	Cat. No.	RL no.	Volume	Cat. No.
25536	XIV	**197**	25568	XIV	**155**
25537		**198**	25569		**156**
25538		**16**	25570		**61**
25539		**151**	25571		**110**
25540		**85**	25572		**60**
25541		**84**	25573		**126**
25542		**106**	25574		**63**
25543		**109**	25575		**35**
25544		**120**	25576		**36**
25545		**121**	25577		**62**
25546		**53**	25578		**34**
25547		**199**	25579		**134**
25548		**157**	25580		**135**
25549		**181**	25581		**140**
25550		**182**	25582		**141**
25551		**113**	25583		**148**
25552		**123**	25584		**146**
25553		**124**	25585		**149**
25554		**66**	25586		**136**
25555		**95**	25587		**142**
25556		**133**	25588		**129**
25557		**107**	25589		**137**
25558		**58**	25590		**138**
25559		**59**	25591		**139**
25560		**115**	25592	XV	**67**
25561		**128**	25593		**178**
25562		**144**	25594		**127**
25563		**57**	25595		**108**
25564		**145**	25596		**154**
25565		**158**	25597		**119**
25566		**116**	25598		**117**
25567		**122**	25599		**118**

RL no.	Volume	Cat. no.	RL no.	Volume	Cat. no.
25600	XV	**15**	25640	XVI	**64**
25601		**8**	25641		**65**
25602		**7**	25642		**47**
25603		**11**	25643		**88**
25604		**6**	25644		**90**
25605		**9**	25645		**104**
25606		**10**	25646		**41**
25607		**5**	25647		**42**
25608		**47**	25648		**101**
25609		**52**	25649		**44**
25610		**51**	25650		**79**
25611		**83**	25651		**80**
25612		**81**	25652		**99**
25613		**26**	25653		**20**
25614		**55**	25654		**24**
25615		**40**	25655		**125**
25616		**19**	25656		**30**
25617		**28**	25657		**111**
25618		**153**	25658		**132**
25619		**38**	25659		**37**
25620		**71**	25660		**100**
25621		**73**	25661		**48**
25622		**43**	25662		**78**
25623		**69**	25663		**82**
25624		**91**	25664		**33**
25625		**105**	25665		**86**
25626		**98**	25666		**87**
25627		**29**	25667		**32**
25628		**114**	25668		**72**
25629		**97**	25669		**92**
25630		**75**	25670		**89**
25631		**56**	25671		**70**
25632		**77**	25672		**49**
25633		**54**	25673		**39**
25634		**21**	25674		**25**
25635		**45**	25675		**143**
25636		**46**	25676		**68**
25637		**50**	25677		**18**
25638		**96**	25678		**22**
25639		**112**	25679		**23**

RL no.	Volume	Cat. no.	RL no.	Volume	Cat. no.
25680	XVI	**2**	25708	XVII	**190**
25681		**1**	25709		**177**
25682		**4**	25710		**131**
25683		**14**	25711		**31**
25684		**3**	25712		**191**
25685		**147**	25713		**192**
25686		**12**	25714		**161**
25687		**17**	25715		**162**
25688		**13**	25716		**163**
25689		**27**	25717		**164**
25690		**94**	25718		**165**
25691		**93**	25719		**166**
25692		**152**	25720		**167**
25693		**130**	25721		**168**
25694		**102**	25722		**169**
25695		**103**	25723		**170**
25696		**76**	25724		**171**
25697	XVII	**183**	25725		**172**
25698		**184**	25726		**193**
25699		**185**	25727		**194**
25700		**150**	25728		**180**
25701		**186**	25729		**173**
25702		**187**	25730		**174**
25703		**159**	25731		**175**
25704		**188**	25732		**176**
25705		**179**	25733		**195**
25706		**189**	25734		**196**
25707		**160**			

CONCORDANCE B

Stelluti Plate numbers and Catalogue numbers

Stelluti 1637 Plate number	Cat. no.
7	**8**
8	**7**
8	**11**
10 (fig. D)	**19**
11 (fig. E)	**117**
11 (fig. F)	**93**
11 (fig. G)	**94**
11 (fig. H)	**76**
12 (fig. H)	**112**
12 (fig. L)	**104**
12 (fig. M)	**95**
12 (fig. N)	**97**
13	**197**
13	**198**
13	**199**

CONCORDANCE C

Seventeenth-century folio numbers and Catalogue numbers

Where catalogue numbers appear in parenthesis, no number is discernible on the drawing but the original number is noted in the seventeenth-century index (Appendix I).

17th-cent. no.	Cat. nos.	17th-cent. no.	Cat. nos.
1	183, 197	35	126, 175
2	184, 198	36	63, 176
3	151, 185	37	35, 195
4	85, 150	38	36, 196
5	84, 186	39	(34), 62
6	106, 187	40	134
7	109, 159	41	135
8	120, 188	42	135
9	121, 179	43	140
10	53, 189	44	141
11	160, 199	45	148
12	157, 190	46	146
13	177, 181	47	149
14	131, 182	48	136
15	31, 113	49	142
16	123, 191	50	129
17	124, 192	51	137
18	(66), 95, 161	52	138
19	133, 162	53	139
20	107, 163	54	67
21	58, 164	55	178
22	59, 165	56	127
23	115, 166	57	108
24	128, 167	58	154
25	144, 168	59	119
26	145, 169, 170	60	117
27	(57), 158	61	118
28	116, 171	62	15
29	122, 172	63	8
30	155, 193	64	7
31	156, 194	65	11
32	61, 180	66	6
33	110, 173	67	9
34	60, 174	68	5, 10

17th-cent. no.	Cat. nos.	17th-cent. no.	Cat. nos.
69	**47**	114	**20**
70	**52**	115	**24**
71	**51**	116	**125**
72	**83**	117	**30**
73	**81**	118	**111**
74	**26**	119	**132**
75	**55**	120	**37**
76	**40**	121	**100**
77	**19**	122	**48**
78	**28**	123	**78**
79	**153**	124	**82**
80	**38**	125	**33**
81	**71**	126	**86**
82	**73**	127	**87**
83	**43**	128	**32**
84	**69**	129	**72**
85	**91**	130	**92**
86	**105**	131	**89**
87	**98**	132	**70**
88	**29**	133	**49**
89	**114**	134	**39**
90	**97**	135	**25**
91	**75**	136	**143**
92	**56**	137	**68**
93	**77**	138	**18**
94	**54**	139	**22**
95	**21**	140	**23**
96	**45**	141	**2**
97	**46**	142	**1**
98	**50**	143	**4**
99	**96**	144	**14**
100	**112**	145	**3**
101	**64**	146	**147**
102	**65**	147	**12**
103	**74**	148	**17**
104	**88**	149	**13**
105	**90**	150	**27**
106	**104**	151	**94**
107	**41**	152	**93**
108	**42**	153	**152**
109	**101**	154	**130**
110	**44**	155	**102**
111	**79**	156	**103**
112	**80**	157	**76**
113	**99**		

In addition to **34**, **57** and **66** (identified above), the following drawing bears no inscribed number: **16**

CONCORDANCE D

Stelluti manuscripts, the *Trattato* and Windsor drawings

Manuscripts H. 173 in Montpellier and It. 1684 in Paris are accompanied by proofs of the engravings for plates 1–9 and 13 of Stelluti's *Trattato*; in addition, H. 173 also contains the proofs of all elements of Plates 10, 11, and 12.[1] The inscription on the important illustration of the piece of *legno ovato* from Rosaro (Plate 7) is handwritten in both these manuscripts. All this tends to support the suggestion made above (pp. 62–3) that these were not just the prime manuscripts, but that they were the ones owned by Cassiano and Peiresc.

The second manuscript in Montpellier, H. 170, on the other hand, contains no engravings, but three drawings (illustrated here as Figs 66–8), with annotations by Cesi on folio 10. These relate to three of the specimens on Stelluti's Plates 10 and 11. Furthermore, this manuscript is bound together not just with Cesi's letter to Francesco Barberini of 1 December 1624 but also with La Ferrière's letter to Peiresc, and a number of other treatises relating to fossil matters.[2] The text of Stelluti's *Trattato* in H. 170 is much closer to the printed version than are the versions in the two manuscripts just discussed, H. 173 and It. 1684. The few differences are fairly minor, with the exception, once more, of the omission of the excursus on ammonites and the long list of different types of figured stones.

At this point in MS H. 170, Stelluti concludes by reaffirming that he has based himself on the writings of Cesi—*riportandomi a quanto n'ha scritto il sudetto Signor Principe per pij piena congnitione di questo parto della Natura* (a passage actually omitted in the printed version)[3]—but adds several lines explaining that

> there remain the two figures drawn in pen which—among the many others not engraved—we have described as especially rare and strange. The first shows some pieces of wood where there are those fibres of thick metal, as shown in the said drawing ... the other figure shows a trunk of wood just as it was found beneath the earth, which was all knotty and the interior part smooth, and like hard bone.[4]

Two of the drawings in this manuscript are indeed related to the pieces of wood shown as Pl. 10, Fig. A and Pl. 11, Fig. F in the *Trattato* (see here Figs 67 and 68).[5] The third drawing (see Fig. 69), on the other hand, does not relate to anything illustrated in the *Trattato*.

Both manuscripts in the Vatican (Barb. Lat. 4355 and Vat. Lat. 8258) contain texts which are almost identical with that of H. 170 in Montpellier. But whereas Barb. Lat. 4355 contains an almost complete set of engravings, it also contains drawn variations (illustrated here as Figs 70 and 71) of the two drawings in H. 170 that were reproduced in part in the *Trattato*. This manuscript must have been made at some point between the initial versions, H. 173 and It. 1684, and the later ones, H. 170 and Vat. Lat. 8258. This latter manuscript contains no engravings but it has other versions of the two drawings in Barb. Lat. 4355 and H. 170, as well as several versions of other works on natural history and fossils.[6]

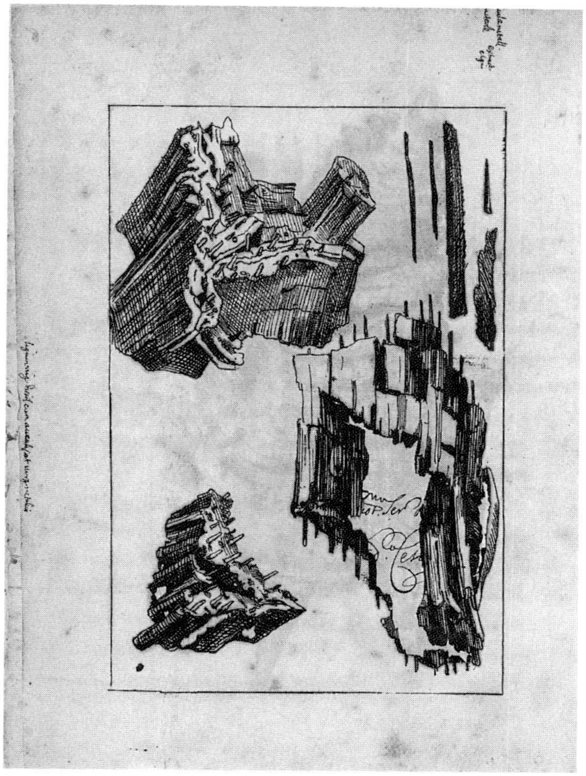

Fig. 67. Montpellier, BEM, MS H.170, fol. 8v: drawings of fossil woods, one for Stelluti 1637, Pl. 10, Fig. A

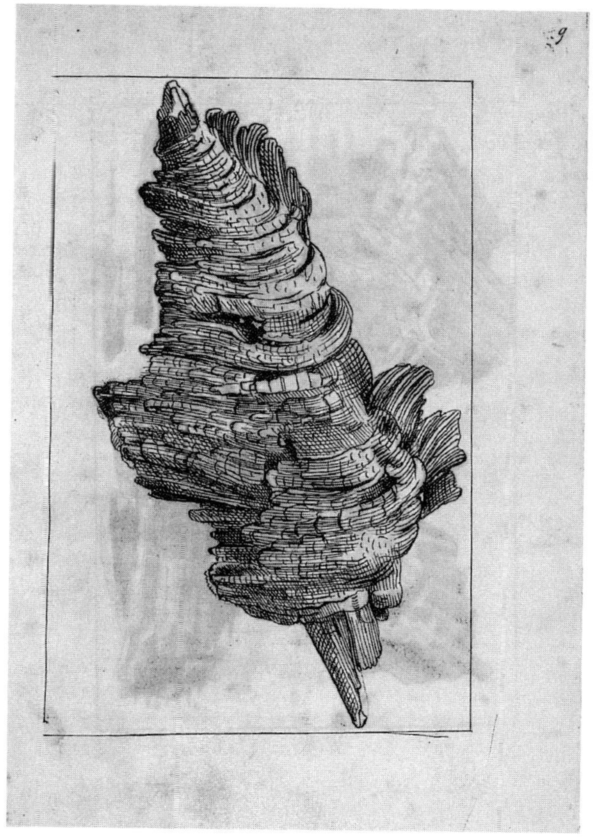

Fig. 68. Montpellier, BEM, MS H.170, fol. 9: drawing for Stelluti 1637, Pl. 11, Fig. F

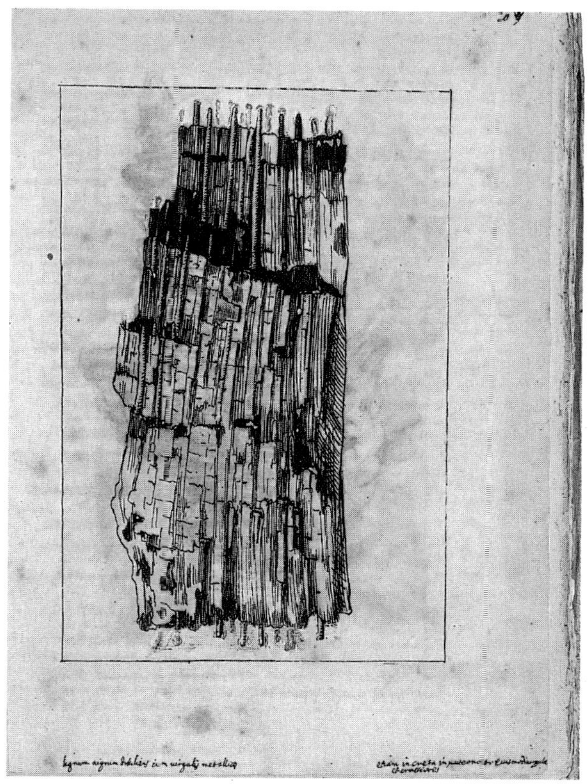

Fig. 69. Montpellier, BEM, MS H.170, fol. 10: drawing of pyritized wood (not in Stelluti 1637)

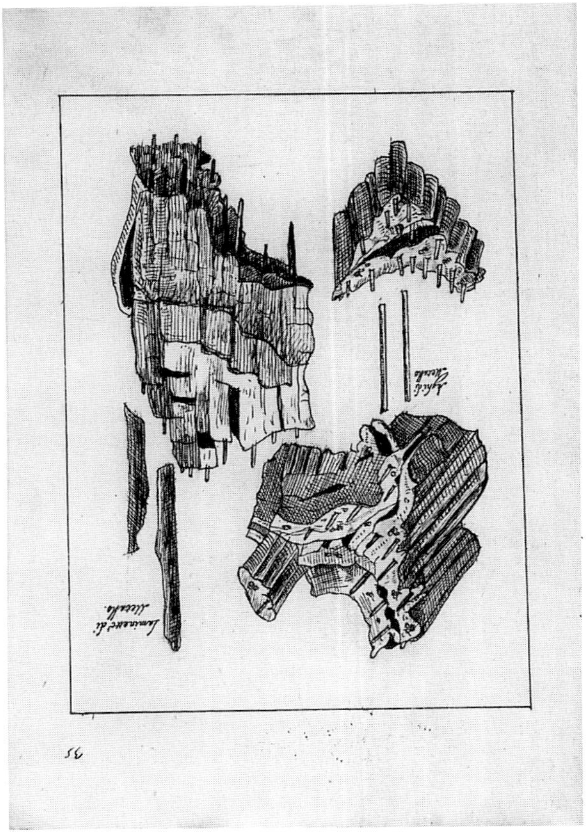

Fig. 70. BAV, Barb. Lat. 4355, fol. 35: drawing of pyritized wood, including study for Stelluti 1637, Pl. 10, Fig. A

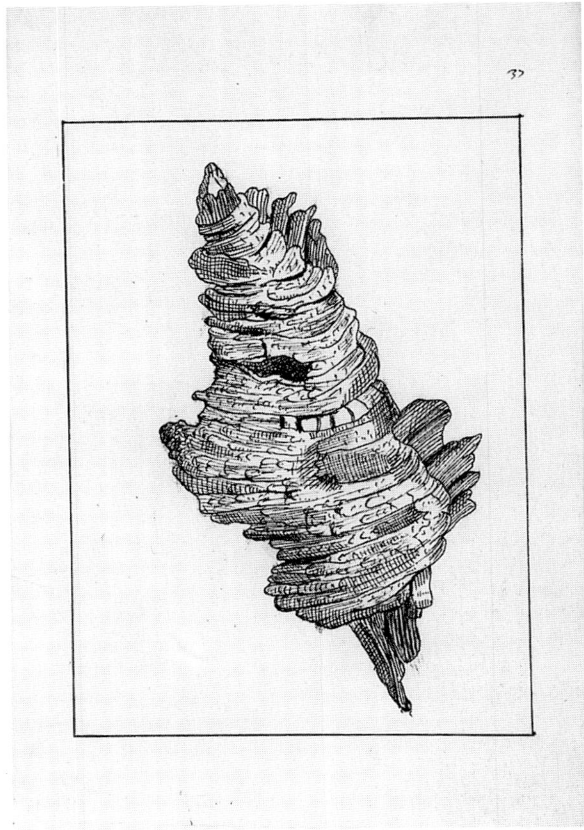

Fig. 71. BAV, Barb. Lat. 4355, fol. 37: showing variants of drawings in H.170

But the differences involved are very small. Manuscripts and printed version were all produced in a period of intense interest in publishing at least something relating to Cesi's controversial investigations into the fossil remains of Acquasparta, Dunarobba, Rosaro and Sismano.

NOTES

1. Montpellier, BEM, MS H 173 thus contains thirteen proofs for the Plates in Stelluti 1637, while Paris, BNF, MS It 1684 has ten. Cf. Appendix II.

2. Other treatises bound with this MS include those on the *succino o ambra gialla*, on a blood-red celestial rain in Sicily, on certain gems, as well as a variety of works of more purely botanical interest; cf. Alessandrini 1978, pp. 241–7 for a summary.

3. Montpellier BEM, MS H. 170, fol. 5v (cf. Stelluti 1637, p. 12).

4. Montpellier, BEM, MS H. 170.

5. However, of the several remarkable drawings on fol. 8 showing rocks with pyritized fibres, only one was used, for Plate 10, Fig. A.

6. For a summary of all these differences, see the table below.

Concordance between the Stelluti manuscripts, the *Trattato* and the Windsor drawings

Stelluti 1637 plate no.	**Montpellier, BEM, H.173** folio no.	**Paris, BN, It. 1684** folio no.	**BAV, Barb. Lat. 4355** folio no.	**Montpellier, BEM, H.170** folio no.	**BAV, Vat. Lat. 8258** folio no.	**Windsor RL no.**	**Cat. no.**
1	8	6v–7	13–14				
2	10	8v–9	16–17				
3	12▲	11▲	21*				
4	11▲	10▲	19*				
5	13	12▲	23				
6	14	13	25				
7	16	14	27			25601	**8**
8	17	15	29			25602 25603	7 **11**
9	15	16	31*				
10	19				24		
10A			35✎ (fig. 17)	8 verso✎ (fig. 14)			
10D						25616	**19**
11	20				25		
11E						25598	**117**
11F			37✎ (fig. 18)	9✎ (fig. 15)		25691	**93**
11G						25690	**94**
11H						25696	**76**
12	21						
12H						25639	**112**
12I						25645	**104**
12L						25555	**95**
12M						25629	**97**
12N							
13	18	17▲	33			25537 23556 25547	**198** **197** **199**
				other drawing fol. 10✎ (fig. 16)			

* in reverse ▲ upside down ✎ drawing

BIBLIOGRAPHY AND ABBREVIATIONS

Institutions

BAV	Vatican, Biblioteca Apostolica
Florence, BNC	Florence, Biblioteca Nazionale Centrale
Montpellier, BEM	Montpellier, Bibliothèque de l'École de Médecine
Naples, BN	Naples, Biblioteca Nazionale
Paris, BN	Paris, Bibliothèque Nationale
Paris, IDF	Paris, Institut de France
Rome, BANL	Rome, Biblioteca della Accademia Nazionale dei Lincei
Windsor, RL	Windsor, Royal Library

Books, articles and manuscripts

Accordi 1975 — B. Accordi, 'Contributions to the history of the geological sciences. Paolo Boccone (1633–1704) — a practically unknown excellent geo-paleontologist of the 17th century', *Geologia Romana* 14, pp. 353–9.

Accordi 1976 — B. Accordi, 'Contributions to the history of the geological sciences. Illustrations of the Kircher Museum naturalistic collections', *Geologia Romana* 15, pp. 113–26.

Accordi 1977 — B. Accordi, 'Contributions to the history of the geological sciences. The Musaeum Calceolariam (XVIth century) of Verona, illustrated in 1622 by Ceruti and Chiocco', *Geologia Romana* 16, pp. 21–54.

Accordi 1981a — B. Accordi, 'Ferrante Imperato (Napoli, 1550–1625) e il suo contributo alla storia della geologia', *Geologia Romana* 20, pp. 43–56.

Accordi 1981b — B. Accordi, 'Tentativi di classificazione delle pietre e delle gemme nei secoli XVI e XVII', *Physis* 23, pp. 311–24.

Agricola 1546 — G. Agricola, *De Natura Fossilium libri X*, Basle (H. Frobenius and N. Episcopius).

Agricola 1556 — G. Agricola, *De Re Metallica libri XII, quibus officia, instrumenta, machinae, ac omnia denique ad metallicam spectantia, non modo luculentissimè describuntur, sed & per effigies, suis locis insertas, adiunctis latinis, germanicisque appellationibus ita ob oculos ponuntur, ut clarius tradi non possint*, Basle (H. Frobenius and N. Episcopius).

Aldrovandi 1648 — U. Aldrovandi, *Musaeum Metallicum in libros IIII distributum*, Bologna (Giovanni Battista Ferrone).

Alessandrini 1978 — A. Alessandrini, *Cimeli Lincei a Montpellier*, Rome: Accademia Nazionale dei Lincei (Indici e sussidi della biblioteca 11).

Alessandrini 1986 — A. Alessandrini *et al.*, *Francesco Stelluti Linceo da Fabriano*, Fabriano.

Altieri Biagi/Basile 1980 — M. Altieri Biagi and B. Basile (eds), *Scienziati del Seicento*, Milan.

Ambrosetti 1995 — P. Ambrosetti *et al.*, 'La foresta fossile di Dunarobbba (Terni, Umbria, Italia Centrale): contesto lithostratigrafico, sedimentologico, palinologico, dendrocronologico e paleomalacologico', *Il Quaternario* 8, pp. 465–508.

Ambrosetti/Cerquaglia 1996 — P. Ambrosetti and Z. Cerquaglia (eds), *La Foresta Fossile di Dunarobba*, 2nd edn, Todi.

Andrews 1980 — H. N. Andrews, *The Fossil Hunters: In Search of Ancient Plants*, Ithaca, New York.

Archivio dal Pozzo — *Il Carteggio di Cassiano dal Pozzo*, Rome, BANL MS, 39 vols. For a catalogue of this, see Nicolò 1991.

Arkell 1956 — W. J. Arkell, *Jurassic Geology of the World*, London.

Ashpitel 1862 — A. Ashpitel, 'Catalogue of the architectural MSS and drawings in the Royal Library at Windsor Castle', unpublished MS, Royal Library, Windsor.

Basilici 1997 — G. Basilici, 'Sedimentary facies in an extentional and deep lacustrine depositional system: the Pliocene Tiberino Basin, Central Italy', *Sedimentary Geology*, 109, pp. 73–94.

Basilici/Gentili 1992 — G. Basilici and S. Gentili, 'La foresta fossile di Dunarobba: una interpretazione sedimentologica', in Cerquaglia 1992, pp. 47–68.

Biagioli 1993 — M. Biagioli, *Galileo, Courtier: The Practice of Science in the Culture of Absolutism*, Chicago.

Biondi 1982 — E. Biondi, '*Taxodioxylon gypsaeum* (Göppert) Kräusel legno fossile rinvenuto nel Miocene dell'Italia centale', *Il Giornale Botanico Italiano* 116, pp. 227–33.

Biondi 1984 — E. Biondi, *Sul 'Trattato del Legno Fossile Minerale' di Francesco Stelluti Accademico Linceo da Fabriano con Ristampa Anastatica dell'Opera'*, Fabriano.

Biondi 1992 — E. Biondi, 'La foresta di Dunarobba: aspetti storici paleontologici e paleoambientali', in Cerquaglia 1992, pp. 227–33.

Biondi/Brugiapaglia 1991 — E. Biondi and E. Brugiapaglia, '*Taxodioxylon gypseum* in the fossil forest of Dunarobba (Umbria, Central Italy)', *Flora Mediterranea* 1, pp. 111–20.

Biondi/Castagnari 1979 — E. Biondi and G. Castagnari, 'Francesco Stelluti — un naturalista fra XVI e XVII secolo', extract from *Natura e Montagna* 2, June.

Blackwell 1991 — R. J. Blackwell, *Galileo, Bellarmine, and the Bible*, Notre Dame, Indiana.

Boncambi 1961 — C. L. Boncambi, 'Geologia e paleontologia dei bacini ligniferi dell'Umbria', *Atti del Convegno Mostra Nazionale delle Ligniti, Perugia 7–11 maggio 1959*, Foligno, Italy (F. Salviati).

Bromehead 1947a — C. E. N. Bromehead, 'A geological museum of the early seventeenth century', *Quarterly Journal of the Geological Society* CIII, pt 2, pp. 65–87.

Bromehead 1947b — C. E. N. Bromehead, 'Aetites or the eagle stone', *Antiquity* XXI, pp. 16–22.

Caley/Richards 1956 — E. R. Caley and J. F. C. Richards (eds), *Theophrastus on Stones*, Greek text, English translation and commentary, Ohio State University, Columbus, Ohio.

Cardanus 1550 — H. Cardanus, *De Subtilitate libri XXI: Liber VII, De Lapidibus,* Nuremberg (I. Petreius).

Cassiano cat. 1993 — *The Paper Museum of Cassiano dal Pozzo*, exhibition catalogue, British Museum, London. *Quaderni Puteani 4*, Milan (Olivetti).

Cassiano cat. 1997 — *Cassiano dal Pozzo's Paper Museum. Drawings from the Royal Collection*, exhibition catalogue, National Gallery of Scotland, Edinburgh (1 April–8 June 1997).

Cerquaglia 1992 — Z. Cerquaglia (ed.), *La Foresta Fossile di Dunarobba*, 1st edn, Todi, Italy [for 2nd edn, see Ambrosetti/Cerquaglia 1996].

Ceruti/Chiocco 1622 — B. Ceruti and A. Chiocco, *Museum Francisci Calceolarii Junioris Veronensis luculenter descriptum, & perfectum. In quo multa ad naturalem, moralem[que] philosophia spectantia, non pauca ad rem medicam pertinentia erudité propnuntur, & explicantur non sine magna rerum exoticarum supellectile, quae artifici plane manu in aes incisae, studiosis exhibentur*, Verona (A. Tamum).

Cesalpino 1583 — A. Cesalpino, *De Plantis libri XVI*, Florence (Giorgio Marescotto).

Cesalpino 1596 — A. Cesalpino, *De Metallicis libri tres*, Rome (Zanetti).

Cesi 1980 — F. Cesi, 'Del natural desiderio di sapere et instituzione de' Lincei per adempimento di esso' in Altieri Biagi/Basile 1980, pp. 39–70.

Colonna 1592 — F. Colonna, *Phytobasanos sive plantarum aliquot historia in qua describuntur diversi generis plantae veriores, ac magis facie, viribusque respondentes antiquorum Theophrasti, Dioscoridis, Plinii, Galeni, aliorumque delineationibus, ab aliis hucusque non anima diversae*, Naples (Oratio Salviani).

Colonna 1606 F. Colonna, *Minus Cognitarum Stirpium aliquot ac etiam rariorum nostro coelo orientium* ΕΚΦΡΑΣΙΣ. *Qua non parca ab Antiquioribus, Theophrastro, Dioscoride, Plinio, Galeno aliisque memoratae declarantur, officinarum usis perquam utiles ... Item, de Aquatilibus, aliisque animalibus quibusdam paucis libellus, eodem Auctore*, Rome (Guglielmo Facciotto) 1606.

Colonna 1616a F. Colonna, *Minus Cognitarum Rariorumque nostro coelo orientium stirpium* ΕΚΦΡΑΣΙΣ. *Qua non paucae ab antiquioribus Theophrasto, Dioscoride, Plinio, Galeno aliisque descriptae, praeter illos etiam in* ΦΥΤΟΒΑΣΑΝΩ *editas disquiruntur ac declarantur. Item de Aquatilibus aliisque nonullis animalibus libellus*, Rome (Giacomo Mascardi) 1616.

Colonna 1616b F. Colonna, *Purpura; hoc est de purpura ab animali testaceo fusa, de hoc ipso animali, aliisque rarioribus testaceis quibusdam* with *De Glossopetris Dissertatio*, Rome (Giacomo Mascardi) 1616.

Conan/Deschesne 1996 J. Conan and O. Deschesne, *Le Bitume à Suse: Collection du Museé du Louvre*, Paris.

Dana 1865 J. D. Dana, *A System of Mineralogy*, 5th edn, New York.

Dati 1664 C. Dati, *Delle Lodi del Commendatore Cassiano dal Pozzo*, Florence (All'Insegna della Stella) 1664.

De Renzi 1992-3 S. de Renzi, 'Storia naturale ed erudizione nella prima età moderna: Johnannes Faber (1574–1629) medico linco', Università degli Studi di Bari, Tesi di Dottorato in Storia della Scienza (VI ciclo).

Desio 1973 A. Desio (ed.), *Geologia dell'Italia*, Turin.

Documentary Culture E. Cropper, G. Perini, F. Solinas (eds), *Documentary Culture. Florence and Rome from Grand-Duke Ferdinand I to Pope Alexander VII*, Papers from a Colloquium held at the Villa Spelman, Florence 1990, Bologna 1992.

Drake 1957 S. Drake (ed.), *Discoveries and Opinions of Galileo*, New York.

Edwards 1931 W. N. Edwards, *Guide to an Exhibition Illustrating the Early History of Palaeontology*, exhibition catalogue, British Museum, London.

Encelius 1557 C. Encelius, *De Re Metallica, hoc est de origine, varietate & natura corporum metallicorum, lapidum, gemmarum atque aliarium quae ex fodinis eruuntur, rerum, ad medicinae usum deseruientium, libri III*, Frankfurt (Hered. Christiani Egenolphi).

Findlen 1990 P. Findlen, 'Jokes of Nature and jokes of knowledge: the playfulness of scientific discourse in early modern Europe', *Renaissance Quarterly* 43, pp. 292–331.

Findlen 1994 P. Findlen, *Possessing Nature. Museums, Collecting, and Scientific Culture in Early Modern Italy*, Berkeley, California.

Finocchiaro 1989 M. A. Finocchiaro, *The Galileo Affair: A Documentary History*, Berkeley, California.

Follieri 1977 M. Follieri, 'Evidence on the Plio-Pleistocene paleofloristic evolution in Central Italy', *Rivista Italiana Paleontologica* 83, pp. 925–30.

Foucault 1973 M. Foucault, *The Order of Things: An Archaeology of the Human Sciences*, New York [originally *Les mots et les choses*, Paris, 1966].

Freedberg 1989a D. Freedberg, 'From Hebrew and gardens to oranges and lemons: Giovanni Battista Ferrari and Cassiano dal Pozzo', Solinas 1989a, pp. 37–72.

Freedberg 1989b D. Freedberg, 'Cassiano, natural historian', *Quaderni Puteani* 1, pp. 10–15.

Freedberg 1989c D. Freedberg, 'Cassiano dal Pozzo's drawings of citrus fruits', *Quaderni Puteani* 1, pp. 16–36.

Freedberg 1992a D. Freedberg, 'Cassiano on the Jewish races', *Quaderni Puteani* 3, pp. 41–56.

Freedberg 1992b D. Freedberg, 'Ferrari on the classification of oranges and lemons', *Documentary Culture*, pp. 287–306.

Freedberg 1993 D. Freedberg, 'Cassiano and the art of natural history', *Quaderni Puteani* 4, pp. 141–222.

Freedberg 1994 D. Freedberg, 'The failure of colour', *Sight and Insight. Essays on Art and Culture in Honour of E. H. Gombrich*, ed. John Onians, London, pp. 245–62.

Freedberg 1996a	D. Freedberg, `Ferrari and the pregnant lemons of Pietrasanta', *Il Giardino delle Esperidi. Gli Agrumi nella Storia, nella Letteratura e nell'Arte*, ed. A. Tagliolini and M. Azzi Visentini, Florence, pp. 41–58.
Freedberg 1996b	D. Freedberg, 'Poussin, Ferrari, Cortone et "l'Aetas Florea"', *Actes du Colloque Poussin*, ed. A. Mérot, Paris, pp. 337–62.
Freedberg 1998	D. Freedberg, 'Iconography between the history of art and the history of science: art, science and the case of the urban bee', in *Picturing Science Producing Art*, ed. P. Galison and C. Jones, London, pp. 272–96.
Freedberg forthcoming	D. Freedberg, *The Eye of the Lynx: Art, Science and Nature in the Age of Galileo*, Chicago.
Freedberg/Baldini 1997	D. Freedberg and E. Baldini, *Citrus Fruit*. B. I. of *The Paper Museum of Cassiano dal Pozzo. A Catalogue Raisonné. Drawings and Prints in the Royal Library at Windsor Castle, the British Museum, the Institut de France and other Collections*, London.
Gabrieli 1959	[G. Gabrieli], 'Il carteggio linceo della vecchia accademia di Federico Cesi (1603–1630)', *Atti della Reale Accademia dei Lincei, Memorie della Classe di Scienze morali, storiche e filologiche*, series 6, Vol. VII, fasc. 1–4 (1938–42), pp. 1–1446.
Gabrieli 1989	G. Gabrieli, *Contributi alla Storia dell'Accademia dei Lincei*, 2 vols., Rome.
Galileo *Opere*	*Le Opere di Galileo Galilei*, ed. A. Favaro, 20 vols., Florence 1890–1909.
Galuzzi 1984	*Novità Celesti e Crisi del Sapere: Atti del Convegno Internazionale di Studi Galileiani*, ed. P. Galluzzi, Florence.
Gayvard-Valy 1994	Y. Gayvard-Valy, *The Story of Fossils: In Search of Vanished Worlds*, London.
Gesner 1558	C. Gesner, *Historiae Animalium liber IIII, qui est de piscium & aquatilium animantium natura*, Zurich (Christoph Froschauer).
Gesner 1565	C. Gesner, *De Rerum Fossilium, Lapidum et Gemmarum maximè, figuris et similitudinibus liber: non soljm medicis, sed omnibus rerum naturae ac philologiae studiosis, utilis & iucundus futurus*, Zurich (Tiguri).
Girotti 1992	O. Girotti, 'Il Bacinotiberino e l'antico tirreno', in Cerquaglia 1992, pp. 29–34.
Gomez 1991	S. Gomez, 'The Bologna stone and the nature of light., *Nuncius* VI:2, pp. 3–30.
Goodman/ Goodman/Colin 1991	D. Goodman, R. Goodman, A. Colin, *The Rise of Scientific Europe, 1500–1800*, London.
Gould 1997	S. J. Gould, 'Seeking to explain sea fossils on mountain ridges, Leonardo developed his theory of the living earth', *Natural History* 1–6 (4), pp. 18–29.
Gould 1998a	S. J. Gould, 'The sharp-eyed Lynx, outfoxed by Nature, Part I', *Natural History*, 5.
Gould 1998b	S. J. Gould, 'The sharp-eyed Lynx, outfoxed by Nature, Part II', *Natural History*, 6.
Haskell 1963	F. Haskell, *Patrons and Painters: A Study of the Relations between Italian Art and Society in the Age of the Baroque*, London.
Hernandez	*see Thesaurus*.
Hooke 1665	R. Hooke, *Micrographia: or some physiological descriptions of minute bodies made by magnifying glasses, with observation and inquiries made thereupon*, London.
Hower 1996	J. Hower, A. C. Scott, B. K. Parekh, A. C. Hutton, 'Coal availability, mining and preparation', *Encyclopaedia of Energy Technology and the Environment*, ed. A. Bisio and S. G. Boots, New York, pp. 603–84.
Imperato 1559	F. Imperato, *Dell' Historia Naturale ... libri XXVIII, nella quale ordinatamente si tratta della diversa condition di miniere, e pietre*, Naples (C. Vitale).
Imperato 1610	F. Imperato, *De Fossilibus Opusculum*, Naples (D. Roncaioli).
Imperato 1628	F. Imperato, *Discorsi Intorno a Diverse Cose Naturali*, Naples (E. Longo).
Inventory A	Manuscript inventory preserved in the Print Room, Royal Library, Windsor, and dating from the end of the eighteenth century: drawn up by, or under the direction of, George III's librarian, Richard Dalton. See p. 17, above.
Knobloch 1993	E. Knobloch, 'Die Planzengesellschaften der Braunkohle', *Fossilen* 6, pp. 357–77.
Kräusel 1947	R. Kräusel, 'Die fossilen Coniferenhölzer', *Paläontographica* 89B, pp. 83–203.

Lhote/Joyal 1989 — *Nicolas-Claude Fabri de Peiresc, Lettres à Cassiano dal Pozzo (1626–1637)*, ed. J.-F. Lhote and D. Joyal, Clermont-Ferrand.

Liceti 1621 — F. Liceti, *De Lucernis Antiquorum Reconditis libri quatuor*, Venice (Evangelista Deuch).

Liceti 1640a — F. Liceti, *De Luminis Natura & Efficientia libri tres*, Udine, Italy (N. Schiratti).

Liceti 1640b — F. Liceti, *Litheosphorus, sive de lapide Bononiensi lucem*, Udine, Italy (N. Schiratti).

Liceti 1641 — F. Liceti, *De Lucidis in Sublimi Ingenuarum Exercitationum liber*, Padua (Cribelliano).

Lumbroso 1874 — G. Lumbroso, 'Notizie sulla vita di Cassiano dal Pozzo', *Miscellanea di Storia Italiana* 15, pp. 131–388.

Martinetto 1994 — E. Martinetto, 'Paleocarpology and the "in situ" ancient plant communities of a few Italian Pliocene fossil forests', in *Studies on Ecology and Palaeoecology of Benthic Communities*. Bollettino della Società Paleontolia Italiana, Spec. Vol. 2, ed. R. Matteucci, M. G. Carboni and J. S. Pignatti, Mucchi, Italy, pp. 189–6.

Martinetto 1996 — E. Martinetto, 'Relazione sull' analisi Paleocarpologica di depositi continentali del Bancino Tiberina E di Pietrafitta' in Ambrosetti/Cerquaglia 1996, pp. 87–95.

Mathiesen 1970 — J. Mathiesen, 'Palaeobotanical investigations into some cormophytic macrofossils from the Neogene Tertiary lignites of Central Jutland', *Det Kongelige Danske Videnskabernes Selskab Bioliske Skrifter* 17, 3, pp. 1–69.

McBurney 1989a — H. McBurney, 'History and contents of the dal Pozzo collection in the Royal Library, Windsor Castle', Solinas 1989a, pp. 75–93.

McBurney 1989b — H. McBurney, 'The later history of Cassiano dal Pozzo's "Museo Cartaceo"', *Burlington Magazine* CXXXI, pp. 549–53.

McBurney 1989c — H. McBurney, 'A brief history of the *Museo Cartaceo*', *Quaderni Puteani* 1, pp. 5–9.

McBurney 1989d — H. McBurney, 'Cassiano dal Pozzo's drawings of birds', *Quaderni Puteani* 1, pp. 37–47.

McBurney 1992a — H. McBurney, 'Cassiano dal Pozzo as scientific commentator: ornithological texts and images from the *Museo Cartaceo*', in *Documentary Culture*, pp. 349–63.

Mercati 1719 — M. Mercati, *Metallotheca. Opus posthumum, auctoritate, & munificentia Clementis Undecimi Pontificis Maximii è tenebris in lucem eductum*, Rome (J. M. Salvioni).

Michaelis 1882 — A. Michaelis, *Ancient Marbles in Great Britain*, trans. C. Fennell, Cambridge.

Morello 1986 — G. Morello, *Federico Cesi e i primi Lincei*, exhibition catalogue, Biblioteca Apostolica Vaticana, Vatican City.

Napoleone 1989 — C. Napoleone, 'Appunti sul "Natural History of Fossils V" della Royal Library di Windsor', Solinas 1989a, pp. 187–98.

Naudé 1650 — G. Naudé, [Mascurat] *Iugement de tout ce qui a esté imprimé contre le cardinal Mazarin, depuis le sixième Janvier, jusques à la declaration du premier Avril mil six cent quarante-neuf*, Paris.

Nicolò 1982 — A. Nicolò, 'Corrispondenza inedita di Francesco Stelluti a Cassiano dal Pozzo nel Carteggio Puteano dell'Accademia dei Lincei', *Atti della Accademia Nazionale dei Lincei, Classe di Scienze Morali, Storiche e Filologiche, Rendiconti* XXVII, fasc. 3–4, pp. 91–9.

Nicolò 1985 — A. Nicolò, 'Francesco Stelluti nelle sue lettere dal 1630 al 1652', *Atti del Convegno Francesco Stelluti*, Fabriano, Italy.

Nicolò 1986 — A. Nicolò, 'Le lettere dal 1630 al 1652 ... Francesco Stelluti e il legno fossile', Alessandrini 1986, pp. 176–80.

Nicolò 1991 — A. Nicolò, *Il Carteggio di Cassiano dal Pozzo: Catalogue*, Florence.

Nicolò/Solinas 1986 — A. Nicolò and F. Solinas, 'Per una analisi del collezionismo Linceo: l'Archivio Linceo 32 e il Museo di Federico Cesi', *Convegno Celebrativo del IV Centenario della Nascita di Federico Cesi (Atti dei Convegni Lincei, 78), Acquasparta: 7–9 ottobre 1985*, Rome, pp. 193–212.

Nicolò/Solinas 1987 — A. Nicolò and F. Solinas, 'Cassiano dal Pozzo: appunti per una cronologia di documenti e disegni (1612–1630)', *Nouvelles de la République des Lettres* II, pp. 59–110.

Page 1990 — C. N. Page, 'Taxodiaceae', *The Families and Genera of Vascular Plants. 1: Pteridophytes and gymnosperms*, ed. K. U. Cramer and P. S. Green, Berlin, pp. 353–61.

Palissy 1580	B. Palissy, *Discours Admirables, de la nature des eaux et fonteines, tant naturelles qu'artificielles, des metaux, des sels & salines, des pierres, des terres, du feu & des emaux*, Paris (Martin le Jeune).
Persio 1613	A. Persio, *De Ratione recte Philosophandi et De Natura Ignis et Caloris*, Rome (Giacomo Mascardi).
Pirotta 1904	R. Pirotta, *Sodales R. Lynceorum Academiae annum CCC. ab eius Institutione concelebrantes Fridirici Cesi auctoris conlegi Opus probatissimum De Plantis ad fidem exemplaris castigatioris studio et cura conlegae Romualdi Pirotta iterum edendum decrevere*, Rome.
Quaderni Puteani 1	*Il Museo Cartaceo di Cassiano dal Pozzo, Cassiano Naturalista. Quaderni Puteani* 1, Milan (Olivetti) 1989.
Quaderni Puteani 2	*Cassiano dal Pozzo's Paper Museum, Volume I* (Proceedings of a conference held at the British Museum and Warburg Institute, London, 14–15 December 1989). *Quaderni Puteani* 2, Milan (Olivetti) 1992.
Quaderni Puteani 3	*Cassiano dal Pozzo's Paper Museum, Volume II* (Proceedings of a conference held at the British Museum and Warburg Institute, London, 14–15 December 1989). *Quaderni Puteani* 3, Milan (Olivetti) 1992.
Quaderni Puteani 4	*Cassiano cat.* 1993
Ravazzi/ Van der Burgh 1994	C. Ravazzi and J. Van der Burgh, 'Coniferous woods in the early Pleistocene brown coals of the Leffe Basin (Lombardy, Italy): ecological and biostratigraphic inferences', in *Revista Italiana di Paleontologia e Stratigrafiya* 100, pp. 97–620.
Redondi 1982	P. Redondi, 'Galilée aux prises avec les théories aristotéliciennes de la lumière (1610–1640)', *Dix-Septième Siècle* 136, pp. 267–83.
Redondi 1984	P. Redondi, 'La luce "messagio celeste"', *Novità Celesti e Crisi del Sapere: Atti del Convegno Internazionale di Studi Galileiani*, ed. P. Galluzzi, Florence, pp. 177–86.
Redondi 1987	P. Redondi, *Galileo Heretic*, trans. R. Rosenthal, Princeton, New Jersey.
Reeves 1991	E. Reeves, 'Daniel 5 and the "Assayer": Galileo reads the handwriting on the wall', *Journal of Medieval and Renaissance Studies* 21:1, pp. 1–27.
Reeves 1997	E. Reeves, *Painting the Heavens: Art and Science in the Age of Galileo*, Princeton, New Jersey.
Robert *ASR*	C. Robert, *Die antiken Sarkophag-Reliefs*, 3 vols., Berlin 1890–1919.
Roiron 1981	P. Roiron, 'La macroflore Pliocene de Pichegu (Saint-Gilles, Gard)', *Paléobiologie Continentale* 12, pp. 27–42.
Rossi 1978	P. Rossi, 'Galileo Galilei e il libro dei Salmi', *Rivista di Filosofia* 10, pp. 45–71.
Rudwick 1976	M. J. S. Rudwick, 'The emergence of a visual language for geological science, 1760–1840', *History of Science* 14, pp. 149–95.
Rudwick 1985	M. J. S. Rudwick, *The Meaning of Fossils: Episodes in the History of Paleontology*, 2nd edn, Chicago.
Schapiro/Frank 1979	B. Schapiro and R. Frank, *English Scientific Virtuosi in the Sixteenth and Seventeenth Centuries*, Los Angeles, pp. 3–55.
Schnapper 1988	A. Schnapper, *Le Géant, la Licorne, la Tulipe: Collections Françaises au XVIIe Siècle. I. Histoire et histoire naturelle*, Paris.
Schreer 1954	C. Schreer, 'The rise of historical geology in the seventeenth century', *Isis*, XLV, pp. 256–68.
Scott 1990	A. C. Scott, 'Anatomical preservation of fossil plants', *Palaeobiology, A Synthesis*, ed. D. E. G. Briggs and P. R. Crowther, Oxford, pp. 263–66.
Scott 1991	A. C. Scott, 'Coal: its origin and future', *Teaching Earth Science* 16, pp. 24–36.
Shea 1977	W. R. Shea, *Galileo's Intellectual Revolution*, New York (Science History Publications).
Solinas 1989a	*Cassiano dal Pozzo. Atti del Seminario Internazionale di Studi*, ed. F. Solinas, Rome.
Solinas 1989b	F. Solinas, 'Percorsi Puteani: note naturalistiche ed inediti appunti antiquari', in Solinas 1989a, pp. 95–129.

Solinas 1989c	F. Solinas, '*Il Trattato del Legno Fossile Minerale* di Francesco Stelluti e i quattro volumi della Natural History of Fossils nelle raccolte della Biblioteca Reale di Windsor', *Quaderni Puteani* 1, pp. 84–94.
Solinas 1992	F. Solinas, 'Sull'atelier di Cassiano dal Pozzo: metodi di ricerca e documenti inediti', *Quaderni Puteani* 3, pp. 57–76.
Sparti 1992	D. L. Sparti, *Le Collezioni dal Pozzo. Storia di una Famiglia e del suo Museo nella Roma Seicentesca* (Collezionismo e storia dell'arte, studi e fonti, I, Scuola Normale Superiore di Pisa), Modena.
Stelluti 1630	*Persio, tradotto in verso sciolto e dichiarato da Francesco Stelluti. Accademico Linceo da Fabriano*, Rome (Giacomo Mascardi).
Stelluti 1637	F. Stelluti, *Trattato del Legno Fossile Minerale nuovamente scoperto nel quale brevemente si accenna la varia & mutabil natura di detto Legno rappresentatovi con alcune figure, che mostrano il luogo dove nasce, la diversità dell'onde, che in esso si vedono, e le sue così varie e meravigliose forme*, Rome (Vitale Mascardi).
Tamizey de Larroque *Correspondants*	Nicholas Claude Fabri de Peiresc, *Les Correspondants de Peiresc: Lettres Inédites et Annotées*, ed. P. Tamizey de Larroque, 21 vols., Paris 1879–97.
Tamizey de Larroque *Lettres*	Nicholas Claude Fabri de Peiresc, *Lettres de Peiresc*, ed. P. Tamizey de Larroque, *Collection de Documents Inédits sur l'Histoire de France. Série 2. Histoire des lettres et des sciences*, 7 vols., Paris 1888–98.
Theophrastus	*see* Caley/Richards 1956.
Thesaurus	Francisco Hernandez, *Rerum Medicarum Novae Hispaniae Thesaurus seu Plantarum Animalium Mineralium Mexicanorum Historia ex Francisci Hernandez novi orbis Medici Primarii relationibus in ipsa Mexicana urbe conscriptis a Nardo Antonio Reccho Monte Corvinate Cath. Maiest. Medico et Neap. Regni Archiatro Generali Jussu Philippi II Hisp. Ind. etc. Regis Collecta ac in ordinem digesta a Ioannae Terrentio Lynceo Constantiense Germano Pho. ac Medico Notis Illustrata*, Rome (Vitale Mascardi) 1651; also Rome (Giacomo Mascardi) 1628; 1630, and Rome (Vitale Mascardi) 1649.
Torrens 1985	H. Torrens, 'Early collecting in the field of geology' in *The Origin of Museums*, ed. A. McGregor and O. Impey, Oxford.
Van der Burg 1975	W. J. Van der Burg, 'Aetites or eagle stones', *Organorama* 11, 4, pp. 16–19.
Van der Burgh 1987	J. Van der Burgh, 'Miocene floras in the Rhenish Basin and their ecological interpretation', *Review of Palaeobotany and Palynology* 52, pp. 299–366.
Venturi 1982	F. Venturi, *Ammoniti Liassiche dell' Appennino Centrale*, I, Castello, Italy.
Venturi 1986	F. Venturi, *Ammoniti Liassiche dell' Appennino Centrale*, II, Castello, Italy.
Venturi 1996	F. Venturi *et al.*, *Rocce e Fossili dell' Umbria Meridionale*, Terni, Italy.
Volpicelli 1866	P. Volpicelli, 'Ritrovamento dell' Inventario degli oggetti appartenenti all'eredità di Federico Cesi', *Atti dell' Accademia dei Lincei* XIV, pp. 203–5.

LIST OF FIGURES

Fig. Numbers refer to text illustrations. Plate numbers refer to plates in Stelluti's 'Trattato'. Numbers in ***bold*** *type refer to catalogue items.*

COMPARATIVE FIGURES

PHOTOGRAPHIC ACKNOWLEDGEMENTS

All items in the Royal Collection are copyright Her Majesty The Queen and are reproduced by permission of Her Majesty. The field-trip photographs and comparative photographs of geological specimens are copyright Andrew C. Scott. We are also grateful for permission to reproduce items in the following collections: Montpellier, Bibliothèque Interuniversitaire de Montpellier, Section Médecine (Figs 67–69); Rome, Biblioteca dell'Accademia Nazionale dei Lincei (Fig. 2); Vatican City, copyright Biblioteca Apostolica Vaticana, Barb. N. IV. 24, Pls 1–13 (Figs 53–66); Barb. Lat. 4355, fols. 35, 37 (Figs 70, 71); Odardo Girotti (Comp. figs 147 a, b).

GEOLOGICAL INDEX

This index should be used in conjunction with the Glossary on p. 355.
References are to pages and to catalogue numbers which are in **bold** type.
References to Figures and Comparative figures are in *italics*

GENERAL INDEX

Figure numbers are given in *italic*, catalogue numbers in **bold**